High-Power Microwave Sources

The Artech House Microwave Library

Introduction to Microwaves by Fred E. Gardiol
Lossy Transmission Lines by Fred E. Gardiol
Microwaves Made Simple: Principles and Applications by W. Stephen Cheung, Frederick H. Levien et al.
Microwaves Made Simple — the Workbook by Salvatore Algeri, W. Stephen Cheung, and Lawrence A. Stark
Microwave Tubes by A.S. Gilmour, Jr.
Electric Filters by Martin Hasler and Jacques Neirynck
Nonlinear Circuits by Martin Hasler and Jacques Neirynck
Receiving Systems Design by Stephen J. Erst
Microwave Mixers by Stephen A. Maas
Feedback Maximization by B.J. Lurie
High-Power Microwave Sources, Victor L. Granatstein and Igor Alexeff, eds.
Applications of GaAs MESFETs by Robert Soares et al.
Gallium Arsenide Processing Techniques by Ralph E. Williams
GaAs FET Principles and Technology by James V. DiLorenzo and Deen D. Khandelwal
E.-Plane Integrated Circuits, P. Bhartia and P. Pramanick, eds.
Evanescent Mode Microwave Components by George F. Craven and Richard F. Skedd
Filters with Helical and Folded Helical Resonators by Peter Vizmuller
Design Guide for Monopole Elements on Circular Ground Planes by M.M. Weiner et al.
Dielectric Resonators, Darko Kajfez and Pierre Guillon, eds.
Modern Spectrum Analyzer Theory and Applications by Morris Engelson
Preparing and Delivering Effective Technical Briefings by David L. Adamy
Writing and Managing Winning Technical Proposals by Timothy Whalen
Solid-State Microwave Devices by Thomas S. Laverghetta
Microwave Materials and Fabrication Techniques by Thomas S. Laverghetta
Handbook of Microwave Testing by Thomas S. Laverghetta
Microwave Measurements and Techniques by Thomas S. Laverghetta
Linear Active Circuits: Design and Analysis by William Rynone, Jr.
The Design of Impedance-Matching Networks for Radio-Frequency and Microwave Amplifiers by Pieter L.D. Abrie
Microwave Filters, Impedance Matching Networks, and Coupling Structures by G. Matthaei, L. Young, and E.M.T. Jones
Analysis, Design, and Applications of Fin Lines by Bharathi Bhat and Shiban K. Koul
Microwave Engineer's Handbook, 2 Vols., Theodore Saad, ed.
Handbook of Microwave Integrated Circuits by Reinmut K. Hoffmann
Microwave Integrated Circuits, Jeffrey Frey and Kul Bhasin, eds.
Computer-Aided Design of Microwave Circuits by K.C. Gupta, Ramesh Garg, and Rakesh Chadha
Microstrip Lines and Slotlines by K.C. Gupta, R. Garg, and I.J. Bahl
Advanced Mathematics for Practicing Engineers by Kurt Arbenz and Alfred Wohlhauser
Microwave Circuit Design Using Programmable Calculators by J. Lamar Allen and Max Medley, Jr.
Stripline Circuit Design by Harlan Howe, Jr.
Microwave Transmission Line Filters by J.A.G. Malherbe
Microwave Diode Control Devices by Robert V. Garver
Tables for Active Filter Design by Mario Biey and Amedo Premoli
Design Tables for Discrete Time Normalized Low-Pass Filters by Arild Lacroix and Karl-Heinz Witte
Laser Applications by W.V. Smith
Microwave Remote Sensing, 3 Vols., by F.T. Ulaby, R.K. Moore, and A.K. Fung

High-Power Microwave Sources

Victor L. Granatstein
and
Igor Alexeff, Editors

Artech House
Boston • London

Library of Congress Cataloging-in-Publication Data

High-power microwave sources.

"This volume grew out of the three-day minicourse titled: 'Generation of high-power microwaves, millimeter waves, and submillimeter waves,' which was held in conjunction with the 1985 IEEE International Conference on Plasma Science [held in Pittsburgh, Pa.]"—Pref.

Includes bibliographies and index.

1. Oscillators, Microwave. I. Granatstein, V. L. (Victor L.), 1935– . II. Alexeff, Igor. III. IEEE International Conference on Plasma Science (1985 : Pittsburgh, Pa.)

TK7872.07H5 1987 621.381'323 87-19323

ISBN 0-89006-241-2

ARTECH HOUSE, INC.
685 Canton Street
Norwood, MA 02062

International Standard Book Number: 0-89006-241-2
Library of Congress Catalog Card Number: 87-19323

10 9 8 7 6 5 4 3 2 1

Contents

Preface

This volume grew out of the very successful three-day minicourse entitled, "Generation of High-Power Microwaves, Millimeter Waves, and Submillimeter Waves," which was held in conjunction with the 1986 IEEE International Conference on Plasma Science. The chapters were contributed by lecturers at the minicourse, and therefore are somewhat tutorial, although they cover the latest developments in theory, experiment, and technology. Thus, in addition to serving as a reference for research workers and industrial scientists in the fields of microwave generation, propagation, and effects, this volume might also provide suitable material for an advanced graduate course.

Research and development of vacuum electronics microwave sources has experienced a resurgence, following a hiatus in the 1960s and early 1970s. New requirements arose for microwave sources at frequency ranges and power levels beyond the capabilities of current technology. Among these requirements were sources for electron-cyclotron heating of plasmas in controlled thermonuclear fusion studies, millimeter-wave communication and radar, simulators of enhanced electromagnetic pulse effect, and drivers of advanced electron accelerators in high-energy physics research.

The attempt to meet these requirements naturally led to exploring microwave generators at higher values of electron beam energy and current density than had been previously used. In addition to extending greatly the performance of such familiar microwave generators as magnetrons and backward-wave oscillators, wholly novel classes of microwave generators were developed. The virtual cathode oscillator, which was observed in electron beams having current that surpassed the space-charge-limiting value, produced gigawatts of peak power when driven by intense relativistic electron beams.

Microwave sources were also developed, and these depended on interaction between electron beams and fast waves (i.e., waves traveling with phase velocity greater than the speed of light). Such fast-wave devices required no periodically rippled walls or dielectric loading, and could instead use simple hollow-pipe waveguide as a circuit. Fast-wave devices thus had obvious advantages for very high power and millimeter-wavelength operation.

This volume is divided into three parts. Part I, "Introduction and Basic Techniques," contains three chapters. Chapter 1 by Levush and Drobot gives an extensive overview of the field, Chapter 2 by Wharton reviews technology and diagnostic techniques useful in high-power microwave studies, and Chapter 3 by Birdsall describes applicable numerical techniques.

Part II is devoted to fast-wave devices. The first device of this type which is described in Chapter 4 by Baird and Chapter 5 by Granatstein is the gyrotron. Gyrotrons generate coherent radiation at the electron-cyclotron frequency, ω_c, and depend on the relativistic variation of ω_c with electron energy for their operation. Free-electron lasers, which are described in Chapter 6 by Pasour, have wide tuning by varying electron energy, and they have operated over wavelengths ranging from the microwave regime down to the ultraviolet. Finally, Chapter 7 by Burke, Manheimer, and Ott and Chapter 8 by Alexeff describe a new class of devices, the orbitron, which is somewhat analogous to the gyrotron, but utilizes electron orbits in electrostatic fields, rather than in magnetostatic fields.

Part III consists of three pairs of chapters, each pair describing the theory and experiments pertaining to a particular class of sources. Chapter 9 by Lau and Chapter 10 by Benford describe the operation of magnetrons at relativistic electron energies. Chapter 11 by Case and Chapter 12 by Walsh describe microwave generation in structures loaded with dielectric so that the electron velocity can surpass the phase velocity of the wave, thereby producing coherent Cerenkov radiation. This process in dielectically loaded waveguides is closely related to backward-waves oscillators and traveling wave tubes employing waveguides with periodically rippled walls. Chapter 13 by Sullivan, Walsh, and Coutsias and Chapter 14 by Thode describe the virtual cathode oscillators discussed above.

While the coverage of high-power microwave generators is not exhaustive, the editors believe that it gives a representative sampling of the research and development work currently in progress. A wide range of physical mechanisms are exploited in the various microwave generators described in this volume. Furthermore, an extensive range in parameters such as microwave frequency and pulse duration are covered, indicating a wide range of potential applications.

The editors are indebted to Professor Akira Hirose of the University of Saskatchewan, who greatly assisted in making arrangements for the minicourse out of which this volume grew.

Victor L. Granatstein
Igor Alexeff
June 1987

The Authors

Igor Alexeff (co-editor), University of Tennessee, Knoxville, Tennessee

J. Mark Baird, Department of Electrical Engineering, University of Utah, Salt Lake City, Utah

James Benford, Physics International Company, San Leandro, California

Charles K. Birdsall, Electrical Engineering and Computer Science Department, University of California, Berkeley, California

John M. Burke, National Research Council Associate and Plasma Physics Division, Naval Research Laboratory, Washington, DC

William Case, Department of Physics, Grinnell College, Grinnell, Iowa

Evangelos A. Coutsias, Department of Mathematics and Statistics, University of New Mexico, Albuquerque, New Mexico

Adam T. Drobot, Science Applications International Corporation, McLean, Virginia

Victor L. Granatstein (co-editor), Electrical Engineering Department and Laboratory of Plasma and Fusion Energy Studies, University of Maryland, College Park, Maryland

Y. Y. Lau, Plasma Physics Division, Naval Research Laboratory, Washington, DC

Baruch Levush, Laboratory of Plasma and Fusion Energy Studies, University of Maryland, College Park, Maryland

Wallace M. Manheimer, Plasma Physics Division, Naval Research Laboratory, Washington, DC

Edward Ott, Science Applications International Corporation, McLean, Virginia and Departments of Electrical Engineering and Physics and Astronomy, University of Maryland, College Park, Maryland

John A. Pasour, Mission Research Corporation, Alexandria, Virginia

Donald J. Sullivan, Mission Research Corporation, Albuquerque, New Mexico

Lester E. Thode, Los Alamos National Laboratory, Los Alamos, New Mexico

John E. Walsh, Department of Physics and Astronomy, Dartmouth College, Hanover, New Hampshire

C. B. Wharton, Laboratory of Plasma Studies, Cornell University, Ithaca, New York

Part I
INTRODUCTION AND BASIC TECHNIQUES

Chapter 1
GENERATION OF HIGH-POWER MICROWAVES, MILLIMETER AND SUBMILLIMETER WAVES: INTRODUCTION AND OVERVIEW

Baruch Levush

University of Maryland
Laboratory for Plasma and Fusion Energy Studies
College Park, Maryland

Adam T. Drobot

Science Applications International Corporation
McLean, Virginia

1.1 INTRODUCTION

The exploration of new and unique concepts for high-power microwaves, millimeter waves, and submillimeter waves has progressed rapidly in recent years and exhibited the excitement to attract bright, enthusiastic researchers. This is in sharp contrast to the mature state of vacuum electronics a decade ago, when the transition from discovery of fundamental principles to an orientation of concern about production of microwave tubes resulted in diminished support for basic research. The progression from the invention of "radar" tubes in the 1940s to the stage when vacuum electronics had become a "staid" field for research took roughly 20 to 25 years. The recent revival in work on new concepts for microwave sources has been driven by the introduction of new technologies and the requirements of present and future applications. From a very personal point of view, the opportunity to participate in interesting and intellectually stimulating research is attractive, and as the chapters of this book demonstrate, there is considerable room for invention and challenging problems to solve.

The two most notable recent achievements, which stand out as topics of chapters in this book, are the ability to produce *short-wavelength radiation* using electron beams and the ability to produce *extremely high power.* Devices like *cyclotron resonant masers* (CRMs) [1] have generated coherent electromagnetic radiation at wavelengths below 1 cm, and thus have filled the spectral gap that existed between lasers at extremely short wavelengths and conventional microwave devices. Even more remarkable has been the development of the *free-electron laser* (FEL) [2], which may eventually extend to the UV and soft X-ray regime. The capability to produce short-wavelength radiation has been accompanied by the ability to provide appreciable power levels in both steady-state and pulsed operation. At longer wavelengths in the microwave range, researchers working on high-power devices, such as relativistic magnetrons, vircators, and *backward-wave oscillators* (BWOs), have generated short pulses with levels exceeding 1 GW. Progress on more conventional tubes like klystron amplifiers has also resulted in dramatic performance improvements, such as stable long-pulse operation with power levels of 150 MW at SLAC.

The overall reason for the fast paced developments of the last few years may be hard to attribute to a specific cause, but what is apparent is the wide and complex technology base that has contributed to success.

In Figure 1.1, we have schematically depicted the interrelation between various components of millimeter, submillimeter, and high-power devices, and the typical issues given consideration. We shall describe these briefly in this introduction, and refer the reader to chapters in the book for a more thorough exposition. The departure point for designs based on a particular concept is the *wave-particle interaction.* A specific interaction between an electromagnetic mode and an electron beam can be characterized by the efficiency η, the frequency f, and a bandwidth Δf. The interaction can result in a device that is either an amplifier or an oscillator.

In an amplifier, a further distinguishing property is the gain $G(f, P_I)$, which relates the output power to the frequency of the input signal and the input power level P_I. An important objective of current research is the ability to phase-lock oscillators by the use of an input signal or premodulation of the electron beam and to control the phase ϕ of the device. Some of the wave-particle interaction concepts that are undergoing intensive study include:

- *Parametric Devices*—Raman or Compton free-electron lasers and klystrons;

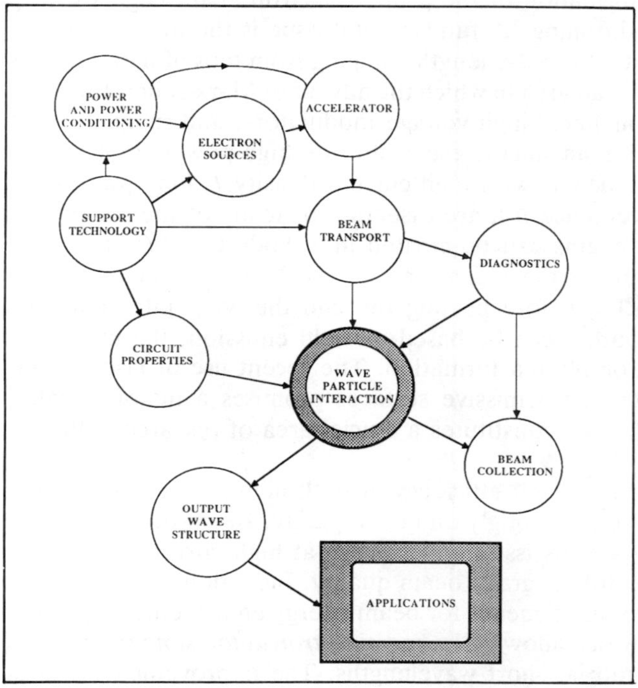

Fig. 1.1 Components of High-Power and Short-Wavelength Electromagnetic Radiation Sources

- *Slow-Wave Devices*—magnetrons, Cerenkov masers, and backward-wave oscillators;
- *Fast-Wave Devices*—cyclotron resonant masers, such as gyrotrons, peniotrons, and CARMs;
- *Plasma Devices*—virtual cathode oscillators, beam-plasma interactions, and orbitrons.

The practical realization of a device with reasonable operating parameters as shown in Figure 1.1, requires the proper electron beam, an appropriate wave circuit, the ability to couple the radiated output, beam collection, and other indispensible support technology.

Wave circuits are used for achieving wave-particle resonances, mode control, and proper matching to beam parameters. The circuit elements often comprise a factor in performance because of breakdown limits, heat dissipation, and spurious mode generation. Advances in vacuum technology, materials, cooling, and the ability to model complicated structures and understand their properties have all led to improvements.

The formation of the proper electron beam begins with power and power conditioning. A fundamental issue is the limit on beam voltage V and current I for pulse lengths ranging from tens of nanoseconds to steady state. This is an area in which the advent of Marx generators, pulsed power transmission lines, high-voltage modulators, and magnetic switching has led to major advances, especially for high-power application. Electron sources for beams with high current density J, long pulse lengths τ, and limited divergence $\Delta\theta$, are essential for many of the concepts mentioned here. The beams can be created in a diode to interact directly with the wave circuit, or the beams can be accelerated to high energy, as in the case of FELs, before passing through the wave-interaction region. The electron sources can be based on field emission, thermionics, secondary processes, or plasma formation. The recent use of lasers in high-voltage diodes with photoemissive surfaces promises a major advance in beam quality [3]. This constitutes a crucial area of research with promising results.

The theoretical efficiency of both high-power and short-wavelength devices depends strongly on beam quality. Beam transport in accelerators is another serious issue, particularly at high currents, where instabilities can significantly degrade beam quality. Induction and RF accelerators are the most common means for beam energy enhancement to high relativistic factors γ, which allow FELs and *cyclotron autoresonance masers* (CARMs) [4] to operate at short wavelengths. The improvements in areas such as magnetics and superconducting structures have again significantly contributed to performance.

For the high-power, short-pulse, and short-wavelength experiments, the diagnosis of what occurs is far from simple to measure. The development of new and innovative diagnostics has played an important role in guiding research and providing the understanding that leads to improvement.

To use the radiated output from a device for a specific application, the output and transmission structure becomes important. Mode conversion at both high powers and short wavelengths is a challenging task, particularly for steady-state operation. The matching of the radiated power to suitable antennas is also an area of research, especially for short pulses.

In addition to handling the radiated power, it is also necessary to deal with the residual energy of the spent electron beam. From a practical point of view, after passage through the interaction region, the beam electrons will still carry considerable energy, which must be deposited over a sufficiently large area to avoid heat load limits or electrical breakdown in the collector. The use of depressed collectors or the recently proposed energy-recovery schemes for FELs may be an additional method to improve overall system efficiency.

By briefly describing the components of an electromagnetic radiation source, as depicted in Figure 1.1, we have tried to convey the broad areas of research that are involved in high-power, short-wavelength devices. In the following sections, we shall discuss potential applications, the general properties of the electromagnetic radiation sources described in this book, and the underlying technology. We conclude with a summary.

1.2 APPLICATIONS

Microwave devices have played an important role in a variety of fields, which include:

- *Basic Physics*
- *Accelerators*
- *Magnetic Confinement Fusion*
- *Communication*
- *Astronomy, Earth and Space Observations*
- *Industrial Uses*
- *Military Applications*

The use of the new capability at short wavelengths and high powers is crucial for applications in several areas, and requirements exceed what can currently be provided by existing sources. We hope that the need for continued research will be obvious from the brief description of applications provided below. The discussion is by no means all-inclusive, and should be regarded as illustrative of potential applications.

1.2.1 Basic Physics

Microwave devices have commonly been used as diagnostics of plasmas and other matter. The techniques used have ranged from determining the attenuation and phase shift in passing a beam of microwaves through a medium to observation of the scattered radiation as a function of angle and frequency shift. The use of lasers at very short wavelengths for scattering from high-density plasmas has resulted in detailed diagnosis of plasma-density profiles, fluctuation levels, and temperatures. The use of new millimeter and submillimeter sources will make it possible to extend the range over which we can perform these diagnostics. The need here is for sources with very narrow line widths and high power.

Tuned millimeter and submillimeter sources also play an important role in atomic and molecular physics as well as biology. The energy levels of molecular states and specific absorption at resonant frequencies can be explored by using these devices. At extremely high powers, the physics of breakdown and plasma formation on extremely short time scales is an active area of research, which may allow simulations of physical conditions that are otherwise unattainable. The reaction of materials and the response of surfaces to short high-power pulses of electromagnetic radiation is also of interest, with applications ranging from basic diagnostics to plasma formation for use in electron-beam or ion-beam devices.

At high-power levels, the interaction of microwaves with plasmas can lead to nonlinear interactions analogous to those observed in laser-plasma experiments. These interactions can be studied to shed light on mechanisms of particle acceleration and simulation of nonlinear phenomena.

1.2.2 Accelerators

The next generation of high-energy physics accelerators and the next frontier in understanding of elementary particles is based on the *super collider*. This is an accelerator that will reach center-of-mass energies sufficient to explore the production and spectroscopy of W and Z particles. The proposals for colliders range from a circular proton-antiproton machine at 20 TeV to a linear electron-positron collider at 1 TeV. The circular collider requires efficient high-power tubes at frequencies below 500 MHz and average powers in the range of 2–5 MW. While existing devices such as klystrons can meet the requirements, any improvement in efficiency and reliability has beneficial results for the cost of operating a super-collider facility. For linear colliders, the most advanced machine is at SLAC, at present consisting of a LINAC structure fed by 240 klystrons with peak power $P = 55$ MW, pulse length τ approximately 5 μs, operating at 2.856 GHz, and producing approximately 17 kW of average power per klystron. For linear colliders to be competitive with circular machines, it is thought that sources at 10 GHz with $P \approx 300$ MW, $\tau = 0.2$ μs, and characteristics that will allow approximately 1000 pulses per second will be necessary. These must be phase-coherent devices, which can be either amplifiers or phase-locked oscillators. The US Department of Energy is also funding the development of the Continuous Electron Beam Accelerator Facility for Nuclear Physics, which requires an average power of greater than 30 kW per tube at the SLAC frequency.

A very active area of research involves high-current accelerators operating with fundamental mode structures. The acceleration of beams with 0.1–10 kA to energies of 100–500 MeV requires the development of high-power (0.2–1.0 GW) sources at operating frequencies ranging from 0.5 to several GHz. Among other uses, these accelerators will be applied to the production of coherent electromagnetic radiation in FELs. Finally, there is an increasing demand for light sources. These are sources of UV or soft X-ray radiation with applications to research and microelectronic lithography. Common devices are based on circular accelerators or microtrons with beam energies from 0.8–5 GeV. Such devices require reliable, high-average-power, efficient microwave tubes.

1.2.3 Magnetic Confinement Fusion

Auxiliary heating of confined plasmas with microwaves and millimeter waves plays an important role in future plans for fusion reactors. Steady-state sources providing several megawatts of power have been contemplated for the following uses: *ion cyclotron resonance heating* (ICRH) at frequencies of $\approx$ 100 MHz; *lower hybrid heating* at 0.5–5 GHz; and *electron cyclotron resonance heating* (ECRH) at above 100 GHz. The use of electromagnetic radiation has spanned a range of applications from plasma formation to diagnostics, heating, current drive, and profile control. In each case, successful experiments have been performed while establishing the need for low-cost and reliable high-power devices. The most dramatic progress has been the development of gyrotrons at 35, 60, and 120 GHz with long-pulse tubes delivering several hundred killowatts of power. Proposals have been made for using FELs as sources for ECRH in high-magnetic-field plasma containment configurations.

1.2.4 Industrial and Military Applications

One of the properties that microwaves possess is the ability to penetrate and deposit energy selectively in appropriate materials. An example of this is the microwave oven, where the frequency is tuned to one of the water-absorption lines. Bulk material heating processes are an example of an application where high average power and reliability play an important role. Curing processes at high frequency, where energy is rapidly deposited in the surface, is another class of applications. Microwave sputtering sources for material processing are in common use and represent an area of considerable interest.

The use of microwaves in isotope separation has resulted in several practical schemes. All of these have required extremely rigid specification on phase and frequency stability. Probably the most common industrial use of microwaves at all frequencies is as a diagnostic for process control. The ability to use higher frequencies at millimeter and submillimeter wavelengths will extend the workable range for process-control diagnostics.

Military uses of microwave devices cover radar, communication, electronic warfare (EW), electronic countermeasures (ECM), and testing. Not all of these applications require either short wavelengths or high power. An interesting area of continued emphasis is efficiency and reliability, and the ability to tailor the frequency gain characteristics over a broad range. The introduction of devices in the millimeter and submillimeter ranges with modest powers is naturally leading to deployment of military components at those frequencies. An open and somewhat controversial question concerns the use of high-power microwaves on military systems. This is currently an area where there is considerable study of the coupling to and interaction with military components.

1.2.5 Space-Based Applications

Satellite communication at 35 and 94 GHz, where transmission windows exist in the atmosphere, are a natural location in the electromagnetic spectrum for future expansion. There are two reasons to contemplate this development. The first is the saturation of available bands at lower frequencies. The second is the increase in bandwidth at the higher frequencies and improvement in directionality. The negative aspect is, of course, the increased attenuation and greater influence of weather. The requirements for these applications are modest powers of less than 10 kW, light weight, high reliability, and stable gain characteristics.

The existence of sources will also make it possible for the astronomers to perform millimeter and submillimeter surveys of the sky, and to identify and map molecular constituents having resonances in that part of the spectrum. Because of superior directionality, high-frequency sources are also preferable for deep-space communication and in radar ranging for planetary sciences. As an example, the Jet Propulsion Laboratory (JPL) is contemplating the use of its Goldstone 70 M radar for mapping of Venus at 35 GHz. By using microwave interferometry techniques, there are applications of microwaves and millimeter waves for terrain mapping, weather prediction, wind-pattern reconstruction, atmospheric profiles, and many other possibilities. Perhaps, a more futuristic application is the beaming of power from solar satellites to earth or from earth to other orbiting space platforms.

1.3 DEVICES

In this section, we discuss the fundamental mechanisms for generation of high-power coherent radiation and the basic principles of various devices.

The basic physical ideas related to the generation of coherent electromagnetic radiation at millimeter and submillimeter wavelengths were developed in the late 1950s [5–6].

In general, such devices amplify or generate coherent electromagnetic radiation by radiative emission from a beam of relativistic electrons. The electrons radiate because they undergo oscillations transverse to the direction of beam motion by the action of an external force. The force can be due to either homogeneous fields, periodic static fields, or crossed static electric and magnetic fields.

The condition for coherent radiation is that the contribution from the electrons reinforces the original emitted electromagnetic radiation in the oscillator or the incident electromagnetic wave in the amplifier. This condition is satisfied if a bunching mechanism exists to create electron density variations of a size comparable with the wavelength of the imposed electromagnetic wave. To achieve such a mechanism, a resonance condition must be satisfied between the periodic motion of the electrons and the radiation wave in the interaction region. The resonance condition usually has the form:

$$\omega - k \cdot v \approx n\Omega, \quad n = 1, 2, \ldots \tag{1.1}$$

where ω and k are the electromagnetic wave frequency and characteristic wavenumber, respectively, v is the beam drift velocity, and Ω is an effective frequency, which is associated with macroscopic oscillatory motion of the electrons.

In the electron *cyclotron resonant maser* (CRM), electromagnetic energy is radiated by relativistic electrons gyrating about an external magnetic field. In this case, the effective frequency Ω corresponds to the relativistic electron cyclotron frequency:

$$\Omega_c = \frac{|e|B_0}{mc\gamma} \tag{1.2}$$

where $|e|$ and m are the respective charge and mass of an electron, c is the speed of light, γ is the relativistic factor, and B_0 is the magnitude of the guide magnetic field. A group of relativistic electrons gyrating in a strong magnetic field will radiate coherently due to bunching caused by the relativistic energy dependence of their gyration frequency. Radiation is emitted because, as an electron loses energy, its relativistic mass decreases and thus gyrates faster. The consequence is that a small amplitude wave's electric field, while extracting energy from the particles, causes them to become bunched in gyrotron phase and reinforces the existing wave electric field. The strength of the magnetic field determines the value of the radiation frequency.

In the case of the periodic magnetic or electric field, the oscillation frequency Ω is proportional to the ratio of the electron beam velocity v to field spatial period λ_w. Thus,

$$\Omega = k_w v, \quad k_w = 2\pi/\lambda_w \tag{1.3}$$

The operating frequency of such devices, an example of which is the free-electron laser, is determined by the condition that an electron in its rest frame "observes" both the radiation and the periodic external force at the same frequency. If the beam is relativistic, the radiation will have a much shorter wavelength than the external force in the laboratory frame. Therefore, FELs are capable of generating electromagnetic waves of, very short wavelength determined by the relativistic Doppler effect. The bunching of the electrons in FELs is due to the perturbation of the beam electrons by the pondermotive wave, a wave caused by "beating" of the electromagnetic wave with the spatially periodic wiggler field.

In crossed-field configurations such as magnetrons, the electrons move along toroidal trajectories with a built-in periodicity. This periodicity is created by a slow-wave structure, which supports several electromagnetic modes. Waves that have phase velocities nearly equal to the electron $E \times B$ drift velocity will interact strongly with the electrons. The forces due to the fields in the electromagnetic wave tend to produce azimuthal electron bunching.

In the orbitron, the electron orbits are controlled by an externally applied electric field, which has no obvious scale length. However, the device is capable of producing coherent radiation with very short wavelengths [7]. There is conjecture that the wavelength is determined by the radius of the on-axis wire.

In the case of CRMs and FELs, unlike most conventional microwave sources and lasers, the radiation wavelength is not determined by the characteristic size of the interaction region. These devices are capable of producing very high power radiation at centimeter, millimeter, and submillimeter wavelengths.

In the remainder of this section, we provide a brief description of CRMs, FELs, slow-wave devices such as the backward-wave oscillator (BWO), and plasma microwave devices such as plasma-filled gyrotrons and vircators. We also review recent achievements and include a current bibliography.

1.3.1 Cyclotron Resonant Masers

Many configurations can be used to produce coherent radiation based on the electron cyclotron maser instability. Accordingly, various types of electron cyclotron masers exist. Often, CRMs are called *gyrotrons*. Voluminous scientific literature exists on the subject, manifesting the intense international activity in the field. A number of review articles are devoted to describing gyrotron-like devices [8–10]. Probably, the best source to obtain information about the most recent achievements in gyrotron research are the collection of articles published in the special issues of the *International Journal of Electronics* on gyrotrons [11].

Figure 1.2 shows the most popular configuration for the cyclotron electron maser, the *axisymmetric gyrotron*. The gyrotron was invented [12] when a CRM first utilized a magnetron-type electron injection gun [13] for production of helical electron beams and an open resonator with diffraction limited power output. Although the principle of CRM operation was understood, previous attempts at practical devices failed because a suitable electron beam was unavailable. In the gyrotron gun, the initial twisting of electron trajectories takes place in the crossed fields near the emitter. In the interaction space, the electrons are guided by a magnetic field. The electrons interact with a transverse electric (TE) mode near the cut-off of a cavity with slowly varying cross section.

In principle, the wavelength is not determined by the geometry and characteristic size of the interaction region, but in actual devices the RF structure is necessary to support the electromagnetic wave. The frequency of the resonant eigenmode must match the relativistic gyrofrequency or one of its harmonics. The structure is also used for mode control.

The theoretically predicted intrinsic efficiency of a gyrotron device can be as high as 70% to 90%. Such high efficiencies can be achieved by optimizing the longitudinal structure of the RF field in the cavity [14] or tapering the longitudinal magnetic field [15]. The theoretical modeling of the gyropeniotron interaction has shown that intrinsic efficiencies close to 100% are possible [16], without tapering the RF or magnetic fields.

Gyrotrons are usually operated at the fundamental of the cyclotron frequency, $\omega = n\Omega_c$, $n = 1$. However, an interaction at harmonics of the cyclotron frequency, $n > 1$, is possible. The predicted efficiency for gyrotrons operating at high harmonics are comparable with those operating at the fundamental frequency [17].

To move the operation of the gyrotron devices toward MW power at *near-millimeter* wavelengths, there must be an increase of the beam

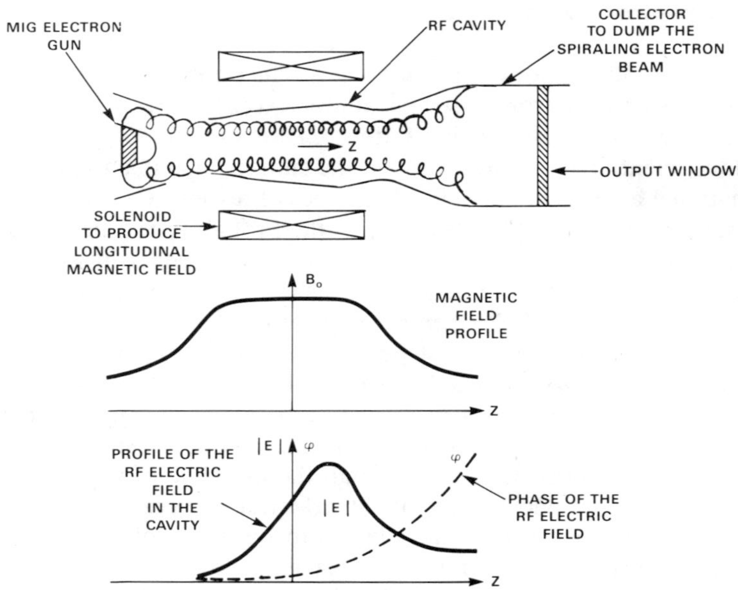

Fig. 1.2 The Gyrotron
Note: Typical configuration for an open resonator gyrotron with a magnetron injection electron gun. The profiles of the resonator cavity and the magnetic field can be tailored to optimize efficiency.

current and the voltage. The cavity size must scale correspondingly to handle the increase in dissipated power by maintaining the wall loading below the threshold for cooling technology. High-current operation in overmoded cavities leads to a problem inherent to all high-power microwave tubes, the supression of parasitic modes. These problems as well as the methods for their solution are the current topics in gyrotron research.

For example, to increase the output power of a gyrotron, we can use cavities with low Q-factors and high beam currents [18, 19]. Another promising approach for MW gyrotrons is the use of a hollow electron beam with a radius close to that of the resonator supporting surface or "whispering gallery" modes [20, 21]. To reduce the Q-factor of such resonators, a cavity with elliptical cross sections is used [22, 23].

The *complex-cavity gyrotrons* represent a new step in the modification of the simple cavity gyromonotron to accommodate high-power requirements [24]. Figure 1.3 shows a Gyrotron with a complex resonator consisting of two strongly coupled cavities.

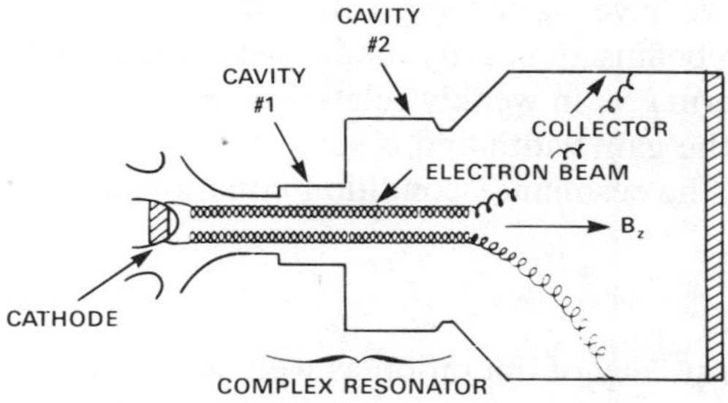

Fig. 1.3 Complex Cavity Gyrotron
Note: This configuration consists of a mistuned input cavity below threshold for oscillation and a closely coupled, open, output cavity.

A more drastic modification of the usual gyrotron configuration is the *quasioptical gyrotron* [25–27], shown in Figure 1.4. This configuration permits operation at higher frequencies and at increased output powers. Specific advantages of the quasioptical gyrotron are natural mode control and separation of the radiated output and the beam collector. The large diffraction losses suffered by the higher order transverse modes in the Fabry-Perot type of optical cavity lead to a mode spectrum that is essentially one-dimensional. Single-mode operation at short wavelengths, corresponding to the fundamental and harmonics of the relativistic gyrofrequency, becomes feasible and is independent of the dimensions of the cavity. [28, 29].

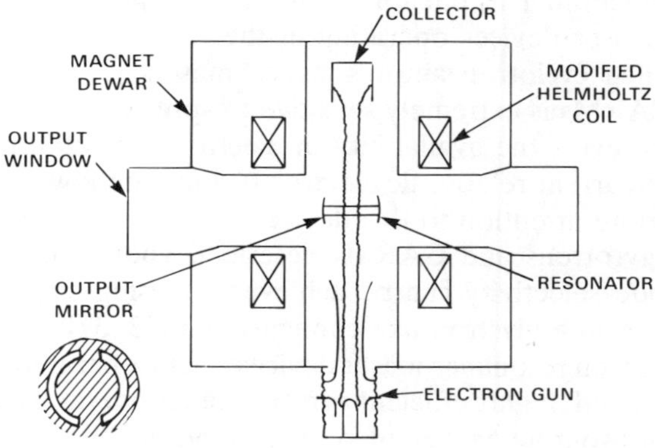

Fig. 1.4 Quasioptical Gyrotron
Note: Typical configuration based on a split Fabry-Perot resonator.

To this point we have discussed gyrotron devices utilizing only weakly relativistic electron beams and cavity modes with small axial wavenumber, $k_\parallel$. The Doppler term $k_\parallel v_\parallel$ in weakly relativistic electron cyclotron masers is on the order of the gain width and is small compared with the radiating frequency. Hence, the resonance condition equation (1.1) is of the form

$$\omega \approx n\Omega_c \tag{1.4}$$

To increase the frequency of operation as well as output radiation power, a relativistic cyclotron resonance maser was suggested [30].

In a gyrotron with a highly relativistic beam, an efficient interaction will lead to an average energy loss on the order of the initial electron energy. The consequence of this is that the change in the gyrofrequency will be much greater than in the mildly relativistic case. It is therefore desirable to identify the condition under which an electron that loses energy remains in synchronism with the RF field. A possibility for achieving such synchronism is to utilize the interaction of electrons with electromagnetic waves propagating at a phase velocity close to the speed of light in the direction of the magnetic field. In this case, the Doppler shift term $k_\parallel v_\parallel$ is large, and the appropriate resonance condition is

$$\omega \approx k_\parallel v_\parallel + n\Omega_c \tag{1.5}$$

If $\beta_{ph} = 1$, where $\beta_{ph} = \omega/k_\parallel c$, the increase in cyclotron frequency exactly compensates the decrease in the Doppler shifted term. Therefore, if the resonance condition is initially fulfilled, it will continue to be satisfied during the interaction. This phenomenon is called *autoresonance* [31], and the cyclotron maser devices operating in the relativistic Doppler-shifted regime are called cyclotron autoresonance masers [32] (CARMs). The efficiency of CARMs is extremely sensitive to spread in the parallel beam velocity $v_\parallel$. However, the availability of electron accelerators capable of producing high-current relativistic electron beams with low velocity spread has attracted more attention to this device.

For the gyrotrons and CARMs described above, higher efficiency and better mode selectivity can be achieved by using one or more pre-bunching cavities in a klystron-like configuration [33–37].

Most cyclotron resonance maser devices can be operated as oscillators or amplifiers. Pulsed, highly efficient gyrotron tubes operating at millimeter wavelengths with output power of several hundred kilowatts have been reported [38–39]. By using pulsed magnetic fields [40], output powers on the order of 100 kW at wavelengths as short as 0.6 mm [41] have been produced. Advances in cooling technology have made it possible to operate a CW gyrotron tube at 75 kW output power at a frequency of 140 GHz [42].

During the past several years, impressive results have been obtained with gyrotrons designed for operation at harmonics of the cyclotron frequency [43]. Experimental results of third harmonic operation have been obtained for a gyropeniotron. This is the first experimental confirmation of the peniotron interaction in a gyrotron device [44]. A traveling wave amplifier (gyro-TWA) was operated with a 3 dB bandwidth of 13% and 18 dB gain at a frequency of 35 GHz [45]. A gyroklystron amplifier with a gain of 40 dB and bandwidth of 0.2% at a frequency of 28 GHz has produced 50 kW radiation output power [46].

Another approach is being developed at the University of Maryland [47] and at UCLA [48], whereby high-frequency radiation is produced with low magnetic fields. This approach is based on the synchronism between axis-encircling electrons and a high-order azimuthal mode.

1.3.2 Free-Electron Lasers

Free-electron lasers have been described by a great number of articles published in a broad spectrum of international scientific journals (see the references at the end of this chapter). In addition, there are a number of review articles on the subject [49] and there have been three special issues of the *IEEE Journal of Quantum Electronics* on FELs [50]. There are seven volumes of workshops and conference proceedings [51] in which hundreds of papers are collected on the topic. Finally, several textbooks have appeared recently, which provide a good introduction to the FEL physics and contain up-to-date bibliographies [52].

The history of the FEL can be traced back to microwave tube development in the 1950s. The tubes developed by Motz [53] and Phillips [54], referred to as the *ubitron,* used the same mechanism as a FEL, but they used mildly relativistic electrons traveling through periodically undulating electric or magnetic fields inside a waveguide.

The modern FEL configuration, shown in Figure 1.5, was suggested by John Madey at Stanford University, where the first set of experiments was carried out [2, 56]. These experiments were the first to use relativistic electrons from a particle accelerator. The use of modern accelerators makes FEL devices particularly attractive for producing intense, coherent radiation at short wavelengths. The limits of what "short" means in this context are being explored at present. It is thought by some that the range of the FEL may span radiation from the soft x-ray portion of the spectrum all the way to microwave frequencies. Additionally, at least in principle,

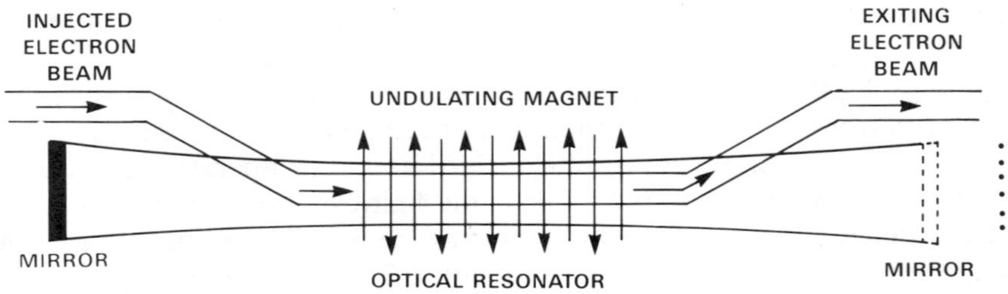

Fig. 1.5 Free-Electron Laser (FEL)
Note: Conceptual configuration of a FEL showing the electron beam, a magnetic undulator, and an optical cavity.

the FEL is continuously tuned by changing the energy of the electron beam.

There are basically two physical regimes for FELs. In the *collective* regime, where the density in the beam is high, the process involves an eigenmode of the beam, such as a slow space-charge wave, the radiation, and the wiggler. This regime is a three-wave interaction. Usually, such intense relativistic electron beams are produced at low energies (1–2 MV), and the radiation is therefore emitted in the far infrared region [57]. In the low-density or "Compton" regime, the interaction is between individual electrons, the radiation field, and the undulator. This regime is a two-wave particle interaction. The electron energy is usually high (10–1000 MeV) and the emission is at shorter wavelengths, ranging from UV to IR.

The first process is sometimes called the high-gain Raman FEL, referring to the fact that the FEL mechanism is governed by stimulated Raman back-scattering [58]. In the first experiments, a low-frequency electromagnetic wave played the role of the undulator [59], the electromagnetic wiggler was later replaced by a periodic-static-magnetic undulator [60]. To enhance the gain of Raman FELs employing high-current-density beams, it has been suggested that the beam must be confined by a strong axial magnetic field [61]. The advantages and disadvantages of such a scheme have been theoretically and experimentally explored [62–64].

The Compton FELs have been configured with single and multiple uses of the electron beam. In single-pass devices, the electron beam is lost after having interacted with the radiation field in the undulator. These FELs are driven by linear accelerators [65, 66], microtrons [67], or electrostatic accelerators. In recirculating-beam oscillators, the electron bunch interacts with the radiation stored in an optical cavity on each pass through the undulator. This effect has been achieved in a storage ring accelerator

[68] in France, at Orsay. By mounting an optical klystron [69] configuration on a storage ring, FEL oscillation has been obtained in the visible range [70].

There are two major approaches to FEL systems driven by linear accelerators:

- *RF-linac-driven FEL* [71]
- *Induction-linac-driven FEL* [72]

RF-driven systems couple a very-high-energy electron beam with low peak current to the FEL, while induction accelerators are used to couple high peak current and lower energy beams. Considerable effort is now being expended on minimizing these differences [73].

The intrinsic efficiency of the Compton FEL interaction is very low, less than a fraction of a percent. In the high-gain Raman regime, efficiency of only a few percent has been achieved [74].

The theoretical analyses of the gain saturation mechanism in the FEL showed that high efficiencies could be achieved by an appropriate tapering of the wiggler period or wiggler amplitude [75–77]. These ideas were further developed by establishing a connection between FEL and RF accelerator physics [77].

The theory of tapered undulators predicts very high efficiency for FELs, in excess of 60%. In the recent induction-linac-driven FEL experiment [78], where this concept was utilized, energy extraction efficiency has reached 40%.

1.3.3 Slow-Wave Devices

A basic feature of slow-wave devices is the extended interaction between an electron beam and an electromagnetic wave. A slow-wave RF structure is provided to maintain the synchronism between the electrons and the electric field of the wave. By slow-wave RF we mean a structure where the phase velocity of radiation field is subluminous. Crossed-field devices, such as the *magnetron* (shown in Figure 1.6), employ a periodic RF structure built on the outer portion of a coaxial configuration. The inner hub serves as a cathode, while the RF structure acts as an anode. The motion of individual electrons consists of a fast gyromotion superimposed on the slow rotation of the electron beam around the cathode. The synchronous electron velocity is determined by the dc electric-to-magnetic field ratio (E_r/B_z).

Experiments in recent years with magnetrons employing relativistic electron beams have demonstrated high efficiency operation at gigawatt power levels [79–84].

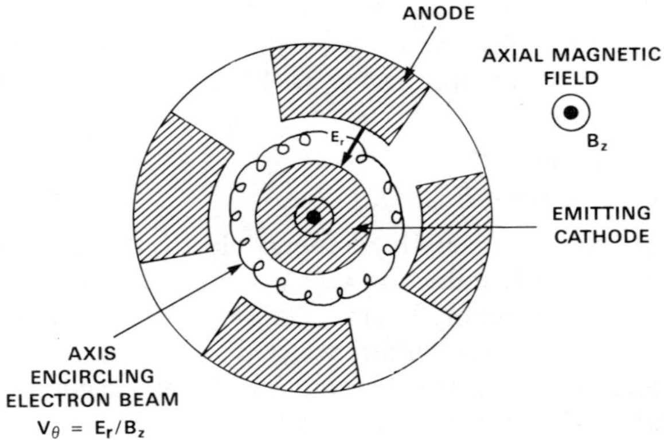

Fig. 1.6 Magnetron
Note: Conceptual configuration for a four-vane, high-power relativistic magnetron.

In slow-wave devices, such as traveling wave tubes, backward-wave oscillators, and Cerenkov masers, the synchronism is maintained by reducing the phase velocity of the electromagnetic wave by means of structures on the RF cavity walls or dielectric loading of waveguides.

The first successful experiments with high-current relativistic electron beams were performed in a BWO operating at centimeter wavelengths [85]. Recent progress in the production of high-quality, high-current relativistic electron beams has resulted in radiation at high-power levels with millimeter wavelengths using other relativistic Cerenkov devices [86–89]. In addition, using the configuration shown in Figure 1.7, further progress has been made with BWOs [90–92].

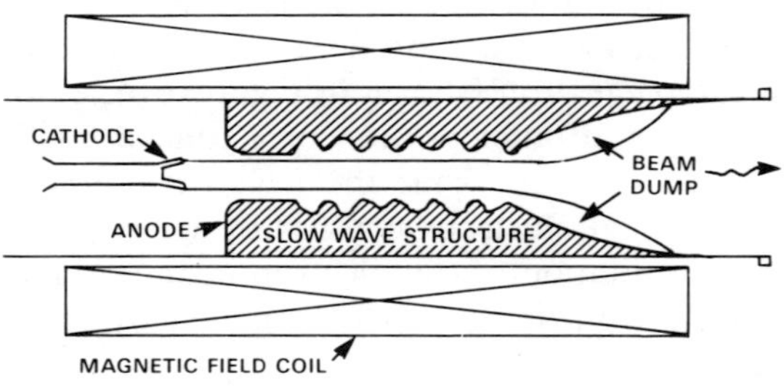

Fig. 1.7 Backward-Wave Oscillator (BWO)
Note: Configuration for a high-power, relativistic electron beam BWO.

1.3.4 Plasma Microwave Devices

In the wide variety of microwave devices that have been investigated in recent years, a factor determining the maximum output power has been the limit on useful beam current. This limitation arises primarily from self-space-charge effects. There are several ways of overcoming the limitation.

One way to do so is to inject a relativistic electron beam into a drift tube in excess of the space-charge-limiting current. In this case, a virtual cathode forms and reflects electrons back toward the accelerator. This is an inherently unstable situation, and it leads to quite efficient production of electromagnetic radiation. The self space charge of the electrons forms a deep electrostatic well, which results in the virtual cathode [93]. Oscillation of the virtual cathode causes strong excitation of the electric field around the potential well region, which, in turn, excites the eigenmodes of the output coupling waveguide. The microwave emission can also be explained by phase bunching of electrons reflecting inside the self-potential well [94]. A typical vircator configuration is shown in Figure 1.8. Power levels between 10 MW and several gigawatts at frequencies in the range of 2–13 GHz, with electron beam power conversion efficiencies between 1.5 and 14% have been reported. Recently, experimental data has been produced at higher frequencies, 18 and 38 GHz, at hundreds of megawatt power levels [95–96].

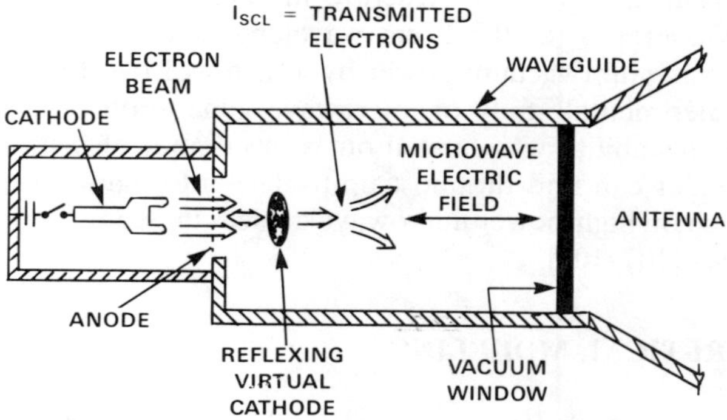

Fig. 1.8 The Vircator
Note: A relativistic configuration indicating a high-current electron diode injecting a beam in excess of the space charge limit into an interaction cavity.

Another way of exceeding space charge limits is by the generation of electromagnetic waves in a plasma. This was predicted when the beam-plasma instability was discovered [97–98]. A plasma microwave generator was not successfully demonstrated until 1973. It took over 20 years of experimental and theoretical effort for sufficient understanding of the beam-plasma instability in finite magnetized plasmas before realization of microwave sources using relativistic beams. This effort was greatly helped by the development of a general theory of plasma microwave generators [101–102].

Plasma microwave sources have a number of advantages over vacuum microwave generators. The plasma in the resonator leads to partial or even complete neutralization of the electron beam's self-space-charge fields. This makes it possible to send electron beams through the resonator at currents much higher that the limiting current. Simultaneously, the spread in the longitudinal velocities due to the self-field-potential well is reduced. The lower velocity spread in the electron beam and the fact that the interaction can be uniformly distributed across the volume of plasma in the resonator can result in a more efficient interaction. Removal of the space-charge-limited current barrier can lead to higher power devices than those in vacuum configurations. Furthermore, it is possible to tune the output frequency by varying the plasma density.

Initial experiments with plasma-filled microwave oscillators were performed by Tkach *et al.* [103]. In the next series of experiments, microwave generation using a slow-wave structure in vacuum and in plasma were compared. Generation at the 3 cm wavelengths reached 600 MW, exceeding the maximum vacuum power by a factor of two to three [104].

The experimental study of microwave generation by a two-stream beam-plasma instability was carried out at a number of institutions [105, 106]. The reader can find the most up-to-date information on the status of the research on high-power microwave devices in two recently published review articles [107–108].

1.4 THEORETICAL MODELING

The experimental effort to produce high-power and short-wavelength radiation using relativistic electron beams has been supported by extensive theoretical research. To describe fully the variety of theoretical models, approaches, and methods is a monumental task that we have not undertaken in this chapter. In general, theoretical models and methods are aimed at either design calculations or addressing fundamental problems. In the latter category, the emphasis has been on the investigation of nonlinear behavior, which occurs in large classes of microwave devices. The theoretical work has been extremely instrumental in challenging the frontiers of present capability and has led to new sources of high-power coherent radiation. Because the contribution of theory has had a profound effect, we will comment on some recent highlights and general concepts.

1.4.1 Linear Theory

Given an equilibrium configuration in which the wave structure and the beam properties are defined, we can calculate the response of the system to infinitesimal perturbations of the radiation field. For many devices, linear theory is a good indication of the way in which the device will operate. The first cut at optimization of operating conditions can usually be obtained by analyzing the dispersion relation arising from the linear calculations. Linear theory can be performed at various levels of approximation. These approximations can consist of neglecting self-field effects, simplifying electron orbit calculations, including a partial field representation, and simplifying the geometry. While these calculations are not always simple, it is now commonplace to perform the analysis without approximation by using numerical methods [109–110].

The results of linear theory can be used to determine the system parameters when instability can lead to growth or gain. Once first-order quantities for the perturbed fields and currents have been obtained, it is also possible to determine start-oscillation conditions on the basis of power balance. Start-oscillation occurs when the energy lost to radiation from the beam electrons is balanced by the finite losses in the wave structure. Thus, from the linear theory, we can obtain the start-oscillation conditions for microwave generators.

1.4.2 Nonlinear Models

Whereas some devices operate in the linear regime [111], in most cases, a stationary nonlinear state is the operating point for efficiently generating high-power coherent radiation [112]. The steady state in microwave devices can be described with the aid of a self-consistent system of nonlinear equations, comprised of equations for the excitation of RF by an electron beam and equations of motion of beam electrons in the high-frequency electromagnetic field. The excitation equations are obtained from the wave equation, written for appropriate boundary conditions, with two simplifying assumptions [113–114]:

- Only one wave in synchronism with the beam electrons exists;
- The wave amplitude and phase is slowly varying in time.

This is called the *single-wave approximation.* The equations of motion of particles in various devices with a high efficiency of electron-wave interaction have different forms. However, in devices with inertial electron bunching (CRMs, CARMs, FELs), the relative variation in the particle energy is small when averaged over a wave period, but the phase bunching caused by this variation is large. The motion of the electrons in such cases can be described by the nonlinear pendulum equation [115–116].

When a beam particle loses energy to the RF field, the particle's motion may no longer maintain synchronism with the wave. This is often a saturation mechanism for energy transfer from the beam to the RF field, and this condition limits the device efficiency. The efficiency of devices such as CRMs, FELs, BWOs, and planar magnetrons can be improved by tailoring of the RF structure, or tapering of the magnetic field.

As mentioned in Section 1.2.1, the theoretically predicted intrinsic efficiency of a gyrotron device can be as high as 70 to 90%. Such high efficiencies can be achieved by optimizing the longitudinal structure of the RF field in the cavity and by tapering the longitudinal magnetic field. These effects were first conceived in theoretical analysis.

The theory of nonstationary operation is also very important for many microwave devices. In particular, it is the goal of most designs to avoid such conditions. In devices driven by very short electron-current pulses, transient effects are obviously important. For some microwave generators, if the current density is above a threshold value, many modes will be excited and no stationary state will exist [117–119]. Therefore, a large fraction of the theoretical activity is concerned with multimode systems. In most devices, the interaction involves two or more time scales. This allows application of powerful mathematical tools, such as multiple-time-scale analyses [120] or Hamiltonian [121] perturbation techniques.

When different time scales are identified, a well defined mathematical procedure for averaging over fast time scales is used to obtain a set of reduced nonlinear equations on the slow time scale. For example, in CRMs the fast time scale is associated with the cyclotron period, and the slow time scale is proportional to the inverse of the difference between the wave frequency and the relativistic cyclotron frequency. In FELs, the time scale that takes the electron across one wiggler period is much shorter than the time associated with the electron motion in the beat wave.

These techniques have been applied successfully to gain a better physical insight into the bunching mechanisms of different coherent microwave sources and to provide an excellent tool for the design of microwave devices.

1.4.3 Numerical Simulation

Realistic modeling of nonlinear behavior, based on purely analytical techniques, is limited. As we can see from many of the chapters in this book, numerical simulations have come to play an increasingly important role as a tool for exploring new concepts and for design of specific devices. The most prevalent approaches are:

1. *Test-particle method;*
2. *Full-scale particle-in-cell simulations.*

The test-particle method introduced by Nordsieck [122] assumes that a single wave or mode represents the radiated electromagnetic field and test particles are advanced in the self-consistent field of the wave. By relying on the single-wave approximation, a set of nonlinear equations for the wave amplitude and phase can be derived, which generally has the form:

$$\frac{d\vec{A}}{dt} = \sum_i \vec{F}^i_A (\vec{A}, \phi, \vec{x}^i, \vec{v}^i) \tag{1.6a}$$

$$\frac{d\phi}{dt} = \sum_i F^i_\phi (\vec{A}, \phi, \vec{x}^i, \vec{v}^i) \tag{1.6b}$$

where $\vec{A}$ is the amplitude, ϕ is the phase, and F^i_A, F^i_ϕ are contributions from individual test particles, the positions and velocities of which are advanced by

$$\frac{d(v^i_\gamma)}{dt} = \frac{q}{m} [\vec{E} (\vec{A}, \phi, t) + \frac{\vec{v}^i}{c} \times \vec{B}(\vec{A}, \phi, t)] \tag{1.7a}$$

$$\frac{d\vec{x}^i}{dt} = \vec{v}^i \tag{1.7b}$$

While Nordsieck's initial application was in electrostatic limit and provided a one-dimensional treatment of a TWT, the method has proved so robust that it has become the basis for modeling of CRMs, CARMs, and FELs. The basic idea has been improved to the point where it takes advantage of averaging techniques developed in multiple-time-scale analysis methods and can account for three-dimensional effects under some circumstances.

For the more intractable problems where self-space-charge effects, multiple waves, complicated geometries, and initial conditions play an important role, full-scale particle-in-cell (PIC) simulations have proved invaluable. A large body of literature exists that describes the basis of coding for PIC simulations [123–130]. The advantage of such codes is that, for the high-power microwave devices described in this book, detailed numerical simulations can be performed at an extremely primitive level requiring almost no approximation. The codes have provided detailed insight into the nonlinear phenomena that occur in microwave devices and have contributed greatly to the design of successful experiments. The PIC simulations are fairly complicated, and they are used as numerical analogues of laboratory experiments. The proliferation of high-speed computers should make such codes increasingly available for both research and design.

1.5 TECHNOLOGY

The generation of millimeter waves, submillimeter waves, and high-power microwaves represents different technological problems. The high-power devices operating at hundreds of megawatts usually have short pulses, typically on the order of tens of nanoseconds, and perhaps as long as a few microseconds. Such high-power devices are fundamentally pulsed devices. Experimentally, most of them are also single-shot devices with low repetition rate. Exceptions to this are tubes, such as the SLAC klystrons, which have gone through a rigorous engineering development process and can be reliably operated for thousands of hours with repetition rates of 360 pulses/s.

The physical conditions encountered in pulsed power devices in this extreme parametric range include high voltages and field gradients on the order of MV/cm and also high-energy-density deposition in the walls of the devices. In this regime, the treatment and processing of materials, the properties of insulators, and even the existence of components such as vacuum windows or coupling structures are as crucial as the interesting physics describing the radiative processes. Probably, the evolution of this technology will permit us to answer the question of defining the limits on high power.

Under high-field-gradient conditions, materials become emitters, and this can lead to the formation of surface plasmas. The motion of the plasma determines the time scale on which a device may function as intended. An example of this is the relativistic magnetron, where gap closure limits the RF pulse to tens of nanoseconds. In a similar way, the duration of the electron source operation is affected by plasma motion in linear devices with field emission cathodes. More crucial is the effect of diode geometry and materials on beam quality. As described in Section 1.3, beam quality is an important factor in realizing optimal conditions for radiation. A case in point is the gyrotron, which was not considered to be a successful concept until the magnetron injection gun was used as the beam source.

Pulsed devices, while providing high-power levels (>1 GW), do not usually provide large average powers. The challenge of producing high-average-power, high-peak-power devices possess interesting technological problems in cooling, materials, and switching technology. The ETA and ATA accelerators at Lawrence Livermore National Laboratory are examples of devices that operate in a mutlipulse mode. This area of technology will have a tremendous effect on concepts such as high-gain FELs and has already led to impressive results. Similar advances can be expected from RF accelerator development for high-current, high-energy microwave devices.

Another area of technology that is important for such devices is magnetics. The use of magnetically insulated transmission lines for power feeds to high-current diodes, the use of magnetic fields for containing and guiding beams, the tapering of fields for energy enhancement, and the construction of wigglers and undulators, all play an increasingly important role in high-power microwave sources. The practical attainment of beam conditions that correspond to those necessary for a particular interaction is tied to the technological limits. An example is the period and peak magnetic field in a wiggler. The scattered radiation in a FEL has a wavelength $\lambda = \lambda_w/2\gamma^2$, where λ_w is the wiggler wavelength and γ is the relativistic factor. For a given γ, shorter wavelength radiation can be produced if wigglers with short periods are possible. Because the strength of interaction also depends on the magnetic field at the wiggler tips, the difficult piece of technology is to produce simultaneously high-magnetic-field, short-wavelength wigglers.

For millimeter and submillimeter waves, the technological issues center on the development of components. The realization of sources that produce considerable power at these wavelengths is recent, and the need for components such as couplers, convertors, and windows is only now being taken seriously. In CW devices, cooling technology is again a serious barrier, where active loading is limited to several kW/cm^2. This limits the power to modest levels unless devices that operate reliably in highly overmoded conditions can be constructed.

At short wavelengths, quasioptical techniques appear to be very attractive, and the marriage of vacuum electronics and optics may result in further breakthroughs. An example of this is the quasioptical gyrotron, which uses a Fabry-Perot cavity. The treatment of the radiation field in FELs also requires the use of optical elements.

1.6 SUMMARY

There have been a number of dramatic developments in high-power microwave sources as well as millimeter and submillimeter devices during the last few years. Laboratory experiments have produced power levels in excess of 1 GW on a regular basis, and predictions have been made that this will be improved by more than an order of magnitude. At short wavelengths, the spectral gap in devices that produced even modest powers at wavelengths below 1 cm has been filled. FELs have managed to produce amplification in the visible portion of the spectrum, and may eventually produce soft X-rays. Near-term experiments promise an exciting number of improvements in device performance and technology. There are numerous applications that will need persistent effort and research to achieve what may now appear to be stringent requirements. In retrospect, devices that are now common and well understood were inconceivable less than a decade ago. There is ample opportunity for progress in this field!

ACKNOWLEDGMENT

We would like to acknowledge the support of the Naval Research Laboratory for the work of B. Levush, and the support of the SAIC executive science and technology council for the work of A.T. Drobot. Additionally, we would like to thank Dr. Victor Granatstein for his encouragement.

REFERENCES

A considerable amount of research not cited here has been published in the following journals:

Physical Review Letters; Physical Review, A: General Physics; Physics of Fluids; Applied Physics Letters; Journal of Applied Physics; Applied Physics B; IEEE Transactions on Quantum Electronics; Optical Communication Particle Accelerators; Soviet Physics: JETP, JTP, Letters to JETP, Letters to JTP; Journal of Quantum Electronics.

1. J.L. Hirshfield, and J.M. Wachtel, *Phys. Rev. Lett.* **12**, 533 (1964).
2. D.A.G Deacon, L.R. Elias, J.M.J. Madey, G.J. Ramian, H.A. Schwettman, and T.I. Smith, *Phys. Rev. Lett.* **38**, 892 (1977).
3. C.H. Lee, P.E. Oettinger, E.R. Pugh, R. Klinkowstein, J.H. Jacob, J.S. Fraser, and R.L. Sheffield, *IEEE Trans. Nuc. Sci.* **NS-32**, 3045 (1985).

4. M.I. Petelin, *Sov. Radioph. and Quant. Electr.* **17**, 686 (1974).
5. R.Q. Twiss, *Australian J. Phys.* **II**, 564 (1958); J. Schneider, *Phys. Rev. Lett.* **2**, 504 (1959).
6. A.V. Gapanov, *Izv. VUZ., Radiofiz.* **2**, 450 (1959); H. Motz, *J. Appl. Phys.* **22**, 527, (1951).
7. A. Alexeff and F. Dyer, *Phys. Rev. Lett.* **43**, 351 (1980).
8. V.A. Flyagin, A.V. Gaponov, M. I. Petelin, and V.K. Yulpator, *IEEE Trans. Microwave Theory Techn.* **MTT-25**, 514 (1977).
9. J.L. Hirshfield and V.L. Granatstein, *IEEE Trans. Microwave Theory Techn.* **MTT-25**, 522 (1977).
10. R.S. Symons and H.R. Jory, in C. Marton, ed., *Advances in Electronics and Electron Physics,* (Academic Press, New York, 1981).
11. Special Issue on Gyrotrons, *Int. J. Electr.* **51** (1981); Special Issue on Gyrotrons, *Int. J. Electr.* **53** (1982); Special Issue on Gyrotrons, *Int. J. Electr.* **57** (1984); Special Issue on Gyrotrons, *Int. J. Electr.* **61** (1986).
12. A.V. Gaponov, M.I. Petelin, and V.K. Yulpatov, *Radiophys. Quant. Electr.* **10**, 794 (1967).
13. A.L. Goldenberg and T.B. Pankratova, *Electr. Ser. 1 Electrtoniks SVch.* **4**, 81 (1971).
14. A.A. Kuraev, "Theory and Optimization of Microwave Electron Devices (in Russian)" (Izd. Nauka i Technika, Minsk, 1979).
15. P. Sprangle and A.T. Drobot, *IEEE Trans. Microwave Theory Techn.* **MTT-25**, 528 (1977).
16. P. Vitello, *Int. J. Infrar. and Mill. Waves* **8** (1987).
17. B.G. Danly and R.J. Temkin, *Phys. Fluids* **29**, 561 (1985).
18. V.L. Bratman and M.I. Petelin, *Radiophys. Quant. Electron.* **18**, 1136 (1975).
19. A.W. Fliflet, M.E. Read, K.R. Chu, and R. Seeley, *Int. J. Electron.* **53**, 505 (1982).
20. Yu.K. Bykov, A.L. Goldenburg, L.V. Nikolaev, M.M. Ofitserov, and M.I. Petelin, *Radiophys. Quant. Electr.* **18**, 1141 (1975).
21. K.E. Kreisher and R.J. Temkin, *Int. J. Infr. and Mill. Waves* **2**, 175 (1981).
22. I.I. Antakov, V.A. Gintsburg, E.V. Zasypkin, and E.V. Sokolov, *Elektronnaya Tekhniks Ser. 1 Elektron. SVch* **8**, 20 (1975).
23. S.W. McDonald, J.M. Finn, M.E. Read, and W.M. Manheimer, *Int. J. Electron.* **61**, 795 (1986).
24. Y. Carmel, K.R. Chu, M.E. Read, A.K. Ganguly, D. Dialetis, R. Seeley, T.S. Levine, and V.L. Granatstein, *Phys. Rev. Lett.* **50**, 112 (1983).

25. P. Sprangle, J.L. Vomvoridis, and W.M. Manheimer, *Phys. Rev.* **A23**, 3127 (1981).
26. T.A. Hargreaves, K.J. Kim, J.M. McAdoo, S.Y. Park, R.D. Seeley, and M.E. Read, *Int. J. Electron* **57**, 977 (1984).
27. N.A. Ebrahim, Z. Liang, and J.L. Hirshfield, *Phys. Rev. Lett.* **49**, 1556 (1982).
28. A. Bondenson, B. Levush, W.M. Manheimer, and E. Ott, *Int. J. Electr.* **53**, 547 (1982).
29. A. Bondenson, W.M. Manheimer, and E. Ott, in K.J. Button, ed., *Infrared and Millimeter Waves*, (Academic Press, New York, 1983), Vol. 9, p. 309.
30. V.L. Bratman, N.S. Ginzburg, G.S. Nusinovich, M.I. Petelin, and P.S. Strelkov, *Int. J. Electr.* **51**, 541 (1981).
31. A.A. Kolomenskii and A.N. Lebedev, *DAN SSSR* **145**, 1259 (1962).
32. V.L. Bratman, C.G. Denison, N.S. Ginzburg, and M.I. Petelin, *IEEE J. Quant. Electr.* **QE-19**, 282 (1983).
33. W.M. Bollen, A.H. McCurdy, B. Arfin, R.K. Parker, and A.K. Ganguly, *IEEE Trans. Plasma Sc.* **PS-13**, 417 (1985).
34. K.R. Chu, V.L. Granatstein, P.E. Latham, W. Lawson, and C.D. Striffler, *IEEE Trans. Plasma Sc.* **PS-13**, 424 (1985).
35. T.M. Tran, B.G. Danly, K.E. Kreischer, J.B. Schutkeker, and R.J. Temkin, *Phys. Fluids* **29**, 1274 (1986).
36. W.M. Bollen, A.H. McCurdy, B. Arfin, R.K. Parker, and A.K. Ganguly, *IEEE Trans. Plasma Sc.* **PS-13**, 417 (1985).
37. K.R. Chu, V.L. Granatstein, P.E. Latham, W. Lawson, and C.D. Striffler, *IEEE Trans. Plasma Sc.* **PS-13**, 424 (1985).
38. K.E. Kreischer, J.B. Schutkeker, B.G. Danly, W.J. Mulligan, and R.J. Temkin, *Int. J. Electr.* **57**, 835 (1984).
39. G. Mourier, G. Faillon, and P. Gapin, *Int. J. Electr.* **61**, 715 (1986).
40. V.A. Flyagin, A.G. Luchinin, and G.S. Nusinovich, *Int. J. Infr. and Mill. Waves* **3**, 765 (1982).
41. A.G. Luchinin, G.S. Nusinovich, and V.A. Flyagin, *J. Techn. Phys.* **53**, 1629, (1983).
42. K. Felch, R. Bier, L.J. Graig, H. Hue, L. Ives, M. Jory, L. Lorez, and S. Spang, *Int. J. Electr.* **61**, 701 (1986).
43. J.L. Byerly, B.G. Danly, K.E. Kreischer, R.J. Temkin, W.J. Mulligan, and P. Woskoboinikow, *Int. J. Electr.* **57**, 1033 (1984).
44. S. Ono, M. Ansai, N. Sato, K. Yokoo, K. Henmi, T. Idehara, T. Tachikawa, I. Okazaki, T. Okamoto, 11*th* Int. Conf. on Infrared and Millimeter Waves, Tirrenia, Italy, Oct. 1986. G. Moruzzi, ed., *Conf. Digest,* p. 37.
45. L. Barnett, Y.Y. Lau, K.K. Chu, and V.L. Granatstein, *IEEE Trans. Electr. Dev.* **ED-28**, 872 (1981).

46. H. Jory, S. Hegji, J. Shirely, and R. Symons, *Microwave J.* **21**, 30 (1978).
47. W.W. Destler, R. Kulkarni, C.D. Striffler, and R.L. Weiler, *J. Appl. Phys.* **54**, 4152 (1983).
48. K.R. Chu, D.S. Furuno, N.C. Luhman, D.B. McDermott, P. Vitello, and K.Ko, *IEEE Trans. Plasma Sc.* **PS-13**, 435 (1985).
49. C. Pellegrini, *IEEE Trans. Nucl. Sc.* **NS-26**, 3791 (1979); D. Prosnitz, "Free Electron Lasers," in M.J. Weber, ed., *CRC Handbook of Laser Science and Technology* (CRC Press, Boca Raton, FL, 1982), Vol. 1, p. 425; V.L. Granatstein, R.K. Parker, and P. Sprangle, in M.J. Weber, ed., *CRC Handbook of Laser Science and Technology*, (CRC Press, Boca Raton, FL, 1982), Vol. 1, p. 441; T.C. Marshall, S.P. Schlesinger, and D.B. McDermott, in L. Marton and C. Marton, eds., Advances in Electronics and Electron Physics (Academic Press, New York, 1980); N.M. Kroll, P. Morton, and M.N. Rosenbluth, *IEEE J. Quant. Electr.* **QE-17**, 1436 (1981); W.B. Colson and A.M. Sessler, *Ann. Rev. Nucl. Sc.* **35**, 25 (1985).
50. *IEEE J. Quant. Electr.* **QE-17**, 8 (1981); *IEEE J. Quant. Electr.* **QE-19**, 3 (1983); *IEEE J. Quant. Electr.* **QE-21**, 7 (1985).
51. S.F. Jacobs, M. Sargent, and M.O. Scully, eds., "Novel Sources of Coherent Radiation," *Physics of Quantum Electronics.,* Vol. 5, (Addison-Wesley, Reading, MA, 1978); S.F. Jacobs, H.S. Pilloff, M. Sargent, and M.O. Scully, R. Spitzer, eds., "Free Electron Generators of Coherent Radiation," *Physics of Quantum Electronics,* Vol. 7, (Addison-Wesley, Reading, MA, 1980); S.F. Jacobs, H.S. Pilloff, M. Sargent, M.O. Scully, and R. Spitzer, eds., "Free Electron Generators of Coherent Radiation," *Physics of Quantum Electronics*, Vol. 8, (Addison-Wesley, Reading, MA, 1982); S.F. Jacobs, H.S. Pilloff, M. Sargent, M.O. Scully, and R. Spitzer, eds., "Free Electron Generators of Coherent Radiation," *Physics of Quantum Electronics*, Vol. 9, (Addison-Wesley, Reading, MA, 1982); Free Electron Laser Conference, *Journal de Physique, Colloq.* **C1**, Suppl. No. 2, 44 (1983); C.A. Brau, S.F. Jacobs, and M.O. Scully, eds., "Free Electron Generators of Coherent Radiation," SPIE **453** (1984).
52. R.B. Miller, *An Introduction to the Physics of Intense Charge Particle Beams*, (Plenum, New York, 1982); T.C. Marshall, *Free Electron Laser*, Macmillan, New York, 1985).
53. H. Motz, *J. Appl. Phys.* **22**, 527 (1951). H. Motz, W. Thone, and R.N. Whitehorst, *J. Appl. Phys.* **24**, 826 (1953).
54. R.M. Phillips, *Trans. IRE Electron Dev.* **7**, 231 (1960).
55. J.M.J. Madey, *J. Appl. Phys.* **42**, 1906 (1971).
56. L.R. Elias, W.M. Fairbank, J.M.J. Madey, H.A. Schwettman, and T.I. Smith, *Phys. Rev. Lett.* **36**, 710 (1976).

57. D.B. McDermott, T.C. Marshall, S.P. Schlesinger, R.K. Parker, and V.L. Granatstein, *Phys. Rev. Lett.* **41**, 1368 (1978).
58. P. Sprangle, R.A. Smith, and V.L. Granatstein, "Sources of Radiation," in K.B. Button, ed., *Infrared and Millimeter Waves*, Vol. 1 (Academic Press, New York, 1979), p. 279.
59. V.L. Granatstein, S.P. Schlesinger, M. Herndon, R.K. Parker, and J.A. Pasour, *Appl. Phys. Lett.* **30**, 384 (1977).
60. P.C. Efthimion and S.P. Schlesinger, *Phys. Rev. A.* **16**, 633 (1977).
61. L. Friedland, *Phys. Fluids* **23**, 2376 (1980).
62. H.P. Freund and A.K. Ganguly, *Phys. Rev. A.* **28**, 3438 (1983).
63. S.M. Gold, W.M. Black, H.P. Freund, V.L. Granatstein, R.H. Jackson, P. Efthimion, and A.K. Kinkead, *Phys. Fluids* **26**, 2683 (1983).
64. J. Fajans, G. Bekefi, Y.Z. Yin, and B. Lax, *Phys. Fluids* **28**, 1995 (1985).
65. T.J. Orzechowski, B. Anderson, W.M. Fawley, D. Prosnitz, E.T. Scharlemann, S. Yarema, D.B. Hopkins, A.C. Paul, A.M. Sessler, and J.S. Wurtele, *Phys. Rev. Lett.* **54**, 889 (1985).
66. R.W. Warren, *et al., Nucl. Instrum. Methods Phys. Res.* **A237**, 180 (1985).
67. E.D. Shaw and C.K.N. Patel, in S.F. Jacobs, *et al.*, eds., *Physics of Quantum Electronics, Vol. 9,* (Addison Wesley, Reading, MA, 1982); C.M. Tang, P. Sprangle, S. Penner, B.M. Kincaid, and R.R. Freeman, *Nucl. Instrum. Methods Phys. Res.* **A250**, 278 (1986).
68. M. Billardon, P. Elleaume, J.M. Ortega, C. Bazin, M. Bergher, M. Velghe, Y. Petroff, D.A.G. Deacon, K.E. Robinson and T.M.J. Madey, *Phys. Rev. Lett.* **51**, 1652 (1983).
69. N.A. Vinokurov and A.N. Strinsky (preprint), *INP 77-59*, (Novosibibsk, 1977).
70. G.A. Kornyukuin, *et al., Nucl. Instr. Meth. Phys. Res.* **A237**, 281 (1985); M. Billardon, *et. al., Nucl. Instr. Meth. Phys. Res.* **A237**, 244 (1985).
71. C.A. Brau, *IEEE J. Quantum Electronics* **QE-21**, 824 (1985).
72. T.J. Orzechowski, B. Anderson, W.M. Fawley, D. Prosnitz, E.T. Scharlemann, S. Yarema, D.B. Hopkins, A.C. Paul, A.M. Sessler, and J.S. Wurtele, *Phys. Rev. Lett.* **QE-21**, 831 (1985).
73. C.W. Roberson, *IEEE J. Quantum Electronics* **QE-21**, 860 (1985).
74. R.K. Parker, R.M. Jackson, S.H. Gold, H.P. Freund, V.L. Granatstein, P.C. Efthimion, M. Herndon, and A.K. Kinkead, *Phys. Rev. Lett.* **48**, 238 (1982).
75. A.T. Lin and J.M. Dawson, *Phys. Rev. Lett.* **42**, 1670 (1979).
76. P.M. Sprangle, C.M. Tang, and W.M. Manheimer, *Phys. Rev. Lett.* **43**, 1932 (1979).

77. N.M. Kroll, P. Morton, and M.N. Rosenbluth, *IEEE J. Quantum Electronics.* **QE-17**, 1436 (1981).
78. T.J. Orzechowski, B.R. Anderson, J.C. Clark, W.M. Fawley, A.C. Paul, D. Prosnitz, E.T. Sharlemann, and S.M. Yarema, *Phys. Rev. Lett.* **57**, 2172 (1986).
79. A. Palevsky and G. Bekefi, *Phys. Fluids* **22**, 986 (1979).
80. R.A. Close, A. Palevsky, and G. Bekefi, *J. Appl. Phys.* **54**, 4187 (1983).
81. N.F. Kovalev, E.G. Krastelev, M.I. Kuznetsov, A.M. Maine, M.M. Ofitserov, V.A. Papdichev, M.I. Fuks, and L.N. Chekanova, *Sov. Tech. Phys. Lett.* **6**, 197 (1980).
82. I.Z. Gleizer, A.N. Didenko, A.S. Sulakshin, G.P. Formant, and V.I. Isvestrov, *Sov. Tech. Phys. Lett.* **6**, 19 (1980).
83. W.W. Destler, R. Kulkarni, C.D. Striffler, and R.L. Werler, *J. Appl. Phys.* **54**, 7 (1983).
84. J. Benford, M. Sze, T. Young, D. Bromley, and G. Proulx, *IEEE Trans. plasma Sc.* **PS-13**, 538 (1985).
85. Y. Carmel, J. Ivers, R.E. Kribel, and J. Nation, *Phys. Rev. Lett.* **33**, 1278 (1974); N.F. Kovalev, *et al., JETP Lett.* **18**, 138 (1973).
86. V.L. Bratman, *et al., Pisma Zh. Tekhn. Fiz.* **10**, 807 (1984).
87. N.I. Zaitser, *et al., Pisma Zh. Tekhn. Fiz.* **8**, 911 (1982).
88. S.D. Polevin, *Pisma Zh. Tekhn. Fiz.* **10**, 1269 (1984).
89. E. Garate, C. Shaughnessy, and J. Walsh, *Proc., 8th Int. Free-Electron Laser Conf.*, Glasgow, Scotland, Sept. 1986.
90. G.G. Denisov, A.V. Smorgensky, V.P. Gubanov, S.D. Korovin, V.V. Rostor, and M.I. Yaladin, *Int. J. Infr.* and *Mill. Waves* **5**, 1389 (1984).
91. R.A. Kehs, A. Bromborsky, B.G. Ruth, S.E. Graybill, W.W. Destler, Y. Carmel, M.C. Wang, *IEEE Trans. Plasma Sc.* **PS-13**, 559 (1985).
92. S.P. Bugaev, V.I. Kanavets, A.I. Klinov, V.I. Koshelev, and V.A. Cheropenin, *Pisma. Zh. Tekhn. Fiz.* **10**, 1229 (1984).
93. C.K. Birdsall and W.B. Bridges, *J. Appl. Phys. R.* **32**, 2611 (1961).
94. R.A. Mahaffey, P. Sprangle, J. Golden, and C.A. Kapetanakos, *Phys. Rev. Lett.* **39**, 843 (1977).
95. A.N. Didenko, Ya.E. Krasik, S.F. Perelygin, and G.P. Fomenko, *Sov. Tech. Phys. Lett.* **5**, 128 (1979).
96. A.L. Peratt, C.M. Snell and L.E. Thode, *IEEE Trans. Plasma Sc.* **PS-13**, 498 (1985).
97. A.I. Akhiezer and Ya.B. Fainberg, *Dokl. Akad. Nauk SSR* **65**, 555 (1949).
98. D. Bohm and E. Gross, *Phys. Rev.* **75**, 1851 (1949).

99. A.K. Berezin, Ya.B., Fainberg, L.I. Bolotin, *et al., Sov. Phys. JETP* **36**, 453 (1973).
100. M.S. Rabinovich, A.A. Rukhadze, P.S. Strelkov and A.G. Shkrarunets, *Zh. Eksp. Teor. Fiz.* **58**, 1219 (1975).
101. L.S. Bogdankevich, M.C. Kuzelev, and A.A. Rukhadze, *Sov. J. Plasma Phys.* **5**, 51 (1979).
102. L.S. Bogdankevich, M.V. Kuzelev, and A.A. Rukhadze, *Sov. Phys. Usp.* **24**, 1 (1981).
103. Yu.V. Tkach, Ya.B. Fainberg, I.I. Magda, V.D. Shapiro, V.I. Shevchenko, A.I. Zukov, Ye.A. Lemberg, I.N. Mondrus, and N.P. Gadetskiy, *Zh. ETEP* **16**, 368 (1972).
104. YuV. Tkach, Ya.B. Fainberg, I.I. Magda, I.I. Gaponenko, G.V. Skachek, S.S. Pushkarev, N.I. Gadetskiy, and A.A. Belukha, *Sov. Phys. Plasma* **1**, 81 (1975).
105. M.V. Kuzelev, F. Kh. Mukhametzyanov, and M.S. Shkvarunets, *Sov. Phys. JETP* **56**, 780 (1982).
106. K.G. Kato, G. Benford, and D. Tzach, *Phys. Fluids* **26**, 3636 (1983).
107. L. Feinstein, and K. Felch, *IEEE Trans. Electron Dev.* **ED-34**, 461 (1987).
108. V.L. Bratman, G.G. Denison, M.M. Ofitserov, S.D. Korovin, S.D. Polevin, and V.V. Rostor, *IEEE Trans. Plasma Sc.* **PS-15**, 2 (1987).
109. T. Antonsen, W.M. Manheimer, and B. Levush, *Int. J. Electr.* **61**, 823 (1986).
110. W.P. Marable, P. Sprangle, and C.M. Tang, *Nucl. Instr. Meth. Phys. Res.* **A250**, 159 (1986).
111. P. Sprangle, C.M. Tang, and C.W. Roberson, *Nucl. Instr. Meth. Phys. Res.* **A39**, (1985).
112. A.V. Gaponov, V.A. Flyagin, A.L. Goldenberg, G.S. Nusinovich, Sk. E. Tsimring, V.G. Usov, and S.N. Vlasov, *Inst. T. Electr.* **51**, 277 (1981).
113. P. Sprangle, C.M. Tang, and W.M. Manheimer, *Phys. Rev.* **A21**, 302 (1980).
114. V.L. Bratman, N.S. Ginzburg, and M.I. Petelin, *Sov. Phys. JETP* **49**, 469 (1979).
115. W.B. Colson, *Phys. Lett.* **64A**, 190 (1977).
116. V.L. Bratman, N.S. Ginzburg, and M.I. Petelin, *Optics Comm.* **30**, 409 (1979).
117. Ya.L. Bogomolov, V.L. Bratman, N.S. Ginzburg, M.I. Petelin, and A.D. Yunakovsky, *Optics Commun.* **36**, 209 (1981).
118. N.S. Ginzburg, G.S. Nusinovich, and N.A. Zavolsky, Int. *J. of Electr.* **61**, 881 (1986).
119. N.S. Ginzburg, *Sov. Phys. Tech. Phys.* **31**, 571 (1986).

120. R.C. Davidson, *Methods in Nonlinear Plasma Theory,* (Academic Press, New York, 1971).
121. A.J. Lichtenberg and M.A. Lieberman, "Regular and Stochastic Motion," *Applied Mathematical Sciences*, Vol. 38, (Springer-Verlag, Berlin, 1983).
122. A. Nordsieck, *Proc. IRE* **41**, 630 (1953).
123. R.W. Hockney and J.W. Eastwood, *Computer Simulation Using Particles* (McGraw-Hill, New York, 1981).
124. A.B. Langdon and B.F. Lasinski, in J. Killeen, ed., *Methods of Computational Physics,* (John Killeen, ed.) (Academic Press, New York, 1976), Vol. 16, pp. 327-366.
125. B. Adler and S. Fernbach, eds., *Methods in Computational Physics,* Vol. 9 (Academic Press, New York, 1970). *See* R.W. Hockney, pp. 136-212; R.L. Morse, pp. 213-240; C.K. Birdsall, A.B. Langdon, and H. Okuda, pp. 241-258.
126. J.P. Boris and R.A. Shanny, eds., *Proc. 4th Conf. Numerical Simulation of Plasma*, Naval Research Laboratory, Washington, DC, Nov. 1970.
127. John M. Dawson, *Rev. Mod. Phys.* **55** (2), (1983).
128. A. Palevsky and A.T. Drobot, "Application of Electromagnetic P.I.C. Codes to Microwave Devices," *Proc. 9th Conf. Numerical Simulation of Plasmas,* Northwestern University, Evanston, IL, June-July, 1980.
129. A. Mankofsky, L. Seftor, C.L. Chang, A. Mondelli, and A.T. Drobot, "ARGUS—A General Three-Dimensional Plasma Simulation Model," *Bull. Am. Phys. Soc.* **28** (8), 1124 (1983).
130. C.K. Birdsall and A.B. Langdon, in *Plasma Physics via Computer Simulation* (McGraw-Hill, New York, 1985).

Chapter 2
HIGH-POWER MICROWAVE TECHNOLOGY AND DIAGNOSTICS

C. B. Wharton

Laboratory of Plasma Studies
Cornell University
Ithaca, New York

2.1 INTRODUCTION*

Microwave generators employing intense relativistic electron beams are capable of generating very large powers (exceeding 100 MW) for short time durations (tens of nanoseconds), on a single-shot or slow-repetition-rate basis. Such generators offer the promise of producing extremely high powers (exceeding 5 GW), but some unusual problems are encountered in relation to their operating conditions. For example, an oscillator producing 200 MW during a 50 ns pulse emits only 10 J of energy, yet creates electromagnetic fields of hundreds of kV/cm, capable of causing breakdown in ordinary waveguides. For that reason, oversized waveguides are used, which are usually cylindrical in cross section, and able to support several possible modes. The investigator may want to excite only one particular mode, possibly requiring the use of a mode discriminator or transition. However, these devices generally contain structures that lower the breakdown threshold, and are thus ruled out in high-power investigations. So, the system is left to its own devices, propagating whatever mode is generated in the source, very often a transverse magnetic (TM) mode that couples to the axial electron beam. TM modes may also be

*This chapter is based on a lecture presented at the IEEE Plasma Sciences Conference, Saskatoon, Saskatchewan, May 21, 1986.

preferred because they are generally able to propagate higher power levels than TE modes before breakdown occurs.

Standard diagnostic instruments for oversized TM-mode cylindrical waveguides are just now beginning to appear. A large portion of this chapter is devoted to diagnostics, including discussion of some instruments that are relatively new to the field.

2.2 SOURCES OF HIGH-CURRENT RELATIVISTIC ELECTRON BEAMS

Field-emission cold cathodes are able to emit large currents when pulsed potentials of hundreds to thousands of kilovolts are applied, on a single-pulse or slow-repetition-rate basis. The pulsed power technology supporting high-current beam devices is now highly developed [1], and reliable systems may be purchased from several manufacturers, or designed and fabricated to well defined specifications by the investigator. Other authors treat this subject in depth [2].

Figure 2.1 shows schematically a pulsed system powering a diode that provides an electron beam in a microwave backward-wave oscillator (BWO) structure. The *Marx bank* consists of a stack of capacitors charged in parallel and discharged on command in series to power the coaxial pulse-forming line to a high voltage, typically 250 kV to 3 MV.

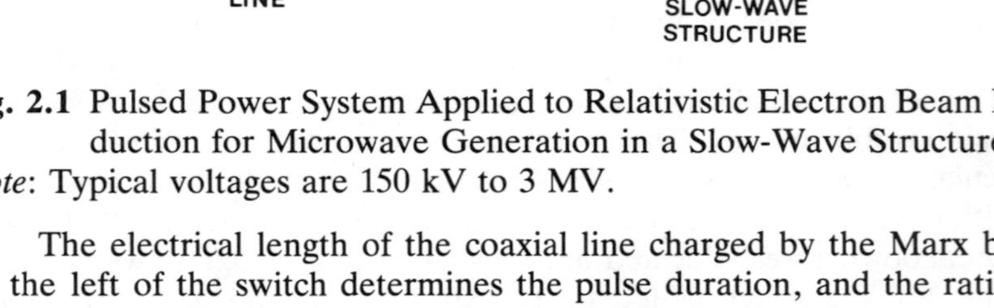

Fig. 2.1 Pulsed Power System Applied to Relativistic Electron Beam Production for Microwave Generation in a Slow-Wave Structure
Note: Typical voltages are 150 kV to 3 MV.

The electrical length of the coaxial line charged by the Marx bank on the left of the switch determines the pulse duration, and the ratio of outer-to-inner diameters determines the impedance:

$$\tau = l/\sqrt{LC} = l/\sqrt{\mu\epsilon_0 K} \quad \text{(s)} \tag{2.1}$$

$$Z = \sqrt{L/C} = \frac{1}{2\pi}\sqrt{\frac{\mu}{\epsilon_0 K}}\ln\frac{b}{a} \quad (\Omega) \tag{2.2}$$

where b and a are the outer and inner diameters, l is the length, and K is the dielectric constant of the insulation.

Distilled, deionized water is often used for insulation between conductors because its dielectric constant of $K = 81$ permits the line to be nine times shorter than if it were filled with air.

The length of pulse line on the right of the switch constitutes a capacitive load to decrease the voltage that appears on the diode during the microsecond or so of charging time of the line by the Marx bank, a procedure called *prepulse suppression*. The diode emits very little current during the prepulse time if the suppression is sufficient.

Field emission diodes are now well understood, and their applications are widespread [1, 2]. The emission current density is limited by space charge, its value being given by the Child-Langmuir equation (MKS units):

$$J = \frac{4\epsilon_0}{9}\sqrt{\frac{2e}{m}}\,V^{3/2}/d^2 \quad (\text{A/m}^2) \tag{2.3}$$

where V is the applied voltage and d is the anode-cathode gap.

Many diodes use a tungsten-coated or graphite cathode and a thin (.0005 to .005 in) titanium or aluminum foil anode, with the beam being extracted through the foil. At currents of 25,000 A or less, the foil survives for many pulses before it needs replacement. At higher currents, an in-vacuum foil changer is usually employed to allow faster changing of ruptured foils, while maintaining clean vacuum conditions.

In addition to the problem of burning out, a foil introduces another problem; it scatters the beam electrons. When the beam is being guided by a strong magnetic field, the scattering has the effect of introducing a transverse temperature. For wave-growth experiments, the best growth rates are obtained with a cold beam [3], requiring as thin a foil as possible.

A third problem inherent with foil-anode diodes is *diode closure,* the effect of plasma, produced by electron collisions, drifting out into the gap. The effective closing of the gap decreases the impedance, causing runaway electron current (or collapsing voltage) on a time scale of 50 to 100 ns. Thus, the application of longer duration pulses is pointless because the diode becomes shorted after the closure time. Damage to the electrodes and excessive loading of the pulse source may result from attempts to increase the pulse length much beyond the closure time limit.

A diode without any foil at all is also possible when the beam is strongly magnetized, in which case the electrons are shot through a hole in the anode, as shown in Figure 2.2. Foilless diodes usually produce hollow beams because the strongest fields are at the outer edge of the cylindrical cathode. A sharp edge can lead to very thin annular beams. Foilless diodes, in principle, can be pulsed at a high repetition rate with microsecond pulses, but the practical limits of present systems are a few thousand pulses in total.

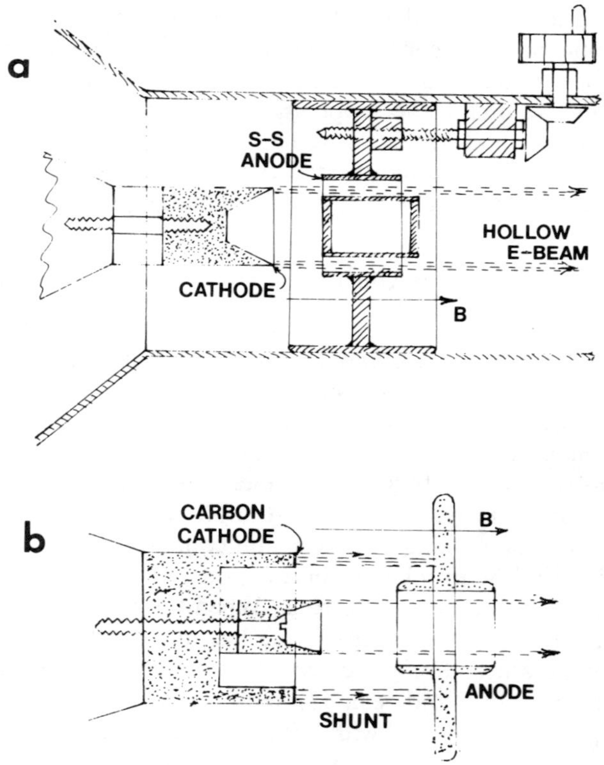

Fig. 2.2 Foilless Diodes for Production of Relativistic Electron Beams: (a) Magnetically Insulated Diode Having Variable Anode-Cathode Spacing; (b) Shunt-Beam Diode Having Outer Cathode Ring that Passes Most of the Current and Inner Cathode that Produces the Beam

Hollow beams may suffer from various instabilities, especially filamentation, in which case the smooth cylinder breaks up into field-aligned columns [5]. The consequences are the generation of spurious modes and frequencies, and an overall degradation of output [6]. Very strong magnetic fields, in excess of 35 kG, together with the close proximity of conducting walls, help to control this instability [6].

Slow-wave generators have beam impedances of 10–100 Ω, while most available pulsed power sources have much lower internal impedances, typically 4–8 Ω. To avoid a mismatch, a shunting cathode structure is often employed, as shown in Figure 2.2(b). The outside cathode ring is spaced from the anode plate such that most of the current is dumped, while the inner cathode ring emits electrons that pass through the hole in the anode plate [4]. The relative proportions are determined by adjusting the gap of the outer cathode ring. The shunt portion of the diode also acts as a voltage regulator, tending to flatten the pulse, at least up until the onset of closure. Use of a high-impedance line, not requiring the shunt, would lessen the closure problem because the magnetic insulation inhibits formation of the anode plasma. A passive resistive shunt can also be used to provide regulation.

The pulsed power technology for other types of microwave generators, such as relativistic magnetrons, vircators, and FELs, is quite similar to the technology discussed above, except that each device has its own requirements for electron generation and propagation. Our discussion here will be limited to rectilinear beams in slow-wave structures.

2.3 BEAM INTERACTIONS WITH SLOW-WAVE STRUCTURES

High-power BWO devices have been investigated for over a decade [7], yielding results that are encouraging, but at the same time disappointing. Efficiencies as high as 35% at moderate levels have been reported [8], but, as power levels have been increased toward the gigawatt level, the outputs saturate and the frequency spectra become broadened and noisy, probably because of beam breakup and turbulence [9]. Thus, although terawatt beams are available, the maximum microwave powers so far claimed (albeit not very well documented with reliable diagnostics) are 1–5 GW [5]. In fact, as experimenters improve their measurement capabilities, it seems that the announced power claims decrease, implying that we need to pay closer attention to the methods and instruments used to make diagnostic measurements.

Conventional, nonrelativistic BWO structures are usually tape-wound helices, which are closely spaced to obtain the slow-wave velocities that match the beam velocity (10). With relativistic electron beams, having velocities of 0.8 c to 0.9 c, the ripple spacing is centimeters instead of millimeters, and the structures must be large and smooth to inhibit surface breakdown. A helix is not optimum in this case for at least two reasons: (1) the relatively large pitch creates a nonaxially symmetrical mode, while most experiments seek TM_{0n} modes; (2) the sharp edges of the tape cause electric field breakdown at much lower fields than the smooth structures to be discussed next.

The smooth structures sketched in Figures 2.1 and 2.2 are typical of most contemporary experiments employing relativistic electron beams at high-power levels. In most cases, the diameters are large enough to permit several modes to propagate, although only one may have sufficient gain for excitation.

The BWO interaction is between the slow-space-charge wave on the beam and the backward wave of the structure, as shown in an X-band example in Figure 2.3, causing wave growth in the upstream direction on the beam [9]. Thus, the microwave output is conventionally taken from the input end of the structure. However, at high powers, where breakdown in the coupling line is imminent, experimenters have found that the waves can be reflected from an impedance discontinuity at the input end and extracted from the same end as the beam exit. Decreasing the magnetic field at the end of the solenoid dumps the beam onto the waveguide wall without further interaction with the wave.

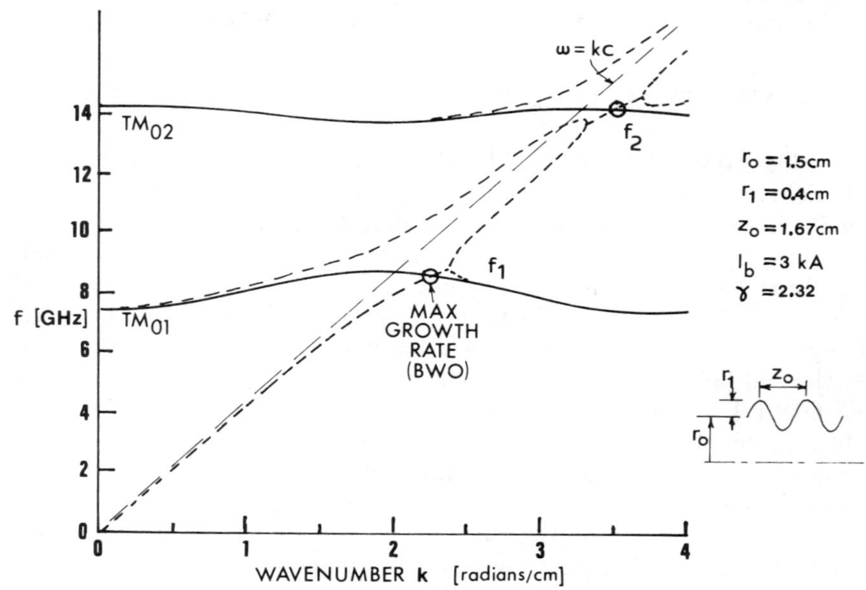

Fig. 2.3 Dispersion Curves for the Two Lowest Modes, Showing Beam Coupling to Backward Waves

Large magnetic fields are required to provide guidance to the electrons and to insulate radially the cathode from the anode. However, for some cases, as the magnetic field is increased, the microwave power output decreases to a very low value at conditions corresponding to wave-wave resonances, and then increases again. Field strengths of 25 to 50 kG are generally above the minimum point, which means that very sturdy solenoid systems, driven by moderate power sources (usually pulsed capacitor banks) are required.

It is possible to fabricate high-power coupling structures for the input end, instead of extracting power as discussed above, and the system shown in Figure 2.4 has a number of advantages. First, the large amplitude wave need not be reflected back through the interaction region from which it began to grow, eliminating one stray feedback loop and decreasing the introduction of noise into the wave. Second, the beam dump is not the wall of the waveguide, where erosion becomes serious after a period of operation. Third, the output coupling can be to TE-mode waves, which may be desirable for radiation because the pattern is not hollow in the center, as is that of a TM mode.

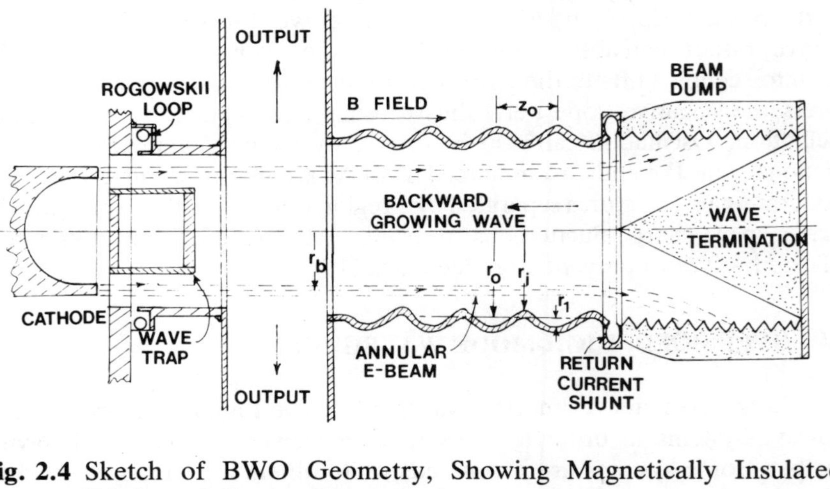

Fig. 2.4 Sketch of BWO Geometry, Showing Magnetically Insulated Diode and Current Loops

The Rogowski loop measures the total electron-beam current entering the interaction region, while the return current shunt (which could also be a Rogowski loop) measures the current that transits the region without being intercepted by the walls due to wave-scattering or other beam-divergence effects. The downstream current monitor can also be divided into several loops around the periphery to detect evidence of filamentation.

A convenient method to fabricate a Rogowski loop is to wind a spiral of narrow polyvinyl chloride (PVC) tape around a section of semirigid coaxial line, and submerge it in an etching solution to leave a metal spiral, supported by the polytetrafluoroethylene (PTFE) insulation. The sensitivity is determined by the diameter and the number of turns, and the current return is provided by the center conductor.

X-ray energy analysis can be applied to beam-wave systems to determine the effective gamma γ of the beam, both before and after interaction with the structure. A 1 mm^2 target of thin titanium foil, supported by a quartz fiber directly in the beam, has been successfully used for this purpose [3], with the results indicating an average energy decrease of about 75 keV per electron for one particular plasma interaction.

Traveling wave tube amplifiers can be made with exactly the same techniques as are used for BWOs, except that the beam interacts with the forward wave of a structure. This means that the periodicity must be more closely spaced in order to obtain a larger longitudinal wavenumber (i.e., smaller wave velocity). The interaction region shown in Figure 2.3 is then on the positive-slope part of the dispersion curve, for which there is convective, rather than absolute, instability. As the beam current is increased, the intersection shifts to the right, due to increasing space charge, until it enters the negative-slope part, thus causing the amplifier to break into oscillation. Oscillation can also be caused by feedback due to wave reflections from the ends. Commercial TWTAs are usually divided into two moderate-gain structures separated by a region that passes the beam (which carries space-charge fluctuations from the first structure), but attenuates reflected waves to prevent stray feedback [10].

2.4 OVERSIDED, HIGH-MODE WAVEGUIDES

Large diameter, cylindrical waveguides have found wide use for microwave systems in order to transmit large power without breakdown. Further, some systems are evacuated, while others are pressurized with sulfur-hexafluoride to increase their power-handling ability. TM modes generally are excited by linear beam structures, the particular mode number being determined by several factors. TM_{01} and TM_{03} are commonly reported modes for BWOs, while modes up to TM_{06} are reported for vircators [11]. Figure 2.5 sketches the field patterns for four common modes, where the shaded regions indicate cross sections of maximum power flow. The magnitude of the electric field depends on the cut-off

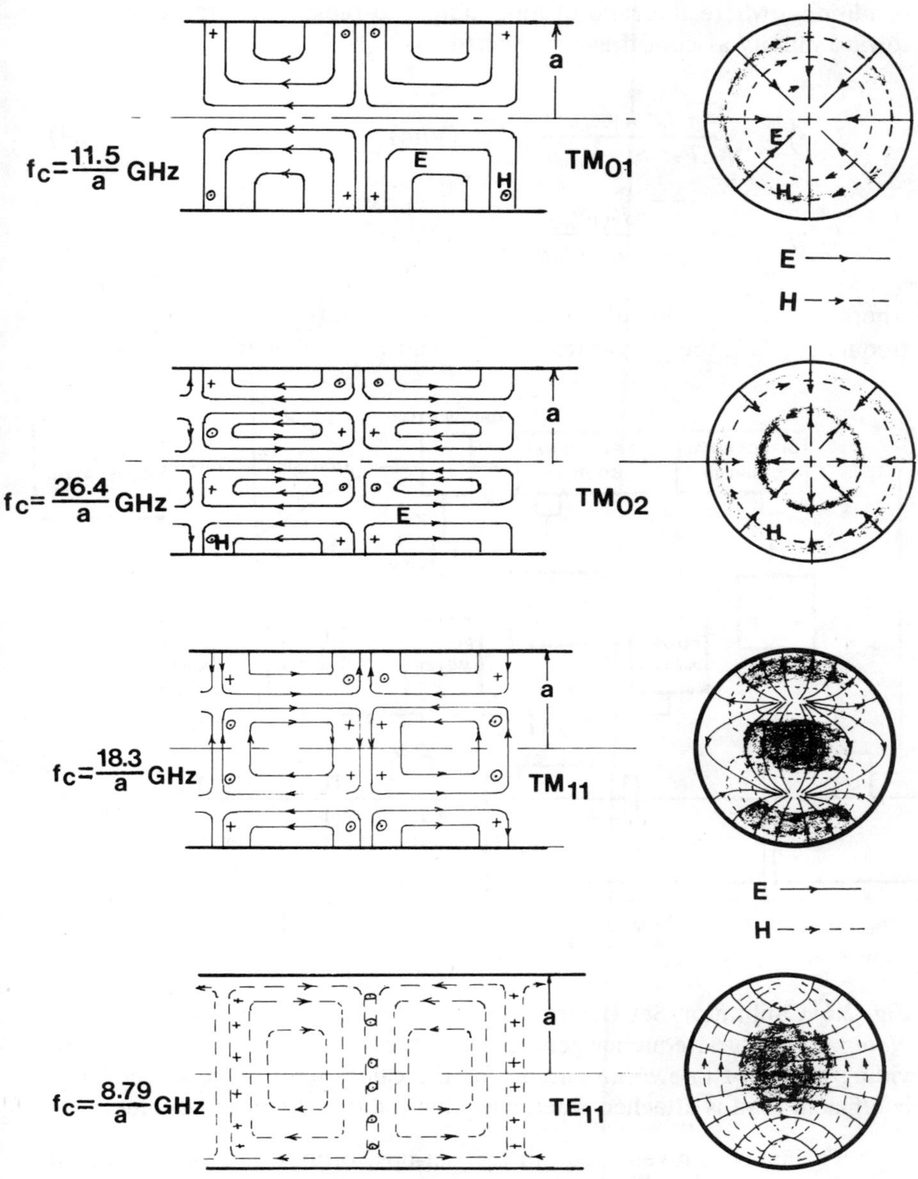

Fig. 2.5 Common Waveguide Field Configurations in Circular Pipe
Note: Shaded regions represent maximum power flow; a = the radius of the pipe.

conditions, where the ratio of longitudinal to radial field strengths tends toward infinity as cut-off is approached:

$$E_z(r) = \sqrt{\frac{4n}{\pi F}} \frac{f_c}{f} \frac{J_0(K_{mn}r)}{aJ_0'(K_{mn}a)} \sqrt{P_\tau} \quad \text{(V/m)} \tag{2.4}$$

$$E_r(r) = -j\sqrt{\frac{4n}{\pi F}} \frac{J_0'(K_{mn}r)}{aJ_0'(K_{mn}a)} \sqrt{P_\tau} \quad \text{(V/m)} \tag{2.5}$$

where k_{mn} is the root of the Bessel function $J_m(k_{mn}r)$, f_c is the cut-off frequency, P_τ is the propagated power, and $F = \sqrt{1 - f_c^2/f^2}$.

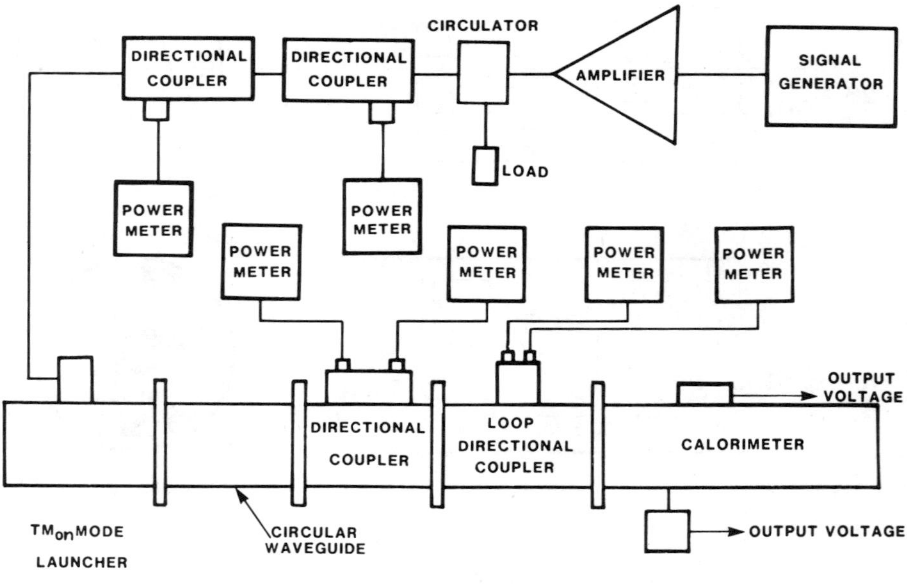

Fig. 2.6 Calibration Set-Up for Oversided Waveguide Instruments
Note: A variable-frequency generator feeds coaxial-line components, providing pulses of known amplitude for the calorimeter. A low-reflection-coefficient load is attached when checking the directivity of couplers.

To launch a given mode in test set-ups, probes or loops placed at locations of field maxima for one mode and at zeros for another mode will provide discrimination. The schematic diagram of a typical test set-up used for calibration of directional couplers and power meters is shown in Figure 2.6. A comprehensive discussion of these instruments is given in the following sections.

A graphic, although qualitative, display of mode patterns can be obtained by placing a sheet of absorbing liquid crystal over the end of the waveguide, and observing the color patterns through a screen that shields the observer from the radiation. Continuous-wave (CW) power of 1–5 W is required for this observation. Antenna patterns, taken in either the far field or the near field give a more quantitative, but more time-consuming, plot.

2.5 SIGNAL SAMPLING

Directional couplers can be used to sample the forward and reflected waves in a waveguide. For applications at hundreds of megawatts, the coupling coefficient must be very small, and also the coupling slots or holes must not degrade the waveguide's performance. In oversized TM modes, the transverse magnetic field is maximum near the wall and the phase velocity is nearly the free-space value, properties that permit coupling into a TEM mode in coaxial line. Such a coupler is sketched in Figure 2.7, where an eight-hole Chebyschev junction is indicated [12]. Coupling coefficients of −65 to −90 dB are appropriate for sampling large pulsed powers, but the calibration of such weakly coupling junctions is difficult, and usually a compromise of −60 dB coupling plus a 20 dB attenuator in each side arm is made. Directivity of 20–30 dB is adequate to permit measurement of even small reflection coefficients in the waveguide, such as due to flange misalignment, vacuum windows, and calorimeters. Window flash-over and other forms of breakdown are easily discernible in the reflection signal. In principle, the couplers should be able to discriminate between modes because they have different guide wavelengths. In practice, however, the wavelengths are not very different because they all are far from cut-off. Thus, when waves are propagating in more than one mode, there may be beating between them in the coupled signal. Even so, the directivity eliminates the fluctations due to standing waves, an effect that often perturbs the signals received by simple probe samplers.

A second type of directional coupler that has been found to be very useful in high-power experiments is the directional loop, which is sketched in Figure 2.8. The loop is oriented so that magnetic lines induce differential signals at its terminals, while electric field lines terminating on its surface induce in-phase signals. When the geometry is properly designed so that the magnitudes of the two voltages are equal, the loop is sensitive to waves traveling in one direction in the circular waveguide, while insensitive to those traveling in the other direction because their signals just cancel.

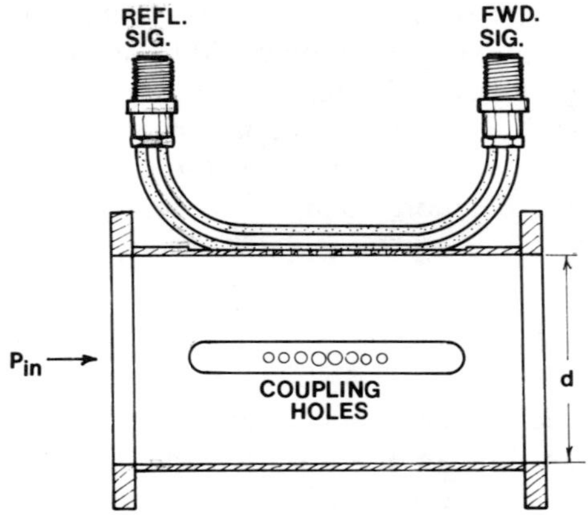

Fig. 2.7 Multihole Directional Coupler
Note: The coupling is between TM_{0n} modes in oversized waveguide and semirigid coaxial cable side arms. Chebyschev relationship of hole sizes gives flat frequency response over the working band.

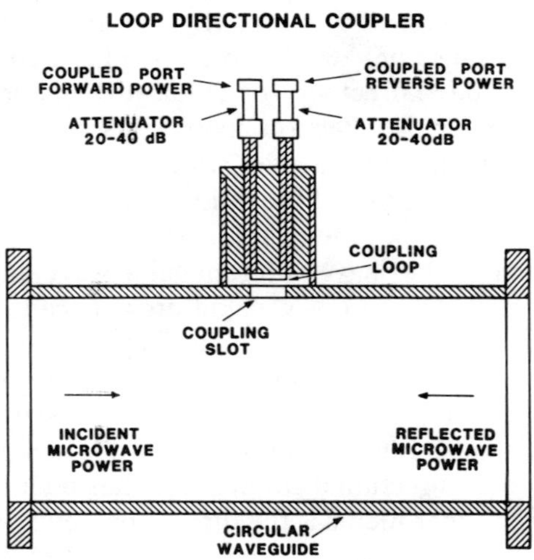

Fig. 2.8 Loop Directional Coupler in Oversized Circular Waveguide
Note: Magnetic and capacitive coupling cancel for reflected waves, giving directivity for over an octave of bandwidth.

2.6 FREQUENCY MEASUREMENT

Methods to measure the frequency of short-duration, high-power microwave pulses are quite different from those used for other microwave signals. For example, sweeping spectrum analyzers have little utility here because there is usually only one pulse, or at most a very few, to analyze. Two other types of spectrum analyzers apply for such signals, one using an array of tuned filters or a grating that feeds multiple detectors, the other using a long dispersive line that converts frequency to delay time.

Figure 2.9 shows an arrangement commonly used for both measurements when viewing radiation patterns [9]. A long run of waveguide operating near cut-off can be used for the dispersive line because the group velocity varies from *c* down to zero over the frequency spectrum. Figure 2.10 plots delay time *versus* frequency for an example of 66 m of X-band waveguide. Unfortunately, neither the dispersion nor the attenuation are linear with frequency in a simple waveguide, resulting in poor discrimination at the high end of the band, and large attenuation at the low end. Thus, for exploratory measurements, where the range of radiated frequencies is poorly known, it is often necessary to use more than one dispersive line to cover an octave or so of frequencies. The shorter is the initial pulse, the better is the frequency resolution at the output. A major advantage of this method is that only one detector is needed at each end of the line, allowing for easy calibration.

The method of using multiple filters and detectors permits spot frequencies to be monitored [13], and any range of frequencies can be covered for which horns, filters, and detectors are available. Multiple-aperture coupled bandpass filters are usually satisfactory if they have at least 40 dB of rejection, but the filters must be tested for spurious responses outside the design band. Low-pass or high-pass filters may be required along with the bandpass unit to suppress these responses. Gratings may also suffer from the same problems due to higher order modes, requiring filters to be inserted in the detector lines.

A third type of frequency measurement, which is most useful after the range has been narrowed, uses a heterodyne detector. The incoming signal is mixed with a local oscillator, and the beat signal is observed directly on a high frequency or digitizing oscilloscope. Software written for digitizing oscilloscopes (e.g., the Tektronix 7912AD transient digitizer) permit Fourier analysis and direct plotting of the beat frequency spectrum.

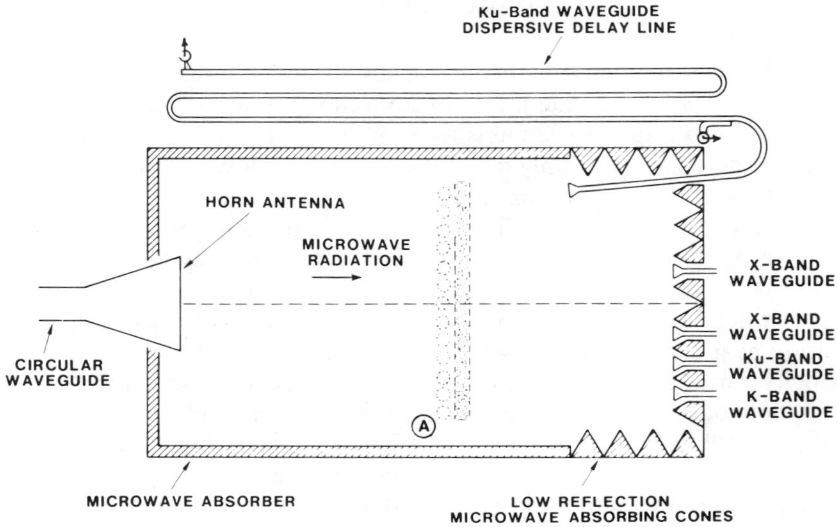

Fig. 2.9 Microwave Radiation Test Chamber
Note: Multiple pickup horns may be used with filters to determine frequencies, or arranged spatially to determine radiation pattern. Fluorescent tubes in array at A are illuminated where the power density is large, showing near-field pattern. Long dispersive line converts frequency to time delay.

2.7 POWER MEASUREMENT

Direct power measurement of single microwave pulses is difficult, especially when involving power levels of hundreds of megawatts. A waveguide attenuator of 90–100 dB, followed by a calibrated fast detector could be considered as a direct measurement. Signal sampling, either with couplers (see Sec. 2.5) or directly within the waveguide provides a second approach if the field distribution and the mode (or modes) are known. A third approach is to measure the total energy deposited by the pulse, the time dependence of which is independently determined by fast signal sampling.

2.7.1 Direct Power Measurement

The direct power method is straightforward, but requires calibration of both the attenuator and the detector in the mode of interest. Cone

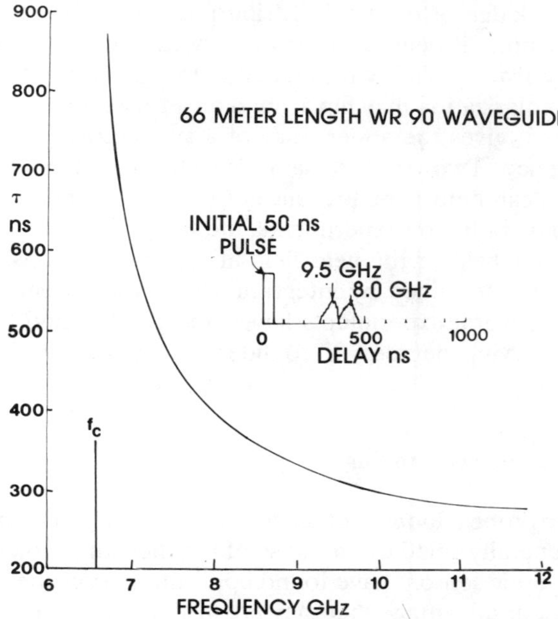

Fig. 2.10 Dispersion of Long Waveguide with Sample Signal Having Two Frequency Components

attenuators have insertion loss that is fairly independent of mode type and frequency, and large values of attenuation can be provided. However, careful calibration of such large values of attenuation requires special equipment. Detectors must be calibrated with pulses of known power because their pulse characteristics may be quite different from the CW response, especially over a wide dynamic range (extending well above the square-law level) and with a 50 Ω termination, which is required to obtain good pulse response. Silicon (Si) or gallium arsenide (GaAs) point contacts and Schottky-barrier diodes having sufficiently fast response times are commercially available in either coaxial SMA or waveguide mounts.

Cryogenically cooled detectors, such as the Putley type or simple crystals, have an advantage of large bandwidth and wide dynamic range, but most do not have sufficiently fast response times and require liquid helium cooling, making these detectors less attractive than conventional ones.

2.7.2 External Power Sampling

Signal sampling, using either simple probes or directional couplers, relies on knowledge of the field distribution, sometimes as a function of time. For example, if beam filamentation occurs, a dominant TM_{01} mode may suddenly switch to a TM_{11} mode, due to the asymmetry of the beam. Further, the polarization may flip around on a very short time scale, which, when averaged, gives the appearance of a symmetrical mode, but at the wrong frequency. Two or more sampling power meters at different azimuthal angles can detect the presence of such polarization flips.

In antenna radiation experiments, the sampling probes may be placed in either the far field or the near field at various locations, and the total power may be determined by integration over the measured or inferred field pattern. Several experiments have used this method [7, 9], yielding results with varying degrees of confidence. Figure 2.9 shows a typical arrangement.

2.7.3 Internal Power Probing

Metallic probes, loops, or thermisters inserted into high-power waveguides are generally ruled out because of breakdown problems. However, various fiber optic sensors have found application as probes. Such sensors have the obvious advantage that they are insulators, introducing minimum perturbation into the system and providing nonconductive data lines from the experiment to the recording instruments in the screen room. Some examples are sketched in Figure 2.11.

Figure 2.11(a) indicates rotation of the plane of polarization in a single-mode fiber by using an optically active crystal, such as KDP or lithium niobate, due to the RF electric field. Similar probes with centimeter-sized KDP crystals have been used in front of S-band horn radiators to measure the E-field intensity [14]. Because a very fast oscilloscope was used, the oscillations themselves were visible, providing a direct optical measurement of the microwave frequency as well.

Figure 2.11(b) represents the rotation of polarization in a single-mode fiber by a strong magnetic field due to the Verdet effect, a method now finding application for the measurement of large pulse currents in high-voltage pulse lines, where insulation of sensors is a problem [2]. RF magnetic fields are not strong enough to apply this method.

Figure 2.11(c) represents a heat-absorbing liquid crystal for which optical transmission depends on temperature. A section of the sensor is coated with microwave absorber so that its temperature depends on the time-averaged power density of the wave at that location. Multimode fibers may be used for this application, thus simplifying the system.

Figure 2.11(d) represents the change in attenuation, length, and refractive index of a single-mode fiber due to temperature. The trend in

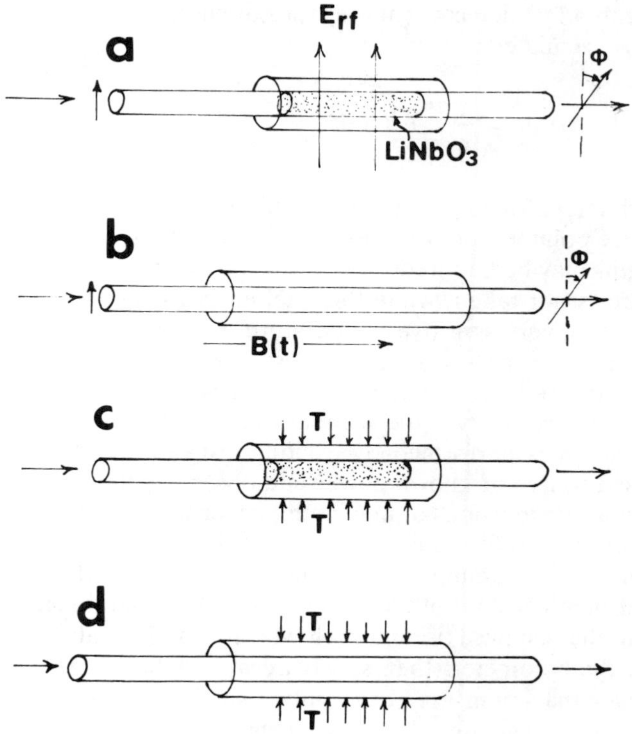

Fig. 2.11 Fiber Optic Sensors for High-Power Microwave and Pulsed Power Applications
Note: 900 nm to 1000 nm is a commonly used wavelength.

recently developed fibers is to minimize these effects, so samples of obsolete fibers must be sought to optimize this sensor. The outside of the fiber is coated with microwave absorber. The change in refractive index or fiber length produces a phase shift, which can be measured with an interferometer, while the change in attenuation can be measured by a sensitive balanced bridge.

2.8 CALORIMETRY

The total energy contained in a microwave pulse or a train of pulses can be measured with considerable precision by using a calorimeter. To determine the power of the wave, the time dependence is simultaneously recorded with a fast detector (of the types discussed in Sec. 2.7.1) and the value $P(t)$ is calculated:

$$W = \int_0^t P(t)\,\mathrm{d}t = \frac{1}{\eta}\int_0^t E^2(t)\,\mathrm{d}t \tag{2.6}$$

Calorimeters for single high-level pulses operate under unusual conditions, as we pointed out in Section 2.1, in that the instantaneous electric field strengths may be hundreds of kilovolts per centimeter, causing breakdown if care is not taken, while the total energy deposited is only a few joules, requiring very sensitive temperature measurements.

At modest power levels, many investigators successfully use sensitive bolometers attached to resistive spears or cones terminating the energized waveguide, either directly in the vacuum system or at atmospheric pressure [15]. Calibration is performed by exciting the desired waveguide mode with a pulsed source of known power and observing the response. Figure 2.5 shows that the locations of maximum power flux depend on the mode, requiring separate calibrations for each. If the bolometer is attached at the location of a maximum, the response rises rapidly, then decays with the thermal time constant of the system. If the bolometer is attached near a maximum, the response rises more slowly, as heat flows along the resistive element to the bolometer, then slowly decays. A bolometer attached far from a power maximum may have only a small response, depending on the relative time constants of heat flow. Thus, if several bolometer elements are attached, we can obtain some information on mode patterns. At high levels of operation, the bolometers or the wires connected to them cause sparks or breakdown, and the instrument is destroyed in some cases. Electrical pickup from the pulsed sources is also a serious problem, requiring very careful shielding of both the elements and the data cables.

A calorimeter, which is insensitive to mode type and has no electrical connections, is shown in Figure 2.12. Microwave power enters through a half-wave window, having less than 5% reflection and is absorbed by a sheet of lossy fabric (space cloth) having the same impedance as the waveguide (200 Ω in this case) [16]. To provide total absorption, the waveguide is terminated $\lambda_g/4$ behind the lossy sheet. The waveguide is quite oversized to permit high-power propagation so that the lowest three modes have nearly the same guide wavelength. The instrument is thus fairly broadband, operating from 6.5 GHz to 12.5 GHz with a voltage standing wave ratio (VSWR) of less than 1.2. The space cloth is heated by the microwaves, and the space cloth transfers its heat to the air contained in the foam-insulated enclosure. The air expands according to Charles' law, raising the column of water in the manometer to a height determined by the mass of

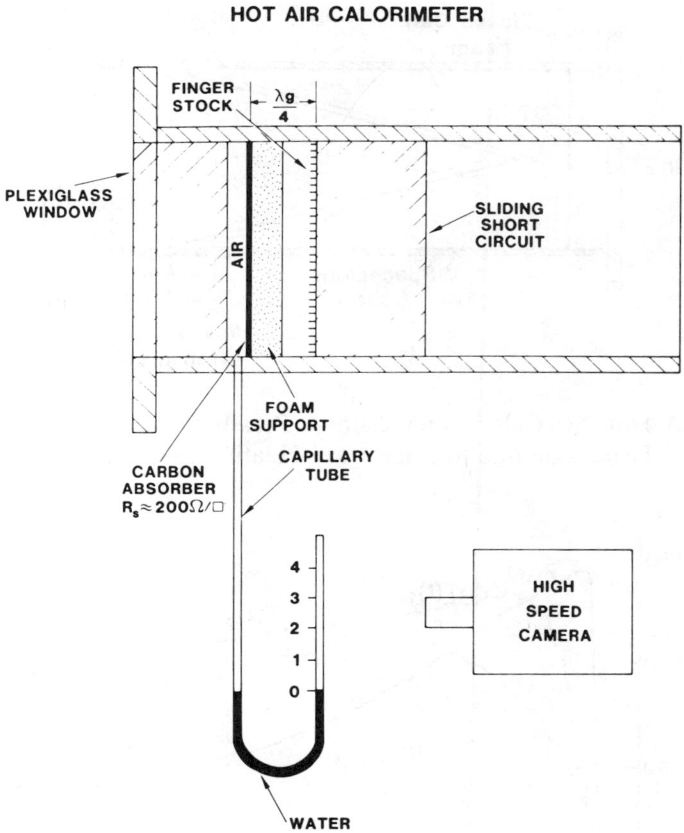

Fig. 2.12 Hot Air Calorimeter Using a Water Manometer to Measure Air Expansion due to Microwave Heating of an Absorber

the absorber and the relative volumes of the chamber and the glass tube. In operation, 10 joules typically raises the column by 15 cm. The total excursion can be recorded with open-shutter photography or a rapid movie camera, neither method being very sensitive to stray pickup [16].

An obvious improvement is to replace the clumsy manometer with a miniature differential pressure gauge, having direct electrical readout, as sketched in Figure 2.13. Now, the expansion is at constant volume, permitting the pressure to be raised to increase the sensitivity. Also, sulfur hexafluoride gas will increase the breakdown threshold, permitting installation of a thermistor for comparative readings. Figure 2.14 shows

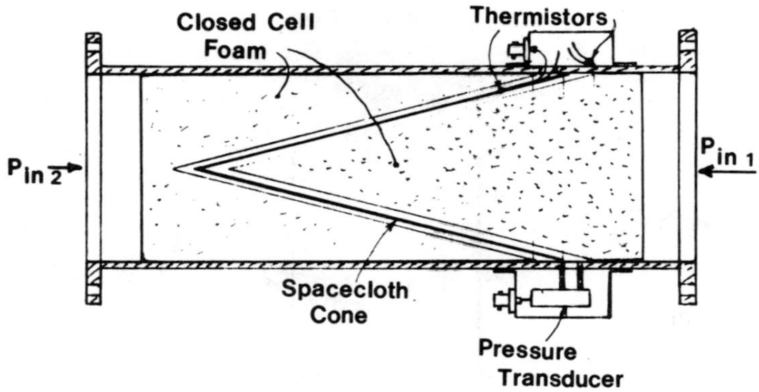

Fig. 2.13 Hot Air Calorimeter Using a Pressure Gauge to Measure Air Expansion due to Microwave Heating of an Absorber

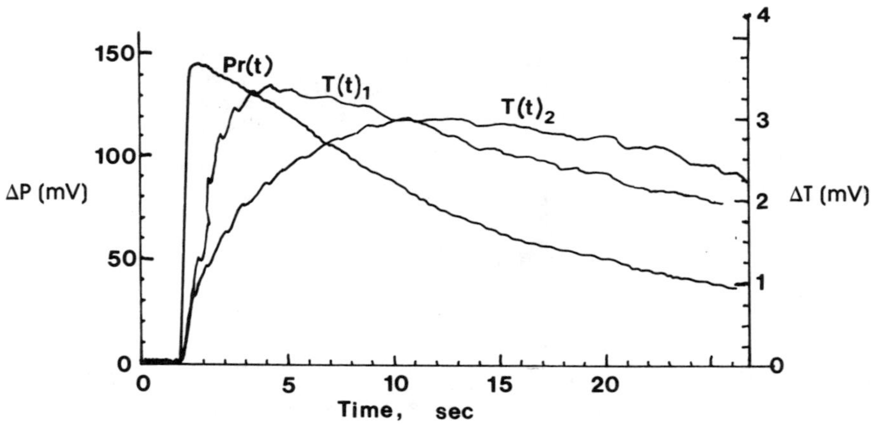

Fig. 2.14 Signal Response from the Calorimeter in Figure 2.13 for a Short Pulse of Microwave Energy Containing Eight Joules.

typical data from both sensors for an eight-joule shot. The thermistor signal is seen to continue to rise after the pulse, as heat flows into its location. The pressure instrument is insensitive to mode type, as long as the absorber is well isolated from the walls, so that no thermal shunts interfere with the heat transfer to the gas. However, this instrument suffers from electrical pickup, requiring careful shielding.

Another calorimeter that is free from problems of stray pickup is shown in Figure 2.15. Here, the absorber is placed at a shallow angle, as in a cone or spear, in Figure 2.15(a) to allow an infrared camera to view

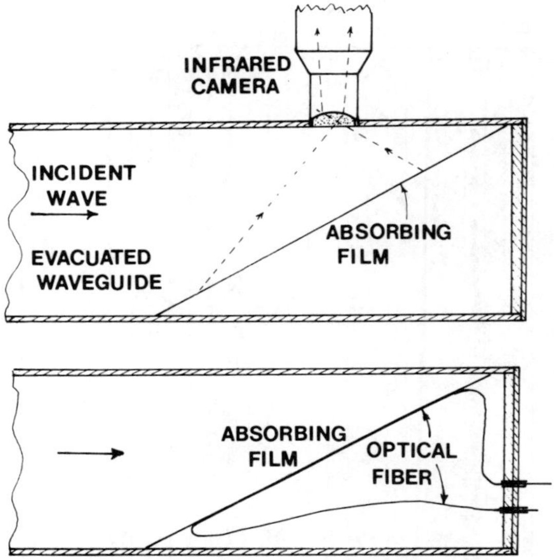

Fig. 2.15 Broadband Calorimeter Having Infrared or Fiber Optic Sensors

the heated surface. The image in the camera shows the shape of the mode pattern, somewhat as the liquid crystal viewer mentioned in Section 2.4, except that here it can be made quantitative with proper calibration. The method of measuring the absorber temperature with a fiber optic sensor, indicated in Figure 2.15(b) could equally be used with the 90° absorber of Figure 2.13.

2.9 RADIATION PATTERNS

Antennas fed by appropriate mode launchers can be calibrated at low powers in an antenna range, showing the hollow center for TM_{0n} modes and giving the gain of the radiator.

At high powers, a direct imaging method has been applied, consisting of a bank of 20 or so long fluorescent tubes mounted in front of the radiator, as indicated by dotted lines and labelled A in Figure 2.9. A typical illumination pattern for a TM_{01} mode generated by the system of [9] is sketched in Figure 2.16, operating at a level of 100–200 MW. The ionization is able to spread vertically in the tubes, while it is constrained in the horizontal direction by the walls. Even so, the patterns for TM_{0n} and TM_{1n} are easy to distinguish.

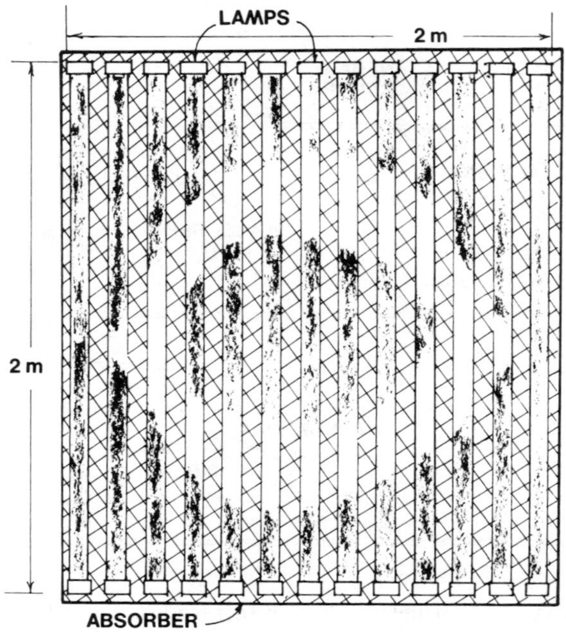

Fig. 2.16 Array of Long Fluorescent Tubes Mounted in Front of Microwave Radiator
Note: glow due to ionization showing the radiation pattern.

To quantify the high-power radiation measurements, glow balls (small spheres of glass filled to a chosen pressure of gas) can be placed in the field, and their illumination intensity will indicate the E-field strength [17]. Glow balls ionize in a time of less than 2 ns in high fields, allowing their light output to be monitored with fast photomultipliers to indicate rapid fluctuations in wave pattern. The fluorescent tube shows a time-averaged image, which will be only a blur if there is mode switching.

REFERENCES

1. R.B. Miller, *Intense Charged-Particle Beams* (Plenum, New York, 1982).
2. K. Prestwich, "Pulsed Power Technology," *Proc. IEEE Plasma Sciences Conf.*, Saskatoon, Saskatchewan, May 1986.
3. M.A. Greenspan, *et al., Phys. Fluids* **23** (1), 205 (1980); J.D. Sethian, and C.A. Ekdahl, *Phys. Rev. Lett.* **42** (11), 711 (1979).

4. Yu.F. Bondar, *Sov. J. Plasma Phys.* **8** (5), 528 (1982).
5. G. Benford, *Plasma Physics* **15**, 483 (1973).
6. S.P. Bugayev, *et al., J. Tech. Phys. Lett.* **9** (22), 1385 (1983); **10** (20), 26 (1984).
7. N.F. Kovalev, *et al., JETP Lett.* **18**, 138 (1973); Y. Carmel, *et al., Phys. Rev. Lett.* **33**, 1278 (1974); V.S. Ivanov, *et al., Sov. Phys. Tech. Phys.* **26**, 580 (1982).
8. A.S. El′chaninov, *Sov. Tech. Phys. Lett.* **6**, 191 (1980).
9. G.T. Leifeste, *et al., J. Appl. Phys.* **59** (4), 1367 (1986); J.A. Swegle, *et al., Phys. Fluids* **28**, 2882 (1985).
10. A.S. Gilmour, *Microwave Tubes* (Artech House, Dedham, MA, 1986).
11. M.C.Clark, Sandia Laboratories, private communication.
12. L.M. Earley, W.P. Ballard, and C.B. Wharton, *IEEE Trans. Nuclear Sci.* **NS32** (5), 2921 (1985).
13. N.F. Kovalev, *et al., Sov. Tech Phys. Lett.* **6**, 197 (1980).
14. J. Chang, *et al., J. Appl. Phys.* **57**, 4843 (1985).
15. P. Efthimion, P.R. Smith, and S.P. Schlesinger, *Rev. Sci. Instr.* **47**, 1059 (1976).
16. L.M. Earley, *et al.,* "A Comprehensive Approach for Diagnosing Intense, Single-Pulse Microwave Sources," Sandia Laboratories Report, SAND85-2141J (1985).
17. P. Felsental, and J.M. Proud, *Phys. Rev.* **139A**, A1796 (1965).

Chapter 3
PARTICLE SIMULATION TECHNIQUES FOR BEAM DEVICES

Charles K. Birdsall

Electrical Engineering and Computer Science Department
University of California
Berkeley, California

3.1 INTRODUCTION

Simulation using particles for electron-beam devices began a long time ago. Examples for traveling wave tubes in the 1950s are Nordsieck [28], Tien, Walker, and Woluntis [39], Rowe [36], Poulter [35]. These simulation models more or less followed a number of electrons (e.g., 32 disks or rings) in one wavelength as these particles moved through the slow-wave structure. These simulations pointed the way toward improved efficiencies, but they were in less than perfect agreement with experiments. Hess [16] added numerous changes such as phase velocity tapering and multielement *depressed* collectors. Magnetrons were well treated in the 1960s, for example, by Yu, Kooyers, and Buneman [40], ending up with relatively good agreement with experiments, as well as clearing up numerous scientific points, such as why the electrons in the smooth bore magnetron (no RF circuit) exhibited anomalously high electron temperatures. Modern simulations have greatly benefitted by improvements in numerical techniques, programming, and, of course, much faster computers. An example for a magnetron is Palevsky [32].

The current professional codes in use are multidimensional, electromagnetic, and relativistic, using realistic (close to laboratory) boundary conditions, with particle injection and collection. Examples are ZOHAR

(Langdon and Lasinski, [24]), MASK (Palevsky *et al.* [33]), and ARGUS (Mankofsky *et al.* [27]). Many, if not all, of these codes were developed for fusion plasma applications, with both electrons and ions, with some attention to boundaries. There are now many professional applications to klystrons, magnetrons, traveling wave tubes, gyrotrons, accelerators, free-electron lasers, and vircators; these are very large codes, typically of 40,000 lines, usually run by groups of professionals.

The beauty and quality of many particle simulations comes from a combination of properties, such as

a. working from first principles, using the Newton-Lorentz equation of motion to move particles, and using Maxwell's equations to solve for the fields;
b. having noninvasive diagnostics to measure the usual quantities, such as fields, potentials, currents, and densities as functions of time and space $(t, \mathbf{x})$; in addition, particle simulations can provide velocities distributions at $(\mathbf{x}, t)$ as well as Fourier decompositions in $(\mathbf{x}, t)$;
c. being fully nonlinear, simulation is not restricted to relatively small amplitudes (although well used to verify linear theory), hence capable of following growth in (x, t) from linear through large amplitudes (e.g., saturation of amplifiers or instabilities);
d. having complete accounting for spatial anisotropies and inhomogeneities;
e. having the ability to employ real boundaries, albeit idealized on occasion.

Simulations are used to complement both theory and experiment, and *to stand on their own.* Thus, simulations are widely used by theoreticians in *verifying* linear and nonlinear theory, adding feedback for improvement in theory. Simulations are useful to experimentalists in complementing explanations obtained from their observations. In addition, simulations have proved to be very useful as a *tool for exploration* into new parameters, in advance of the comparable theory or experiments. Simulations have also been a tool for *discovery,* trying out new ideas, or (very importantly!) recognizing new and unexpected behavior. For example, discovering electrostatic virtual cathode oscillations in our simulations was a surprise to W. B. Bridges and myself in 1959, when we were studying the noise properties of a potential minimum (to be sure, we had some anticipation from theory). Observation of the potential minimum oscillating through zero (cathode potential) with a regular period added the final step in explaining the planar diode limiting current mysteries of the previous decades (then called the "6L6 effect"). Now, electromagnetic vircators generate terawatts. This miniservon is to say that particle simulations are both an asset to theory and experiment *and* have a role of their own.

The purpose here is to introduce some of the methods of particle simulations and to make application to simple programs that can be run in a few minutes on personal computers, but using a periodic model (very simple), rather than one with a slow-wave circuit or cavities. The applications presented are the *cold two-stream instability* and the *cold beam-plasma instability*.

We use material directly from Birdsall and Langdon [4] with acknowledgment of the copyright by McGraw-Hill Company. We recommend our full text to those looking for more detail, as well as the text by Hockney and Eastwood [17]. Please see also the introductory article by Denavit and Kruer [12], and the review article by Dawson [11].

Our work in Berkeley is supported by the US Department of Energy (DOE) and by the US Navy Office of Naval Research (ONR). We are grateful for recent help from Dr. Dan Hitchcock at DOE and from Dr. Charles Roberson at ONR.

Our procedure is to follow a widely used program ES1, an *electrostatic one-dimensional code,* complete with initial conditions and diagnostics to tell us about the physics, as well as to check on the numerics. This is preparatory to running the program from tens to thousands of time steps on trial runs. We may find that we were not as wise as we thought on the first run. Hence, we will modify the program, improve the initial conditions, add some new diagnostics, and try again. After a few such iterations as this, we may have what we started out to find, the essence of the physics. Of course, along the way, we probably will find that we want more solutions to the dispersion equation for the waves under study, better estimates of the nonlinear behavior to be observed, or other information.

Our initial system for study has charged particles in both self and applied electrostatic fields, and an applied magnetic field. This initial use of an electrostatic model, which follows the historical development of plasma particle simulation, is perhaps the easiest of all models to understand, and also leads directly to fully electromagnetic models.

3.2 THE ELECTROSTATIC MODEL

The model (Figure 3.1) consists of charged particles moving about due to forces of their own and applied fields. The physics comes from two parts, the fields produced by the particles and the motion produced by the

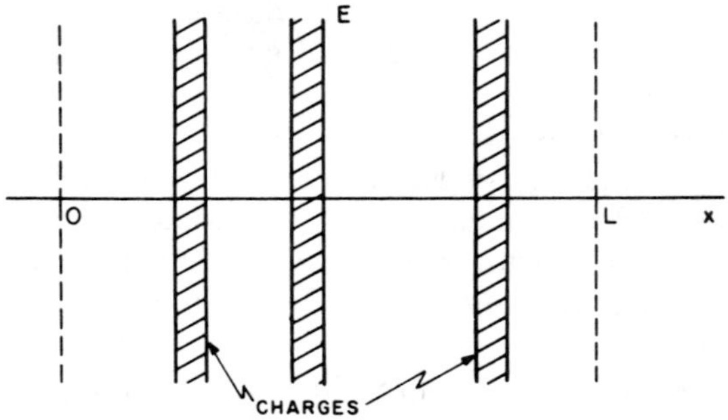

Fig. 3.1 A One-Dimensional Model Consisting of Many Sheet Charges with Self and Applied Electric Fields **E** Directed Along the Coordinate x

Note: There are no variations in y or z.

forces (or fields). The fields are calculated by *Maxwell's equations* from the known positions of all of the particles and their velocities, and the forces on the particles are found by using the electric and magnetic fields in the *Newton-Lorentz* equation of motion. We calculate the fields from the initial charge and current densities, then move the particles (small distances) and recalculate the fields due to the particles at their new positions and velocities. This procedure is repeated for many time steps.

The difference from a laboratory plasma is that simulations proceed discontinuously in time *step by step,* using *digital* rather than analog computation. We must show care in developing *numerical methods* that provide sufficient *accuracy and stability* to make the simulations useful for many characteristic cycles of the plasma, whether these be plasma, cyclotron, or hybrid (or whatever) periods of the ions or electrons. We use a *mathematical spatial grid,* as shown in Figure 3.2, usually fine enough to resolve a *Debye length,* in order to measure the charge density, and then calculate the electric field **E**.

There are some exceptions to these generalities. For example, we may calculate the electric field in one-dimensional problems relatively easily without using a spatial grid. However, the grid provides a *smoothing effect* by not resolving spatial fluctuations that are smaller than the grid size. An exact field calculation would keep everything, which is usually more than we want.

The use of temporal and spatial grids, which are mathematical and not physical, causes concern about accuracy and may create what we will term *nonphysics.* Suffice it to say that the possibility of nonphysical effects

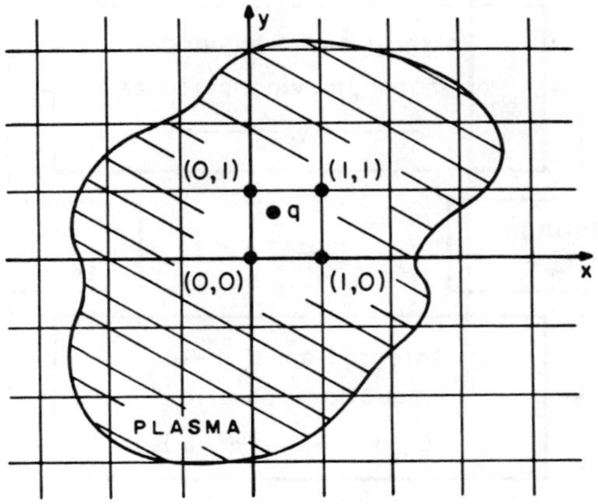

Fig. 3.2 A Mathematical Grid is Set into the Plasma Region in order to Measure Charge and Current Densities ρ, **J**

Note: From these we will obtain the electric and magnetic fields **E** and **B** on the grid. A charged particle q at (x, y) will typically be counted in terms of ρ at the nearby grid points (0,0), (1,0), (1,1), (0,1) and in terms of **J** at the faces between these points. The force on q will also be obtained from the fields at these nearby points.

may restrict our choices of parameters on occasion, but generally these effects can be avoided. Inaccuracies will always be with us and simply must be made small.

3.3 THE COMPUTATIONAL CYCLE

At each step in time, the program solves for the fields from the particles and then moves the particles (this cycle is shown in Figure 3.3). There may be tens of steps in a characteristic period of the plasma and there may be tens of periods in a typical run, which adds up to hundreds or thousands of time steps in a given run.

The cycle starts at $t = 0$, with some appropriate initial conditions on the particle positions and velocities. The computer runs to the number of time steps that it is told. Various *diagnostics* are printed out at the end of the run. Some are in the form of *snapshots* at particular times, such as densities, fields, or velocity distributions, and some are in the form of *time histories,* such as energy *versus* time. These graphs are the record from

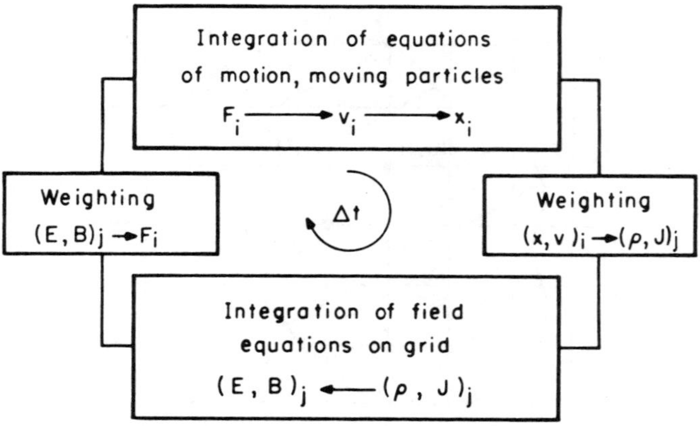

Fig. 3.3 A Typical Cycle, One Time Step, in a Particle Simulation Program
Note: The particles are numbered $i = 1, 2, \ldots, NP$; the grid indices are j, which become vectors in two and three dimensions.

which we obtain the physics of the simulation. Numbers *per se* are very seldom the object of a plasma simulation.

The particles are processed through the boxes shown in Figure 3.3, much as the fields and forces are created in the actual plasma. Let us follow a cycle by starting from initial values of positions and velocities. Keep in mind that hundreds or thousands (or 10^6 and up in two and three dimensions) of particles are being processed.

The particle quantities, such as velocity and position, are known at the particle and may take on all values in **v** and **x** space, called *phase space*. The name of the particle is given by index i, such as v_i and x_i. The field quantities will be obtained only on the spatial grid, known only at discrete points in space labeled with index j such as E_j. The ties from the particle position and velocities to the field quantities are made by first calculating the charge and current densities on the grid. This step requires stating how to produce the grid densities from the particle positions and velocities. This process of charge and current assignment implies some *weighting* to the grid points that is dependent on particle position. Once the densities are established on the grid, then we will use various methods to obtain the electric and magnetic fields. With the fields known on the grid, but with the particles scattered around within the grid, we interpolate the fields from the grid to the particles in order to apply the force *at the particle* by again performing a weighting function.

How do we distinguish particles? What information is stored for the particles and the fields? The particles may be known by the way in which they are stored in the computer memory; they may be in some kind of order, with only their present velocity and position stored ($\mathbf{v}_i$, $\mathbf{x}_i$); their values of charge q_i and mass m_i may be put elsewhere (*e.g.*, with only two species, electrons and ions, q and m would change only once in running all of the particles through in a time step). The fields are known at the grid points and stored, probably indexed in an array, so that they can be recalled with ease. There will almost always be *many more particles than grid points,* and external storage allows use of more particles than might fit into the computer's fast memory, along with the field quantities. Because the particles are integrated independently, only a few need be in fast memory at a time. The field quantities probably will be retained in fast memory because they have to be recalled randomly in most present methods.

3.4 INTEGRATION OF THE EQUATIONS OF MOTION

A problem may call for 10,000 particles to be run for 1000 time steps. This means that the equations of motion must be integrated $10{,}000 \times 1000 = 10^7$ times. We want to use as fast a method as possible, and still retain acceptable accuracy. The time per particle per step currently is on the order of microseconds. Explicitly, a 10^4 particle program of 10^3 time steps running at 10 μs/particle/step on a machine with a $720 per hour charge would cost $20 for the 100 s run.

One commonly used integration is called the *leap-frog method.* The two first-order differential equations to be integrated separately for each particle are

$$m\frac{d\mathbf{v}}{dt} = \mathbf{F} \tag{3.1}$$

$$\frac{d\mathbf{x}}{dt} = \mathbf{v} \tag{3.2}$$

where $\mathbf{F}$ is the force. These equations are replaced by the *finite-difference equations*

$$m\,\frac{\mathbf{v}_{\text{new}} - \mathbf{v}_{\text{old}}}{\Delta t} = \mathbf{F}_{\text{old}} \tag{3.3}$$

$$\frac{x_{\text{new}} - \mathbf{x}_{\text{old}}}{\Delta t} = \mathbf{v}_{\text{new}} \tag{3.4}$$

The flow in time and notation are shown in Figure 3.4 which makes clear the *time centering*. The computer will advance $\mathbf{v}_t$ and $\mathbf{x}_t$ to $\mathbf{v}_{t+\Delta t}$, $\mathbf{x}_{t+\Delta t}$, although $\mathbf{v}$ and $\mathbf{x}$ are not known at the same time. The user must show care in at least two ways. First, initial conditions for particle velocities and positions given at $t = 0$ must be changed to fit in the flow; we push $\mathbf{v}(0)$ back to $\mathbf{v}(-\Delta t/2)$ using the force $\mathbf{F}$ calculated at $t = 0$. Second, the energies calculated from $\mathbf{v}$ (kinetic) and $\mathbf{x}$ (potential, or field) must be adjusted to appear at the same time.

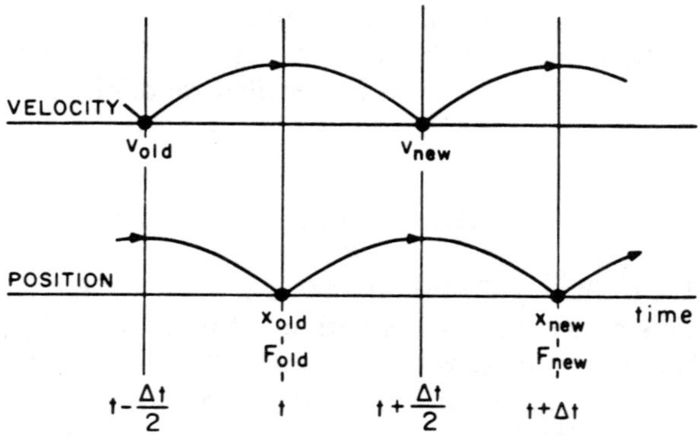

Fig. 3.4 Sketch of Leap-Frog Integration Method Showing the Time Centering of Force **F** while advancing **v**, and that of **v** while advancing **x**.

The leap-frog method has error, which vanishes as $\Delta t \to 0$. We will use a leap-frog integrator in nearly all of our programs because it is both simple (easy to understand, with minimum storage) and surprisingly accurate.

Let us try a simple model to illustrate the leap-frog method. The model is that of a simple harmonic oscillator, which is described by the second-order differential equation:

$$\frac{d^2x}{dt^2} = -\omega_0^2 x \tag{3.5}$$

This equation has solutions:

$$x(t, t_0) = A(t_0) \cos\omega_0 t + B(t_0) \sin\omega_0 t \tag{3.6}$$

The finite-difference equations of the leap-frog method are

$$\frac{dx}{dt} \rightarrow \frac{x_t - x_{t-\Delta t}}{\Delta t} = v_{t-\Delta t/2} \tag{3.7}$$

$$\frac{dx}{dt} \rightarrow \frac{x_{t+\Delta t} - x_t}{\Delta t} = v_{t+\Delta t/2} \tag{3.8}$$

$$\frac{dv}{dt} \rightarrow \frac{v_{t+\Delta t/2} - v_{t-\Delta t/2}}{\Delta t} = \frac{x_{t+\Delta t} - 2x_t + x_{t-\Delta t}}{\Delta t^2} \tag{3.9}$$

We start computation at some time and work foward (e.g., with x_t and $v_{t-\Delta t/2}$ given). First, knowing x_t, we find the force and solve for $v_{t+\Delta t/2}$ from the acceleration equation, then$x_{t+\Delta t}$ is solved from the velocity equation. This gives a new x and the process is repeated time and time again. The equations are time centered for stability.

We are now interested in substituting these finite-difference approximations into a homogeneous equation of motion, then solving it analytically, and comparing the result with the physical solution. The homogeneous equation to be solved is

$$\frac{x_{t+\Delta t} - 2x_t + x_{t-\Delta t}}{\Delta t^2} = -\omega_0^2 x_t \tag{3.10}$$

This is a standard finite-difference equation, which can be readily solved by assuming solutions of the form:

$$x_t = A_{\mathrm{e}}^{-i\omega t} \tag{3.11}$$

A is an initial value and ω is the unknown. Substituting this (and $x_{t-\Delta t} = A_{\mathrm{e}}^{-i\omega(t-\Delta t)}$, *et cetera*) into (3.10), we obtain

$$\sin\left(\omega\frac{\Delta t}{2}\right) = \pm\omega_0\frac{\Delta t}{2} \tag{3.12}$$

which is plotted in Figure 3.5. We see that for $\omega_0\Delta t/2 \ll 1$, $\omega \approx \omega_0$, as desired. However, we see that as $\omega_0\Delta t$ increases beyond 2, the (initially wholly) real solution for ω becomes complex, with growing and decaying roots, which indicates *numerical instability*.

Let us find the magnitude of the *phase error* for small $\omega_0\Delta t$. (The amplitude error is nil for $\omega_0\Delta t < 2$.) For $\omega_0\Delta t \ll 1$, we find

$$\omega\frac{\Delta t}{2} \approx \omega_0\frac{\Delta t}{2}\left[1 + \frac{1}{6}\left(\omega_0\frac{\Delta t}{2}\right)^2 + \ldots\right] \tag{3.13}$$

which shows a *quadratic error term,* as desired. The cumulative phase after N steps is $\omega N\Delta t$, so that the cumulative phase error is, approximately,

$$\omega_0 N\Delta t\,\frac{1}{6}\left(\omega_0\frac{\Delta t}{2}\right)^2 = \frac{1}{24}N(\omega_0\Delta t)^3 \tag{3.14}$$

For this error to equal, let us say, 1/24 radian, then

$$N = \frac{1}{(\omega_0\Delta t)^3}, \text{ which } \begin{cases} \text{for } \omega_0\Delta t = 0.1 & \text{is 1000 steps, about 16 cycles} \\ \text{for } \omega_0\Delta t = 0.3 & \text{is 37 steps, about 2 cycles} \end{cases}$$

For this error to equal, let us say, 1 radian, then

$$N = \frac{24}{(\omega_0\Delta t)^3}, \text{ which } \begin{cases} \text{for } \omega_0\Delta t = 0.1 & \text{is 24,000 steps, about 384 cycles} \\ \text{for } \omega_0\Delta t = 0.3 & \text{is 890 steps, about 42 cycles} \end{cases}$$

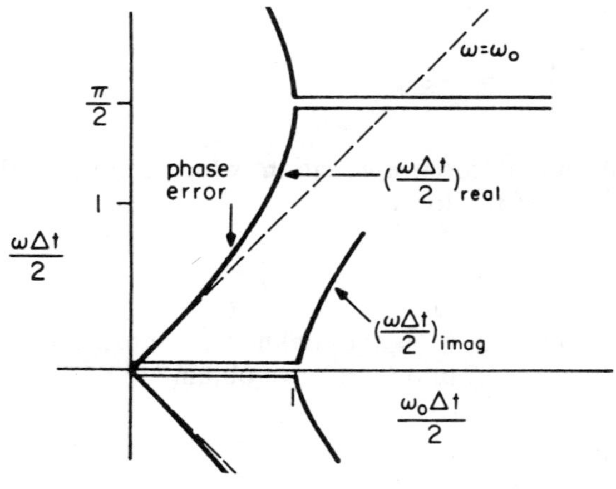

Fig. 3.5 Solution for ω in Terms of ω_0 for Simple Harmonic Motion from the Leap-Frog Finite-Difference Equation

Note: Frequency ω agrees with ω_0 for small $\omega_0\Delta t$, but it is larger than ω_0 as $\omega_0\Delta t$ increases. For $\omega_0\Delta t/2 > 1$, the solution becomes complex with growing and decaying roots for ω (this is *numerical instability*); one trademark is the breakup into odd and even steps, the manifestation of the π phase shift. Observe the double root at $\omega = 0, \pi/2$.

The points are (1) restricting the allowable error to small values limits the number of steps (and cycles) and (2) increasing the step increases the error as the cube of the step size. A common compromise is $\omega_0 \Delta t \approx 0.2$; normal usage is 1000 to 10,000 steps, with useful physics sometimes extending to even more steps.

The force **F** has two parts:

$$\mathbf{F} = \mathbf{F}_{\text{electric}} + \mathbf{F}_{\text{magnetic}} \tag{3.15}$$

$$= q\mathbf{E} + q(\mathbf{v} \times \mathbf{B}) \tag{3.16}$$

Here, the electric field **E** and magnetic field **B** are to be calculated *at the particle*. Hence, using a spatial grid, we must interpolate **E** and **B** from the grid to the particle. We will do so in the same way as we determined the charge density (at the grid points) from the particle positions.

For our one-dimensional purpose, now let us consider the particle displacement to be along x, and that we have velocities v_x and v_y, with a uniform static magnetic field B_0, along z. The $q(\mathbf{v} \times \mathbf{B})$ force, as shown in Figure 3.6, is simply a *rotation* of **v** (i.e., **v** does not change in magnitude). However, the $q\mathbf{E} = \hat{x}qE_x$ force does alter the magnitude of **v** (or v_x); $E_y = 0$. Hence, a physically reasonable scheme which is centered in time is as follows (with t' and t'' as dummy variables, $t - \Delta t/2 < t' < t'' < t + \Delta t/2$).

Half acceleration:

$$\begin{aligned} v_x(t') &= v_x\left(t - \frac{\Delta t}{2}\right) + \left(\frac{q}{m}\right) E_x(t)\left(\frac{\Delta t}{2}\right) \\ v_y(t') &= v_y\left(t - \frac{\Delta t}{2}\right) \end{aligned} \tag{3.17}$$

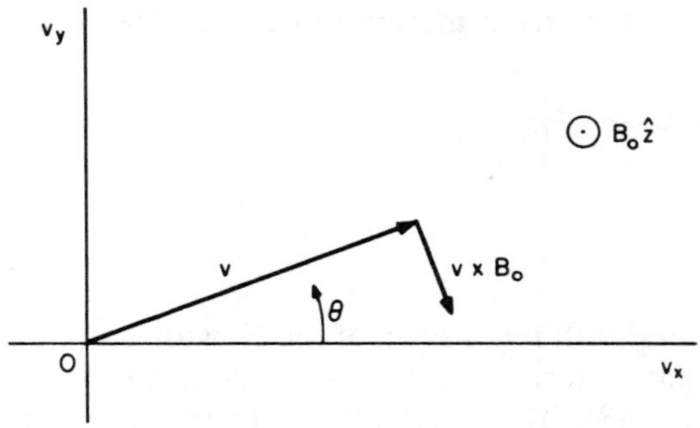

Fig. 3.6 The v_x, v_y Plane, Showing the $q(\mathbf{v} \times \mathbf{B})$ Force Normal to **v**, which Results in a Rotation of **v**, with no Change in Speed with $\dot{\theta} < 0$ for $(q/m) > 0$, $B_0 > 0$

Rotation:

$$\begin{pmatrix} v_x(t'') \\ v_y(t'') \end{pmatrix} = \begin{pmatrix} \cos\omega_c\Delta t & \sin\omega_c\Delta t \\ -\sin\omega_c\Delta t & \cos\omega_c\Delta t \end{pmatrix} \begin{pmatrix} v_x(t') \\ v_y(t') \end{pmatrix} \tag{3.18}$$

Half acceleration:

$$\begin{aligned} v_x\left(t + \frac{\Delta t}{2}\right) &= v_x(t'') + \left(\frac{q}{m}\right) E_x(t) \left(\frac{\Delta t}{2}\right) \\ v_y\left(t + \frac{\Delta t}{2}\right) &= v_y(t'') \end{aligned} \tag{3.19}$$

The angle of rotation is

$$\Delta\theta = -\omega_c\Delta t \; (i.e., \; \dot{\theta} = -\omega_c) \tag{3.20}$$

as desired, where the *cyclotron frequency* (radians/second) as defined by

$$\omega_c \equiv \left(\frac{q}{m}\right) B_0 \tag{3.21}$$

which carries the sign of q and of B_0.

3.5 NEWTON-LORENTZ FORCE

3.5.1 Three-Dimensional v × B Integrator

The particle equations of motion to be integrated are

$$m\frac{d\mathbf{v}}{dt} = q(\mathbf{E} + \mathbf{v} \times \mathbf{B}) \tag{3.22}$$

$$\frac{d\mathbf{x}}{dt} = \mathbf{v} \tag{3.23}$$

We desire a centered-difference form of the Newton-Lorentz equations of motion. The magnetic term is centered by averaging $\mathbf{v}_{t-\Delta t/2}$ and $\mathbf{v}_{t+\Delta t/2}$, following Buneman [8]. The other terms are treated as before. Hence (3.22) becomes

$$\frac{\mathbf{v}_{t+\Delta t/2} - \mathbf{v}_{t-\Delta t/2}}{\Delta t} = \frac{q}{m}\left[\mathbf{E} + \frac{\mathbf{v}_{t+\Delta t/2} + \mathbf{v}_{t-\Delta t/2}}{2} \times \mathbf{B}\right] \tag{3.24}$$

This vector equation for $\mathbf{v}_{t+\Delta t/2}$ can be solved as three simultaneous scalar equations, one for each component. Instead, we choose to obtain a simpler solution using several steps.

Another method separates the electric and magnetic forces completely (Boris [5]). Substitute the following into (3.24):

$$\mathbf{v}_{t-\Delta t/2} = \mathbf{v}^- - \frac{q\mathbf{E}}{m}\frac{\Delta t}{2} \tag{3.25}$$

$$\mathbf{v}_{t+\Delta t/2} = \mathbf{v}^+ + \frac{q\mathbf{E}}{m}\frac{\Delta t}{2} \tag{3.26}$$

then $\mathbf{E}$ cancels entirely (not just $\mathbf{E}_\perp$), which leaves

$$\frac{\mathbf{v}^+ - \mathbf{v}^-}{\Delta t} = \frac{q}{2m}(\mathbf{v}^+ + \mathbf{v}^-) \times \mathbf{B} \tag{3.27}$$

Equation (3.27) is a rotation. The steps to compute here are (1) add half the electric impulse to $\mathbf{v}_{t-\Delta t/2}$ using (3.25) to obtain $\mathbf{v}^-$, (2) rotate according to (3.27) to obtain $\mathbf{v}^+$, and (3) add the remaining half of the electric impulse (3.26) to obtain $\mathbf{v}_{t+\Delta t/2}$. These are the same steps, differently motivated, as in the preceding section. Separation of parallel and perpendicular components is not needed with Boris' method, and the relativistic generalization is straightforward.

Finally, we check the angle of rotation θ, which we expect to be close to $\omega_c\Delta t = qB\Delta t/m$. We see that

$$\left|\tan\frac{\theta}{2}\right| = \frac{|\mathbf{v}^+ - \mathbf{v}^-|}{|\mathbf{v}^+ + \mathbf{v}^-|} = \frac{qB}{m}\frac{\Delta t}{2} = \frac{\omega_c\Delta t}{2} \tag{3.28}$$

where we have used (3.26) in the last step. Hence, the difference equation (3.26) produces a rotation through the angle:

$$\theta = 2\arctan\left(\frac{qB}{m}\frac{\Delta t}{2}\right) = \omega_c\Delta t\left(1 - \frac{(\omega_c\Delta t)^2}{12} + \ldots\right) \tag{3.29}$$

which has less than one percent error for $\omega_c\Delta t < 0.35$.

3.6 INTEGRATION OF THE FIELD EQUATIONS

Now, starting from the charge and current densities as assigned to the grid-points, we obtain the electric and magnetic fields, in general, from

Maxwell's equations, using ρ and J as sources. Here, we take this step for an electrostatic problem (meaning $\nabla \times \mathbf{E} = -\partial\mathbf{B}/\partial t \approx 0$, so that $\mathbf{E} = -\nabla\phi$) in one dimension x.

The differential equations to be solved are

$$\mathbf{E} = -\nabla\phi \text{ or } E_x = -\frac{\partial\phi}{\partial x} \tag{3.30}$$

$$\nabla \cdot \mathbf{E} = \frac{\rho}{\epsilon_0} \text{ or } \frac{\partial E_x}{\partial x} = \frac{\rho}{\epsilon_0} \tag{3.31}$$

which are combined to obtain *Poisson's equation*:

$$\nabla^2\phi = -\frac{\rho}{\epsilon_0} \text{ or } \frac{\partial^2\phi}{\partial x^2} = -\frac{\rho}{\epsilon_0} \tag{3.32}$$

One approach is to solve the finite-difference equations (3.30) and (3.32) using the grid shown in Figure 3.7 as

$$E_j = \frac{\phi_{j-1} - \phi_{j+1}}{2\Delta x} \tag{3.33}$$

$$\frac{\phi_{j-1} - 2\phi_j + \phi_{j+1}}{(\Delta x)^2} = -\frac{\rho_j}{\epsilon_0} \tag{3.34}$$

This last may be written in matrix form as

$$\mathbf{A}\phi = -\frac{(\Delta x)^2}{\epsilon_0}\rho \tag{3.35}$$

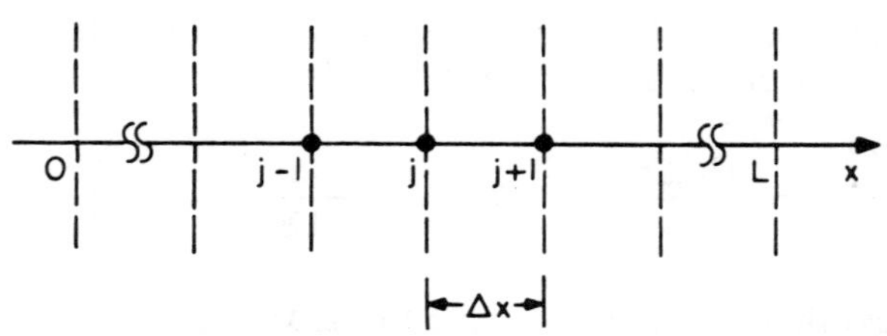

Fig. 3.7 One-Dimensional Numerical Grid with Grid Planes Located at $X_j = j\Delta x$, uniformly spaced

Note: The charge density ρ, the potential ϕ, and the electric field E_x will be obtained only at the X_j points.

We are to use the ρ_j values, known from the x_i values, to obtain the unknown ϕ_j values and then the E_j values, for j, from 0 to $L/\Delta x$ (roughly), where L is the length of the system of NG points. By using the known boundary conditions at $x = 0, L$, and all of the ρ_j values, there will be as many equations as unknowns (the ϕ). Hence, the problem is solvable and there are numerous methods of solution.

3.7 PARTICLE AND FORCE WEIGHTING

3.7.1 Connection Between Grid And Particle Quantities

We need to calculate the charge density on the discrete grid points from the continuous particle positions, and (after the fields are obtained) to calculate the force at the particles from the fields on the grid points. In Figure 3.3, these calculations are called weighting, which implies some form of interpolation among the grid points nearest the particle. It is desirable to use the same weighting in both density and force calculations in order to avoid a self-force (i.e., a particle accelerates itself).

First-order weighting smooths the density and field fluctuations, which reduces the noise (relative to zero-order weighting), but requires additional expense in accessing two grid points for each particle, twice per step. We may view this step as either an improvement in using finite-size particles or one of better interpolation. The charged particles seem to be finite-sized rigid *clouds*, which may pass freely through each other. We call the model *cloud-in-cell*, or CIC (Birdsall and Fuss [2]). If we take the *nominal cloud* to be of uniform density and Δx wide, as shown in Figure 3.8 (the so-called square cloud), then the grid assignment is self-evident, using zero-order weighting for each element. Thus, for total cloud charge of q_c, the part assigned to j is

$$q_j = q_c \frac{\Delta x - (x_i - X_j)}{\Delta x} = q_c \frac{X_{j+1} - x_i}{\Delta x} \tag{3.36}$$

and the part assigned to $j + 1$ is

$$q_{j+1} = q_c \frac{x_i - X_j}{\Delta x} \tag{3.37}$$

The net effect is to produce a *triangular* particle shape $S(x)$ that has width $2\Delta x$. In computation, the nearest left-hand grid point j is located

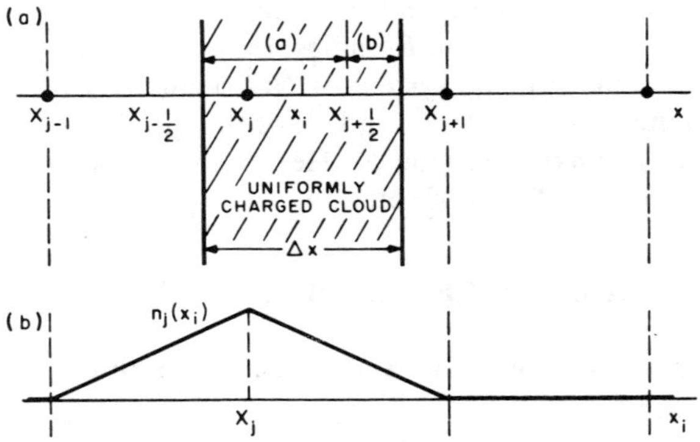

Fig. 3.8 (a) First-Order Particle Weighting, or Cloud-in-Cell Model (CIC) (*Note*: The nominal finite-sized charged particle, or cloud, is one cell wide, with center at x_i. This weighting puts that part of the cloud which is in the jth cell at X_j, fraction (a), and that part which is in the $(j + 1)$th cell at X_{j+1}, fraction (b). This weighting is the same as applying zero-order interpolation to each elemental part); (b) Grid Density $n_j(x_i)$ at Point x_i as the Particle Moves Past X_j (Again displaying the effective particle shape $S(x)$.)

first, so that $x_i > X_j$ always. Then, we calculate the weights and assign the charges. Note that assignment of a point charge at x_i to its nearest grid points by linear interpolation would produce the same result. This viewpoint is called *particle-in-cell,* or PIC, modeling. As a cloud moves through the grid, it contributes to density much more smoothly than with zero-order weight, as we can see from Figure 3.8. Hence, the resultant plasma density and field will have much less noise and be more acceptable for most plasma simulation problems.

The *field* or *force weighting* operates in the same manner. The *first-order force* (CIC-PIC) in fact comes from linear interpolation, exactly as in the charge assignment. For a particle at x_i,

$$E(x_i) = \left[\frac{X_{j+1} - x_i}{\Delta x}\right] E_j + \left[\frac{x_i - X_j}{\Delta x}\right] E_{j+1} \tag{3.38}$$

The first-order weighting consumes more computer time per particle than does the zero-order weighting. However, for a given noise level, CIC allows both a coarser grid and fewer particles than NGP, and thus regains some of the required computer time per particle. We will also see that higher-order weighting and smoothing tend to reduce nonphysical effects.

3.8 CHOICE OF DIAGNOSTICS

The object of simulation is to gain insight into the physics of plasmas. The object of any simulation run, or series of runs, is to study oscillations, waves, instabilities, heating, radiation, or transport, *et cetera,* but usually a limited number of these phenomena in any one set of runs. Hence, the simulation program's output should consist of information that shows the physics and provide specific information about the phenomena under study.

One could ask that all of the information generated ($\mathbf{x}$, $\mathbf{v}$, ρ, ϕ, $\mathbf{E}$, *et cetera*, at every step) be put in storage (e.g., magnetic tape) or printed out, and then diagnosed separately, either by hand or by computer. This can and has been done for a thermal (Maxwellian) plasma, as an example, where the simulator wished to do the diagnostics at his or her leisure, after the run, and more than once. However, the most common output is *graphical,* and consists of plots made during the run (snapshots) and at the end (time histories), complete with Fourier analyses in time and space, printed out on film or paper. No other information will be saved, so the plots chosen must contain the physics of interest. To accommodate hindsight, saving more than the minimum perceived as necessary is usually prudent.

The first page should contain the name of the program, the version being used, and it must also have the initial conditions in considerable detail. The next pages may be snapshot plots made at intervals during the run, such as $t = 0$, $t = 60\Delta t$, and $t = 120\Delta t$. Information of interest in these snapshots might be, for particles:

a. Phase space, v_x *versus* x;
b. Velocity space, v_y *versus* v_x;
c. Density in velocity, $f(v)$ *versus* v, or $f(v^2)$ *versus* v^2, or $\ln[f(v^2)]$ *versus* v^2.

For grid quantities, we might plot:

a. Charge density $\rho(x)$ *versus* x or particle density $n(x)$ *versus* x;
b. Potential $\phi(x)$ *versus* x;
c. Field $E(x)$ *versus* x;
d. Distribution of electrostatic energy $\frac{1}{2}\rho_k\phi_k^*$ *versus* k.

The result at the end of a run will consist of plots of histories of various quantities, such as:

a. Electrostatic energy $\Sigma_k \frac{1}{2}\rho_k\phi_k^*$;
b. Particle kinetic energy by species $\Sigma_i \frac{1}{2}m_i v_i^2$;
c. Particle drift energy $\Sigma_i \frac{1}{2}m_i \langle v_i\rangle^2$;

d. Particle thermal energy $\Sigma_i \frac{1}{2}m_i(\langle v_i^2\rangle - \langle v_i\rangle^2)$;

e. Total energy, electrostatic plus particle (usually with the zero suppressed, so as to magnify loss of energy conservation);

f. Mode plots, $\frac{1}{2}\rho_k\phi_k^*$, for each k, possibly with Fourier analyses of each, with plots *versus* ω.

These lists give an idea of a one-dimensional program output. In a typical run of 1000 steps, we may ask for snapshots every $60\Delta t$ of the seven particle and grid plots (126 plots), and at the end for the five energy plots, and let us say, 32 mode plots. This is a total of $1 + 126 + 5 + 32 = 164$ plots, which fits nicely on a 192-page 10×15 cm microfiche.

Many other quantities may be of interest. In studying diffusion, we wish to follow the mean square deviation of particle velocities or positions, so that $[v_i - v_i(t = 0)]^2$ or $[x_i - x_i(t = 0)]^2$ must be stored, and then plotted. In studying waves, we may wish to resolve frequencies, so that a Fourier analysis in time would be performed on some quantity like $\phi(k, t)$, perhaps following the run (called *postprocessing*), necessitating storage of the quantity. The dynamics of phase, velocity, or distribution function space may be of interest, in which case a *movie* might be made, probably from snapshots every Δt or $2\Delta t$.

In this way, the simulator, like an experimenter in the laboratory, will accumulate sufficient data to verify the correctness of the physics and to provide the desired insight into the particular plasma problem.

3.9 INTRODUCTION TO ES1 (ONE-DIMENSIONAL ELECTROSTATIC CODE)

We are ready to take a look at a particular program. This look is in preparation for using this program (or any other) on one-dimensional electrostatic problems.

The program ES1 (electrostatic, one dimension) was written by A. B. Langdon for this author's course in Berkeley, in 1972. Although ES1 was designed to be a teaching tool, it has formed a base for programs used in industry. ES1 is available to all users of the National Magnetic Fusion Energy Computer Center, and can be easily adapted to other computer systems.

The program flows very much like the scheme shown earlier, where the main loop steps were shown, without the initial step. Now, we state the scheme in computer terms, use the names of computer *variables* and *subroutines,* bunch the steps together somewhat differently, and add the initial step.

First, the program reads the input data, then initializes the plotting routine HISTRY. Next, ES1 calls subroutines INIT, FIELDS (which calls the Fourier transform routine and its inverse), and SETV (which calls ACCEL), in that order. These routines are used once to set up the proper initial conditions, given $\mathbf{x}(0)$, $\mathbf{v}(0)$ for all particles, provided $\mathbf{x}(0)$, $\mathbf{v}(-\Delta t/2)$ as starting conditions for the main loop. Actually, more is done, as follows:

FIRST Connect input file. Create output file for "tape 3" and plots.
Read data
HISTRY Later plots quantities (like the various energies) *versus* time.
INIT Change input variables (like ω_p, q/m) into computer variables (like q, m). Calculate initial particle positions, including perturbations $x(0)$. Calculate initial particle velocities, such as cold or Maxwellian, including perturbations $v(0)$.
SETRHO Convert x to $x/\Delta x$, computer variable. Accumulate charge on the grid, with weighting.
FIELDS Using charge on the grid, solve Poisson's equation for ϕ. Use CPFT, RPFT2, RPFT12, CPFT, the Fourier transform subroutines. Calculate field energy from $\Sigma\rho_k\phi_k^*$. Difference ϕ to obtain E on grid. Make field snapshots.
SETV Change $v(0)$ to $v(-\Delta t/2)$, using $E(0)$ in ACCEL. Convert v to $v\Delta t/\Delta x$, computer variable.

Then, the program continues for as many time steps as are requested in the main loop, which calls subroutines ACCEL, MOVE, and FIELDS in order, once each step. The details for each of these tasks are listed below, which is like the block diagram of Figure 3.3 with the weighting moved into the subroutines ACCEL and MOVE.

Enter main loop from initializing steps, with $x(0)$, $v(-\Delta t/2)$, $E(0)$:

PLOTXV, Particle diagnostics.
etcetera
ACCEL Convert E to $(q/m)E\Delta t^2/\Delta x$, called A, a computer variable. Advance velocity one time step, using weighted E (or A). Calculate momentum and kinetic energy. Repeat for each species.
MOVE Advance position one time step. Accumulate charge density on grid, with weighting. Repeat for each species.
HISTRY Saves quantities for plotting versus time.
Advance time step counter
FIELDS (Already described.)
Return to start of main loop for total of NT steps

After running NT time steps we exit from the main loop through HISTRY and LAST. Thus,

HISTRY	Make plots of quantities versus time.
LAST	Close files and terminate execution.

3.10 TWO-STREAM INSTABILITY: LINEAR ANALYSIS

The model consists of two opposing streams of charged particles as sketched in Figure 3.9. Models with relative motion between two sets or streams of charged particles have been studied in great detail since papers by Haeff [15] and Pierce [34]. Detailed knowledge of the nonlinear behavior of opposing streams came much later, from the simulations done by Dawson [10]. The fluid analog was given much earlier, such as that by Heinrich Hertz in the 1880s (see comprenhesive books on hydrodynamics and acoustics, such as Lamb [21] or Rayleigh [37].

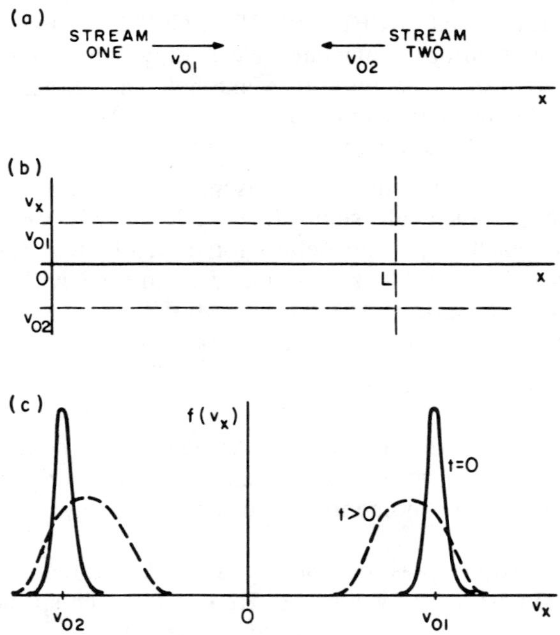

Fig. 3.9 (a) Two Opposing Streams as Seen in the Laboratory; (b) The Streams in Phase Space at the Start of the Problem, $t = 0$; (c) The Streams in Velocity Space at $t = 0$ and $t > 0$

We can readily see that an opposing stream system must be unstable. When two streams move through each other by one wavelength in one cycle of the plasma frequency, a density perturbation on one stream is reinforced by the forces due to bunching of particles in the other stream and *vice versa*. Hence, $\Delta n_1 \propto n_1$, so that the perturbation *grows exponentially in time.*

The longitudinal linear dielectric function for two independent cold streams may be obtained by applying the equations of motion and continuity separately for each stream, and adding the currents of each in the field equation. The result is

$$\frac{1}{\epsilon_0}\,\epsilon(\omega, k) = 1 - \frac{\omega_{p1}^2}{(\omega - \mathbf{k}\cdot\mathbf{v}_{01})^2} - \frac{\omega_{p2}^2}{(\omega - \mathbf{k}\cdot\mathbf{v}_{02})^2} \tag{3.39}$$

for two streams with drift velocities $\mathbf{v}_{01}$ and $\mathbf{v}_{02}$. This result is also obtainable directly from the usual Vlasov-Poisson set by letting the velocity distribution be two delta functions:

$$f_0(\mathbf{v}) = A\delta(\mathbf{v} - \mathbf{v}_{01}) + B\delta(\mathbf{v} - \mathbf{v}_{02}) \tag{3.40}$$

A system of N independent cold streams produces a sum over streams or species s:

$$\frac{1}{\epsilon_0}\,\epsilon(\omega, k) = 1 - \sum_{s=1}^{N} \frac{\omega_{ps}^2}{(\omega - \mathbf{k}\cdot\mathbf{v}_{0s})^2} \tag{3.41}$$

The solutions for complex ω, assuming real k (i.e., an absolute instability, growth in time only, no convection in space), opposing streams of equal strength, $\omega_{p1} = \omega_{p2} \equiv \omega_p$, $v_{01} = -v_{02} \equiv v_0$, is found from $\epsilon(\omega, k) = 0$, which is quartic in ω with four independent solutions. These are

$$\omega = \pm[k^2v_0^2 + \omega_p^2 \pm \omega_p(4k^2v_0^2 + \omega_p^2)^{1/2}]^{1/2} \tag{3.42}$$

for which

$$0 < \frac{kv_0}{\omega_p} < \sqrt{2}\begin{cases}\text{two roots are real}\\ \text{two roots are imaginary}\end{cases} \tag{3.43}$$

$$\sqrt{2} < \frac{kv_0}{\omega_p}, \quad \text{all four roots are real} \tag{3.44}$$

$$\frac{kv_0}{\omega_p} = \frac{\sqrt{3}}{2}, \quad \omega_{\text{imaginary}} = \frac{\omega_p}{2}, \quad \text{maximum growth rate} \tag{3.45}$$

This behavior is sketched in Figure 3.10, and the growth ($\omega_{\text{imaginary}}$) is given in more detail in Figure 3.11. In this model, where there is growth ($\omega_{\text{imaginary}} > 0$), we find that $\omega_{\text{real}} = 0$; that is, there is no oscillatory part associated with the growth, a situation which is not generally true.

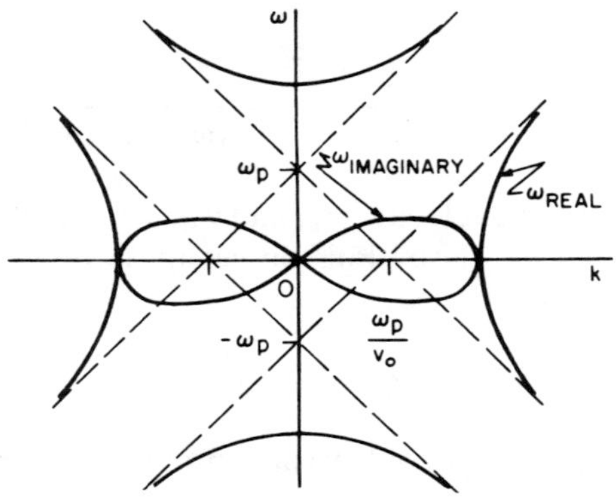

Fig. 3.10 Dispersion ($\omega - k$) Diagram for Two Equal Opposing Streams, Real k, Complex ω

Note: The uncoupled space-charge waves are shown dashed. For each value of k, there are four values of ω that correspond to four linearly independent waves.

A point of Figure 3.11 is to clarify the existence of a *minimum unstable length L of the system*. In this model (normalized):

$$\frac{\omega_p L}{v_0} > \frac{2\pi}{\sqrt{2}} \text{ (unstable)} \tag{3.46}$$

in order to obtain growth. This is the same as (3.43), using $L = 2\pi/k_0$, where k_0 is the smallest wavenumber in the system.

Growth that begins at small amplitude continues until the streaming is destroyed. Indeed, the distribution becomes nearly Maxwellian. Hence, we say that "the colliding streams have *thermalized*," although not by collisions. Instead, collective effects build up large electric fields at long

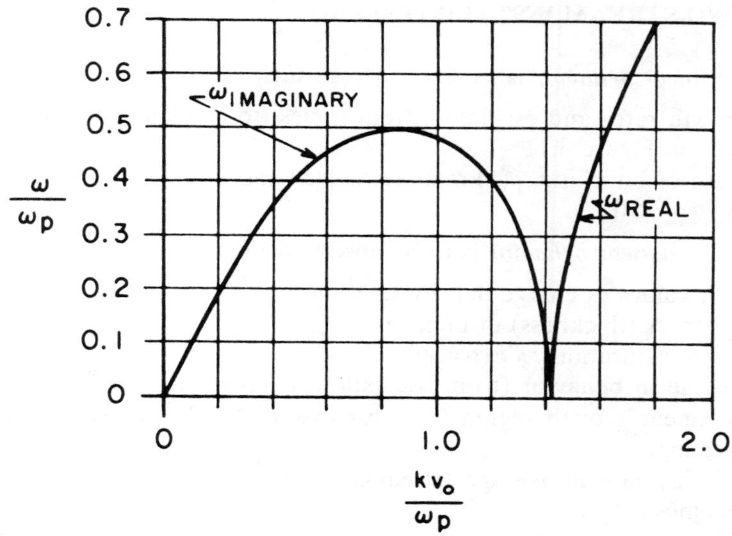

Fig. 3.11 Growth Rate $\omega_{imaginary}$ for Two Opposing Streams

wavelengths ($\lambda \gg$ particle spacing) and these *scatter the particles* in phase space.

As the instability grows, two changes are readily observed in $f(v)$ as indicated for $t > 0$ in Figure 3.9. The *width* of each beam increases (measured directly on an $f(v)$ plot or by $(\overline{v^2} - \overline{v}^2)$ of one stream), which is taken as an increase in the *temperature* of each beam (but perhaps carelessly so, for if the electric field were suddenly shut off—and you should try this—the spread might decrease). The drift or mean velocity $\overline{v}$ decreases.

We might expect, as $v_{thermal}$ increases and v_{drift} decreases, that the conditions for linear growth would cease to be met (see Stringer [38], who shows the growth threshold for electron-electron streams to be $v_{drift} \approx 1.3 v_{thermal}$) and the exponential growth would stop. However, at this time, the conditions for linearity are largely violated, with perturbed charge densities comparable to the zero-order density. Particles in one stream are about to pass their neighbors and wrap into *vortices in phase space*; that is, the particles become trapped. Hence, the growth need not stop, although we might be tempted to look for a change in character of the growth (e.g., away from exponential in time) at the time when v_t exceeds $\overline{v}/1.3$. Keep this in mind in your project. Of course, ES1 can readily be run with warm beams, so look for growth with $v_0 = 2v_t$, but stability $v_0 = v_t$.

3.11 TWO-STREAM INSTABILITY: PROJECT

The *linear behavior* is to be observed, such as:

a. Growth rates and oscillation frequencies, to be checked with those given in Figure 3.11;
b. Phase of bunching, $\rho(x)$ relative to $\phi(x)$ and $E(x)$, for $q_1q_2 > 0$ and for $q_1q_2 < 0$.

The *nonlinear behavior* is to be observed, such as:

a. Peak values of charge densities;
b. Tightness (thickness) of bunches;
c. PE (at saturation)$/KE(t = 0)$;
d. Change in behavior (from oscillation to growth) of modes outside the linear growth region (i.e., for $kv_0/\omega_p > \sqrt{2}$) due to nonlinear coupling;
e. The decrease in average (streaming) velocities, $\langle v_1 \rangle$ and $\langle v_2 \rangle$ as PE becomes large;
f. The increase in velocity spread as given by $[\langle v^2 \rangle - \langle v \rangle^2]$, a measure of the change from cold to warm plasma;
g. The difference in $(PE)_{max}$ between the electron-electron ($q_1q_2 > 0$) and the electron-positron ($q_1q_2 < 0$) two-streaming cases.

A list of initial values follows to aid in getting you started: NSP = 2, L = 2π, DT = 0.2, NT = 300, NG = 32, IW = 2, EPSI = 1.0, A1, A2 = 0, plots every $50\Delta t$, mode plots 1, 2, 3. For species 1, N = 128, WP = 1.0, WC = 0.0, QM = -1.0, V0= 1.0, MODE = 1, X1 = 0.001 (the other values are zero). For species 2, use V0 = -1.0. Note that these values put the first mode at $k_0v_0/\omega_p = 1$, in the region of linear growth and the second mode at 2, outside the linear growth region, where ω is real. Hence, when mode 1 (which is growing) becomes large enough to produce harmonics at nk_0, $n = 2, 3, \ldots$, then mode 2 begins to grow. The interesting part here is that these *nonlinearities* appear at amplitudes (for both mode 1 and 2) that are *far below their maximum values* (at saturation). This should give you some concern for the "boundary" between linearity and nonlinearity.

You might choose to use larger initial displacements, for example, 10 or even 100 times larger, but run fewer steps (e.g., 50), just to see how the program runs and to obtain some idea of growth rates, saturation amplitudes, and whether you want more or less frequent diagnostics. Of course, such a run probably is nonlinear at almost $t = 0$, and is to be used in order to plan better and longer runs at smaller excitation. You might also consider moving k_0 to smaller values to allow for more growing modes.

You may observe changes in saturation amplitude as the k values become more dense, pointing up some deficiencies of single-mode excitation, or use of discrete k values inherent in simulation.

If you wish further challenge, go back to the linear cold-beam theory and work out the initial values of displacements and velocities (perturbation amplitudes) needed to excite only the growing wave. See if there is a change in the nonlinear behavior from that for excitation of a single mode (all four waves) or many modes (e.g., all modes excited, with small but random values of the amplitudes). Try to excite only the *decaying* mode. What terminates the decay?

In contrasting like-sign and opposite-sign results, return to the linear cold-beam theory and obtain the relative phases of ρ_1, x_1, ϕ_1, and E_1 of the growing wave. Apply this information to explain the differences in phase space (v_x, x) plots of the $e - e$ and $e - p$ runs.

3.12 TWO-STREAM INSTABILITY: SELECTED RESULTS

Simply as examples of results, we present a few results of two-stream runs made with ES1. First, let us look at an *electron-electron* run. At $t = 0$, the streams have drift velocities ± 1.0 and are perturbed slightly in position. Figure 3.12 shows the evolution of the instability. Also, the grid charge density, the potential, and the electric field stay sinusoidal in space until about time $t = 15$, when the perturbation becomes quite large and the charge density has two peaks (not shown). Then, with a large electric field retarding the stream, some particles stop at about $t = 17$ and reverse as seen at about time $t = 18$. A phase-space vortex thus forms and persists. Note that the large fields also accelerate some particles out to $|v| \approx 3$, or 9 times their original kinetic energy. At the end, the program plots the energies *versus* time. The electrostatic or field energy reaches a peak of about 40% of total initial kinetic energy, which is almost wholly in mode 1 ($k_1 v_0/\omega_0 = 1$), with exponential growth over many decades at very close to the theoretical rate as measured on a semilog plot (not shown). Not shown are modes 2 and 3, which are linearly stable, but grow due to nonlinear coupling (at two and three times $\omega_{\text{imaginary}}$ of mode 1), starting at about $t = 8$, $t = 10$, when mode 1 energy is still about 1000 and 100 times smaller than at saturation.

Second, let us look at the *electron-positron* colliding stream model, which has the same dispersion relation as the electron-electron model, but has different phases among ρ, ϕ, and E for the growing wave. Figure 3.13 shows some of the results. We see an entirely different form of bunching than with like-sign particles at $t = 15.9$, when positive and negative bunches

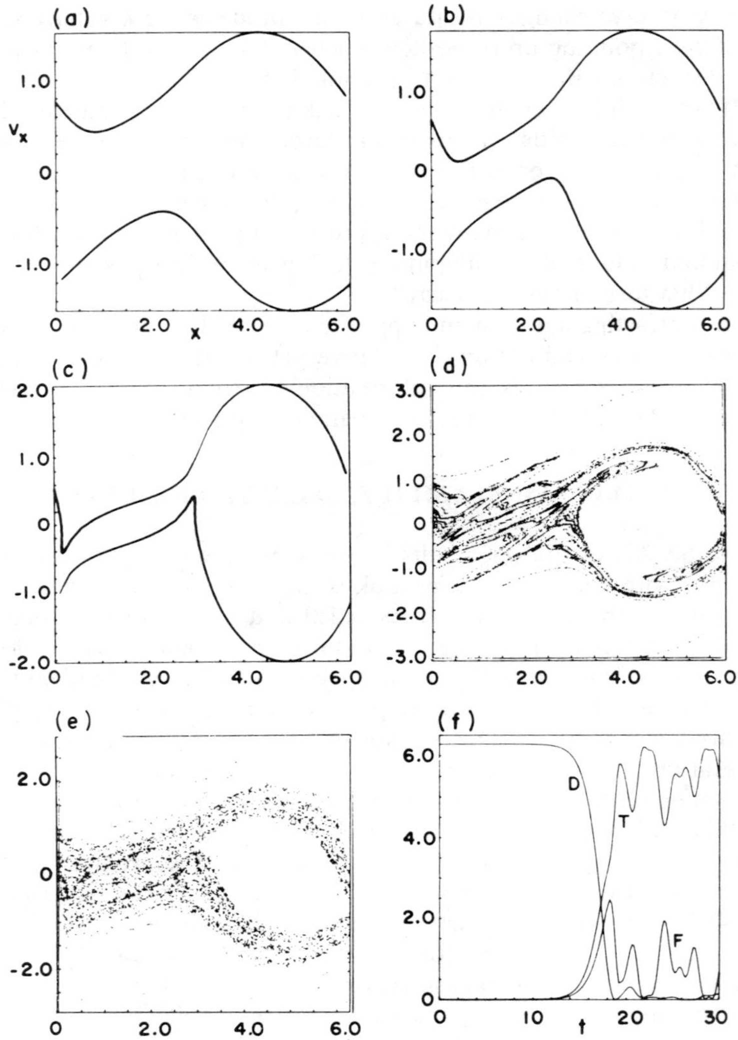

Fig. 3.12 Evolution of Electron-Electron Two-Stream Instability in $v_x - x$ Phase Space Plots

Note: a, b, c, d, and e at $t = 16, 17, 18, 34$, and 60 (see changes in v_x scale). Also shown are the energy histories, drift (D), thermal (T) and field (F), for $t = 0$ to 30 (in f). The initial velocities are ± 1.0. There are 4096 particles in each stream; the grid has 32 cells, $\omega_{p1} = 1 = \omega_{p2}$.

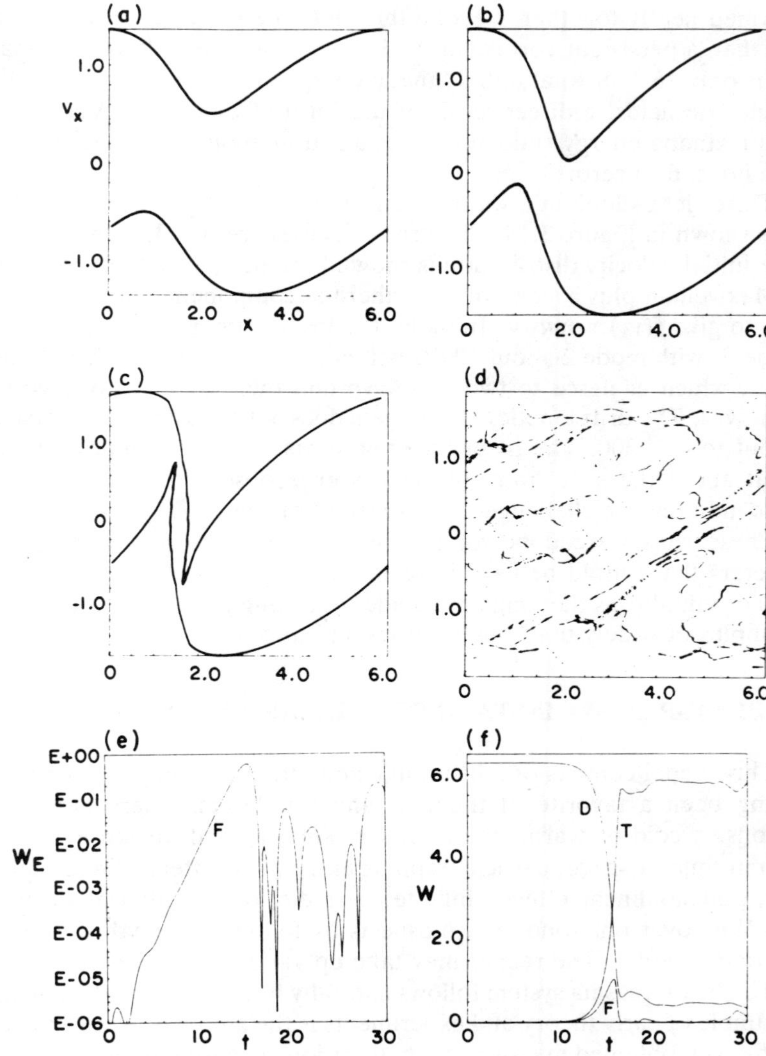

Fig. 3.13 Evolution of Electron-Positron Two-Stream Instability in v_x – x Phase-Space Plots (a, b, c, and d at t = 14, 15, 16, and 33), Mode 1 Field Energy History (e), and Energy Histories (f)

Note: The parameters are the same as in Figure 3.12.

are formed nearly together. Beyond this time, there is a lot of scrambling rather than a persistent vortex. In the history plots, the field energy maximum is only 16% of total initial kinetic energy and has relatively smaller peak electric fields, a direct result of the form of bunching. Also, while the drift kinetic energy suddenly drops again in about one plasma cycle, it does not fall to zero.

Third, let us look at a warm electron-electron two-stream instability run, as shown in Figure 3.14. The drift velocities are ± 1.0, with $v_{te} = v_0/2$. The initial velocity distribution is shown both in $v_x - x$ space (a quiet-start-Maxwellian plus a small random thermal component) and integrated over x to give $f(v_x)$ *versus* v. Initially, the instability grows exponentially in mode 3, with mode 2 about 10 dB behind, and forms three phase-space vortices, which coalesce to two as shown and then finally to one vortex (not shown). The drift kinetic energy again does not go to zero, but persists, even out to $t = 300$. The reader may note that, because modes 2 and 3 grow at about the same rate, the coalescence observed may be due to initial conditions—a challenge to be pursued further.

These runs were done for fun, without the care and variation in parameters that should be used in scientific study. The runs serve as examples of initial runs that might be made in starting a thorough study, and as examples of widely used observations and diagnostics.

3.13 BEAM-PLASMA INSTABILITY: LINEAR ANALYSIS

This model consists of a beam injected into a stationary plasma and has long been a favorite of theorists and simulators. There are many variations: a cold or warm stream and plasma; a weak or strong beam; growth in time or space; particle trapping and passing effects; linear, quasilinear, and nonlinear effects; infinite, periodic, or bounded models; *etcetera*. We cover only one or two aspects by following growth in time in our periodic model. The reader may take up variations as projects.

The beam-plasma system follows naturally from the two-stream work. An example of early theory and experiments is the amplifier work of Boyd *et al.* [6], who obtained microwave amplification at centimeter wavelengths by using an electron beam injected into a long gas-discharge column. The experimenters modulated the beam before it went into the plasma and stripped off the modulation after the beam left the plasma. The device aroused great interest for producing waves at millimeter wavelengths because the beam and plasma were intimately mixed, and did not require a slow-wave circuit, which was tiny and vulnerable to grazing by an intense electron beam. More recent work by *Gentle and Lohr* [13, 14] emphasizes nonlinear behavior. The model is still fascinating.

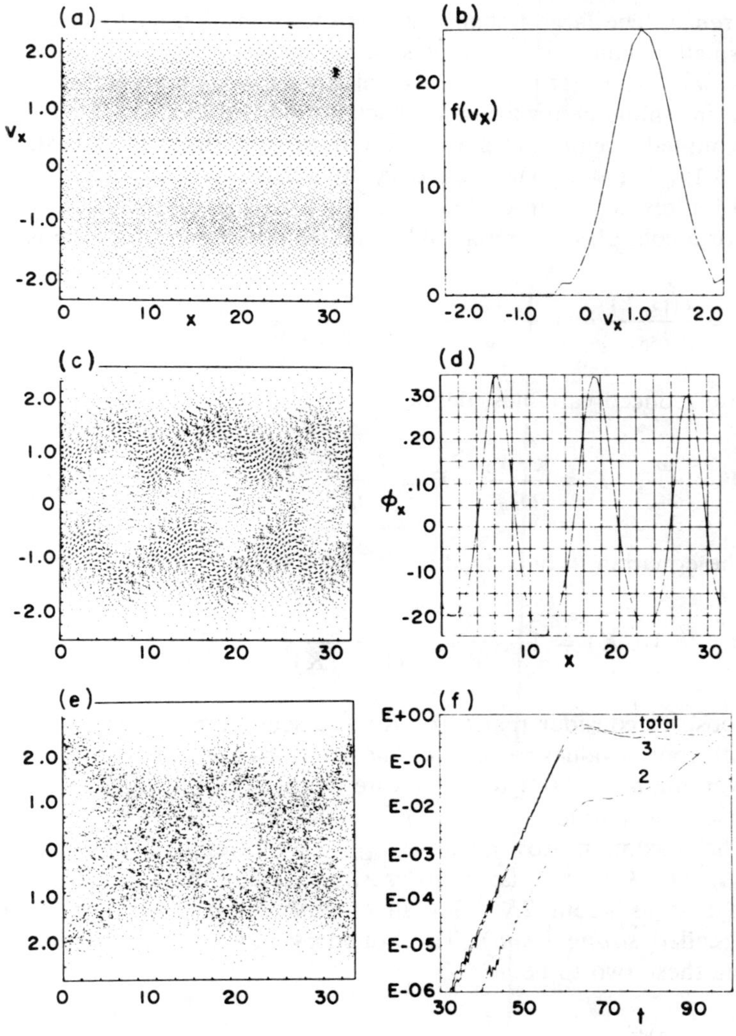

Fig. 3.14 Evolution of Warm Electron-Electron Two-Stream Instability with $v_x - x$ Phase-Space Plots (a, c, and e at $t = 0$, 60, and 100), Initial Velocity Distribution of One Stream $f(v_x)$ *versus* v_x (in b), Potential $\phi(x)$ *versus* x (in d at $t = 60$), and Field Energy History (in f; total, modes 2 and 3, $t = 0$ to 100).

The physical behavior is an example of double streaming as presented earlier. Those ideas can be lifted and used here by going to a new frame of reference. The largest change is that the beam density n_{b0} is generally much smaller than that of the plasma n_{p0} such that $\omega_{pb}^2 \ll \omega_{pp}^2$, the *weak-beam model.* The first interactions which we follow are electron-electron with an immobile neutralizing ion background (i.e., $m_i/m_e \to \infty$). Hence, for uncoupled stream and plasma, the dispersion diagram is as shown in Figure 3.15. The dominant frequency obviously is ω_{pp} and the wave numbers of interest are near $k = \omega_{pp}/v_0$.

For a cold plasma and a cold beam, the dispersion relation is

$$0 = \frac{\epsilon(\omega, k)}{\epsilon_0} = 1 - \frac{\omega_{pp}^2}{\omega^2} - \frac{\omega_{pb}^2}{(\omega - \mathbf{k} \cdot \mathbf{v}_0)^2} \tag{3.47}$$

If we use reduced variables, defined by

$$W \equiv \frac{\omega}{\omega_{pp}}, \quad K \equiv \frac{\mathbf{k} \cdot \mathbf{v}_0}{\omega_{pp}}, \quad R = \left(\frac{\omega_{pb}}{\omega_{pp}}\right)^2 \tag{3.48}$$

then, the equation to be solved is

$$0 = F(W, K) = 1 - \frac{1}{W^2} - \frac{R}{(W - K)^2} \tag{3.49}$$

We choose to consider real k (or K) and seek complex values of ω (or W), with four ω values to each k. The solutions are shown in Figure 3.16 for R varying from 0.001 to 0.1. (Note the utility of having a root solver and a plotting routine.)

The maximum growth rate $\omega_{\text{imaginary}}$ is plotted in Figure 3.17 along with ω_{real} and k at the maximum growth. We see that $(\omega_{\text{imag}})_{\max}/\omega_{pp}$ increases first as about $2R^{1/3}/3$ (called *weak beam*) and then as about $R^{1/4}/2$ (called *strong beam*). The boundary is found graphically, or by equating these two to be at

$$R = \left(\frac{3}{4}\right)^{12} = 0.032 \tag{3.50}$$

as noted on the figure. This boundary is not to be taken seriously, but regarded merely as some kind of transition region from *three-wave interaction* $\omega_i \propto (n_0)^{1/3}$, Figure 3.16(a) and (b), and *four-wave interaction,* $\omega_i \propto (n_{b0})^{1/4}$, Figure 3.16(c).

For *weak beams,* we see that maximum growth occurs very close to intersection B in Figure 3.15 $\omega_{\text{real}} \approx \omega_{pp}$, $k \approx \omega_{pp}/v_0$. Note that even a

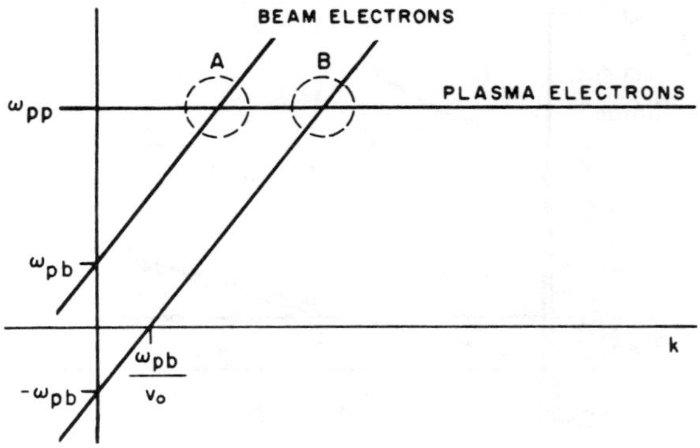

Fig. 3.15 Sketch of Dispersion Diagram for the *Weak-Beam Model*, with Beam of Density ω_{pb} Injected at Velocity v_0 into a Plasma of Density ω_{pp}, Uncoupled
Note: Coupling between the beam space-charge waves and plasma oscillations is expected at *A* and *B*.

very weak beam, let us say, with 10^{-4} the density of the plasma, sets up measurable growth. In this case, the growth is 1.8 dB/plasma cycle, or a factor of 10 after 5.5 plasma cycles. For more exact results, see Bers [1]. For strong beams, the interactions at *A* and *B* in Figure 3.15 move apart widely to *C* and *D* in Figure 3.18, and the three-wave weak-beam interaction becomes a four-wave interaction with $\omega_{\text{real}} \approx \omega_{pp}$, $k \approx \omega_{pb}/v_0$. This regime is physically puzzling with respect to neutralization, and so deserves comment. This implies a beam electron density no longer trivial relative to background plasma electron density, which is usually taken as equal to the background ion density and is now insufficient for system neutrality. Hence, the plasma ion density must exceed the plasma electron density by

$$\frac{n_{pi}}{n_{pe}} = \frac{1 + R}{1 - \dfrac{m_e}{m_i}} \tag{3.51}$$

This is quite different from our initial discussion because the beam is no longer just slightly perturbing a neutral plasma, but is actually providing many of the electrons.

Of course, *the background could be all be ions,* such that $\omega_{pp}^2 = (q/m_i)(n_{b0}q/\epsilon_0)$, which forces $\omega_{pp}^2 \ll \omega_{pb}^2$. This model of electrons streaming

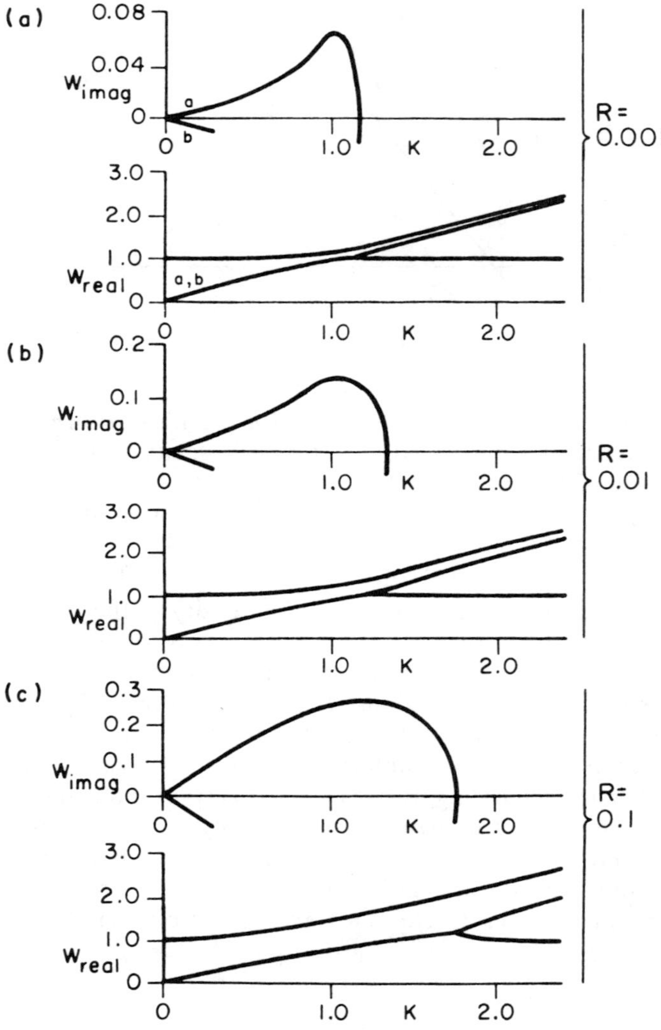

Fig. 3.16 Real and Imaginary Roots of the Dispersion Relation (3.49): (a) for $R = 0.001$; (b) for $R = 0.01$; (c) for $R = 0.1$.

Note: The complex roots are paired by the letters (a, b) in (b); the other roots are wholly real. The root at $W_{\text{real}} \approx -1.0$ is not shown.

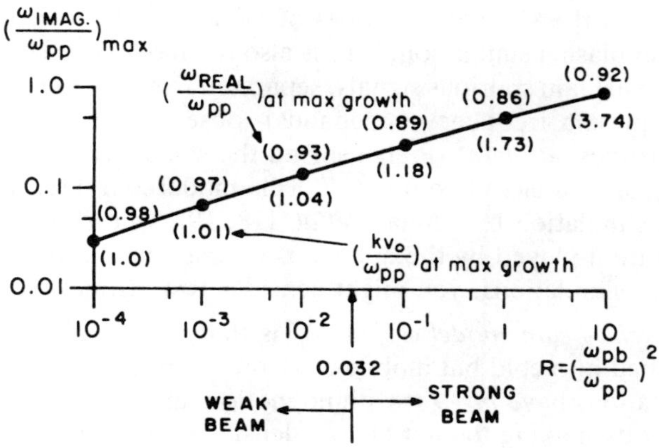

Fig. 3.17 Maximum Growth Rate for Cold-Beam-Plasma System as a Function of the Beam-to-Plasma Density Ratio

Note: The density ratio $R = 0.032$ is taken as the boundary between weak and strong beams, the break point between growth at $R^{1/3}$ and $R^{1/4}$.

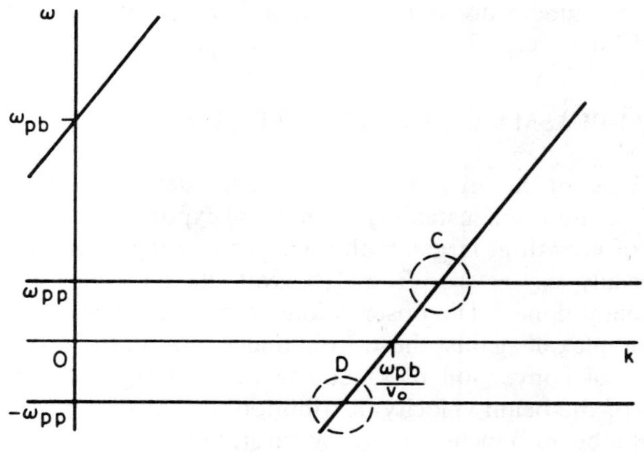

Fig. 3.18 Sketch of Dispersion-Dispersion Diagram for the *Strong-Beam Model,* Uncoupled

Note: Coupling between the beam and plasma is expected at C and D, roughly two two-wave couplings.

through ions is the simulation model of Buneman [8], one of the early milestones in plasma simulation. This is also the model used by Pierce [34] in seeking to explain spurious signals, separated from the microwave signal by the ion plasma frequency (sidebands), observed in microwave tubes (triodes, tetrodes, *etcetera*). Pierce solved the same dispersion relation as (3.48), but for real ω and complex k. This model is treated in nonlinear theory and simulation by Ishihara *et al.* [18, 19], who show the key role in the saturation played by the appearance of spatial harmonics.

In your simulations, you might consider two distinct models:

a. The *weak-beam model* $\omega_{pb}^2 \ll \omega_{pp}^2$ is that of a weak electron beam launched into cold but mobile electrons of much higher density. The plasma ions have $m_i \to \infty$ with no motion, and neutralization is done simply by making the net charge density zero in the simulation program.
b. The *strong-beam model* $\omega_{pb}^2 \gg \omega_{pp}^2$ is that of an electron beam launched into cold but mobile ions of equal density, but with $m_i \gg m_e$.

Thus, for model a, the model is electron-electron; for model b, it changes to electron-ion. Of course, you are free to choose other models as well. As the beam temperature is increased, the nature of the roots changes (O'Neil and Malmberg [31]).

3.14 BEAM-PLASMA INSTABILITY: PROJECT

The object of the project is to observe the development of the weak $R = 0.001$ beam-plasma instability from initial exponential growth through to the end of growth at large amplitude. (The strong beam-plasma instability, especially $\omega_{pb} = \omega_{pp}$, $R = 1$, is very close to the two-stream instability already done.) The observations to be made are the checking of values of complex ω against those from linear theory, the saturation level ("efficiency" of conversion of kinetic to field energy), the slowing and broadening of the beam velocity distribution, and the bounce or trapping motion of the beam bunches following saturation.

For $R \ll 1$, the main plasma (in contrast to the weak beam) remains linear by various tests (e.g., n_1/n_0 remains $\ll 1$). Therefore, the plasma particles could be replaced by a linearized fluid, or treated by a linear susceptibility. The particle-fluid model proved useful in a related model (magnetized ring plasma), where the noise of the plasma had obscured the desired interaction (Lee and Birdsall, [25]). We use another trick:

because only the plasma frequency of a linearly-responding species matters, we are free to choose a small value for q/m to ensure that the plasma remains linear, which permits us to use a small number of plasma particles (as few as one per cell!). The fixed background ions are put in ES1 by default.

Let the ratio $(\omega_{pb}/\omega_{pp})^2 \equiv R = 0.001$ in the weak-beam region ($R \leqslant 0.032$). The beam becomes highly distorted in $v_x - x$ phase space, wrapping itself into a vortex. Therefore, for adequate accuracy, far more particles are required than for the plasma. Hence, try $N = 512$ for the beam and $N = 64$ for the plasma, with $NG = 64$ cells. Use $\omega_{pp} = 1.0$ and $\omega_{pb} = (1000)^{-1/2}$, with beam $v_0 = 1$. For a test run, place the first mode near the peak of the growth rate, $kv_0/\omega_{pp} = 1.01$, where $\omega/\omega_{pp} \approx 0.97 + i0.006$. Hence, set $k_0 = 1$, $L = 2\pi$. For later runs, put the tenth mode at this peak, using $L = 20\pi$, and excite all modes initially using VT1. Excitation of the beam at $X1 = 0.01$ (to be decided on basis of best excitation of the growing wave) produces several decades of exponential growth, ending in saturation by time $t = 115$. The use of $\Delta t = 0.2$ requires NT = 600. Running for a longer time allows viewing of bounce, or trapping motion.

Check the nonlinear predictions of Kainer *et al.* [20]. Predict the potential when trapping is supposed to begin. Does it do this according to your phase-space plots? Use phase-space plots made at saturation and several subsequent peaks and valleys (this takes a preliminary run to determine the times to make the plots) in order to verify (or not) the early trapping and subsequent expected mixing. Is the beam $f(v)$ spreading and slowing development as predicted by Kainer *et al.* [20]? Are there some very hot particles, as are best seen in $\log f(v)$ *versus* v or v^2 plots? Follow the decrease of beam $\langle v \rangle$ and increase of $\langle v^2 \rangle$ in time.

There are more variations worth noting, which you may try. First, suppose only the beam is allowed to move, with the plasma held immobile. Then, only the fast and slow space-charge waves are expected, with no growth. However, there is a nonphysical instability produced by aliasing, as shown analytically by Langdon [22, 23] and Chen *et al.* [9] and verified by them and Okuda [29, 30]. This instability causes heating of the beam to a level of $\lambda_D/\Delta x \approx 0.05$ (where $\lambda_D = v_t/\omega_p$ of the beam), and then stops, leaving a stable but slightly warm (and noisy) beam, as shown by Birdsall and Maron [3]. The maximum growth rate is $0.2\omega_{pb} = 0.2\omega_{pp}R^{1/2}$, which is roughly $\omega_{pp}/150$ for $R = 0.001$, about a factor of 10 less than the expected beam-plasma growth rate. Hence, this self-heating effect may be viewed here as transient.

Second, examine what happens if the plasma is excited at relatively large amplitude, for example, by using $X1$ on the order of the unperturbed particle spacing. Does the beam-plasma interaction take place?

Third, ties may be made to linear and nonlinear experiments, where an electron gun launches a cold beam into a plasma. The simulation may be altered to fit the experiment more closely, where the beam electrons are continuously injected at $x = 0$ and collected at $x = L^-$ (not returned to $x = 0^+$, as in the usual periodic model) and the plasma electrons are reflected, in some careful manner, at $x = 0^+$, $x = L^-$. The excitation may be a velocity or density modulation of the beam by a sinusoidal (in time) source at $x = 0^+$. The level of modulation and length L should be chosen to produce saturation at $x < L$. These changes fit experiments more closely where spatial growth (real ω and complex k) is expected. See Briggs [7, p. 76] on the distinction among absolute, convective, and "neither" instabilities.

Selected results for $R = 0.001$ are shown in Figure 3.19. The $v_x - x$ phase space plots were chosen to coincide with peaks of field energy (hence, valleys of beam kinetic energy) at $t \approx 146$, 202, and 258, and valleys at $t \approx 173$, 230, and 286. Thus, the peak times should exhibit bunching at $v < v_0 = 1.0$, and valley times should exhibit bunching at $v > v_0$. The reader may see that this is roughly the case. The use of ϕ_{max} in $\omega_{trap} = k(q\phi_{max}/m)^{1/2}$ produces $\tau_{trap} \approx 56$, very close to that observed. Note that the vortex takes on considerable structure so that simple-bounce ideas fail.

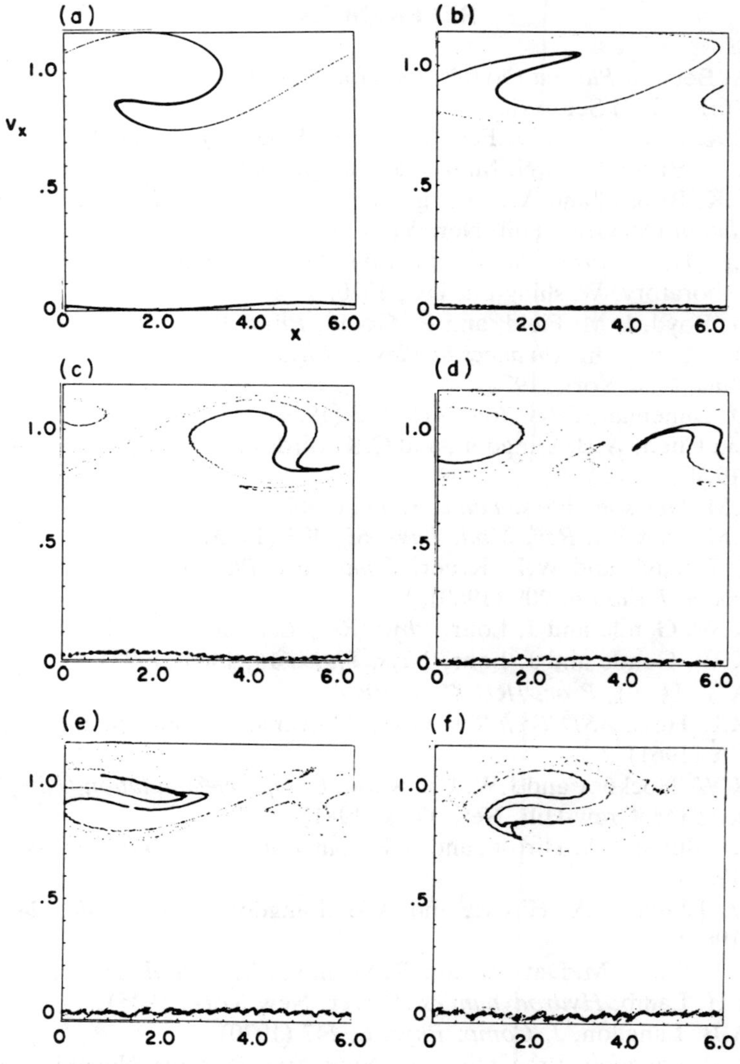

Fig. 3.19 Evolution of Weak-Beam Plasma Instability in $v_x - x$ Phase Space with Strength $R = 0.001 = (\omega_{pb}/\omega_{pp})^2$ and Beam Velocity $v_0 = 1.0$.

Note: The left-hand set occurs very close to the first, second, and third peaks of field energy and the right-hand set occurs very close to the subsequent valleys, showing the bounce motion and mixing. The sequence is $t = 146, 173, 202, 230, 258, 286$.

REFERENCES

1. A. Bers, in *Plasma Physics, Les Houches,* (Gordon and Breach, New York, 1972) Sec. xxii.
2. C.K. Birdsall and D. Fuss, *J. Comp. Phys.* **3**, 494 (1969).
3. C.K. Birdsall and N. Maron, *J. Comp. Phys.* **36**, 1 (1980).
4. C.K. Birdsall and A.B. Langdon, *Plasma Physics via Computer Simulation* (McGraw-Hill, New York, 1985).
5. J.P. Boris, *Proc. 4th Conf. Num. Sim. Plasmas,* Naval Research Laboratory, Washington, DC, 1970.
6. G. Boyd, L.M. Field, and R. Gould, *Phys. Rev.* **109** (1958).
7. R.J. Briggs, in *Advances in Plasma Physics,* Vol. 4 (John Wiley and Sons, New York, 1971).
8. O. Buneman, *Phys. Rev.* **115**, 503 (1959).
9. Liu Chen, A.B. Langdon, and C.K. Birdsall, *J. Comp. Phys.* **14**, 200 (1974).
10. J.M. Dawson, *Phys. Fluids* **5**, 445 (1962).
11. J.M. Dawson, *Rev. Mod. Phys.* **55**, 403 (1983).
12. J. Denavit and W.L. Kruer, *Comments, Plasma Physics and Controlled Fusion* **6**, 209 (1980).
13. K.W. Gentle and J. Lohr, *Phys. Rev. Lett.* **30**, 75 (1973).
14. K.W. Gentle and J. Lohr, *Phys. Fluids* **16**, 1464 (1973).
15. A.V. Haeff, *Proc. IRE* **37**, 4 (1949).
16. R.L. Hess, *ASD Tech Rep. 61–51*, University of California, Berkeley, CA (1961).
17. R.W. Hockney and J.W. Eastwood, *Computer Simulation Using Particles* (McGraw-Hill, New York, 1981).
18. O. Ishihara, A. Hirose, and A.B. Langdon, *Phys. Rev. Lett.* **44**, 1404 (1980).
19. O. Ishihara, A. Hirose, and A.B. Langdon, *Phys. Fluids* **24**, 452 (1981).
20. S. Kainer, J.M. Dawson, and R. Shanny, *Phys. Fluids* **15**, 493 (1972).
21. S.H. Lamb, *Hydrodynamics* (Dover, New York, 1945).
22. A.B. Langdon, *J. Comp. Phys.* **6**, 247 (1970).
23. A.B. Langdon, *Proc. 4th Conf. Num. Sim. Plasmas,* Naval Research Laboratory, Washington, DC (1970).
24. A.B. Langdon and B. Lasinski, in J.K. Killeen, ed., *Methods in Computational Physics* (Academic Press, New York, 1976), Vol. 16, p. 327.
25. J.K. Lee and C.K. Birdsall, *Phys. Fluids* **22**, 1306 (1979).
26. J.K. Lee and C.K. Birdsall, *Phys. Fluids* **22**, 1315 (1979).

27. A. Mankofsky, L. Seftor, A.A. Mondelli, and A.T. Drobot, *Bull. APS* **28**, 1124 (1983).
28. A. Nordsieck, *Proc. IRE* **41**, 630 (1953).
29. H. Okuda, *Phys. Fluids* **15**, 1268 (1972).
30. H. Okuda, *J. Comp. Phys.* **10**, 475 (1972).
31. T.M. O'Neil and J.H. Malmberg, *Phys. Fluids* **11**, 1754 (1968).
32. A. Palevsky, Ph.D. Thesis, Department of Physics, Massachusetts Institute of Technology, Cambridge, MA, 1980.
33. A. Palevsky, G. Bekefi, and A.T. Drobot, *J. Appl. Phys.* **52**, 4938 (1981).
34. J.R. Pierce, *J. Appl. Phys.* **19**, 231 (1948).
35. H. Poulter, *ERL Tech. Rep. No. 73,* Stanford University, Stanford, CA, 1954.
36. J.E. Rowe, *IRE Trans.* **ED-3**, 39 (1956).
37. B. Rayleigh, *The Theory of Sound* (Dover, New York, 1945).
38. T.E. Stringer, *J.Nucl. Energy, Pt. C* (Plasma Physics) **C6**, 267 (1964).
39. P.K. Tien, L.A. Walker, and V. Woluntis, *Proc. IRE* **43**, 260 (1955).
40. S.P. Yu, G.P. Kooyers, and O. Buneman, *J. Appl. Phys.* **36**, 2550 (1965).

Part II
FAST-WAVE DEVICES

Chapter 4
GYROTRON THEORY

J. Mark Baird

Department of Electrical Engineering
University of Utah
Salt Lake City, Utah

4.1 INTRODUCTION

The archival literature on gyrotron theory now extends to several hundred papers so that it is not possible in a single chapter to provide a thorough review of the subject. We therefore have elected to provide a limited tutorial review with the following major thrusts:

Section 4.2 provides a phenomenological description of the cyclotron resonance MASER (CRM) mechanism and its characteristics relative to the competing Weibel and peniotron mechanisms.

Section 4.3 is devoted to a review of the major models and analytical approaches used for analysis and design of gyrotron devices.

Section 4.4 provides a review of magnetron injection gun (MIG) design and other electron-beam forming systems for gyrotrons.

There are two excellent reviews in the literature which focus on the historical development of the gyrotron [1, 2]. Briefly, the first publications on the CRM mechanism came from the independent work of three different researchers in 1958–59. Twiss [3], in Australia, and Schneider [4], in the USA, discovered the CRM mechanism from the quantum mechanical point of view in which the energy dependence of the cyclotron frequency permits MASER operation [5]. Gaponov, in the USSR, used a classical physics approach [6]. The mechanism was experimentally verified in the 1960s, in both the US [7] and the USSR [8].

The major developmental breakthroughs that yielded high average power millimeter-wave oscillators (gyrotrons) were first published in the USSR in the mid-1970s [9, 10]. These breakthroughs consisted of using tapered, open-ended waveguide cavities (discussed in Section 4.3.4) in conjunction with a beam-forming system based on the *magnetron injection gun* (MIG). The gyrotron MIG (discussed in Section 4.4) forms an electron beam with most of the energy in cyclotron motion.

Prior to the Soviet publications, a parallel effort at the Naval Research Laboratories (NRL) in Washington, DC had focused on cold cathode, megavolt-level CRM devices in the early 1970s [11]. After publication of the major Soviet breakthroughs in 1974, NRL quickly became the leading source of theoretical development work on gyrotron devices in the US [12–16]. Since that time, contributions have come from many different organizations, too numerous to name here.

For the reader who desires to pursue the extensive gyrotron literature, the following list of journals contain the large majority of the published work:

IEEE Transactions on Electron Devices

IEEE Transactions on Microwave Theory and Techniques

Vol. MTT-25, June 1978, is a special issue on the subject and is a good starting point.

IEEE Transactions on Plasma Science

Vol. PS-13, Dec. 1985 is a special issue on high-power microwaves.

International Journal of Electronics

This Journal has been especially active with four special issues on gyrotrons: Vol. 51, No. 4, Oct. 1981, Vol. 53, No. 6, Dec. 1982, Vol. 57, No. 6, Dec. 1984, and Vol. 61, No. 6, Dec. 1986.

International Journal of Infrared and Millimeter Waves

Radiophysics and Quantum Electronics

Izv. VUZ and *Radiofizika*

English translations of these two Soviet Journals are available.

4.1.1 General Characteristics of Gyrotrons

Gyrotrons can operate throughout the microwave and millimeter-wave spectra and beyond. Their inherent region of domination, however, is in the millimeter-wave region. In the microwave region, they receive stiff competition from conventional microwave tubes, which most often can provide better efficiency or wider bandwidths at the demanded power levels. In the submillimeter-wave range, gyrotrons suffer from the need to work at harmonics of the cyclotron frequency because of the inherent limitations of the superconducting solenoids that provide the magnetic field. In this region, it appears that FELs will be the principal competitor.

Gyrotron oscillators can produce RF power levels above 100 kW CW and 1 MW pulsed at 100 GHz (3 mm wavelength). The efficiency of these devices operating in the fundamental cyclotron mode are inherently in the range of 30%. With efficiency enhancement (not including beam energy recovery as yet), the efficiency extends above 50%. As a result, gyrotrons are important devices for RF heating applications such as *electron cyclotron resonance heating* (ECRH) in plasmas for fusion studies.

Typical gyrotron oscillators are built as shown schematically in Figure 4.1. A strong externally applied magnetic field supports the cyclotron motion of the electrons in the beam. Note that the interaction region is placed at a point of maximum longitudinal magnetic field. The magnetic field in this region must be very stable and precisely formed (close to a uniform field) so that the cyclotron frequency (or one of its harmonics) will closely match the frequency of the RF fields in the beam frame of reference.

The interaction region consists of an open ended waveguide cavity with a transverse cross section such as that shown in Figure 4.1(b). The RF fields in this region interact with the orbital cyclotron motion of electrons in the beam and convert the orbital kinetic energy into RF output. The electrons in the beam, therefore, must have a strong transverse velocity $v_\perp$ as well as the usual longitudinal beam velocity v_z. For the gyrotron depicted in Figure 4.1, most of this transverse velocity comes from the magnetic mirror effect produced by the increasing magnetic field leading up to the interaction region. The final ratio of the transverse to longitudinal velocity $\alpha = v_\perp/v_z$ in the interaction region is typically between 1 and 2 for gyrotrons that use MIGs with thermionic cathodes. For relativistic beam gyrotrons with field emission cathodes, this ratio is usually less than unity because of the difficulty in forming the beam.

Gyrotron amplifiers are not as well developed as oscillators, but they also promise to exceed contemporary amplifier power levels throughout the millimeter-wave band. The amplifiers can take the form of either a gyro-TWT or a gyroklystron configuration. In the gyro-TWT, the interaction cavity shown in Figure 4.1 is replaced by a length of interaction waveguide with an RF input waveguide at the leading end to supply the input signal. In the gyroklystron, the beam passes through a series of waveguide cavities, the first of which is supplied with the input signal, and the last looks very much like an oscillator cavity.

For introductory purposes, we mention two design variables related to the electron gun that strongly affect the RF interaction efficiency of

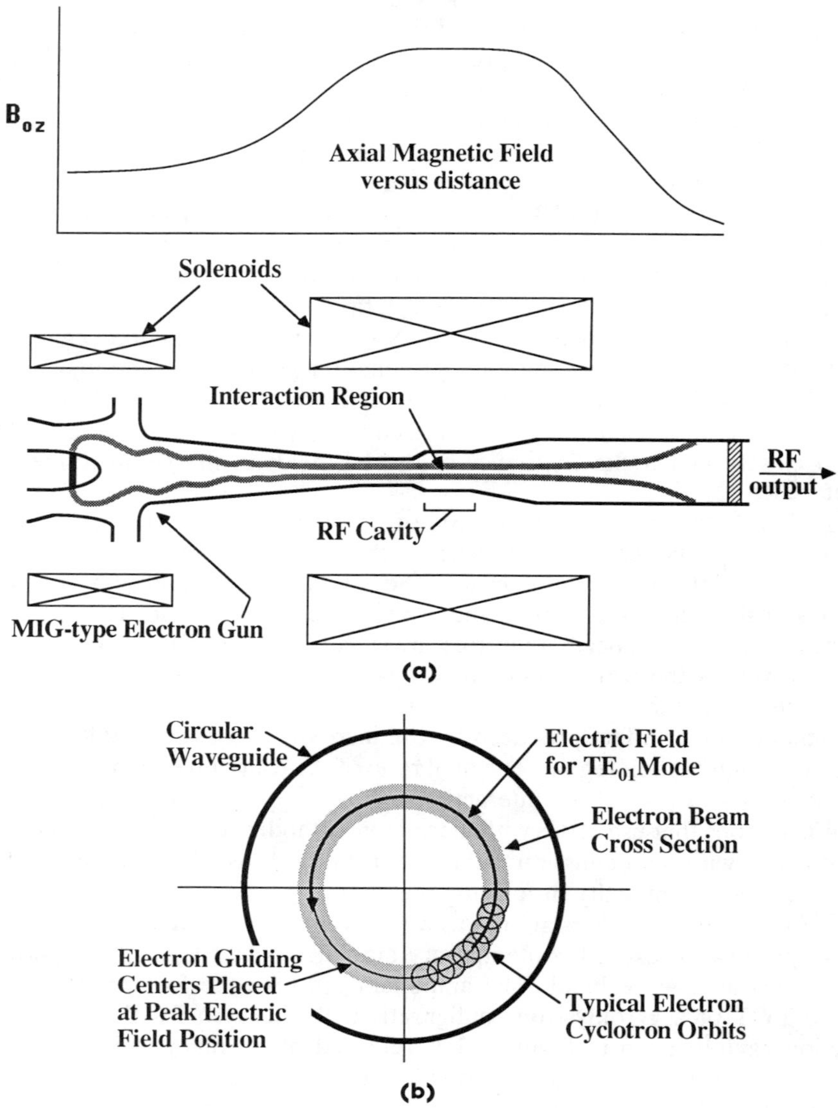

Fig. 4.1 Schematic Diagrams Showing Typical Gyrotron Geometry: (a) Longitudinal Cross Section Showing the Magnetic Field Solenoids and the Field Profile, the MIG (Electron Gun), the RF Interaction Waveguide Cavity, and the RF Output Waveguide; (b) Beam-Cavity Cross Section in the Interaction Region for a Circular TE_{01} Operating Mode

gyrotron devices. The first is the spatial distribution of the electron beam, which is illustrated in Figure 4.1(b). In general, the electrons in the beam follow intermixed helical paths about guiding centers (orbit centers), which are distributed in the transverse plane. However, to provide a strong interaction, the guiding centers must be located at positions where the transverse electric field is strong. For the TE_{01} waveguide mode illustrated, this gives a hollow ring distribution pattern, which nicely matches the natural beam pattern of the electron MIG. For other waveguide modes, matching the beam pattern to the field pattern may be more difficult.

The second gun-related phenomenon that effects RF interaction efficiency is the longitudinal velocity spread $\Delta v_z/v_z$ in the beam. For short interaction regions, such as that used in the gyrotron oscillators, high efficiency can be maintained with velocity spreads up to 15–20%. Longer interaction devices like gyro-TWT or gyroklystron amplifiers may require velocity spreads of 5% or below. Five-percent velocity spread is about the limit of what has been achieved in typical high-power gyrotron beams.

The problems associated with achieving low velocity spread and the proper beam geometry constitute one of the major challenges in building cyclotron beam devices. This is the reason for including a discussion of gyrotron beam-forming techniques in this chapter.

4.1.2 Dispersion Diagrams of Fast Cyclotron Mode Interactions

Gyrotrons operate through a fast cyclotron wave interaction. For those accustomed to classifying electron beam devices according to their dispersion diagrams, these interactions take place in or near the fast-wave region (phase velocity greater than the speed of light) as shown in Figure 4.2. They are therefore classified as a fast-wave devices. The exception to this is the interaction depicted in Figure 4.2(d), which takes place slightly below the speed-of-light line. This is the Weibel* interaction (discussed in the next section), which still falls into the same class of fast-cyclotron-wave devices. For each case shown in Figure 4.2, the fast cyclotron wave on the beam (or one of its harmonics) interacts with a waveguide mode in or near the fast-wave region of the dispersion diagram.

The reader may note that the potential for using fast-wave interactions to generate millimeter-wave power has been generally recognized since the 1960s [17]. The reason is that the RF supporting structures can be simple waveguides, and are therefore practical to build down to wavelengths of 1 mm or below. Ordinary microwave tubes use periodic circuits

*The name "Weibel" mode is only one of several names being used in the literature for this interaction mechanism. Other names are *slow-cyclotron-wave* mode, *whistler* mode, and *eigenmode*.

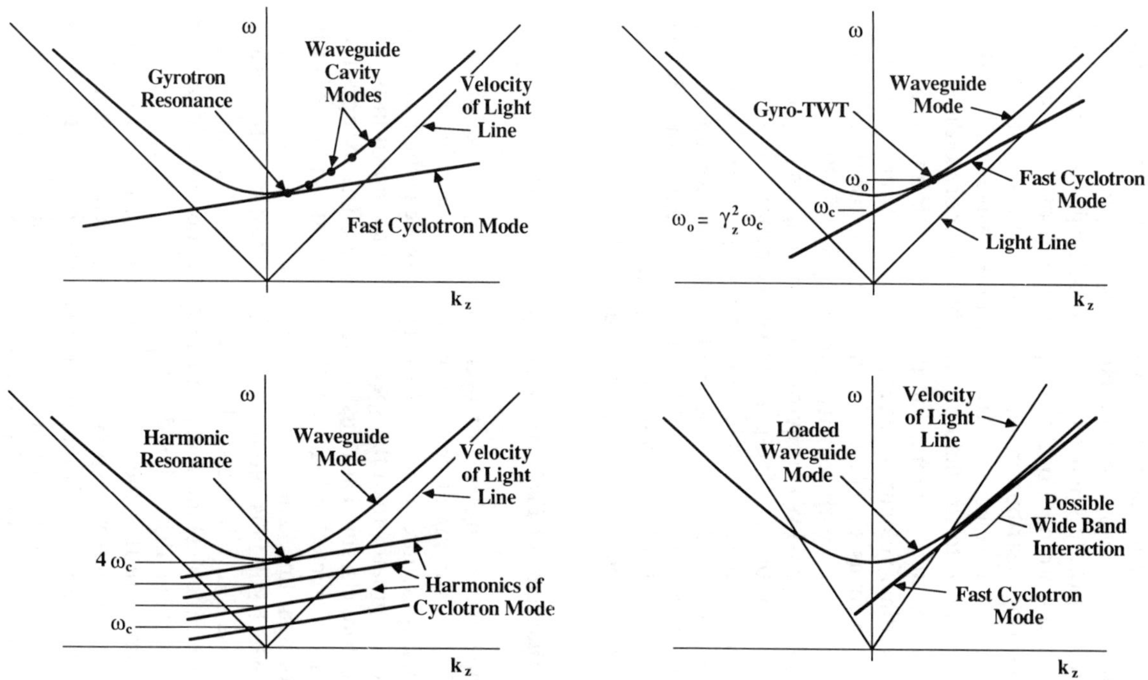

Fig. 4.2 Dispersion Diagrams Showing the Region of Cyclotron Wave Interaction between a Waveguide Mode and the Fast Cyclotron Mode of the Beam for (a) a Gyrotron Oscillator, (b) a Gyro-TWT (Traveling Wave Tube Interaction), (c) an Interaction with Higher Harmonics of the Cyclotron Wave, and (d) an Interaction at a Phase Velocity Slightly Below the Velocity of Light, where Synchronism Occurs over a Much Wider Bandwidth

to slow down the RF wave for interaction with a relatively slow-moving beam. At millimeter wavelengths, these periodic circuits are difficult to build and very fragile. Fast-wave devices, as noted, use a periodic beam that has harmonic modes passing through the fast-wave region to interact with waveguide modes.

4.2 A PHENOMENOLOGICAL DESCRIPTION OF CYCLOTRON WAVE INTERACTIONS

In this section we provide a description of the physical mechanisms by which cyclotron wave interactions convert beam energy into RF output. Although the CRM mechanism utilized in gyrotrons is the principal subject, we also discuss and compare the Weibel [18, 19] and peniotron [20–22] mechanisms, which share dispersion diagrams similar to those shown in Figure 4.2. Each of these three growth mechanisms operates by extracting energy from the transverse cyclotron motion of the electrons. The three differ significantly, however, in the manner by which they accomplish energy extraction. Under certain conditions, the mechanisms interact with each other, and therefore need to be discussed and understood as a group.

A Single Electron Beamlet

To investigate the mechanisms more closely, we begin by focusing on a single beamlet or ensemble of electrons, all of which have nearly the same guiding center as illustrated in Figure 4.3. This is the simplest of the beam models that completely describe cyclotron interactions in the absence of space-charge effects. The beam in any cyclotron wave device can be built up from an overlapping distribution of many such beamlets. The RF space charge usually plays a minor perturbing role, which will also be discussed briefly.

If we place the z-axis of our coordinate system at the guiding center of a single beamlet, all of the electrons will initially be orbiting the z-axis at a constant Larmor radius r_L given by

$$r_L = \frac{v_\perp}{\omega_c} \tag{4.1}$$

Here, ω_c is the relativistic cyclotron frequency:

$$\omega_c = \frac{eB_0}{m\gamma} \tag{4.2}$$

where e/m is the magnitude of the charge-to-mass ratio of the electrons, B_0 is the externally applied magnetic field, and γ is the relativistic factor which can be written in terms of either the total velocity $v = |\mathbf{v}|$, the total momentum $p = |\mathbf{p}|$, or the beam voltage V_0.

$$\gamma = \frac{1}{\sqrt{1 - \frac{v^2}{c^2}}} = \sqrt{1 + \frac{p^2}{m^2c^2}} = 1 + \frac{e}{mc^2} V_0 \tag{4.3}$$

By using Figure 4.3 as the starting point, we may now discuss each of the three cyclotron mechanisms in more detail. Because the CRM and the Weibel processes are always working against each other at the same time, we proceed by discussing these two together. Then, we discuss and compare the relatively different peniotron mechanism.

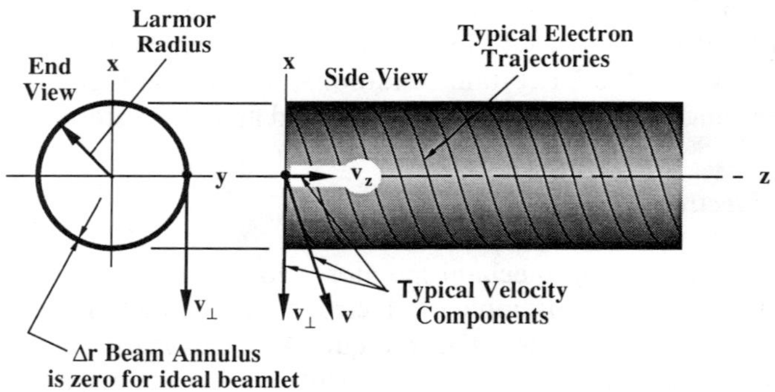

Fig. 4.3 Diagram Illustrating a Single "Beamlet" of Electrons
Note: All are orbiting about the same guiding center along the z-axis. This is the basic unit for understanding cyclotron beam interactions.

4.2.1 Physical Mechanisms of the CRM and Weibel Instabilities

The CRM and Weibel instability mechanisms share a common two-step process:

Step 1 consists of the action of the RF fields which bunch the electrons of a single beamlet together in phase so that they are no longer uniformly distributed around the orbit.

Step 2 consists of positioning the phase bunches in a phase with respect to the RF field so that the electrons as a group loose energy to the field.

These two steps progress simultaneously in actual devices. At the same time that the electron bunching is proceeding, a slight detuning between the RF and cyclotron frequencies is causing a progressive phase shift, which brings the bunched electrons into a phase where energy extraction can occur. The efficiency of the interactions can be substantially altered and improved by controlling the timing between these two steps through tapering of the RF electric field or the dc magnetic field.

Examples of the phase bunching produced by the action of the RF fields in gyrotron and Weibel devices are given in Figures 4.4 and 4.5. These figures were created by tracking the trajectories of the electrons in a single beamlet under the influence of the RF fields of a uniform right-hand circularly polarized traveling wave moving in the same direction as the beam. The beam parameters in the two cases shown are $\gamma = 1.1$ and $\alpha = 1.5$ $(=v_{\perp}/v_z)$, and the electric field amplitude in each case was held constant at a level which produced saturation in about 14 to 16 cyclotron orbits. For the CRM case, the phase velocity of the interaction was set at ten times the speed of light (i.e., near cut-off), while for the Weibel case, the phase velocity was slightly less than the speed of light at $v_p/c = 0.8$. These two cases correspond roughly to the dispersion diagrams shown in Figure 4.2(a) and (d).

Figures 4.4 and 4.5 show plots of the relative positions of 50 representative electrons in the beamlet which initially are uniformly distributed around the Larmor orbit. The time t is normalized to the initial cyclotron orbit time, $2\pi/\omega_{c0}$, and two plots are shown at each time interval of $\Delta t = 2$. The first plot is an x-y plot showing the transverse position of the electrons with respect to the direction of the local E-field (arrow). Each electron in this plot has been rotated along with the electric field vector at its z position to create the superimposed plot. If all the electrons are at the same z position (as they are in the case of the gyrotron), this plot also represents the beam cross-sectional geometry. The second "side view" plot is simply a z-y plot of the same data showing the relative movement of the electrons in the z direction. The z-axis in this plot points to the right and the scale is in waveguide wavelengths. In viewing this plot, we must remember that electrons separated by one half-wavelength in z have been rotated by an angle π with respect to each other for plotting purposes.

Note in Figures 4.4 and 4.5 that the electrons first begin to move in phase so as to produce a higher density of electrons (i.e., a bunch) on one side of the ring. In the case of the CRM mechanism, this bunch first begins to form at the phase position of the electric field and then drifts behind in phase for energy extraction. The magnetic tuning in this case was $B_0/B_{\text{sync}} = 0.95$, where B_0 is the longitudinal dc magnetic field, and B_{sync} is the B-field value that would produce exact synchronism between the rotating electric vector and the cyclotron motion of the electrons.

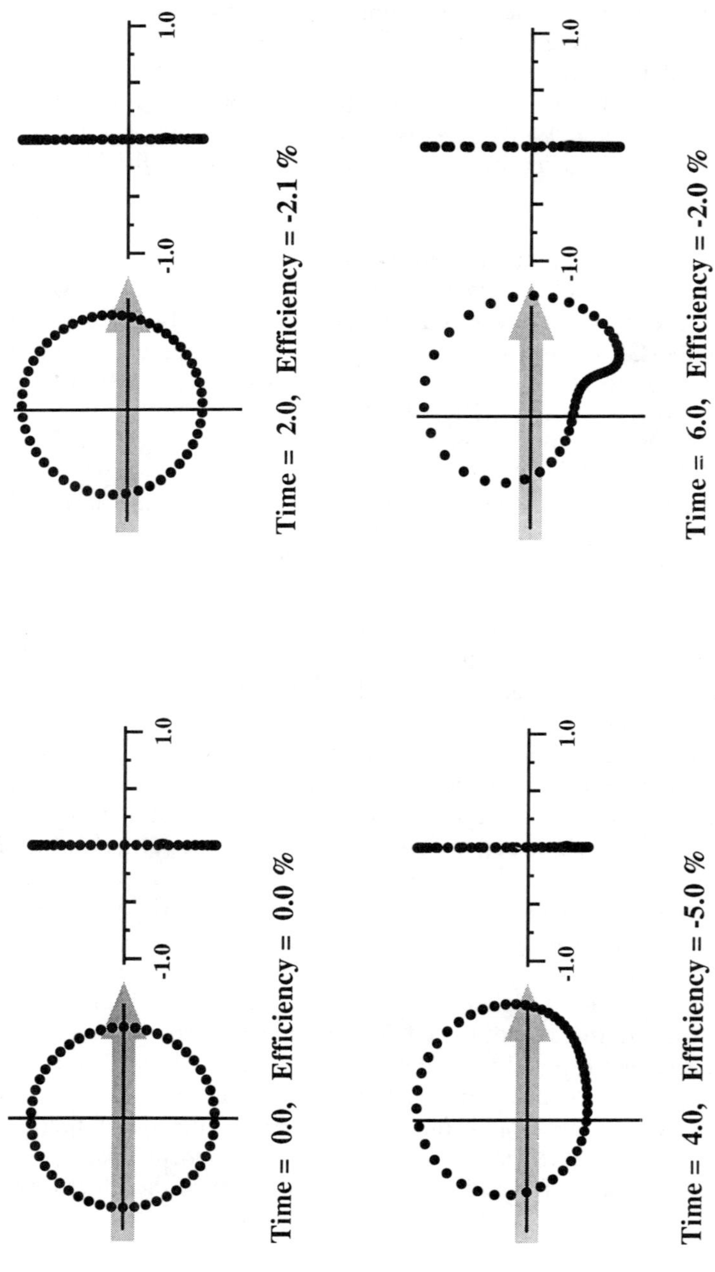
Time = 0.0, Efficiency = 0.0 %
-1.0
1.0
Time = 2.0, Efficiency = -2.1 %
-1.0
1.0
Time = 4.0, Efficiency = -5.0 %
-1.0
1.0
Time = 6.0, Efficiency = -2.0 %
-1.0
1.0

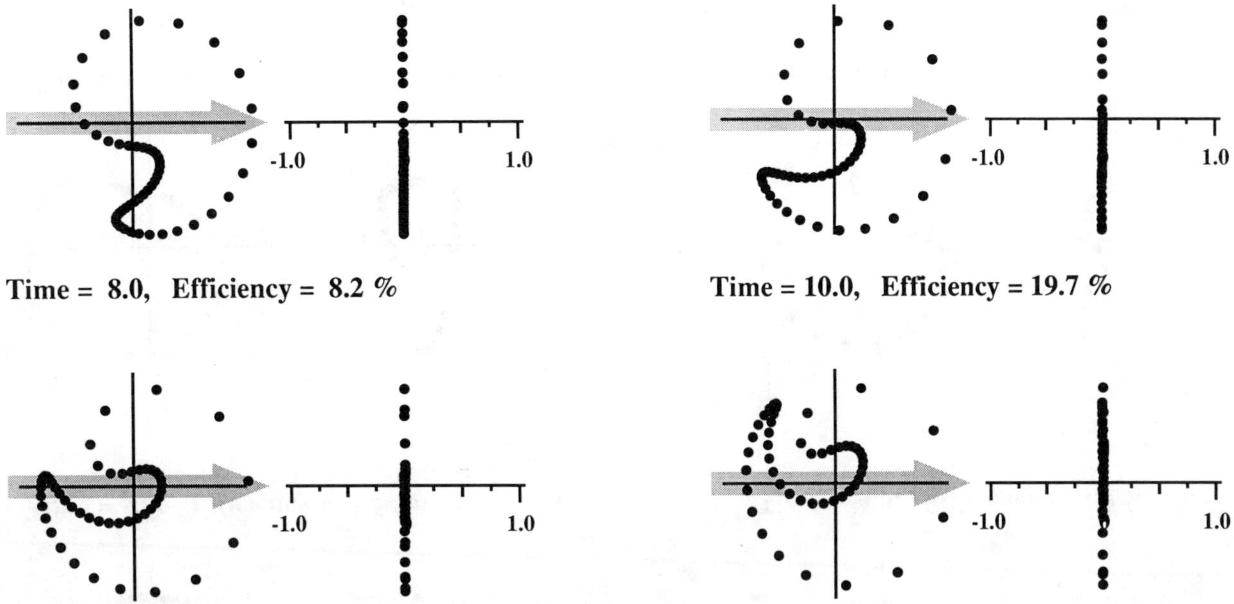

Time = 8.0, Efficiency = 8.2 %

Time = 10.0, Efficiency = 19.7 %

Time = 12.0, Efficiency = 26.8 %

Time = 14.0, Efficiency = 26.3 %

Fig. 4.4 Illustration of the Phase Bunching Produced by the CRM Interaction Mechanism

Note: The time t is in units of $2\pi/\omega_{c0}$, and the plots show the evolution of the x-y positions of the electrons in the beamlet with respect to the phase of the electric field (arrow). The z-y plot shows the corresponding z positions of the electrons.

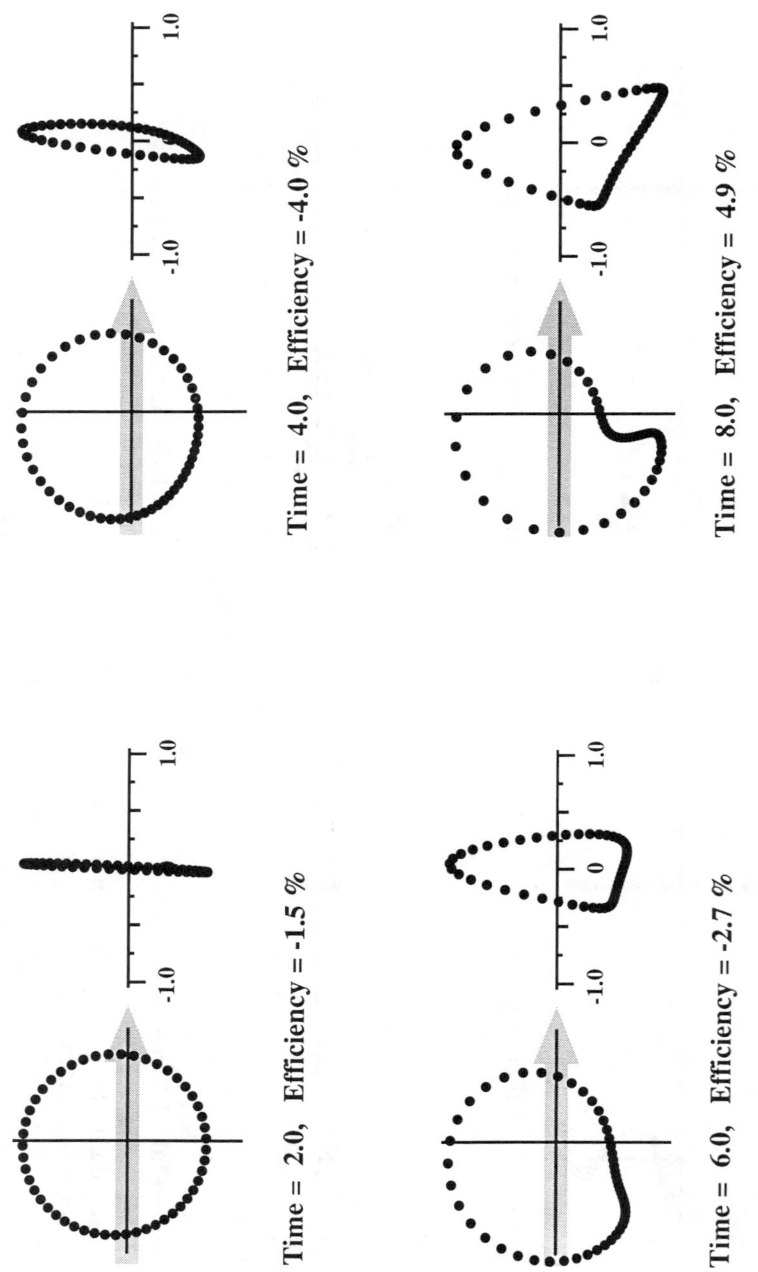
-1.0
0
1.0
Time = 2.0, Efficiency = -1.5 %
Time = 4.0, Efficiency = -4.0 %
Time = 6.0, Efficiency = -2.7 %
Time = 8.0, Efficiency = 4.9 %

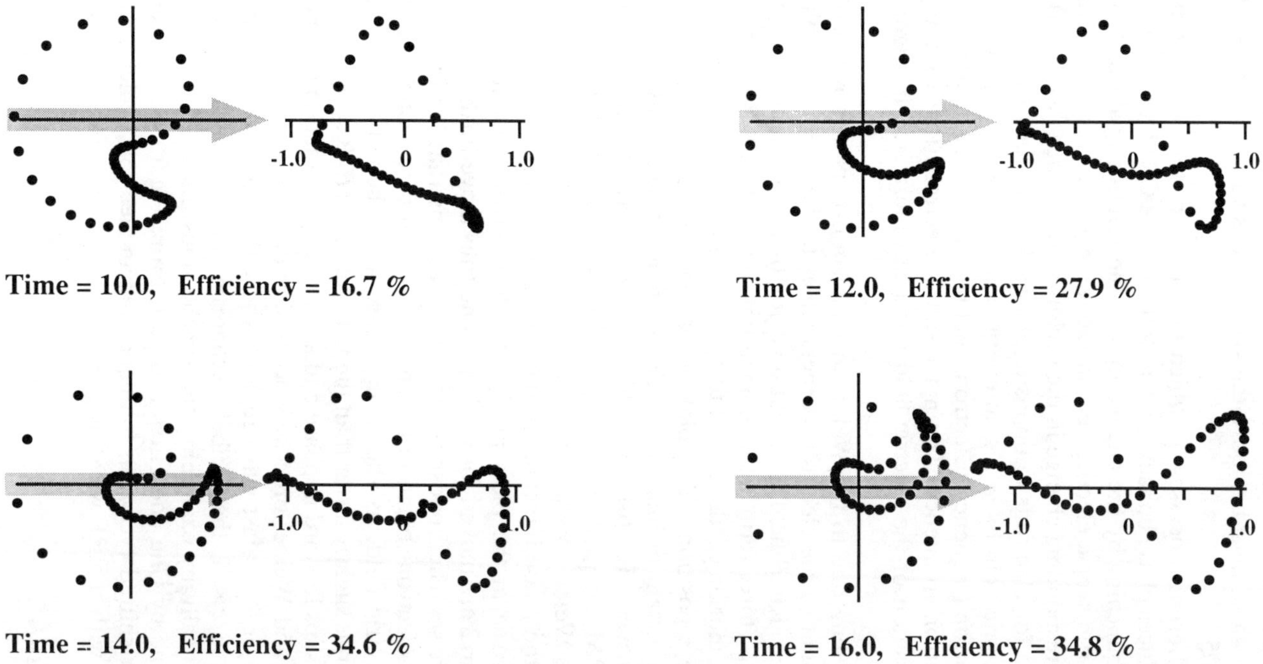

Fig. 4.5 Illustration of the Phase Bunching Produced by the Weibel Interaction Mechanism

Note: The time t is in units of $2\pi/\omega_{c0}$, and the plots show the evolution of the x-y positions of the electrons in the beamlet with respect to the phase of the electric field (arrow). The z-y plot shows the corresponding z positions of the electrons.

In the Weibel case, the bunch first begins to form at a phase shift of π with respect to the electric field vector, and then drifts forward in phase for energy extraction. In this case, the magnetic tuning, which causes the bunch to drift forward, must be above the synchronous value, and was set at $B_0/B_{\text{sync}} = 1.05$.

The two different processes by which the action of the RF field on the beamlet produces phase bunching are as follows. The CRM mechanism produces phase bunching by the change in γ as the electrons gain or lose energy from the transverse electric field in the waveguide. This change in γ produces a corresponding change in the cyclotron frequency (see (4.2)), which is opposite for electrons in opposite phases of the orbit. Electrons that lead the electric field by $\pi/2$ gain energy and have a corresponding decrease in cyclotron frequency. Electrons that lag by $\pi/2$ lose energy and increase in cyclotron frequency. The net result is electrons that drift together at the position of the electric field phase, as illustrated in Figure 4.4.

Phase bunching due to the Weibel mechanism occurs as a result of the axial movement of the electrons perpendicular to the cyclotron orbit. The perpendicular RF B-field is responsible for this movement through the $-e\mathbf{v}_{0\perp} \times \mathbf{B}_{\text{RF}\perp}$ force term in the momentum equation (4.5). In Figures 4.4 and 4.5, the z motion produced by this force is plotted in the companion plot showing the z positions. This plot shows the z distribution that has occurred at each instant of time due to the longitudinal acceleration produced by the transverse RF B-field. Note that there is virtually no z movement in the CRM interaction at $v_p/c = 10$, whereas there is major movement in the Weibel interaction at $v_p/c = 0.8$. It is this movement in z that produces the phase bunching in the Weibel mechanism.

As we can see in the plots in Figure 4.5, electrons that lead the electric field by $\pi/2$ are moved forward in z, and thus move further ahead of the field in phase. The opposite is true for electrons that lag by $\pi/2$. The net result is electrons begin to bunch at a position that is opposite in phase to the electric field, as shown in Figure 4.5. Then, because the cyclotron rotation is slightly faster than the field rotation ($B/B_{\text{sync}} = 1.05$), the phase bunch slips forward in phase and a net energy extraction occurs.

The CRM and Weibel processes occur simultaneously in cyclotron devices. When $v_p/c > 1$ (fast-wave region), the CRM mechanism dominates, and when $v_p/c < 1$, the Weibel mechanism dominates. This can be shown mathematically by examining the synchronous phase of the electrons, as was done by Chu and Hirshfield [18]. From Figure 4.2, we see that the Doppler-shifted cyclotron frequency for an electron, which is to be synchronized with the RF wave, is given by

$$\Omega_D = \frac{\Omega_c}{\gamma} + k_z v_{z0} \tag{4.4}$$

where Ω_c is the nonrelativistic cyclotron frequency, and γ is the relativistic factor. The change in Ω_D as the electron velocity changes through interaction with the RF fields can be determined by taking the time derivative of Ω_D. This derivative can then be evaluated by using the z component of the Lorentz force equation:

$$\frac{d\mathbf{p}}{dt} = -e(\mathbf{E} + \mathbf{v} \times \mathbf{B}) \tag{4.5}$$

along with the energy equation:

$$mc^2 \frac{d\gamma}{dt} = -e\mathbf{v} \cdot \mathbf{E} \tag{4.6}$$

and Faraday's law relating the transverse E and B fields for a TE waveguide mode:

$$\mathbf{B}_\perp = \frac{k_z}{\omega} \hat{\mathbf{z}} \times \mathbf{E}_\perp \tag{4.7}$$

The final result is

$$\frac{d\Omega_D}{dt} = \frac{-ek_z^2}{m\gamma\omega} \left[\overset{(1)}{1} - \overset{(2)}{\frac{\omega^2}{k_z^2 c^2}} \right] \mathbf{v}_\perp \cdot \mathbf{E}_\perp \tag{4.8}$$

Equation (4.8) has two terms, which are designated (1) and (2). Term (1) comes from the Lorentz force term $-e(\mathbf{v}_\perp \times \mathbf{B}_\perp)$, which moves electrons in the z direction. This is the force that produces the phase bunching associated with the Weibel mechanism. The second term (2) comes from the time rate of change of γ (i.e., the energy change due to acceleration or deceleration of the electron). Term (2) produces the phase bunching associated with the CRM mechanism.

To interpret (4.8), we apply it to two test electrons (a) and (b), shown in Figure 4.6. These test electrons are in the orbital positions at which the factor $\mathbf{v}_\perp \cdot \mathbf{E}_\perp$ in the equation is maximum. Note that the sign of $\mathbf{v}_\perp \cdot \mathbf{E}_\perp$ in (4.8) is opposite for the two test electrons, indicating that they

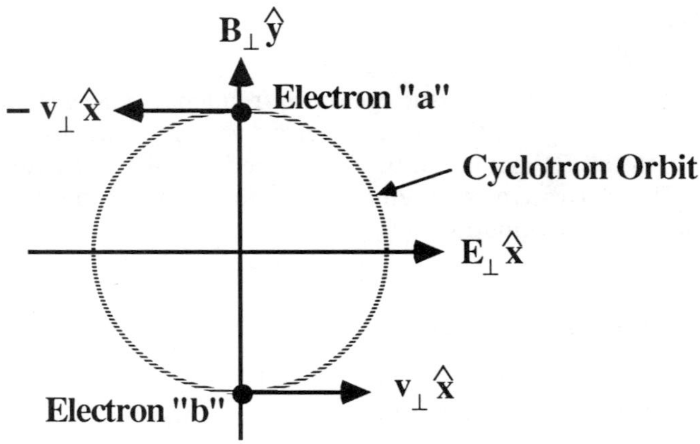

Fig. 4.6 Illustration of Two Test Electrons in a Cyclotron Orbit in Positions of Maximum $\mathbf{v}_\perp \cdot \mathbf{E}_\perp$
Note: This diagram is used to interpret Eq. (4.8).

will begin to drift toward each other around the orbital path. Thus, phase bunching will occur.

The rate and the direction of the phase bunching depends on the relative size of term (2). If $\omega > k_z c$, then term (2) will be larger than term (1), and the phase bunching associated with the CRM mechanism dominates. Alternatively, when $\omega < k_z c$, $d\Omega_D/dt$ reverses sign for the two test electrons, the phase bunching proceeds toward the opposite side of the orbit path, and the Weibel mechanism is dominant. Thus, the CRM mechanism dominates in the fast wave region, and the Weibel mechanism dominates when $v_p < c$.

Relative Wavelength Viewpoint

An alternative reason for why the two mechanisms dominate in these respective regions can be obtained by examining the electron forces with respect to the relative wavelengths. Figure 4.7 shows two rings of electrons separated by a distance equal to one-half of the guide wavelength. The forces on electrons placed at the previous two test points are shown by vector arrows in three-dimensional perspective. Note that in the fast-wave region, where guide wavelength is long, the z directed forces are not only smaller, but they also need to move the electrons by a greater distance in order to produce any significant phase change. Note, too, that the z forces at the two points on a given ring are oppositely directed, having the effect

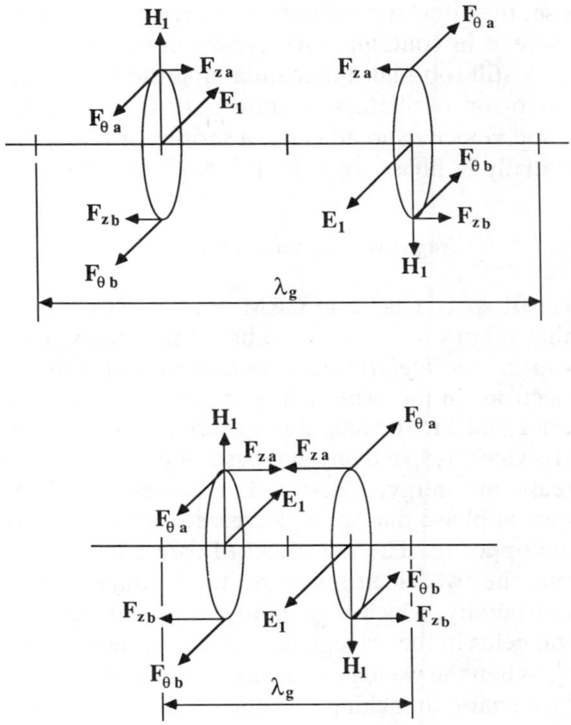

Fig. 4.7 A Wavelength View of CRM *versus* Weibel Phase Bunching Showing the Relative Strengths and Direction of Forces

Note: In (a), long RF wavelengths help the CRM mechanism to dominate in the fast-wave region. In (b), short wavelength helps the Weibel mechanism to dominate in the slow-wave region. Both cases cause phase bunching, which permits RF energy transfer.

of tilting the rings toward each other to produce the electron bunching. In the slow-wave region, where the guide wavelength is shorter, the z directed forces are not only larger, but they only need to move the electrons a short distance in order to tilt the rings and produce phase bunching.

The shorter wavelength that makes the Weibel mechanism possible has also made it difficult to develop working devices [19, 23]. As we can see from Figure 4.2(d), this interaction appears very attractive from a bandwidth point of view because synchronism can be maintained over a considerable frequency range. The problem is that the $\Delta v_z/v_z$ velocity spread in the beam must be very low when the bunching wavelength is short; otherwise, the bunching is dissipated very quickly and no gain occurs. The velocity spread in contemporary cyclotron beam devices is typically 5–15%, which is still too high for maintaining the Weibel interaction. In comparison, gyrotron oscillators operate near cut-off, where the wavelength is long and velocity spread is not a serious problem because the RF fields are essentially in phase over the full cavity length.

Space Charge and the Negative Mass Instability

The effect of space charge in CRM devices is to enhance the space-charge bunching. Figure 4.8 illustrates how this occurs. Once the bunch has begun to form, the electric field associated with the bunch starts to effect other electrons in the orbit. Those electrons in the positions shown as test electrons 1 and 2 in the diagram, experience forces due to the space-charge E-field, which respectively increase and decrease their energies. With an increase in energy, electron 1 increases slightly in radius and moves backward in phase due to its decreased cyclotron frequency. Electron 2 does the opposite. The net effect of forces *away* from the bunch, then, is to bring the two electrons *closer* to the bunch; hence, the name *negative mass* instability, which is given to this process [24]. In the absence of normal mode fields in the waveguide, this bunching process can proceed on its own [25]. When the waveguide fields are present, this effect increases the rate of space charge bunching and slightly improves the gain [26].

Note that when many beamlets, such as those shown in Figure 4.8, are distributed side by side over an area to form an electron beam, the space-charge effect just described can be greatly altered due to the bunching on adjacent beamlets. In the limiting case of a uniform distribution of beamlets over all space, which are acted upon by a uniform plane wave, there is no net space charge bunching, although the phase bunching proceeds on each beamlet as illustrated in Figures 4.4 to 4.8.

Harmonic CRM Interactions

The CRM mechanism can also operate at harmonics of the cyclotron frequency if the electromagnetic field has a corresponding harmonic number so that the field and beam rotate together in synchronism. To illustrate this, we can approximate the transverse fields in a waveguide near the guiding center of the electron beamlet by a field expansion in terms of multipole fields of the form [1]:

$$\mathbf{E}_n = nA_n r^{n-1}\,[\cos(\omega t - k_z z - n\phi)\hat{\mathbf{r}} + \sin(\omega t - k_z z - n\phi)\hat{\phi}] \tag{4.9}$$

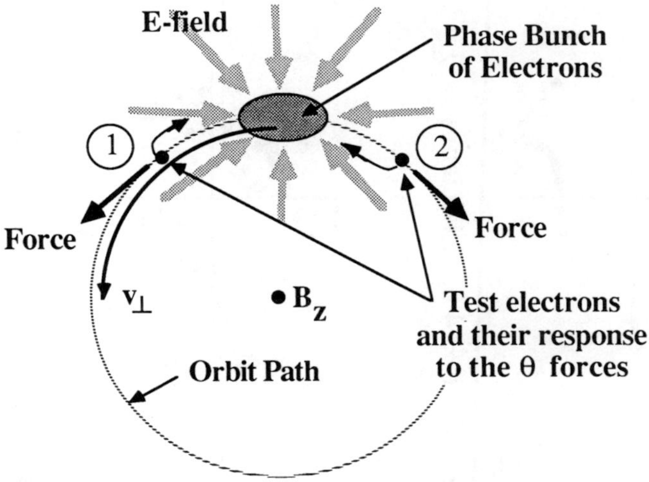

Fig. 4.8 Diagram Showing How the RF Space Charge due to Bunching of Electrons in a Beamlet Enhances the Bunching Process

were (r, ϕ) are the polar coordinates in the beamlet coordinate system, n is the angular harmonic number, and A_n is a constant. Note that the simple right-hand circularly polarized field used previously is just the $n = 1$ mode of this field (i.e., a uniform field which rotates once each RF cycle).

If we now specify the $n = 2$ field mode while simultaneously synchronizing the second harmonic of the cyclotron frequency with the Doppler-shifted RF field, so that $2\omega_c = \omega - k_z v_{z0}$, then the orbital electrons and the RF field still rotate in synchronism. The result is that two phase bunches are developed, as shown in Figure 4.9. As might be anticipated, the number of bunches formed at higher harmonics is equal to the harmonic number.

Because of the high magnetic fields required for millimeter-wave gyrotrons ($\approx 1.2T$ at 35 GHz and $\approx 3.5T$ at 100 GHz), it is very attractive to attempt to use harmonic interactions and take advantage of the magnetic field reduction by the factor $1/n$. However, because the electrons in the phase bunches typically continue to spread in phase faster than the energy of the bunches can be extracted, the saturated efficiency of the harmonic interactions decreases rapidly with increasing harmonic number, typically falling below 10% at the third or fourth harmonic. Although Gyrotrons have enjoyed considerable success at the second harmonic [10] for higher harmonic interactions the peniotron mechanism, which has a predicted third harmonic saturated efficiency greater than 50% [22, 27], appears more attractive at this time. This is discussed further in the next section.

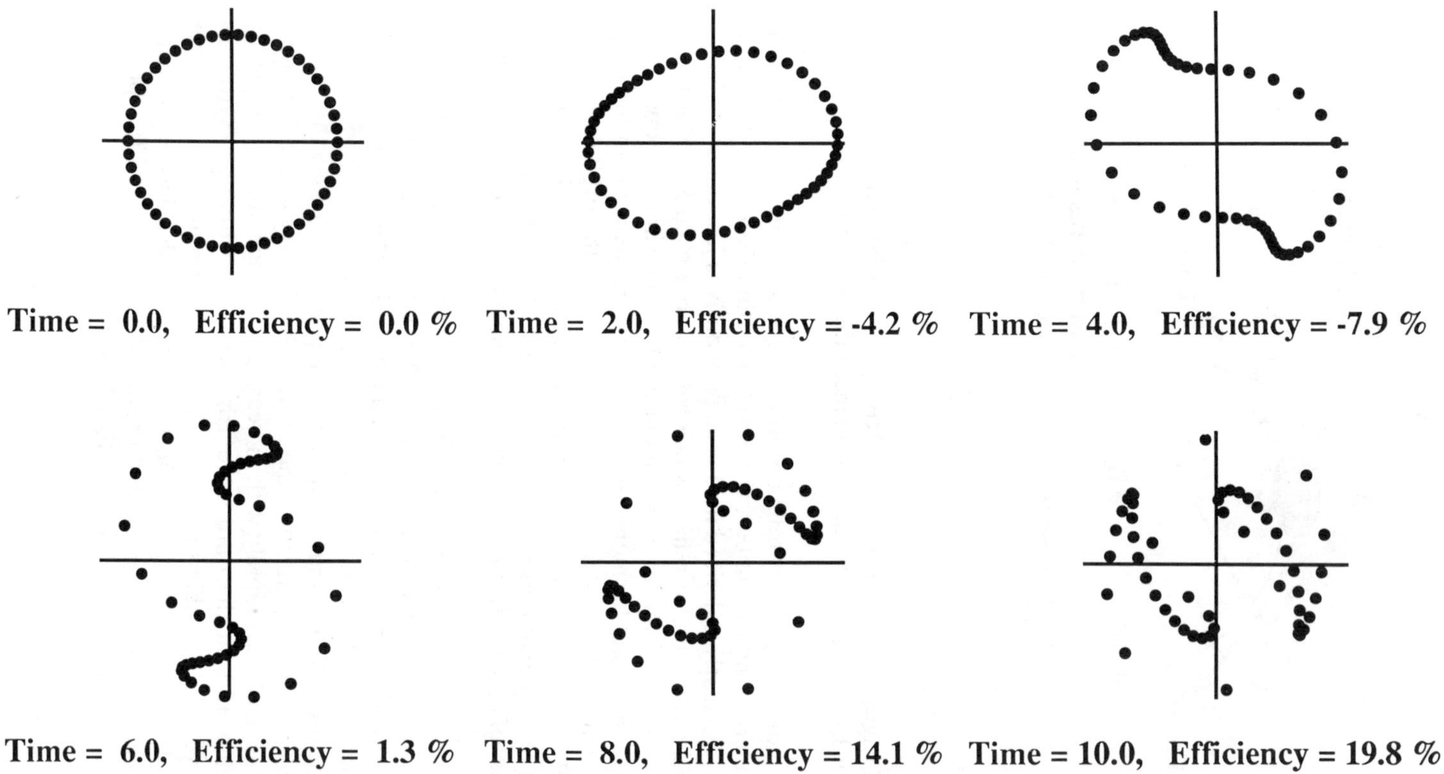

Fig. 4.9 Cross-Sectional Plots Showing the CRM Phase Bunching on a Beamlet Interacting at the Second Harmonic of the Cyclotron Frequency

Autoresonance

Before concluding the phenomenology associated with the CRM and Weibel mechanisms, we point out one additional condition in (4.8) that is of interest. When $\omega \approx k_z c$, little or no phase change occurs, although the energy of the electrons is changing. This is the phenomenon known as *autoresonance* and is utilized by a class of relativistic devices known as *cyclotron autoresonant masers* or CARMs. Once an electron beam has been phase bunched, this phenomenon can be used to extract large amounts of energy from a beam without losing synchronism. As ω_c increases (due to extraction of beam energy), autoresonant tuning causes v_z to decrease (due to the $B_\perp$ field forces) such that $\omega = \omega_c + v_z k_z$ remains constant. Thus, the total energy of the beam is available for RF conversion, not only the transverse component. This fact makes the auto resonant interaction attractive when working with highly relativistic beams in which the largest component of velocity is v_z due to the difficulty of forming the beam. CARM devices are currently an important area of investigation [28, 29].

4.2.2 The Phenomenology of Peniotron Interactions

The peniotron interaction mechanism was first described by Ono, *et al.*, in Japan, in 1962 [30] at about the same time as the discovery of the CRM mechanism. With the resurgence of research on cyclotron devices since the 1970s, investigation of the peniotron mechanism was also revived and in the early 1980s working millimeter-wave oscillators and amplifiers were demonstrated in the US [31]. The peniotron interaction is still an active area of research because of its potential to provide high efficiency at high harmonics of the cyclotron frequency [21, 27, 32, 33]. Its potential therefore is to provide millimeter-wave power without the need for superconducting magnets.

The peniotron mechanism differs significantly from that of the CRM and Weibel mechanisms because phase bunching is not a part of the process by which energy is extracted. The peniotron mechanism is characterized by a drift of the electron guiding centers in directions which cause each electron to loose energy to the transverse electric field. This guiding center drift and energy loss can only occur when the field contains a strong right-hand circularly polarized harmonic with angular mode number n that is related to the cyclotron harmonic number s by $n = s + 1$. The frequency tuning with respect to the Doppler-shifted RF frequency is therefore given by

$$\omega_c = \frac{\omega - k_z v_{z0}}{n - 1}$$

With this tuning, the rotating electric field harmonic slips behind the orbital electrons in phase by 2π each orbit. Therefore, the electrons in a beamlet see one complete cycle of the RF field during each orbit.

The guiding center drift which results from this cycling RF electric field is produced by the radial variation of the field. The best way to understand this drift is see how test electrons respond to a specific field. For example, Figure 4.10 shows an illustration of the fields of a right hand circularly polarized TE_{21} waveguide mode with an electron beamlet placed at the center of the guide. Note that the electric field is zero at the center, and increases radially over the central portion of the guide, where the electron beam interacts. Note also that for this $n = 2$ case, the electric fields reverse every $\pi/2$ in angle, and rotate at one-half the rotation speed of the electrons ($s = n - 1 = 1$). Although this particular illustration does not yield a reduction in magnetic field, it is a good example from the standpoint of understanding the phenomenology of the interaction. It can also be exceptionally efficient as we will illustrate.

Because there are four field reversals, we will examine the initial behavior of the four test electrons, which are placed on the beamlet at the positions of maximum E_ϕ as illustrated in Figure 4.10 and shown expanded in Figure 4.11. Test electrons (1) and (3) are initially decelerated so that their orbits become smaller than original. Thus, when the electrons reach the π phase point of their cyclotron orbits, where they are reaccelerated by the opposite phase of the field (which has rotated through $\pi/2$ during the same time interval), the electrons are in a lower field region due to the radial variation of the electric field. As a result, the acceleration they receive at the π point will not be as great as the initial deceleration. The net result for test electrons (1) and (2) is that the orbits are smaller (loss of energy) and the guiding centers (denoted by small crosses) have moved in the outward direction with respect to their initial phase positions, as shown in Figure 4.11.

For test electrons (2) and (4), the guiding center drift pattern is different. These electrons are initially accelerated so that their orbits are larger than original. Thus, when they reach the π phase point of their orbits, where they are now decelerated, they are in a higher field region. The deceleration that the electrons receive at this point is therefore larger than the original acceleration and their new orbits will carry them back to their starting phase at a smaller radius than at the start. Test electrons (2) and (4) therefore have experienced guiding center drifts in the opposite direction to their original starting phase and have also received a net deceleration.

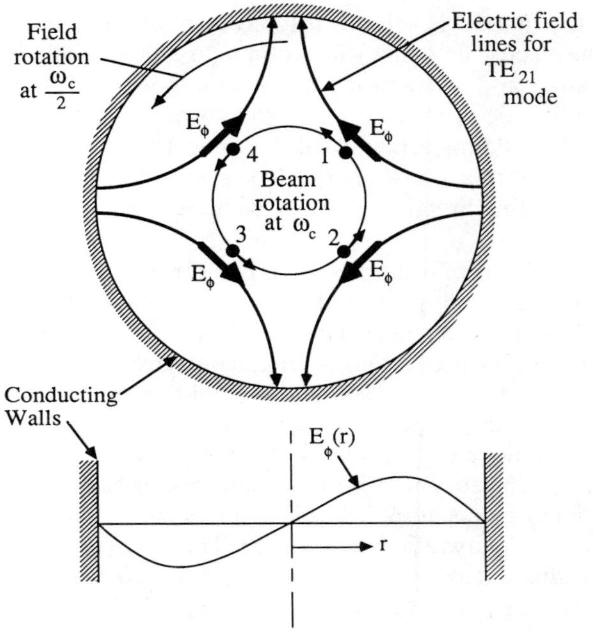

Fig. 4.10 Illustration of the Fields of a Right-Hand Circularly Polarized TE_{21} Waveguide Mode with a Superimposed Electron Beamlet Along the Axis

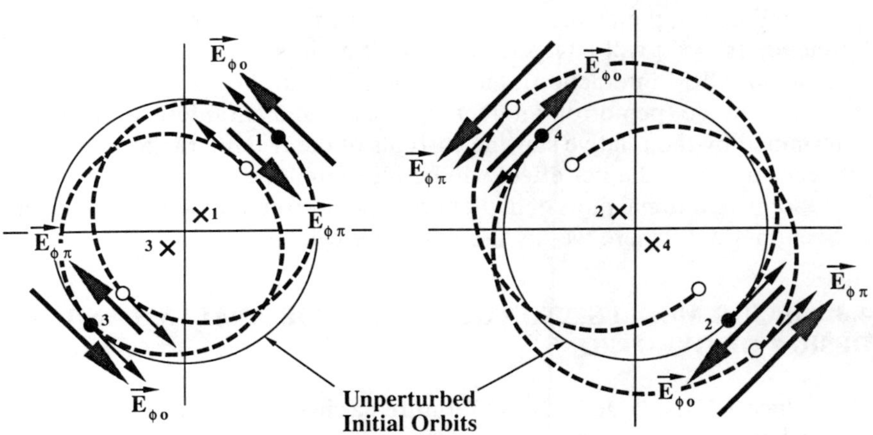

Fig. 4.11 Illustration of the Initial Orbits of the Test Electrons Shown in Figure 4.10

After the first orbit, all four test electrons are now in a position with respect to the electric field that will continue to cause an imbalance between the decelerating and accelerating fields on opposite sides of their orbits. The deceleration field on the outside will always be stronger with the net result that the guiding centers will continue to move outward and the electrons will continue to lose energy. The reader may note that the guiding center drift is in the direction of $\langle \mathbf{E} \rangle \times \mathbf{B}_0$, where $\langle \mathbf{E} \rangle$ is the average E-field over the orbit.

Figure 4.12 shows orbit plots of a peniotron interaction for a ring of eight electrons in a multipole field given by (4.9) with $n = 2$. These plots show the positions of the electrons in the beam cross section with respect to a reference RF phase. Along with each electron, we show the instantaneous orbit size (dotted circles) and the guiding centers (small crosses). Note that the guiding centers spread out uniformly in the radial direction and continue to move outward until the transverse energy is nearly gone. The transverse efficiency in this ideal beam case is close to 100%, but the overall efficiency peaks at 69% because the beam still has its longitudinal velocity (initial velocity ratio was $\alpha = 1.5$). There was essentially no change in the longitudinal velocity in this case because v_p/c was set at 10.

The dispersion diagram for the peniotron interaction is the same as that of the CRM interactions shown in Figure 4.2. This points out the fact that in general both interactions can occur at the same time in a given device. The relative strengths of the two interactions will be determined by the relative strengths of the two harmonic field components on which they depend. For example, if an RF circuit has field components at both $n = 3$ and $n = 4$, then a beam tuned to the third harmonic of the cyclotron frequency ($s = 3$) will have electrons orbiting in synchronism with the $n = 3$ mode (CRM interaction), and slipping forward by 2π with respect to the $n = 4$ mode (peniotron interaction). The dominant interaction will be determined by the relative starting currents of the two interactions. If both proceed together, the net effect will be interference.

Some additional plots of higher order peniotron interactions are given in Section 4.3.2, where we discuss nonlinear calculations.

4.3 MAJOR MODELS AND TOOLS USED FOR ANALYSIS AND DESIGN OF GYROTRONS

Since 1974, a large body of literature has been published dealing with the design and analysis of gyrotron devices. The tools and approaches used in this work vary greatly in their complexity and in their targeted physical geometries, but they stem from a relatively few basic approaches. It is

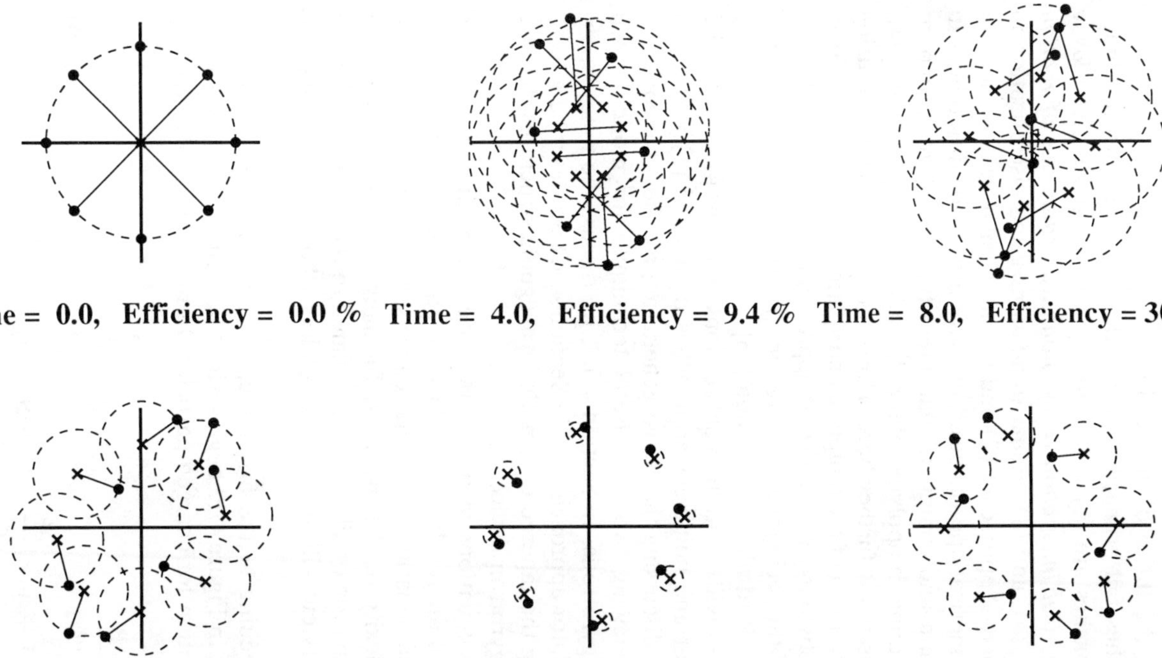

Time = 0.0, Efficiency = 0.0 % Time = 4.0, Efficiency = 9.4 % Time = 8.0, Efficiency = 30.7 %

Time = 12.0, Efficiency = 54.2 % Time = 16.0, Efficiency = 69.0 % Time = 20.0, Efficiency = 62.0 %

Fig. 4.12 Calculated RF Trajectories for a Ring of Eight Peniotron Electrons in a Second Harmonic Electric Field ($n = 2$)

Note: The crosses show the guiding center positions and the circles show the instantaneous orbit sizes. (Beam parameters: $\gamma = 1.1$, $\alpha = 1.5$, $v_p/c = 10$, $B/B_{sync} = 0.97$)

instructive to review these approaches from the standpoint of their usefulness, either to investigate and understand basic physical phenomena, or to predict the results of design features for optimization. These two purposes comprise the thrust of this section.

The approaches fall primarily into the three areas shown in Figure 4.13. These approaches are (1) *kinetic theory* and the *linearized Vlasov equation,* (2) *numerical integration of the nonlinear equations of motion* ("ballistic" equations), and (3) *plasma simulations* using discretization in both time and space. Approaches (2) and (3) are similar in that they are both nonlinear numerical approaches, but there are major differences in their implementation and computational intensity. Only a small amount of work has been done in applying plasma simulations to gyrotrons [34], and we will not discuss it further here. However, for problems requiring a complete understanding of the temporal and spatial evolution of gyrotron interactions, it is obviously an important approach to consider. An excellent description of the techniques involved in the computer simulation of plasmas is given by Birdsall and Langdon [35].

This section is divided into four subsections. Sections 4.3.1 and 4.3.2 deal with the linear and nonlinear approaches designated (1) and (2) in Figure 4.13. After discussing these two general approaches, we conclude this section by reviewing two very useful techniques for modeling RF structures used in gyrotrons. In Section 4.3.3, we discuss the time-dependent coupled oscillator approach, and in Section 4.3.4 we describe a technique for analyzing the tapered-overmoded waveguide structures that are used so widely in gyrotron devices.

Many of the equations in this section are taken directly from the literature to explain the results and to illustrate the type of analysis. Also, the display of these equations gives the reader some insight into the procedure and complexity of the methods. In many cases, space does not permit a full explanation of all the details related to the equations, but we hope that the overview will assist those readers who wish to pursue the subject further.

Before proceeding to the first topic, we note that each of the approaches indicated in Figure 4.13 begins with Maxwell's equations and the Lorentz force equation for charged particles. In mks units, these are

$$\begin{aligned} \nabla \times \mathbf{E} &= -\frac{\partial \mathbf{B}}{\partial t} & \nabla \cdot \mathbf{E} &= \frac{\rho}{\epsilon_0} \\ \nabla \times \mathbf{B} &= \mu_0 \mathbf{J} + \mu_0 \epsilon_0 \frac{\partial \mathbf{E}}{\partial t} & \nabla \cdot \mathbf{B} &= 0 \end{aligned} \tag{4.10}$$

and

$$\frac{d\mathbf{p}}{dt} = q(\mathbf{E} + \mathbf{v} \times \mathbf{B}) \tag{4.11}$$

where **E** and **B** are the electric and magnetic fields, **J** and ρ are the current and space-charge densities, $\mathbf{p} = m\gamma\mathbf{v}$ is the momentum of a particle, and q and m are the signed charge and mass of the particles.

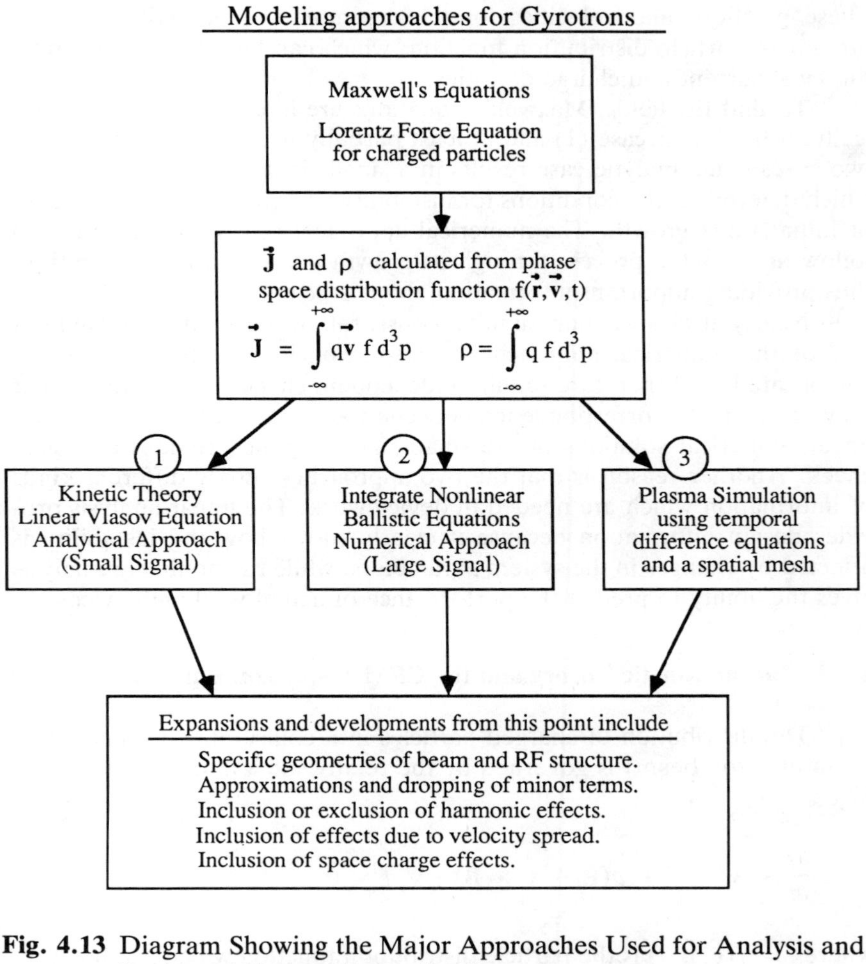

Fig. 4.13 Diagram Showing the Major Approaches Used for Analysis and Design of Gyrocyclotron Devices

The major difference between the linear and nonlinear approaches indicated in Figure 4.13 lies in the manner by which the particle distribution function $f(\mathbf{r}, \mathbf{p}, t)$ is evaluated so that the current and charge densities can be determined from the equations shown in the figure. In case (1), the linearized Vlasov equation (described below) is used to solve analytically for the current and charge densities as functions of the RF fields. In the other two cases, we must choose a representative ensemble of macroparticles, and numerically integrate the nonlinear ballistic equations to find the position and velocity of each particle at the desired time of evaluation. These positions and velocities for the ensemble then constitute an approximate particle distribution function, which can be used to determine the local current and charge densities.

To find the fields, Maxwell's equations are likewise handled analytically in the linear case (1) and at least partially numerically in the other two cases. The analytic case results in a small signal dispersion equation which determines the conditions for instability and gives the characteristics of initial wave growth. The numerical approaches provide the ability to follow an unstable process through its growth phase and into saturation, thus providing important information about efficiency.

Nearly always we must simultaneously follow the analytic and at least one of the numerical approaches. One reason is that the theories are complicated, and therefore require independent checks to help verify that they are correct. Correspondence between the linear theory and the nonlinear numerical solutions in the small signal regime provides one such check. Another reason is that the two approaches supply different kinds of information which are needed in device work. The linear analysis provides design information because it clearly shows how the instability is affected by changes in the system parameters, while the nonlinear analysis gives the ability to predict the performance of actual working devices.

4.3.1 Linear Kinetic Theory and the CRM Dispersion Equation

The distribution of charged particles in a collisionless plasma (such as an electron beam) is governed by the relativistic Vlasov equation [36, 37]:

$$\frac{\partial f}{\partial t} + \mathbf{v} \cdot \nabla f + q(\mathbf{E} + \mathbf{v} \times \mathbf{B}) \cdot \nabla_p f = 0 \tag{4.12}$$

where $f = f(\mathbf{r}, \mathbf{p}, t)$ is the particle distribution function which is a function of time t, configuration space $\mathbf{r} = (x, y, z)$, and momentum space $\mathbf{p} = \gamma m(v_x, v_y, v_z)$. The gradient operators ∇ and ∇_p apply to configuration space and momentum space, respectively. The six-dimensional combination of the **r** and **p** spaces is called phase space $(\mathbf{r}, \mathbf{p}) = (x, y, z, p_x, p_y, p_z)$.

The Vlasov equation is a mathematical statement of conservation of particle points in phase space, and serves the same function in this space as the equations for conservation of particles or charge do in configuration space. The Vlasov equation is derived given the assumptions that the charged particles move under the influence of the Lorentz force and do not collide with each other. These conditions are closely achieved in an electron beam.

By definition, integration of the distribution function over momentum space must return the density distribution in configuration space $n(\mathbf{r}, t)$:

$$n(\mathbf{r}, t) = \int_{-\infty}^{\infty} f(\mathbf{r}, \mathbf{p}, t)\, \mathrm{d}^3 p \tag{4.13}$$

and the macroscopic average of any variable $A(\mathbf{r}, \mathbf{p}, t)$ over momentum space is given by

$$\langle \mathrm{A} \rangle_p = \frac{\int_{-\infty}^{\infty} Af\mathrm{d}^3 p}{\int_{-\infty}^{\infty} f\mathrm{d}^3 p} = \frac{1}{n}\int_{-\infty}^{\infty} Af\mathrm{d}^3 p \tag{4.14}$$

By using these definitions, the more widely known fluid equations can be recovered from the Vlasov equation by taking averages over momentum space [38].

A general linear development of kinetic theory that describes the behavior of waves in plasmas is given by Baldwin, Bernstein, and Weenik [37]. The theory begins by assuming that all dependent variables in (4.10) and (4.11) are the sum of a steady-state part (subscript 0) and a small perturbational part (subscript 1), which represents the wave variable of interest. The equations are then linearized by neglecting second-order perturbation terms, and solved by Fourier and Laplace transform techniques. Although the theory is relatively complex, it is basic to plasma physics and worth the effort to understand.

The small signal linearized Vlasov equation used in this analysis has the form:

$$\begin{aligned}\frac{\partial f_1}{\partial t} + \mathbf{v} \cdot \nabla f_1 + q(\mathbf{E}_0 + \mathbf{v} \times \mathbf{B}_0) \cdot \nabla_p f_1 = \\ -q(\mathbf{E}_1 + \mathbf{v} \times \mathbf{B}_1) \cdot \nabla_p f_0\end{aligned} \tag{4.15}$$

Krall and Trivelpiece [36] show how this equation can be integrated by using the method of characteristics, with the result:

$$f_1 = \int_{\text{Unperturbed Orbits}} -q(\mathbf{E}_1 + \mathbf{v} \times \mathbf{B}_1) \cdot \nabla_p f_0 \, \mathrm{d}t \tag{4.16}$$

in which the integration is carried out over the unperturbed orbits of the particles. The current density and space charge density are then given by

$$\begin{aligned} \mathbf{J}_1 &= \int q\mathbf{v} f_1 \mathrm{d}^3 p \\ \rho_1 &= \int q f_1 \mathrm{d}^3 p \end{aligned} \tag{4.17}$$

and these quantities can be used in Maxwell's equations.

The result of the linear theory is to obtain dispersion equations for various plasma waves that can exist in a magnetized plasma. One of these waves is the well known *right-hand circularly polarized plasma wave* traveling along the magnetic field [38]. It is this plasma wave that exhibits the CRM instability when the initial distribution of electron velocities in the plasma f_0 is properly chosen. A lower frequency branch of this same wave is well known in plasma theory as the *whistler mode*. The instability in this branch is the Weibel instability and hence the alternate use of these names by some authors.

The fact that we can classify the CRM and Weibel mechanisms in terms of instabilities in one of the well known plasma waves demonstrates a value of the kinetic theory approach. It provides a theoretical basis from which to understand and interpret the CRM instability as compared with other plasma waves. Chu and Hirshfield [18] used this theoretical approach to good advantage in discussing the relativistic terms that give rise to the instability and the differences between the CRM and the Weibel instabilities. Another advantage of the kinetic theory approach is the built in method for including the effects of electron velocity spread through the specification of f_0. As we noted previously, velocity spread in the electron beam can significantly reduce the expected gain or efficiency in gyrotrons that have a long interaction region.

The kinetic theory dispersion equation for the right-hand circularly polarized plasma wave has the following general form for an electron plasma in which the ions are assumed to be immobile [18]:

$$\omega^2 - k_z^2c^2 = 2\pi\omega_p^2 \int_0^\infty p_\perp \mathrm{d}p_\perp \int_{-\infty}^\infty \mathrm{d}p_z \frac{f_0}{\gamma}$$

$$\times \left[\frac{\omega - \dfrac{k_z p_z}{\gamma m}}{\omega - \dfrac{k_z p_z}{\gamma m} - \dfrac{\Omega_c}{\gamma}} - \frac{p_\perp^2(\omega^2 - k_z^2c^2)}{2\gamma^2 m^2 c^2 \left(\omega - \dfrac{k_z p_z}{\gamma m} - \dfrac{\Omega_c}{\gamma}\right)^2} \right] \tag{4.18}$$

In this equation, f_0 is the initial unperturbed phase space distribution function, Ω_c is the nonrelativistic cyclotron frequency, and $\omega_p = (e^2n/m\epsilon_0)^{1/2}$ is the plasma frequency. Other parameters are as previously defined. Note that for a cold stationary plasma described by

$$f_0(\mathbf{p}) = \frac{\delta(p_\perp)\delta(p_z)}{2\pi p_\perp} \tag{4.19}$$

(where $\delta(x)$ is the Dirac delta function) we can easily evaluate (4.18) to recover the familiar dispersion equation for the right-hand circularly polarized wave obtained from fluid theory [38]:

$$\frac{c^2k_z^2}{\omega^2} = 1 - \frac{\dfrac{\omega_p^2}{\omega^2}}{1 - \dfrac{\omega_c}{\omega}} \tag{4.20}$$

The simplest form of f_0 that produces the CRM instability is the case in which all of the electrons are uniformly circling in the magnetic field with the same perpendicular momentum $p_{\perp 0}$:

$$f_0 = \frac{\delta(p_\perp - p_{\perp 0})\,\delta(p_z)}{2\pi p_\perp} \tag{4.21}$$

When (4.18) is evaluated with this f_0, the resulting dispersion equation is

$$\omega^2 - k_z^2c^2 = \frac{\omega_p^2}{\gamma_0}\left[\frac{\omega}{\omega - \Omega_c/\gamma_0} - \frac{\beta_{\perp 0}^2(\omega^2 - k_z^2c^2)}{2(\omega - \Omega_c/\gamma_0)^2}\right] \tag{4.22}$$

where $\beta_{\perp 0} = v_{\perp 0}/c$.

Equation (4.22) is of second order in k_z, and the two roots give the forward and backward wave propagation characteristics of the right-hand circularly polarized plane wave in a plasma that has no average drift velocity. Figure 4.14, taken from Chu and Hirshfield, shows the solutions to this dispersion equation for two different values of cyclotron frequency. The figure shows that the CRM mechanism only occurs when the cyclotron frequency exceeds the plasma frequency, whereas the Weibel instability (lower whistler mode branch) is unstable at much lower magnetic field levels.

The more interesting case for gyrotron applications is a drifting plasma in which $p_z = p_{z0}$. For an ideal electron beam, this can be specified by

$$f_0 = \frac{\delta(p_\perp - p_{\perp 0})\,\delta(p_z - p_{z0})}{2\pi p_\perp} \tag{4.23}$$

This distribution of electrons is the same as would be obtained by uniformly distributing the beamlets used in Section 4.2 throughout all space. Using this in (4.18) gives

$$\omega^2 - k_z^2c^2 = \frac{\omega_p^2}{\gamma_0}\left[\frac{\omega - k_zv_{z0}}{\omega - k_zv_{z0} - \Omega_c/\gamma_0} - \frac{\beta_{\perp 0}^2(\omega^2 - k_z^2c^2)}{2(\omega - k_zv_{z0} - \Omega_c/\gamma_0)^2}\right] \tag{4.24}$$

Note that the Doppler frequency shift k_zv_{z0} now appears in three places in (4.24) such that the equation is now of fourth order in k_z. The two added modes are a pair of beam modes in place of the familiar fast cyclotron mode in a nonrelativistic electron stream. The double root for the beam waves comes from the factor $(\omega - k_zv_{z0} - \omega_c)^2$, which results from relativistic factors in the theory and disappears if the initial transverse velocity $\beta_{\perp 0}$ is zero.

Equation (4.24) gives the dispersion relation for a plane-wave CRM instability in an infinite plasma of streaming electrons. For practical applications such as an electron beam in a waveguide, we must return to Maxwell's equations and include the transverse waveguide cut-off wavenumber k_{cn} (for the nth waveguide mode, where n represents a double index) along with the finite transverse geometry of the electron beam. We

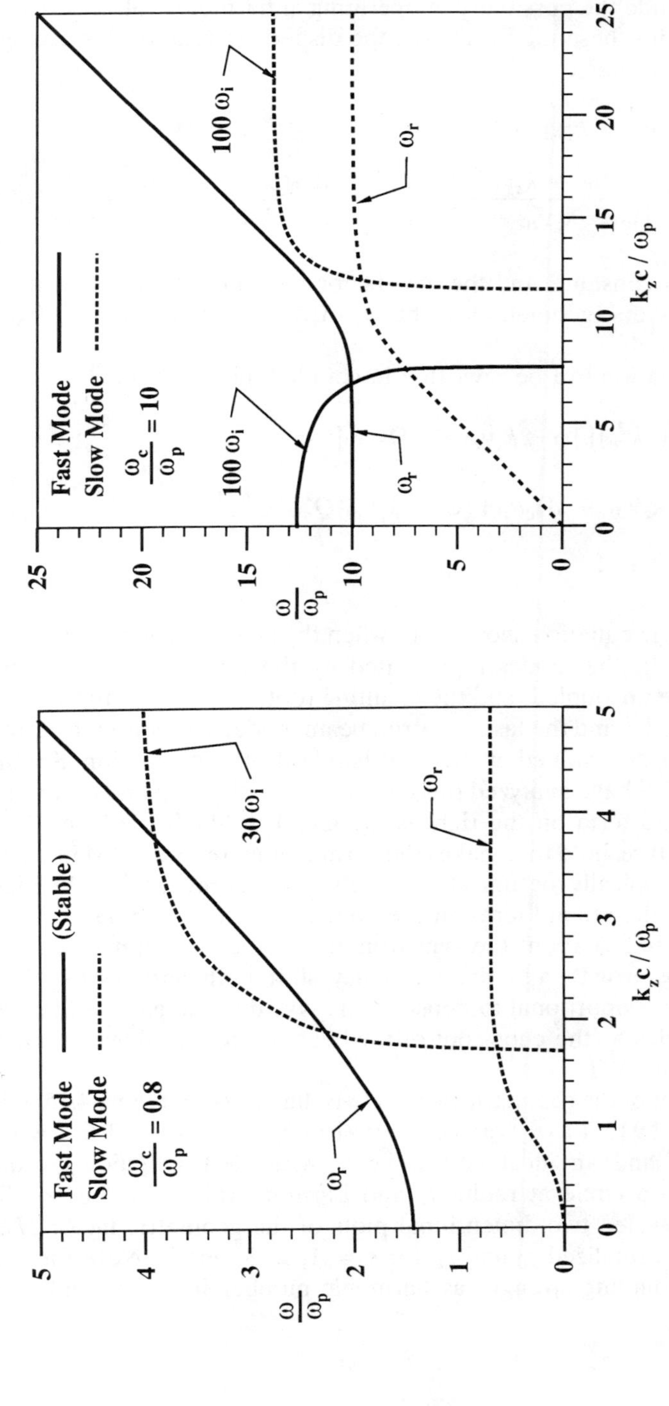

Fig. 4.14 Dispersion Plots of the Right-Hand Circularly Polarized Plasma Modes with the CRM Instability (From Chu and Hirshfield [18])

must also include the possibility of operating at harmonics of the cyclotron frequency. With these modifications, the dispersion relation is once again slightly altered to give

$$\omega^2 - c^2(k_z^2 + k_{cn}^2)$$
$$= C_0 \frac{\omega_p^2}{\gamma_0}\left[\frac{(\omega - k_z v_{z0})Q_{ns}}{\omega - k_z v_{z0} - s\Omega_c/\gamma_0} - \frac{\beta_{\perp 0}^2(\omega^2 - k_z^2c^2)H_{ns}}{2(\omega - k_z v_{z0} - s\Omega_c/\gamma_0)^2}\right] \tag{4.25}$$

where C_0 is a constant, and the new factors H_{ns} and Q_{ns} are functions of the finite transverse geometry and the cyclotron harmonic number s (where $s = 1, 2, 3 \ldots$).

This equation can be rewritten in coupled-mode form as

$$[\omega^2 - c^2(k_z^2 + k_{cn}^2)]\,[\omega - k_z v_{z0} - s\Omega_c/\gamma_0]^2 =$$
$$C_0 \frac{\omega_p^2}{\gamma_0}\left[(\omega - k_z v_{z0} - s\Omega_c/\gamma_0)\,(\omega - k_z v_{z0})Q_{ns} - \frac{\beta_{\perp 0}^2}{2}\,(\omega^2 - k_z^2 c_2)H_{ns}\right] \tag{4.26}$$

This form of the equation shows that, when the beam intensity is negligibly small ($\omega_p^2 \approx 0$), the modes represented by the two left-hand factors in (4.26) become uncoupled, and the resulting roots correspond to the empty waveguide modes and the fast cyclotron beam mode. The coupling between these modes is determined by the right-hand side of the equation. Sprangle and Drobot [12] have analyzed this dispersion relationship and determined that the second term on the right is the CRM destabilizing term. When $\beta_{\perp 0}^2$ reaches a threshold that makes this term dominate, the CRM instability begins. This typically occurs at relatively low energy levels. At higher levels of $\beta_{\perp 0}^2$, the second term on the right completely dominates, and an expansion of (4.26) about the synchronous interaction point reveals that the instability growth and the frequency shift both vary as $(\omega_p^2 H_{ns})^{1/3}$. Because ω_p^2 is proportional to density, this says that the gain of the CRM instability varies as the cube root of the beam current as does the gain in a conventional TWT.

The effect of the geometric factor H_{ns} is illustrated in Figure 4.15, taken from Chu [39]. This figure pertains to circular TE_{0n} waveguide modes, and an ideal hollow beam with electron guiding centers distributed on a circle at radius r_0 and Larmor orbits of radius r_L. The figure shows scaled two-dimensional plots of the geometric factor H_{ns} as functions of normalized r_0 and r_L for $s = 1, 2, 3$, and 4. Note the rapid decrease in coupling strength as harmonic number increases and as the

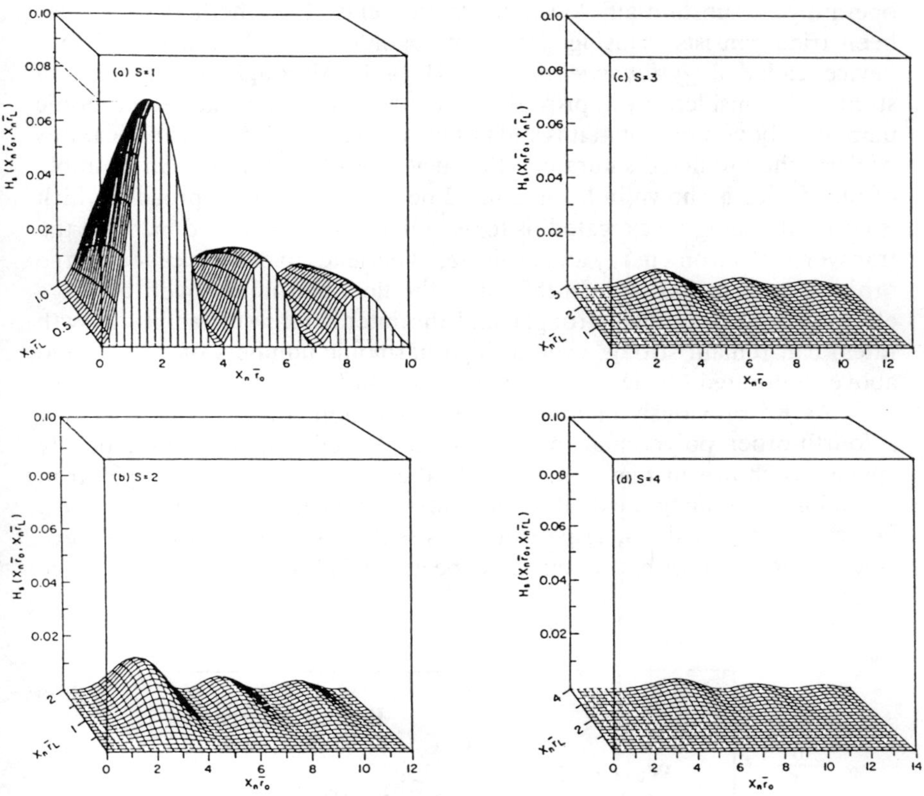

Fig. 4.15 Plots of the Beam-Wave Geometric Coupling Factor H_{ns} for Cyclotron Harmonics $s = 1, 2, 3,$ and 4 (From Chu [39])

beam guiding center is moved from the position of peak interaction strength. For quasirelativistic beams, these figures graphically illustrate the fact that the fundamental interaction ($s = 1$) is the strongest, and the beam and field geometries must be well matched to provide a strong CRM interaction. As interaction strength decreases, the saturated efficiency is also found to decrease. A possible exception to this is the case of the second harmonic, where at least one study has shown that saturated efficiency can remain high under certain conditions [40]. Also, we must note that a fair number of successful second harmonic gyrotrons have been built [10].

Finding techniques to enhance the interaction strength at the higher harmonics is still an attractive area of research because of the benefits of operating at substantially lower magnetic fields. One method which has been tried consists of using slotted waveguide walls. This has lead to a device called a *gyromagnetron* [16, 41, 42]. Although the interaction strength is considerably improved by the magnetron-like wall, at harmonic numbers above $s = 3$ or 4, it is still not very attractive for most applications because the predicted saturated efficiency drops below 10%. An example of this device is shown in Section 4.3.2 below. A second approach, which is currently being investigated, is to use relativistic beams and increase the transverse electron energy and orbit size. This leads to devices called "large orbit" gyrotrons [43–45]. In this case, the field harmonics over the large electron orbit are much stronger and the interactions (instability growth rates) can remain strong, even at high harmonic numbers of $s = 10$ and above. Saturated efficiencies, however, still suffer.

As we previously mentioned, the dispersion equation (4.26) is now a fourth-order polynomial in both ω and k_z. Dispersion plots typically appear as shown in Figure 4.16. Note the splitting of the fast cyclotron mode line containing the double root shown on the left-hand side of (4.26). For gyro-TWT analysis, dispersion plots such as shown in this figure provide important information on the expected gain-bandwidth performance of an amplifier.

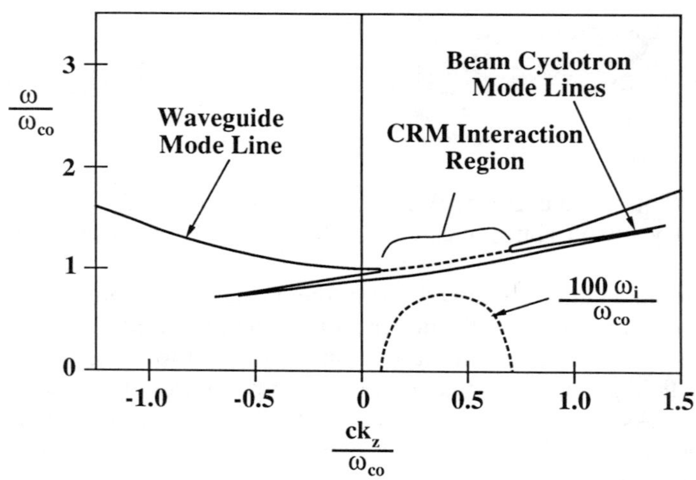

Fig. 4.16 Typical Plots of the CRM Dispersion Equation for Gyro-TWT Tuning

Addition of Velocity Spread to the Dispersion Equation

One of the nice features of the kinetic theory of gyrotrons is the fact that the general dispersion equation given in (4.18) is in terms of an initial velocity distribution f_0. This provides a straightforward method for including the effects of velocity spread in the gain calculations. Velocity spread can be assumed Gaussian [46], or specific data may come from electron-beam measurements or computer simulations. For pointwise velocity spread data, evaluation of the dispersion equation (4.18) is most easily accomplished by treating each data point as a separate beamlet with characteristics specified by (4.23). In this way, the ith beamlet has momentum components $p_{\perp 0i}$ and p_{z0i}, and a plasma frequency of ω_{pi}. By replacing $\omega_p f_0$ in (4.18) by a sum over $\omega_{pi} f_{0i}$, the integrals are easily evaluated and a dispersion equation is obtained, which includes a sum over the beamlets. Our point is that when the effects of velocity spread are important for accurate gain calculations, the linearized Vlasov equation approach gives a method for including these effects in the dispersion equation.

Fluid Models of the CRM Instability

We conclude this section on kinetic theory by noting that it is possible to obtain a simplified version of small signal analysis on a single cyclotron beamlet by using the more widely known fluid equations (which, of course, are contained in the Vlasov equation by taking averages over velocity space). The fluid equations required are the current continuity equation:

$$\frac{\partial \rho}{\partial t} + \nabla \cdot \mathbf{J} = 0 \tag{4.27}$$

where $\rho(\mathbf{r}, t)$ is the charge density, and the force density equation:

$$n \frac{\mathrm{d}\mathbf{p}}{\mathrm{d}t} = \rho \mathbf{E} + \mathbf{J} \times \mathbf{B} \tag{4.28}$$

where the current density $\mathbf{J} = \rho \mathbf{v}$. This fluid approach has been used by Lau [47] and by Shrivastava and Grow [48].

The approach is not as simple as we might hope, however, because a single beamlet has no volumetric current density, and therefore must be represented by a surface current density. To describe the RF surface current perturbations in this case, we must work in polarization variables [49], rather than the usual Eulerian variables, which are common to electron beam analysis for microwave tubes. If we place the beamlet at the center of a circular waveguide and perform small-signal perturbational analysis by use of this technique, it results in the same CRM dispersion equation (4.26). One of the advantages of this approach is the fact that both the CRM and peniotron resonances become clearly evident in the equations.

For example, if we assume that the electric field in a waveguide is right-hand circularly polarized, such that all the fields vary as $\exp[j(\omega t - k_z z - n\theta)]$, where n is the azimuthal harmonic number, then the small signal RF displacement $\mathbf{r}_1$ (polarization variable), can be expressed as [48]:

$$\begin{aligned}\mathbf{r}_1(\mathbf{r}_0, t) = {} & \frac{C_1}{k_2^2}\left[\mathbf{E}_1 + \mathbf{v}_0 \times \mathbf{B}_1 - \frac{(\mathbf{E}_1 \cdot \mathbf{v}_0)\mathbf{v}_0}{c^2}\right] \\ & - \frac{C_1^2}{k_1 k_2^2}\left[\mathbf{B}_0 \times \mathbf{E}_1 + \mathbf{B}_0 \times (\mathbf{v}_0 \times \mathbf{B}_1)\right] \\ & - \frac{C_1^3}{k_1^2 k_2^2}\left(\frac{\mathbf{E}_1 \cdot \mathbf{v}_0}{c^2}\right) B_0^2 \mathbf{v}_0 \\ & + \frac{C_1^3}{k_1^2 k_2^2}\left[\mathbf{E}_1 \cdot \mathbf{B}_0 + \mathbf{v}_0 \times \mathbf{B}_1 \cdot \mathbf{B}_0\right]\mathbf{B}_0\end{aligned} \tag{4.29}$$

where C_1 is a constant $[C_1 = e/(m\gamma_0)]$, and the resonance factors k_1 and k_2 are given by

$$k_1 = j(\omega - k_z v_{z0} - n\omega_c) \tag{4.30}$$

$$k_2^2 = [\omega - k_z v_{z0} - (n + 1)\omega_c]\,[\omega - k_z v_{z0} - (n - 1)\omega_c] \tag{4.31}$$

As we can see from the form of the leading coefficients of each term in (4.29), the RF displacement becomes large (i.e., resonant unstable growth occurs) when either k_1 or k_2^2 go to zero. The resonance $k_1 = 0$ is the synchronous condition for the CRM and Weibel instabilities. In this case, the magnetic tuning is such that the cyclotron harmonic number s is equal to the azimuthal field eigennumber n and the orbital electrons rotate in synchronism with the field. The resonance $k_2^2 = 0$, however, requires a different tuning of the cyclotron frequency. In this case, $s = n + 1$ or $s = n - 1$, which causes the orbital electrons and the rotating electric field to be out of synchronism by one period. This says, in effect, that as an electron moves through one complete orbit, it sees one complete cycle of the RF field. Put another way, the RF phase slips either forward or backward by 2π during each cyclotron orbit. Only the tuning $s = n - 1$, as discussed in Section 4.1.2, transfers the beam energy to the RF fields.

Shrivastava and Grow show that the first and second terms in (4.29) are predominantly peniotron interaction terms, and the third and fourth are predominantly CRM terms. This is consistent with the fact that the relative strengths of these resonances is determined by the powers of the resonant factors in k_1 and k_2.

As we noted in Section 4.1.2, when this tuning condition is coupled with a nonuniform distribution of the field over the orbit's extent, the guiding centers of the electrons in a beamlet move in the direction of the average $\langle \mathbf{E} \rangle \times \mathbf{B}$ and the electrons either loose or gain energy. By carefully choosing the spatial shape of the RF field distribution, we can cause all of the electrons to loose energy, and thereby create an active peniotron interaction. Another example of such a peniotron is given in Section 4.3.2 below.

4.3.2 Nonlinear Analysis via Direct Numerical Integration

As with the linear theory, we begin the nonlinear theory by combining Maxwell's equations to eliminate the H-field, and for TE waveguide modes this results in the transverse wave equation for the electric field in terms of the current and space-charge densities:

$$\left(\nabla^2 + \epsilon_0\mu_0 \frac{\partial^2}{\partial t^2}\right) \mathbf{E}_\perp = \mu_0 \frac{\partial \mathbf{J}_\perp}{\partial t} + \nabla_\perp \left(\frac{\rho}{\epsilon_0}\right) \tag{4.32}$$

For gyrotron interactions, the dominant physical effect described by this equation is the coupling of beam power to the RF waveguide modes through the $q\mathbf{v}_\perp \cdot \mathbf{E}_\perp$ changes in electron energy. Accordingly, the space-charge term can usually be neglected. If we then take the time harmonic Fourier transform of (4.32) and expand the electric field into waveguide modes, the result is

$$\sum_n \left(\frac{\partial^2}{\partial z^2} + k_{zn}^2\right) \mathbf{E}_{\perp n}(\mathbf{r}) = \mathrm{j}\omega\mu_0 \frac{1}{2\pi} \int_0^{2\pi} \mathbf{J}_\perp(\mathbf{r}, t)\, e^{-\mathrm{j}\omega t} \mathrm{d}(\omega t) \tag{4.33}$$

To obtain this equation, we have eliminated $\nabla_\perp^2$ by using the transverse wave equation for the modes in a hollow waveguide:

$$(\nabla_\perp^2 + k_{cn}^2)\,\mathbf{E}_{n\perp} = 0 \tag{4.34}$$

and defined k_{zn} to be the empty waveguide propagation constant for the nth waveguide mode:

$$k_{zn}^2 = \mu_0\epsilon_0\omega^2 - k_{cn}^2 \tag{4.35}$$

where k_{cn} is the cut-off wavenumber and n is again a multiple index. Note that the right-hand side of (4.33) stays in integral form because the nonlinear current density will be determined from the time-dependent equations of motion.

To reduce (4.33) to a coupled set of one-dimensional ordinary differential equations, we proceed by separating the transverse electric field for each mode into amplitude and field parts, such that

$$\mathbf{E}_{\perp n} = V_n(z)\mathbf{e}_n(\mathbf{r}_\perp) \tag{4.36}$$

where $\mathbf{e}_n$ is a field basis function (described in Section 4.3.4 below) and $\mathbf{r}_\perp$ represents the transverse coordinates. Because the basis vectors $\mathbf{e}_n$ are defined to be orthonormal with respect to the integration of $\mathbf{e}_n \cdot \mathbf{e}_m^*$ over the waveguide cross section, we can use this procedure to separate (4.33) and eliminate the transverse variables. The result is

$$\frac{\mathrm{d}^2V_n}{\mathrm{d}z^2} + k_{zn}^2V_n = \mathrm{j}\omega\mu_0\,\frac{1}{2\pi}\int_0^{2\pi}\left[\int_S \mathbf{J}_\perp(\mathbf{r}, t)\cdot\mathbf{e}_n^*(\mathbf{r}_\perp)\,\mathrm{d}S\right]\mathrm{e}^{-\mathrm{j}\omega t}\,\mathrm{d}(\omega t) \tag{4.37}$$

where the right-hand side is now only a function of z due to the integration over the cross section S and the phase period 2π, and the partial derivative has been changed to a total derivative because V_n is only a function of z.

An important extension of (4.37) is the fact that it can often be used for gyrotrons with weakly tapered waveguides. In this case, we simply treat $k_{zn} = k_{zn}(z)$ as a slowly varying function of z because the cut-off wavenumber k_{cn} changes slowly along the guide. The conditions under which this assumption is justified can be rigorously determined by using the tapered waveguide theory given in Section 4.3.4 below. This theory provides a complete description of the effects of tapered waveguide and can be used to augment (4.37) if necessary.

If two or more gyrotron modes are simultaneously excited at the same frequency, each mode can be represented by (4.37) resulting in a coupled set of equations with the transverse beam current density driving each mode and acting as the coupling medium between the modes. In most cases of interest, however, only one mode at a time is excited, and (4.37) then provides the total field amplitude in the gyrotron. This is the case we now continue to develop for purposes of nonlinear analysis.

To solve (4.37) numerically for cyclotron devices, we must determine $\mathbf{J}_\perp$ as a function of space and time. We do this by numerically integrating the ballistic equations of motion for a representative ensemble of electrons covering one period of the RF cycle. For the ith electron in the ensemble, we must therefore integrate

$$\begin{aligned} \frac{d\mathbf{p}_i}{dt} &= -e\left(\mathrm{Re}[\mathbf{E}_{\perp n}e^{j\omega t}] + \frac{\mathbf{p}_i}{m\gamma_i} \times \mathbf{B}_0 + \frac{\mathbf{p}_i}{m\gamma_i} \times \mathrm{Re}[\mathbf{B}_n e^{j\omega t}]\right) \\ \frac{d\mathbf{r}_i}{dt} &= \frac{\mathbf{p}_i}{m\gamma_i} \end{aligned} \tag{4.38}$$

where the relativistic factor γ is calculated as a function of $\mathbf{p}_i$ *as*

$$\gamma_i = \sqrt{1 + \frac{|\mathbf{p}_i|^2}{m^2c^2}}$$

and the RF magnetic field $\mathbf{B}_n$ is determined from the waveguide field relationships to be

$$\mathbf{B}_n = \frac{1}{-j\omega}\nabla_\perp \times \mathbf{E}_{\perp n} + \frac{\mu_0}{Z_n}\hat{\mathbf{z}} \times \mathbf{E}_{\perp n} \tag{4.39}$$

The first term in (4.39) is the B_z component of the RF field, which is nearly always neglected because of its small effect compared to the large dc magnetic field $\mathbf{B}_0$. The second term, however, is the source of the longitudinal acceleration that causes the Weibel instability. As previously illustrated, this term can only be neglected when operation is near the cutoff frequency, where wavelength is long and the wave impedance Z_n is high.

The ensemble of electrons for a "single wave" (single mode) analysis usually takes the form of a distribution of beamlets at representative guiding centers, with each beamlet represented by a single ring of electrons equally spaced at intervals of $\Delta\phi$ around the ring. All of the electrons then enter the interaction region together at the time $t = t_0$. An equivalent ensemble consists of replacing the rings for each beamlet by electrons that are launched into the RF interaction region at equally spaced times Δt

over one RF period and all placed at the same starting position on the rings. For a single-wave RF interaction, these two models are completely equivalent, and the relatively simple ensemble of electrons represent every possible interaction condition in a beam.

By contrast, when a multimode analysis is desired, each beamlet must be represented by many rings of electrons over a representative longitudinal period (e.g., the "beat" period between modes). Because (4.38) represents a set of six ordinary differential equations that must be integrated for each electron, it is easy to see that the computational requirements can become very large in the case of multiple RF modes.

The connection between (4.37) and (4.38) is from the definition of the current density in terms of the phase space distribution function $f(\mathbf{r}, \mathbf{p}, t)$.

$$\mathbf{J}(\mathbf{r}, t) = -e \int_{-\infty}^{+\infty} \mathbf{v}\, f(\mathbf{r}, \mathbf{p}, t)\, \mathrm{d}^3 p \tag{4.40}$$

In nonlinear theory, the positions and momenta of the ensemble of particles are used to represent $f(\mathbf{r}, \mathbf{p}, t)$ through the point distribution form:

$$f(\mathbf{r}, \mathbf{p}, t) = \sum_{i=1}^{N} \delta(\mathbf{r} - \mathbf{r}_i)\, \delta(\mathbf{p} - \mathbf{p}_i) \tag{4.41}$$

where $\delta(x)$ is the Dirac delta function and the sum extends over the ensemble of electron positions and momenta. When equation (4.41) is put into (4.40), we obtain

$$\mathbf{J}(\mathbf{r}, t) = -e \sum_{i=1}^{N} \mathbf{v}_i(t)\, \delta(\mathbf{r} - \mathbf{r}_i(t)) \tag{4.42}$$

and the transverse part of this equation can, in turn, be put into (4.37). The transverse integrals over the waveguide cross section can then be carried out by using the transverse portion of the delta function, leaving only one final integral to evaluate in terms of the final delta function in the z direction. Thus,

$$\frac{\mathrm{d}^2 V_n}{\mathrm{d}z^2} + k_{zn}^2 V_n = -\mathrm{j}\omega\mu_0 e \sum_{i=1}^{N} \frac{1}{2\pi} \int_0^{2\pi} \cdot [\mathbf{v}_i(t) \cdot \mathbf{e}_n^*(\mathbf{r}_{i\perp}(t))\, \delta(z - z_i(t))]\, \mathrm{e}^{-\mathrm{j}\omega t} \mathrm{d}(\omega t) \tag{4.43}$$

and the task is now to interpret the final integral over one phase period with a delta function integrand in terms of the z position.

It is evident, if the particles have a spread in longitudinal velocity, that there is not a one-to-one relationship between time and longitudinal position z. There will either be a spread in the positions z_i at any fixed time, or, if we fix a position z, there will be a spread in the times t_i when the particles arrive at z. It is also evident in our time harmonic formulation that the independent variable z in the right-hand term of this equation is the position at which we desire to evaluate the time harmonic of the current density. To preserve this time harmonic form, we therefore require all of the equations to be evaluated at the same z position.

As the phase integral is carried out over the ensemble of electrons, the only points at which the integrand will have a value are those where $z_i(t) = z$. We must therefore evaluate the integrand for each electron at the time t_i when it arrives at the position z. As long as the spread in these times does not exceed one RF period, the contribution of the ensemble acts properly as a single source term representing the Fourier component of $J_\omega(z)$, which drives the RF wave equation.

We can evaluate the final integral in (4.43) by converting the delta function to ωt by the rule:

$$\delta(z - z_i(t)) = \frac{\delta(t - t_i)}{\left|\dfrac{dz_i(t)}{dt}\right|} = \frac{\omega\delta(\omega t - \omega t_i)}{v_{zi}} \tag{4.44}$$

The final result is

$$\frac{d^2V_n}{dz^2} + k_{zn}^2 V_n = -j\omega\mu_0 e \sum_{i=1}^{N} \frac{\mathbf{v}_{\perp i} \cdot \mathbf{e}_n^*(\mathbf{r}_{i\perp})\, e^{-j\omega t_i}}{v_{zi}\tau} \tag{4.45}$$

where the velocities $\mathbf{v}_i = \mathbf{p}_i/m\gamma$ and the positions $\mathbf{r}_i$ are evaluated at the times t_i when the electrons arrive at the position of the independent variable z, and τ is the RF period. Thus, the right-hand side of (4.45) is a continuous function of z as needed for integration of the equation.

This selection of the variable z as the independent variable requires a change of variable in the ballistic equations (4.38). We must now evaluate the time derivative as

$$\frac{d(\ )}{dt} = \frac{d(\ )}{dz}\frac{dz}{dt} = v_{zi}\frac{d(\ )}{dz} = \frac{p_{zi}}{m\gamma}\frac{d(\ )}{dz} \tag{4.46}$$

and we must invert the z position equation to read

$$\frac{dt_i}{dz} = \frac{m\gamma}{p_{zi}} \tag{4.47}$$

Thus, the times t_i, when the electrons arrive at the position z, become dependent variables for the ensemble. With this understanding, we will continue to write (4.38) in the time-dependent form for the sake of simplicity.

Self-Consistent Solutions in Gyrotrons

Equations (4.38) and (4.45) are used for *self-consistent* large-signal analysis of CRM devices [50, 51, 52]. This implies simply that the equations are solved together as a coupled set of ordinary differential equations. Thus, the effects of the RF fields on the beam and the changes in the RF fields due to the beam are in a self-consistent form. These equations should be used for large-signal analysis of gyro-TWT devices and nonlinear analysis of gyrotron oscillators with low-Q cavities ($Q \leqslant 300$), where the field profile of the working device is perturbed from its empty cavity form.

In special cases where high-Q cavities are used, the shape of the field profile is determined by the RF cavity alone, thus simplifying the solution procedure. In this case, the "fixed" field profile is determined by solving (4.45) with the right-hand term set to zero (empty cavity), and (4.38) is integrated through the cavity using the solution for $V_n(z)$ multiplied by an arbitrary constant amplitude. From beam energy considerations, the resulting beam efficiency, after passing through the cavity, is given by

$$\eta = \frac{\gamma_0 - \frac{1}{N}\sum_{i=1}^{N} \gamma_i}{\gamma_0 - 1} \tag{4.48}$$

where the sum extends over all electrons in the ensemble. This efficiency is used to calculate the RF power extracted from the beam [$P_{\text{RF Beam}} = \eta P_B = \eta V_0 I_0$], which is then compared to the RF output power from the cavity [$P_{\text{RF out}} = \omega W/Q$]. The cavity Q is known from the empty cavity analysis, and the stored energy W is known from the field profile and the field amplitude used in the ballistic calculations. To complete the solution for saturated efficiency, a search is undertaken to determine the field amplitude in the cavity that makes $P_{\text{RF Beam}} = P_{\text{RF out}}$ for various input beam powers.

The reason that this shortcut for high-Q cavities may be preferred over the direct solution of the coupled equations (4.38) and (4.45) is that it simplifies the search procedure required to find the frequency and field amplitude that satisfy the RF boundary conditions. The frequency search involves only the empty cavity so that it is accomplished by using only four ordinary differential equations (i.e., a second-order differential equation with a complex variable). The methods described in Section 4.3.4 show how this is done. This is followed by a field amplitude search (to balance the RF power) by using the $6N$ ordinary differential equations for the beam.

Although the amount of computation involved in this second step is still large due to the number of equations, it is simplified by the fact that it is only a one-dimensional search and readily converges to a solution. In contrast, the self-consistent solution of (4.38) and (4.45) requires a two-dimensional search involving all $6N + 4$ ordinary differential equations.

Self-consistent solution of (4.38) and (4.45) proceeds by integrating the $6N$ differential equations representing the ensemble of beam electrons along with the two first-order complex differential equations contained in (4.45) all together as an initial value problem. The frequency is assumed to be close to the cavity reasonance desired and the electron positions and velocities are distributed so as to represent the initial phase space distribution function f_0. The initial values of the field amplitude V_n and derivative $\mathrm{d}V_n/\mathrm{d}z$ are set by guessing a value for $V_n(0)$ and calculating $\mathrm{d}V_n/\mathrm{d}z$ from the known evanescent field condition at the input to the cavity (due to RF cut-off). This condition is

$$\left[\frac{\mathrm{d}V_n}{\mathrm{d}z} = -\mathrm{j}k_{zn}V_n\right]\Bigg|_{z=0} \tag{4.49}$$

where k_{zn} is calculated below cut-off as

$$k_{zn} = \sqrt{k^2 - k_{cn}^2} = \mathrm{j}\sqrt{k_{cn}^2 - k^2}$$

With these initial conditions, the numerical integration proceeds through the interaction region to a selected point z_{out} in the output waveguide. At this point, we must check to determine if $V_n(z_{\mathrm{out}})$ satisfies the desired RF output boundary condition. This is usually given as

$$\left[\frac{\mathrm{d}V_n}{\mathrm{d}z} = -\mathrm{j}k_{zn}V_n\right]\Bigg|_{z=z_{\mathrm{out}}} \tag{4.50}$$

which represents outgoing power into a matched waveguide load.

Although, initially, the output boundary condition will not be achieved, it is interesting to note that the solutions obtained are still valid for the device under study. The solutions simply need to be interpreted as a condition in which the output waveguide is carrying power in both directions. Physically, the solution represents the operation of the device when an external power source (either an active device or a load reflection) is supplying the incoming power found in the solution.

Because we usually desire to find the operation of the device with a matched RF output load, we must seek a condition in which the final RF field satisfies (4.50). This is usually done by implementing an iterative search procedure in which the initial amplitude $V_n(0)$ and the operating frequency ω (real variable) are changed in response to the output error, and the integration of (4.38) and (4.45) is repeated until the desired output $V_n(z_{\text{out}})$ is achieved. The final time-averaged power output is then given by

$$P_{\text{out}} = \frac{|V_n|^2}{2Z_n} \tag{4.51}$$

where

$$Z_n = \sqrt{\frac{\mu_0}{\epsilon_0}}\frac{k}{k_{zn}}$$

is the wave impedance for the TE waveguide mode. This equation for output power relies on the proper normalization of the field basis functions. The final operating frequency ω, which is found from this iterative search procedure, will be shifted by some small value with respect to the natural frequency of the cavity, and represents the natural point of operation as an oscillator [53].

The Introduction of Slow Variables

For purposes of numerical calculations, (4.38) and (4.45) are most often reduced to differential equations involving slow variables. This simply means that the equations are rewritten in terms of amplitude and phase variables, which vary slowly with respect to the known unperturbed values. Because there are so many choices in this process, the slow variable equations found in the literature vary considerably with the preferences of the individual analysts doing the work. The major phase terms to be eliminated, however, are evident from the differential equations themselves, and we will provide an example of this process based on these terms.

In the absence of coupling between the beam and the RF wave, (4.38) and (4.45) will yield cyclotron motion at frequency ω_c for the particles and wave propagation with wavenumber k_{zn} for the RF waves. Thus, the fast phase variations to be canceled are $\exp(\mathrm{j}\omega_c t)$ and $\exp(-\mathrm{j}k_z z)$, respectively.

One popular method for defining slow variables for the electron trajectories is to rewrite the transverse part of (4.38) in polar form [1]. We start with the transverse part of the force equation which, neglecting small RF B_z field effects, can be written in the form:

$$\frac{\mathrm{d}\mathbf{p}_{\perp i}}{\mathrm{d}t} + \omega_c \mathbf{p}_{\perp i} \times \hat{\mathbf{z}} = -e\,\mathrm{Re}(V_n \mathbf{e}_n\, \mathrm{e}^{\mathrm{j}\omega t}) \tag{4.52}$$

$$\frac{\mathrm{d}\mathbf{r}_{\perp i}}{\mathrm{d}t} = \frac{\mathbf{p}_{\perp i}}{m\gamma} \tag{4.53}$$

These two equations are then separated into their vector components and recombined in polar form to give

$$\frac{\mathrm{d}P_i}{\mathrm{d}t} - \mathrm{j}\omega_c P_i = -e[\mathrm{Re}(V_n e_x\, \mathrm{e}^{\mathrm{j}\omega t}) + \mathrm{j}\,\mathrm{Re}(V_n e_y\, \mathrm{e}^{\mathrm{j}\omega t})] \tag{4.54}$$

$$\frac{\mathrm{d}R_i}{\mathrm{d}t} = \frac{P_i}{m\gamma} \tag{4.55}$$

where the new variables are defined as the transverse position $R = x + \mathrm{j}y$ and momentum $P = p_x + \mathrm{j}p_y$. Note that a constant $R_0 = x_0 + \mathrm{j}y_0$ can be added to each R_i to represent the guiding center of the particle without changing the equations.

The right-hand side of (4.54), which involves the x and y components of the field basis vector, must now be expanded at the guiding center position of each beamlet. As shown in [1], it may be argued that, in general, this expansion takes the form of a right-hand circularly polarized multipole field (2.9) given previously in Section 4.2.1. The mathematical form for the fundamental component of this field can be expressed in rectangular coordinates as

$$\mathbf{e}_n = \mathrm{j}e_n(\hat{\mathbf{x}} - \mathrm{j}\hat{\mathbf{y}}) \tag{4.56}$$

where e_n is a constant amplitude. This particular form fits nicely with (4.54) because we can insert the vector components into the right-hand term, take the real parts as indicated, and reassemble the terms into polar form to obtain

$$\frac{\mathrm{d}P_i}{\mathrm{d}t} - \mathrm{j}\omega_c P_i = -\mathrm{j}eV_n e_n \, \mathrm{e}^{\mathrm{j}\omega t} \tag{4.57}$$

where V_n is the complex field amplitude and e_n is, in general, a complex constant.

To eliminate the $\mathrm{j}\omega_c$ fast phase term from the left-hand side of this equation, we define an electron slow variable P_{si} related to P_i by

$$P_i = P_{si} \, \mathrm{e}^{\mathrm{j}\omega_c t} \tag{4.58}$$

When this is inserted into equation (4.57) the result is

$$\frac{\mathrm{d}P_{si}}{\mathrm{d}t} = -\mathrm{j}eV_n e_n \, \mathrm{e}^{\mathrm{j}(\omega t - \omega_c t)} \tag{4.59}$$

To put (4.45) in slow variable form, we begin by examining the factor $\mathbf{v}_{\perp i} \cdot \mathbf{e}_n^*$ in the right-hand term. When the dot product is taken by using (4.56) and the result is reinterpreted in terms of our new slow variable, we have

$$\frac{\mathrm{d}^2 V_n}{\mathrm{d}z^2} + k_{zn}^2 V_n = -\mathrm{j}\omega\mu_0 e \sum_{i=1}^{N} \frac{e_n^* P_{si} \, \mathrm{e}^{\mathrm{j}(\omega_c t_i - \omega t_i)}}{m\gamma v_{zi}\tau} \tag{4.60}$$

To complete the slow variable transformation of this equation, we must eliminate the fast phase term k_{zn}^2 on the left-hand side. As before, we do this by defining the slow variable V_{sn} in terms of V_n as

$$V_n = V_{sn}(z) \, \mathrm{e}^{-\mathrm{j}k_{zn}z} \tag{4.61}$$

and inserting this into the equation so as to cancel the fast phase term. This time the result is

$$\frac{\mathrm{d}^2 V_{sn}}{\mathrm{d}z^2} - \mathrm{j}2k_{zn}\frac{\mathrm{d}V_{sn}}{\mathrm{d}z} = -\mathrm{j}\omega\mu_0 e \sum_{i=1}^{N} \frac{e_n^* P_{si} \, \mathrm{e}^{-\mathrm{j}(\omega t_i - k_{zn}z - \omega_c t_i)}}{m\gamma v_{zi}\tau} \tag{4.62}$$

Because V_{sn} is a slow variable, the second derivative term in this equation is often negligible, thus reducing it to first order in the slow variable.

If we now use (4.61) to eliminate V_n from (4.59), we obtain

$$\frac{\mathrm{d}P_{si}}{\mathrm{d}t} = -\mathrm{j}eV_{sn}e_n\,\mathrm{e}^{\mathrm{j}(\omega t - k_{zn}z - \omega_c t)} \tag{4.63}$$

and, using the transformation of independent variable given in (4.46), we obtain

$$\frac{\mathrm{d}P_{si}}{\mathrm{d}z} = \frac{-\mathrm{j}eV_{sn}e_n\,\mathrm{e}^{\mathrm{j}(\omega t_i - k_{zn}z - \omega_c t_i)}}{v_{zi}} \tag{4.64}$$

The longitudinal velocities v_{zi} are found from the momenta p_{zi} by integrating the z component of (4.38), which, using (4.39), can be written in terms of the slow variables as

$$\frac{\mathrm{d}p_{zi}}{\mathrm{d}z} = \frac{-e\mu}{Z_n p_{zi}}\,\mathrm{Im}[P_{si}V_{ns}^{*}e_n^{*}\,\mathrm{e}^{-\mathrm{j}(\omega t_i - k_{zn}z - \omega_c t_i)}] \tag{4.65}$$

The times t_i in these equations are found from

$$\frac{\mathrm{d}t_i}{\mathrm{d}z} = \frac{m\gamma}{p_{zi}} \tag{4.66}$$

Note that the coupled equations (4.62) and (4.63), in terms of the slow variables V_{sn} and P_{si} for each electron, only depend on the transverse position through the field expansion coefficient e_n. Also, note that the right-hand sides involve the driving phase factor $(\omega t_i - k_{zn}z - \omega_c t_i)$, which must be close to zero for synchronous tuning of the fundamental cyclotron interaction. Thus, near the synchronous condition of interest, we are assured that our new variables will indeed have a slow phase variation.

As we previously mentioned, a large number of variations of this basic slow variable approach exist in the literature. Often the slow variables are separated into amplitude and phase parts and the differential equations rewritten in terms of real variables. For operation near cut-off, the changes in longitudinal velocity are also often neglected, which results in discarding the longitudinal momentum equation in favor of the simple distance-time relationship $z = v_{z0}t$. A good description of this type of slow variable formulation is given by Fliflet [29].

Nonlinear Calculation of CRM Efficiency

An example of the application of nonlinear techniques to the CRM instability mechanism is given by Sprangle and Drobot [12] in a classic paper, which explores the basic saturation mechanism and the saturated efficiency for a wide range of parameters. These authors found that saturated efficiency always exhibits a maximum with respect to electron energy. The curves calculated are shown in Figures 4.17, 4.18, and 4.19. In each of these figures, the efficiency is plotted in the beam frame of reference (i.e., the transverse efficiency) and must be transformed back to the laboratory frame to get the actual efficiency, which will be lower.

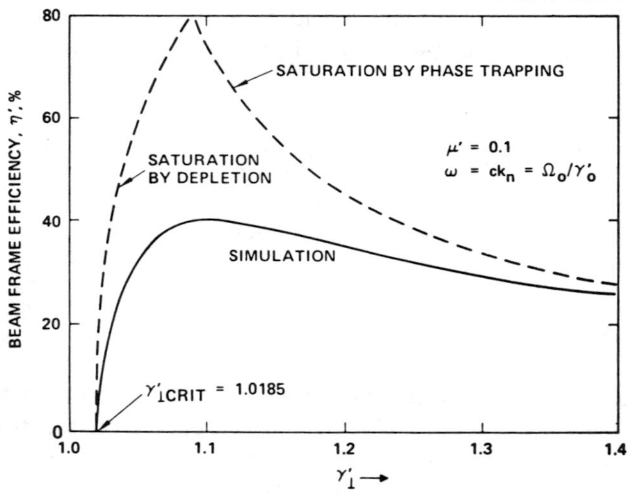

Fig. 4.17 Plot Showing the Typical CRM Saturated Efficiency *versus* Beam Energy in the Beam Frame of Reference, with Two Analytically Derived Estimates of Saturated Efficiency at Low and High Energies (Dashed Curves) [12]

Figure 4.17 shows a typical plot of saturated efficiency *versus* beam energy with superimposed plots (dashed lines) of analytical estimates of the saturation process. One mechanism of saturation is called energy depletion and simply comes from a criterion of estimating the point where the instability will stop, due to the reduction in the average transverse velocity below the critical level of instability in the small-signal dispersion

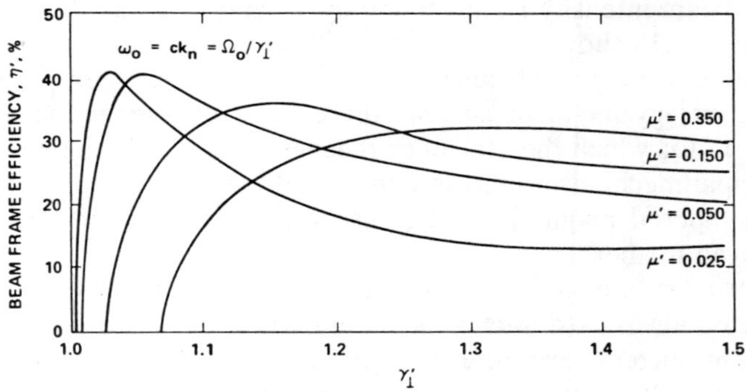

Fig. 4.18 Efficiency *versus* Beam Energy in the Beam Frame with Normalized Current Density μ as a Parameter [12]
Note: The magnetic field was synchronously tuned.

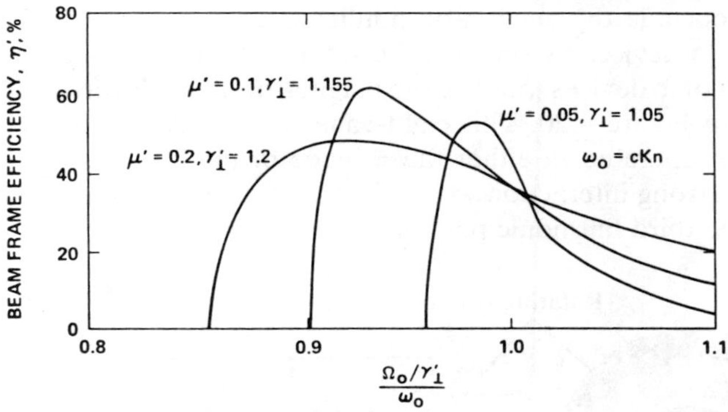

Fig. 4.19 Efficiency *versus* Magnetic Field Tuning for Several Different Currents and Beam Energies [12]
Note: Again, the energy is in the beam frame.

relationship. (See (4.26) in Section 4.3.1.) The other mechanism of saturation is called phase trapping and comes from an estimate of the number of particles that lose sufficient energy to become trapped in phase synchronism with the electric field. In Figure 4.17 the actual saturation level predicted by the numerical simulation shows a peaked curve, which fits between the two asymptotic estimates.

Figures 4.18 and 4.19 then show trade-offs of efficiency *versus* normalized beam intensity μ, beam energy $\gamma_\perp$ and magnetic tuning $\Omega_0 = eB_0/m$. Figure 4.18 shows the saturated efficiency as a function of energy for synchronous tuning and beam current as a parameter. Figure 4.19 shows that, at each energy and beam current, there is a level of magnetic detuning ($B/B_{sync} < 1$) for which the efficiency reaches a peak. Magnetic detuning is the principal mechanism used to enhance efficiency. In more advanced schemes, a tapered magnetic field is also used to improve the saturated efficiency even further.

The primary message of these CRM parameter trade-off studies is that efficiency always exhibits a maximum with respect to beam energy for any beam current level and magnetic tuning. It is therefore wise to consider beam voltage as a part of any design trade-off. There will always be a particular beam energy at which the best possible efficiency will be attainable.

Example of Large-Signal Calculations of Gyromagnetron and Peniotron Efficiencies

We conclude this section on nonlinear computation of efficiency in gyrocyclotron devices by showing the saturated efficiencies calculated for higher harmonic devices that use the magnetron-like RF waveguide circuit illustrated in Figure 4.20. This eight-vane circuit has a strong fourth harmonic when operated in either the π mode or the 2π mode. It therefore produces a strong interaction with either the fourth harmonic CRM mechanism or the third harmonic peniotron mechanism.

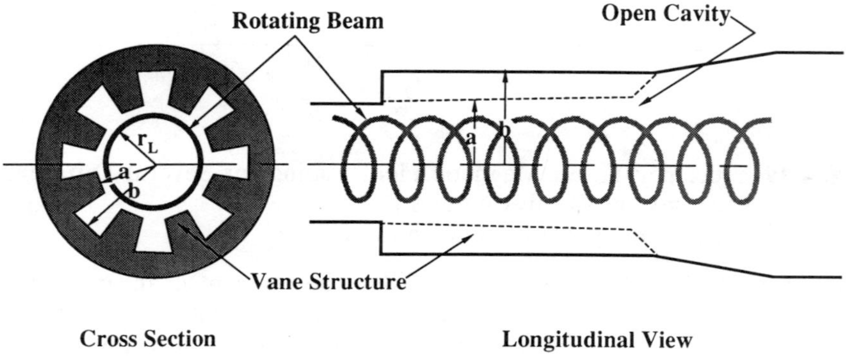

Fig. 4.20 Illustration of Magnetron-like Waveguide Circuit and Beam Used for Harmonic Gyrotron and Peniotron Interactions

The results of calculations on these two different devices are shown in Figures 4.21 through 4.23 [52]. Figure 4.21 shows some typical beam cross sections for the CRM and Peniotron calculations, and Figure 4.22 shows typical peniotron orbits for different phase electrons. In the case of the third harmonic peniotron, it is evident that a significant amount of the transverse energy is given up to the RF field. Finally, Figure 4.23 shows plots of the predicted saturated efficiencies for the two devices. Note that, even when a substantial spread is introduced in the guiding centers of the beam, the predicted peniotron efficiency is still high enough to be attractive for high-power millimeter-wave device development.

4.3.3 Time-Dependent Cavity Analysis

A time-dependent cavity analysis provided by Slater [54] has proved useful in gyrotron theory. This formulation describes the interaction between the RF cavity fields and the current density of a charged particle beam. The analysis has been used very effectively for design and analysis of gyrotrons at MIT [55, 56] and constitutes an approach worth noting.

In this approach, the cavity modes of an ideal cavity are used as a basis set for expansion of Maxwell's equations. The ideal cavity may have any desired shape with two different types of boundary conditions: a perfectly conducting area S on which $\hat{\mathbf{n}} \times \mathbf{E} = 0$ and the remaining area S' on which $\hat{\mathbf{n}} \times \mathbf{H} = 0$. The surface S' corresponds to an insulating area such as an open hole in the cavity wall connected to a cut-off waveguide so that no energy escapes through the hole.

The analysis begins by normalizing the cavity mode fields $\mathbf{E}_n(\mathbf{r})$ and $\mathbf{H}_n(\mathbf{r})$ such that

$$\int_V \mathbf{E}_n \cdot \mathbf{E}_m \, dV = \int_V \mathbf{H}_n \cdot \mathbf{H}_m \, dV = \delta_{nm} \tag{4.67}$$

where V is the volume of the cavity, and m and n are composite mode indices. The general fields in a nonideal cavity can then be expanded in terms of these basis field functions as

$$\begin{aligned} \mathbf{E}(\mathbf{r}, t) &= \sum_n p_n(t)\mathbf{E}_n(\mathbf{r}) \\ \mathbf{H}(\mathbf{r}, t) &= \sum_n q_n(t)\mathbf{H}_n(\mathbf{r}) \end{aligned} \tag{4.68}$$

where the time-dependent expansion coefficients $p_n(t)$ and $q_n(t)$ represent the amplitudes of the field modes.

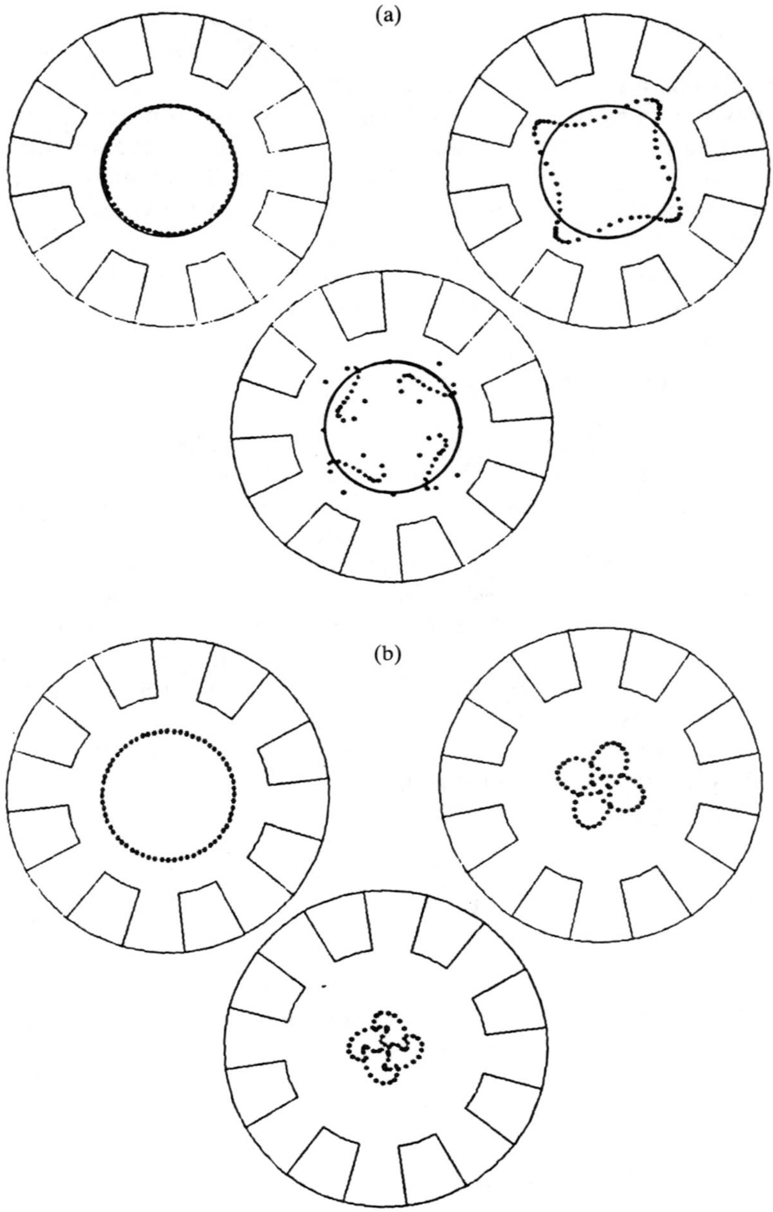

Fig. 4.21 Typical Beam Cross Sections Encountered in the Simulation of (a) a Fourth Harmonic Gyrotron Interaction, and (b) a Third Harmonic Peniotron Interaction (From Rha [52])

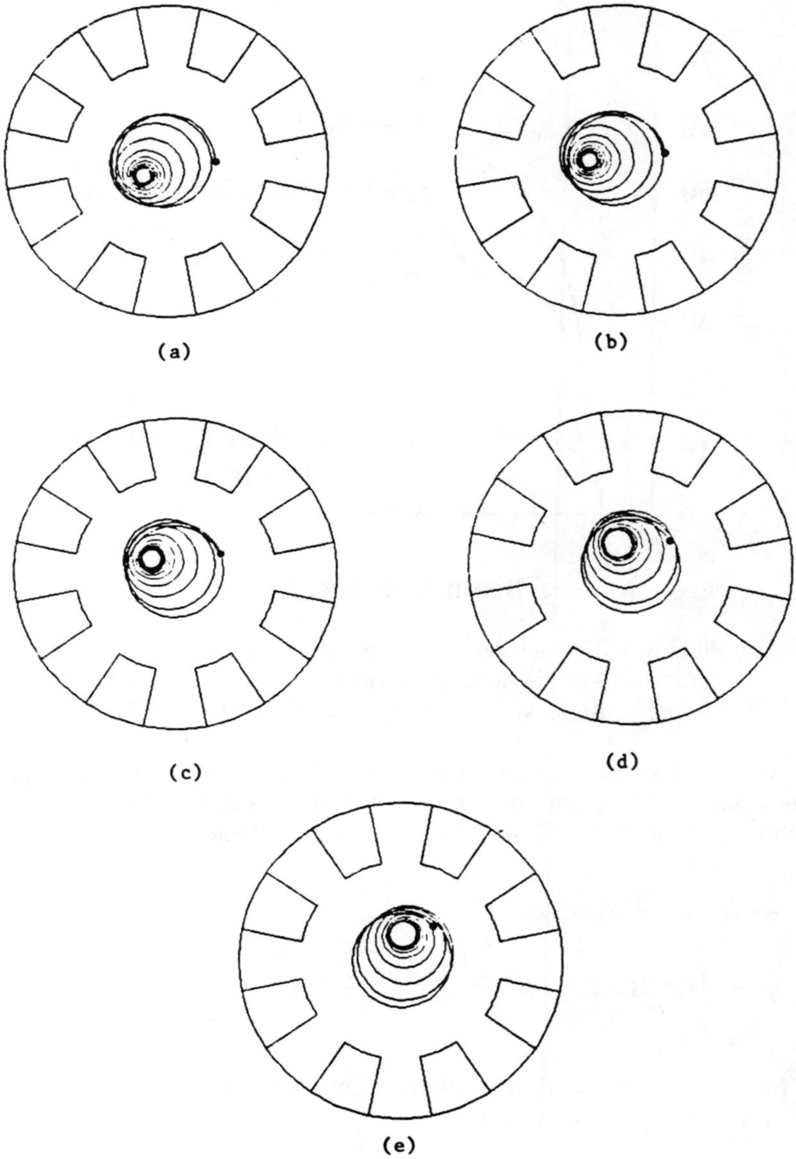

Fig. 4.22 Plots of the Orbital Trajectories for Five Peniotron Electrons Starting at Different RF Phase Positions (From Rha [52])

Note: All electrons in this configuration lose most of their transverse energy to the RF field.

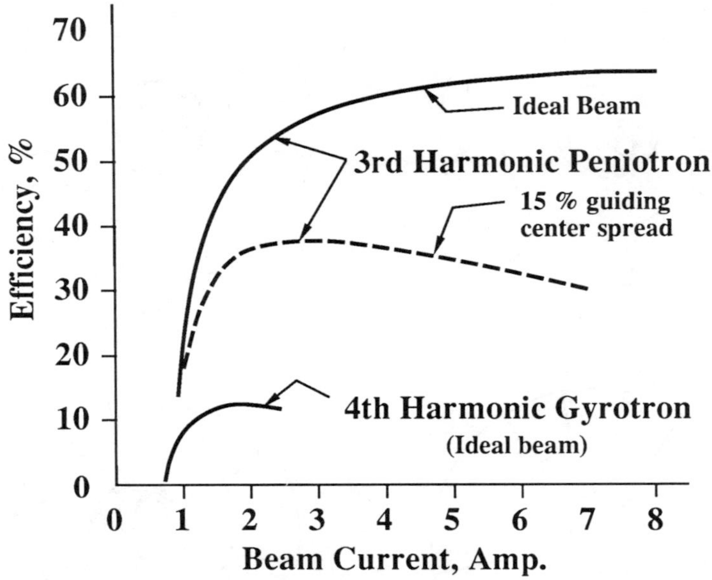

Fig. 4.23 Plots of Efficiency *versus* Beam Current for the Third Harmonic Peniotron and the Fourth Harmonic Gyrotron in an Eight-Vane Magnetron-like Waveguide Circuit (From Rha [52])

When Maxwell's equations are vector-dot-multiplied by an appropriate basis field function and integrated over the cavity volume, it is not difficult to recognize and insert the expansion coefficients:

$$\begin{aligned} p_n &= \int_V \mathbf{E} \cdot \mathbf{E}_n \mathrm{d}V \\ q_n &= \int_V \mathbf{H} \cdot \mathbf{H}_n \mathrm{d}V \end{aligned} \tag{4.69}$$

and to combine the resulting equations into a set of coupled harmonic oscillator equations. The result is

$$\frac{1}{c^2}\frac{\mathrm{d}^2 p_n}{\mathrm{d}t^2} + k_n^2 p_n = -\mu_0 \frac{\mathrm{d}}{\mathrm{d}t}\left[\int_V \mathbf{J} \cdot \mathbf{E}_n \mathrm{d}V - \int_{S'} (\hat{\mathbf{n}} \times \mathbf{H}) \cdot \mathbf{E}_n \mathrm{d}S\right] - k_n \int_S (\hat{\mathbf{n}} \times \mathbf{E}) \cdot \mathbf{H}_n \mathrm{d}S \tag{4.70}$$

To obtain this form, we must use integration by parts and the divergence theorem, which gives rise to the surface integrals shown. As we previously noted, when the cavity to be analyzed is ideal, then $\hat{\mathbf{n}} \times \mathbf{E} = 0$ over S, $\hat{\mathbf{n}} \times \mathbf{H} = 0$ over S', and the surface integrals in (4.70) are zero.

The left-hand side of (4.70) describes cavity oscillations for the nth cavity mode, and, in the absence of the perturbing terms on the right-hand side, we see that oscillations occur at the eigenmode frequency of the ideal cavity $\omega = \omega_n = ck_n$. The first perturbing term on the right-hand side is the driving term due to the current density $\mathbf{J}$ within the cavity. The integral of $\mathbf{J} \cdot \mathbf{E}_n$ over the cavity volume represents the total power exchange between the beam and the RF cavity fields at any instant of time. When the cavity modes are closely spaced in frequency, this term provides strong coupling between the modes. In this case, each mode is described by an equation of the form (4.70) with the same beam density $\mathbf{J}$ in each equation, and beam modulations from one mode can affect other modes.

The two surface integrals on the right-hand side of (4.70) are used in gyrotron applications to describe the power escaping from the cavity. The second term describes the power propagating out of the open end of the gyrotron cavity represented by the surface S', and the third term describes the power being absorbed in the resistive walls of the cavity over the surface S.

For applications to gyrotron cavities, the two surface integrals in (4.70) can be rewritten in terms of the cavity Q values associated with their respective power losses. If we define Q_{Dn} to be the diffraction Q associated with the output power of the cavity, and Q_{0nn} to be the intrinsic Q associated with the power lost in the resistive walls, then (4.70) can be rewritten as

$$\frac{1}{c^2}\frac{\mathrm{d}^2 p_n}{\mathrm{d}t^2} + k_n^2 p_n = -\mu_0 \frac{\mathrm{d}}{\mathrm{d}t}\left[\int_V \mathbf{J} \cdot \mathbf{E}_n \mathrm{d}V + \epsilon_0 \frac{\omega_n}{Q_{Dn}} p_n\right] - (i-1)\frac{k_n^2}{Q_{0nn}} p_n \tag{4.71}$$

This form of the equation is obtained by neglecting a small coupling between modes due to the resistive walls.

Equations (4.71) can be used for a variety of analytical purposes. Gyrotron design equations, which give the steady-state power and the frequency detuning, can be obtained by assuming a steady-state solution of the form $p_n(t) = p_{n0}\exp(\mathrm{j}\omega t)$, where p_{n0} is the constant field amplitude and the operating frequency ω is real. The change in frequency due to magnetic detuning is caused by the interaction between the natural frequency of the cavity and the cyclotron frequency of the electron beam. When the real and imaginary parts of (4.71) are equated for the steady-state case, the result is an energy balance equation relating p_{n0} to the volumetric integral of $\mathbf{J} \cdot \mathbf{E}_n$:

$$\frac{\omega_n}{Q_{Tn}} = \frac{1}{\epsilon_0} \operatorname{Re}\left[\frac{1}{p_{0n}} \int_V \mathbf{J} \cdot \mathbf{E}_n \mathrm{d}V\right] \tag{4.72}$$

and an equation for the detuned frequency of operation ω in terms of other known variables:

$$(\omega_n')^2 - \omega^2 = \frac{\omega}{\epsilon_0} \operatorname{Im}\left[\frac{1}{p_{0n}} \int_V \mathbf{J} \cdot \mathbf{E}_n \mathrm{d}V\right] \tag{4.73}$$

In these equations,

$$\frac{1}{Q_{Tn}} = \frac{1}{Q_{Dn}} + \frac{1}{Q_{0nn}} \tag{4.74}$$

and

$$\omega_n' = \omega_n \left(1 - \frac{1}{Q_{0nn}}\right)^{1/2} \tag{4.75}$$

As with other theoretical approaches, the current density $\mathbf{J}$ in (4.72) and (4.73) can now be evaluated in two different ways: by linear small signal theory or by nonlinear electron trajectory methods. Examples of the results obtained from these two different approaches are illustrated in Figures 4.24 and 4.25 from [55, 56].

The starting current for oscillation and the frequency detuning in Figure 4.24 come from using sinusoidal (S) and Gaussian (G) shaped longitudinal cavity fields, and an ideal annular gyrotron beam to evaluate the perturbed velocity distribution f_1 as a function of the unperturbed trajectories (as in (4.16) Section 4.3.1). When $\mathbf{J}$ is expressed as in (4.17) and the integrals are evaluated, the result is an equation for the small-signal current intensity, which will sustain oscillations (i.e., the starting current).

Figure 4.25 comes from efficiency optimization studies using nonlinear beam calculations to determine the current density $\mathbf{J}$ in (4.72) [56].

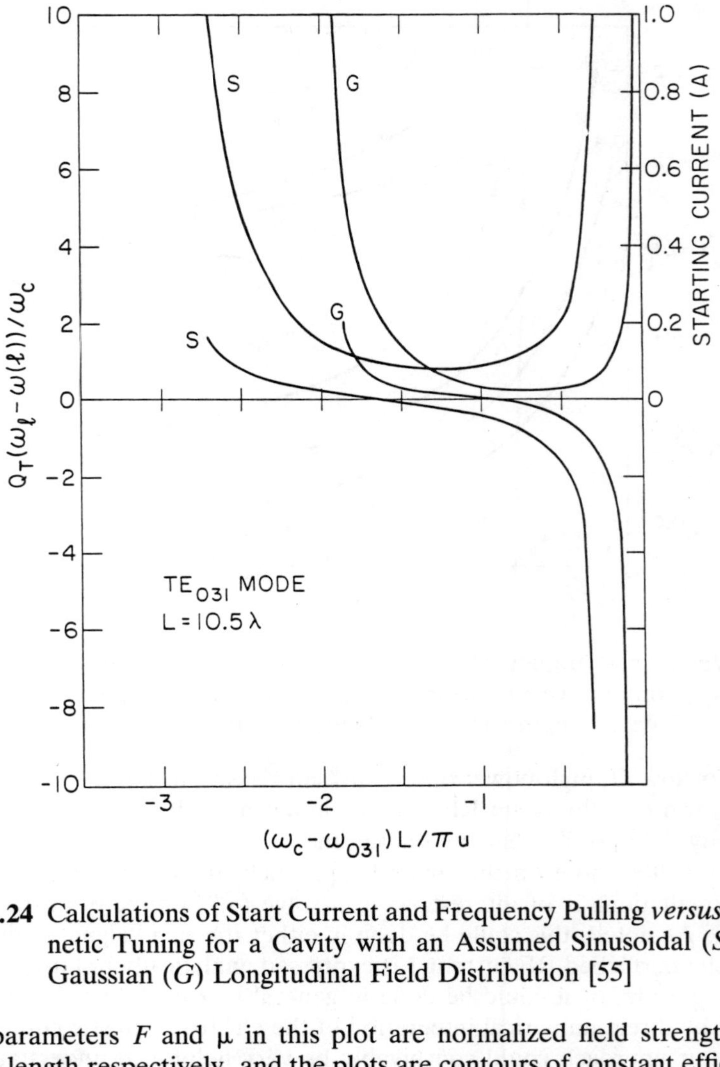

Fig. 4.24 Calculations of Start Current and Frequency Pulling *versus* Magnetic Tuning for a Cavity with an Assumed Sinusoidal (S) and Gaussian (G) Longitudinal Field Distribution [55]

The parameters F and μ in this plot are normalized field strength and cavity length respectively, and the plots are contours of constant efficiency when magnetic tuning is optimized. Plots such as this have great value for design purposes. When overlaid by the constraints of maximum cavity wall heating for various modes and beam voltages, the plots become a very effective tool for assessing the feasibility and predicting the performance of high-power gyrotron oscillators. The single drawback of the approach is the fact that the cavity field profile is fixed in this analysis. As previously

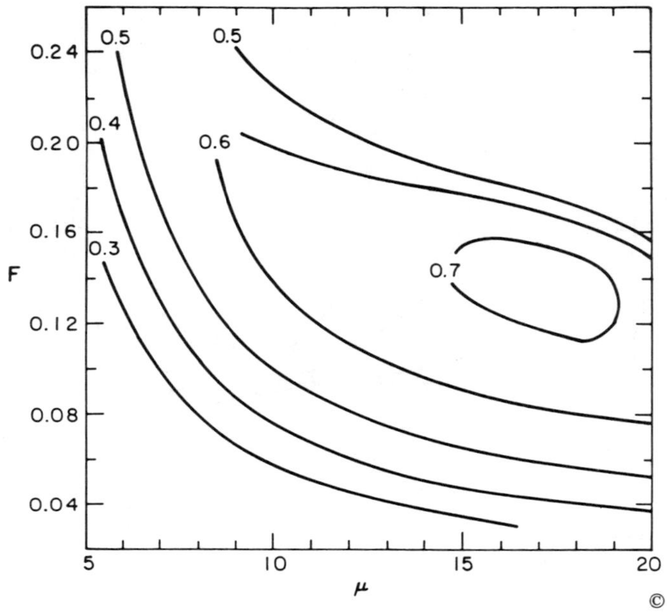

Fig. 4.25 Three-Parameter Nonlinear Analysis Showing Plots of Efficiency Contours *versus* Normalized Field Strength (F) and Normalized Cavity Length (μ) when Magnetic Tuning is Optimized [56]

noted for low Q, high-power operation, self-consistent interaction studies have shown that the beam-RF interaction has a small effect in reshaping the cavity field profile near the output end.

As a final note on this analysis approach, if the time evolution of cavity oscillations is of interest, we can solve (4.71) directly in the time domain. As noted, this could be done in either the small signal regime in which the perturbed RF current $\mathbf{J}$ is expressed analytically in terms of the field amplitude, or it could be done in general by calculating the current density by direct numerical integration of the electron ballistic equations. This latter approach could conceivably be very helpful in understanding and solving time-dependent gyrotron anomalies, such as the generation of spurious modes during pulse turn-on or unstable switching between modes during operation.

4.3.4 Analysis of Overmoded Waveguide Cavities and Components

High-power millimeter-wave gyrotrons require large overmoded waveguide cavities and transmission lines. The analysis of cavities for gyrotron applications is complicated by the fact that the open-ended cavities must not only meet specified design values of Q, but also include arbitrary wall tapers and wall steps to improve efficiency and enhance mode control. At least one recent paper that explores a new generalized approach to tapered waveguide systems has appeared in response to this need [57]. In general, however, the techniques found in the archival literature (although a little dusty and often unappreciated) have proved to be very adequate for the task. Two specialized books are available, which treat many of these techniques [58, 59].

The focus of this section is to describe a technique introduced by Reiter [60] and Solymar [61] in 1959 for reducing Maxwell's equations in tapered overmoded waveguides with arbitrary cross-sectional geometry to a manageable set of coupled transmission-line equations. Solymar gives explicit formulas for determining the intermode scattering coefficients due to the tapered waveguide walls. These formulas are determined at each waveguide cross section by the cross-sectional geometry and the wall taper, and are expressed as line integrals around the waveguide boundary. This theory has been very successfully used not only to analyze gyrotron cavities, but also some rather sophisticated mode converters [62–64] and waveguide transitions [65]. In the opinion of the author, the potential of this theory for analyzing multimode gyrotron interactions is yet to be fully exploited. It could, perhaps, explain some of the spurious mode content in the output of high-power gyrotrons now in operation.

The papers by Reiter and Solymar are quite condensed, but the theory has been more fully explained and expanded in a report by the author [66]. In the cited report, the theory is also extended to include the effects of nonideal waveguide walls (i.e., walls with losses, side wall ports, *et cetera*). This theory has proved to be quite versatile in analyzing a number of waveguide transitions and couplers. Where the transverse wave functions are difficult to obtain because of complicated waveguide cross sections, we have even had good results by using numerical techniques to simultaneously find the transverse mode functions and evaluate Solymar's scattering coefficients [67, 68].

Field Expansion Basis Vectors and the Mode Scattering Equations

The theory rests on the use of transverse field expansions in terms of orthonormal basis vectors, which, in turn, are derived from the wave potential functions from TE and TM field analysis [69, 49]. This type of basis vector was introduced in 1951 by Marcuvitz [70] as a technique for reducing Maxwell's equations in uniform waveguides of arbitrary cross section to transmission-line form. Expansions of the transverse fields are given by

$$\mathbf{E}_\perp = \sum_n V_n \mathbf{e}_n$$
$$\mathbf{H}_\perp = \sum_n I_n \mathbf{h}_n \tag{4.76}$$

where the expansion coefficients V_n and I_n are the voltage and current amplitudes of the transmission line modes (units of volts and amperes, respectively), and the transverse basis vectors $\mathbf{e}_n$ and $\mathbf{h}_n$ are defined by

$$\mathbf{e}_n = \begin{Bmatrix} -\nabla_\perp \Psi_n \times \hat{z} \\ -\nabla_\perp \Psi_n \end{Bmatrix}$$
$$\mathbf{h}_n = \begin{Bmatrix} -\nabla_\perp \Psi_n \\ \nabla_\perp \Psi_n \times \hat{z} \end{Bmatrix} \tag{4.77}$$

The functions Ψ_n are the solutions of the two-dimensional transverse Helmholtz equation:

$$(\nabla_\perp^2 + k_{cn}^2)\Psi_n = 0 \tag{4.78}$$

with boundary conditions:

$$\begin{Bmatrix} \dfrac{\partial \Psi_n}{\partial n} \\ \Psi_n \end{Bmatrix} = 0$$

The large braces { } in the above equations designate alternate forms of the quantities for TE and TM modes (top and bottom, respectively), $\nabla_\perp$ is the transverse part of the ∇ operator, and $\partial/\partial n$ at the boundary is the outward directed gradient operator normal to both the cross-sectional boundary and the z propagation direction. With these definitions and the proper normalization of the functions Ψ_n, the usual orthonormal inner product for the basis vectors can be defined as

$$\int_S \mathbf{e}_n \cdot \mathbf{e}_k^* \mathrm{d}S = \int_S \mathbf{h}_n \cdot \mathbf{h}_k^* \mathrm{d}S = \delta_{nk} \tag{4.79}$$

where S is the cross-sectional area of the waveguide. With this formulation, the time-averaged power in the nth waveguide mode is given by

$$\overline{P}_n = \frac{1}{2}\,\mathrm{Re}(V_n I_n^*) \tag{4.80}$$

Reiter [60] showed that when the waveguide is tapered so that the basis vectors and the cross-sectional area must be treated as functions of the longitudinal coordinate z, it is still possible—without approximations—to reduce Maxwell's equations to transmission-line form. Under these more general conditions, coupling terms appear, which cause intermode coupling of the following form:

$$\begin{aligned} \frac{\mathrm{d}V_n}{\mathrm{d}z} &= -\mathrm{j}k_{zn}Z_n I_n + \sum_k T_{kn}V_k \\ \frac{\mathrm{d}I_n}{\mathrm{d}z} &= -\mathrm{j}k_{zn}Y_n V_n - \sum_k T_{nk}^* I_k \end{aligned} \tag{4.81}$$

where the uncoupled mode propagation constants are given by

$$k_{zn} = \sqrt{k^2 - k_{cn}^2} \tag{4.82}$$

with $k = \omega/c$, and the uncoupled TE and TM mode wave impedances are given by

$$Z_n = \frac{1}{Y_n} = \sqrt{\frac{\mu}{\epsilon}} \begin{Bmatrix} \dfrac{k}{k_{zn}} \\ \dfrac{k_{zn}}{k} \end{Bmatrix} \tag{4.83}$$

The mode coupling, or "transfer," coefficients in (4.81) are defined as

$$\begin{aligned} T_{kn} &= \int_S \mathbf{e}_k \cdot \frac{\partial \mathbf{e}_n^*}{\partial z}\,\mathrm{d}S \\ T_{nk}^* &= \int_S \frac{\partial \mathbf{e}_k}{\partial z} \cdot \mathbf{e}_n^*\,\mathrm{d}S \end{aligned} \tag{4.84}$$

and Solymar [61] reduces these expressions to line integrals around the waveguide boundary with integrands involving the wall taper angle and the transverse wave functions Ψ_n and Ψ_k. Solymar's line integral forms for the coupling coefficients are given in Appendix 4.A at the end of this chapter.

When modes are above cut-off, it is common to make use of the normal modes of a uniform waveguide a_n^+ and a_n^- (forward and backward signals), which are related to the voltage and current variables by the transformation [49]:

$$\begin{bmatrix} V_n \\ I_n \end{bmatrix} = \begin{bmatrix} \sqrt{2Z_n} & \sqrt{2Z_n} \\ \sqrt{2Y_n} & -\sqrt{2Y_n} \end{bmatrix} \begin{bmatrix} a_n^+ \\ a_n^- \end{bmatrix} \tag{4.85}$$

With this transformation, the time-averaged power (4.80) becomes

$$\overline{P}_n = |a_n^+|^2 - |a_n^-|^2 \tag{4.86}$$

Above cut-off, this transformation is very useful to find relationships between voltage and current for boundary conditions involving the amplitude and direction of power flow. For boundary conditions below cut-off, the transformation (4.85) breaks down, and it is necessary to revert to the differential equations directly to determine the voltage and current boundary conditions for evanescently growing or decaying solutions.

Solutions of the Mode Scattering Equations for Mixed Boundary Value Problems

The coupled pair of equations (4.81) are in a form that can be readily implemented on the computer, and therefore lend themselves to a wide variety of surprisingly sophisticated analyses of tapered waveguide systems. The first thing to realize is that (4.81) is simply a coupled set of first-order ordinary differential equations, which, if the initial values are specified, are in the standard form for numerical integration [71]:

$$\frac{\mathrm{d}\mathbf{X}}{\mathrm{d}z} = \mathbf{F}(\mathbf{X}, z)$$

Even more advantageously, (4.81) are linear in the dependent variables, and therefore can be written in the form:

$$\frac{\mathrm{d}\mathbf{X}}{\mathrm{d}z} = \bar{\bar{\mathbf{M}}}\mathbf{X} \tag{4.87}$$

where $\mathbf{X}$ is a $2N$ dimensional vector of the voltages and currents for the N waveguide modes included, and $\bar{\bar{\mathbf{M}}}$ is a $2N \times 2N$ square matrix with elements as functions of z for a tapered waveguide system.

The solution of (4.87) for a gyrotron cavity provides not only the field profiles, the resonant frequencies, and the Q values for the various cavity modes, but also includes the effects of mode coupling due to the cavity wall tapers. If it were fully exploited, for example, analysis of a circular TE_{0n} cavity mode would also include the output power in other spurious TE_{0m} $(m < n)$ modes, which are generated by wall tapers in the cavity region (especially at the input and output ends). Although (4.87) has been used extensively for analysis of mode conversion in tapered transmission lines, cavity analysis to date appears to have been limited to single-mode analysis.

Because the system of (4.87) is linear, a transition matrix $\bar{\bar{\mathbf{T}}}_x$ can be found that links the variables on each side of the cavity and permits a solution in terms of arbitrary boundary conditions. We begin by recognizing that a general solution can always be written as a linear combination of $2N$ independent solutions. If we designate these independent solutions by $\mathbf{X}_i$, then a general solution is

$$\mathbf{X}(z) = \sum_{i=1}^{2N} C_i \mathbf{X}_i(z) = \bar{\bar{\mathbf{W}}}(z)\mathbf{C} \tag{4.88}$$

where $\bar{\bar{\mathbf{W}}}$ is the matrix whose columns are the independent solutions $\mathbf{X}_i(z)$ and $\mathbf{C}$ is the vector of constant coefficients. If the matrix $\bar{\bar{\mathbf{W}}}$ is known at both ends of the cavity or waveguide section, (4.88) can be used as follows to obtain the transition matrix $\bar{\bar{\mathbf{T}}}_x$. Let z_1 and z_2 be the left and right boundaries, respectively, which define the length of waveguide. Then,

$$\mathbf{X}(z_1) = \bar{\bar{\mathbf{W}}}(z_1)\mathbf{C}$$
$$\mathbf{X}(z_2) = \bar{\bar{\mathbf{W}}}(z_2)\mathbf{C}$$

and eliminating $\mathbf{C}$ gives

$$\mathbf{X}(z_2) = \bar{\bar{\mathbf{W}}}(z_2)\bar{\bar{\mathbf{W}}}^{-1}(z_1)\mathbf{X}(z_1) \equiv \bar{\bar{\mathbf{T}}}_x \mathbf{X}(z_1) \tag{4.89}$$

Equation (4.89) defines $\bar{\bar{\mathbf{T}}}_x$ in terms of the $\bar{\bar{\mathbf{W}}}$ matrices at each end of the waveguide section, and provides a matrix relationship between the waveguide variables at either end of the section.

To find the transition matrix it is therefore necessary to determine the matrix $\bar{\bar{\mathbf{W}}}$. For a difficult waveguide section where many waveguide modes are required to give an accurate description of the fields, a set of independent solutions for (4.87) are most easily obtained by direct numerical integration [71]. Once the matrix $\bar{\bar{\mathbf{M}}}(z)$ has been determined from

(4.81) (using Solymar's integral expressions for the coupling coefficients), a first choice for $2N$ independent initial condition vectors at the position z_1 might be the columns of the identity matrix, or an appropriate diagonal matrix. By using these initial conditions, (4.87) is then numerically integrated $2N$ times to find the corresponding solutions at the position z_2. Note from the form of (4.88) and (4.89) that these $2N$ pairs of vector $\mathbf{X}_i$ values completely define the transition matrix $\bar{\bar{\mathbf{T}}}_x$.

Once the transition matrix for a section of waveguide has been determined, there are a number of possible uses. Equation (4.89) is a linear set of $2N$ equations in $4N$ unknowns, which to be used directly would require that $2N$ of the boundary variables be known. For the mixed boundary value problem, where some of the mode variables are known at each end of the waveguide, (4.89) can be rearranged and partitioned to separate the knowns from the unknowns to obtain a solution. The classical example of this procedure is the scattering matrix formulation in which the variables at each port are written in terms of the normal modes entering and leaving the port. After converting the variables to normal mode form, the equations are then rearranged in the form:

$$\mathbf{b} = \bar{\bar{\mathbf{S}}}\,\mathbf{a} \tag{4.90}$$

where $\mathbf{a}$ is the vector of the incoming mode signals, $\mathbf{b}$ is the vector of outgoing signals, and $\mathbf{S}$ is the scattering matrix.

If the boundary conditions are so complicated as to involve linear relationships between variables (such as waveguide steps producing mode conversion at the boundaries), then (4.89) can be augmented by an additional set of $2N$ boundary condition equations of the form:

$$\bar{\bar{\alpha}}\,\mathbf{X}(z_1) + \bar{\bar{\beta}}\,\mathbf{X}(z_2) = \Sigma \tag{4.91}$$

where $\bar{\bar{\alpha}}$ and $\bar{\bar{\beta}}$ represent square $2N \times 2N$ matrices (usually very sparce and not invertible), and Σ represents driving RF sources for the waveguide system (i.e., fixed values of input or output power). The combined $4N$ set of (4.89) (4.91):

$$\begin{bmatrix} \bar{\bar{\mathbf{T}}} & -\bar{\bar{\mathbf{I}}} \\ \bar{\bar{\alpha}} & \beta \end{bmatrix} \begin{bmatrix} \mathbf{X}(z_1) \\ \mathbf{X}(z_2) \end{bmatrix} = \begin{bmatrix} \mathbf{0} \\ \Sigma \end{bmatrix} \tag{4.92}$$

can then be inverted to solve for the $4N$ variables $\mathbf{X}(z_1)$ and $\mathbf{X}(z_2)$.

Solution for Gyrotron Cavity Modes and Q Values

In the special case where Σ is zero (no incoming RF), (4.92) has a solution only when the determinant of the coefficients is zero. The frequencies at which this is accomplished provide a method for not only determining resonances within the system, but also the Q values. When power is being lost from the system (as in the case of open ended gyrotron cavities), the frequencies of interest are complex and the Q of the cavity can be determined from energy decay considerations as

$$Q = \frac{\mathrm{Re}(\omega)}{2\mathrm{Im}(\omega)} \tag{4.93}$$

A specific example of the above procedure for analysis of the single-mode gyrotron cavity shown in Figure 4.26 would appear as follows for the nth waveguide mode. Equation (4.87) becomes

$$\frac{\mathrm{d}}{\mathrm{d}z}\begin{bmatrix} V_n \\ I_n \end{bmatrix} = \begin{bmatrix} T_{nn} & -\mathrm{j}k_{zn}Z_n \\ -\mathrm{j}k_{zn}Y_n & -T_{nn}^* \end{bmatrix}\begin{bmatrix} V_n \\ I_n \end{bmatrix} \tag{4.94}$$

and the elements of the right-hand matrix are calculated as functions of z from a point about one e-folding distance beyond cut-off on the left to a position well outside the cavity on the right as shown in Figure 4.26.

The applied boundary conditions are $V_n = Z_n I_n$ on the left (which for cut-off modes produces solutions that decay exponentially to the left) and $V_n = Z_n I_n$ on the right (which for propagating modes produces outgoing power on the right). These boundary conditions in (4.92) give

$$\begin{bmatrix} T_{x11} & T_{x12} & -1 & 0 \\ T_{x21} & T_{x22} & 0 & -1 \\ 1 & -Z_n(z_1) & 0 & 0 \\ 0 & 0 & 1 & -Z_n(z_2) \end{bmatrix}\begin{bmatrix} V_n(z_1) \\ I_n(z_1) \\ V_n(z_2) \\ I_n(z_2) \end{bmatrix} = 0 \tag{4.95}$$

where the T_{xij} elements of the transition matrix come from integrating (4.94) through the cavity twice to obtain the two independent solutions required to construct the transition matrix as given in (4.89).

Note that in order to integrate (4.94) through the cut-off region, it is necessary to calculate $k_{zn}Z_n$ and $k_{zn}Y_n$ as products because this eliminates the singularities at $k_{zn} = 0$. Thus, for TE and TM modes,

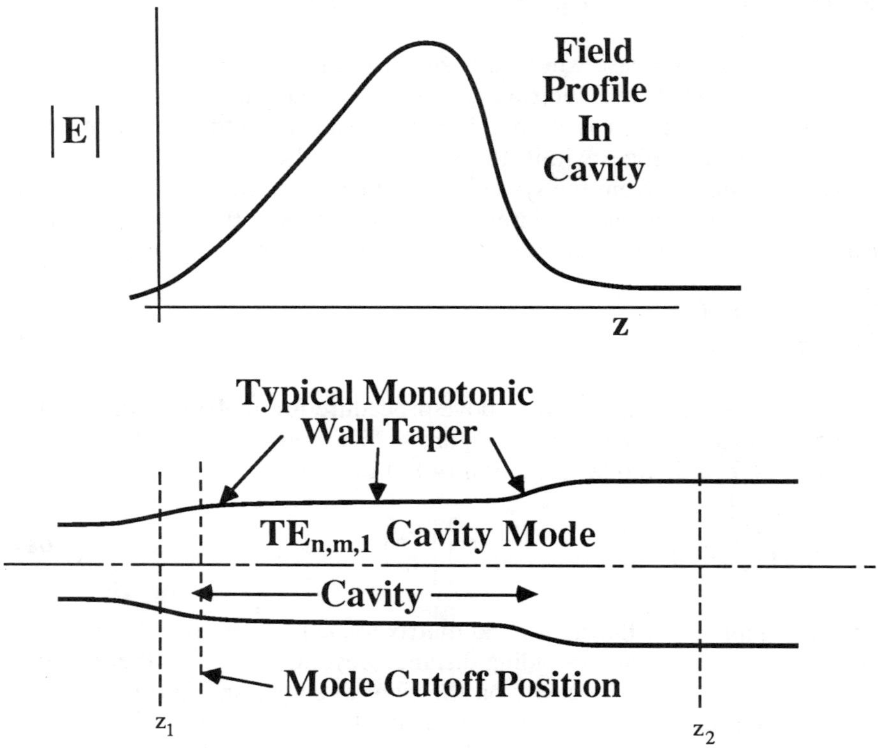

Fig. 4.26 Schematic Drawing of a Typical High-Power Gyrotron Cavity and its Field Amplitude Profile

$$k_{zn}Z_n = \sqrt{\frac{\mu}{\epsilon}}\left\{\begin{matrix} k \\ \left(\frac{k_{zn}^2}{k}\right)\end{matrix}\right\}; \quad k_{zn}Y_n = \sqrt{\frac{\epsilon}{\mu}}\left\{\begin{matrix} \left(\frac{k_{zn}^2}{k}\right) \\ k \end{matrix}\right\} \qquad (4.96)$$

where $k_{zn}^2 = k^2 - k_{cn}^2$. Although (4.95) cannot be evaluated precisely at cut-off because of the singularities in the elements containing Z_n, if the left and right boundaries do not fall directly on a cut-off point, this is not a problem. The values of the wave impedance Z_n in these equations can be correctly obtained both above and below cut-off by calculating k_{zn} using (4.82) and the complex square root function.

A complex search for frequencies which make the determinant of the coefficients of (4.95) go to zero then give the resonances of an arbitrarily shaped gyrotron cavity for the single-mode assumption.

Note that the self-coupling terms T_{nn} in (4.94) can be important for some cavity modes, even for weakly tapered cavities. Equations of the form (4.37) in Section 4.3.2, which are so prevalent in the literature, neglect these self-coupling terms. Care must be taken, therefore, in applying the simplified equation to tapered cavities [72].

As a final note, Figure 4.26 includes a typical gyrotron cavity field profile, which tends to improve the oscillator efficiency. The field peak in this case is moved closer to the output end of the cavity by a small taper angle along the cavity length. This type of field profile has been found to produce improved electron bunching followed by beam energy extraction over a short distance near the field peak. It therefore tends to separate the two steps of the CRM process (see Section 4.2.1). The rapid decrease in the field near the cavity output prevents the beam from reabsorbing the field energy.

4.4 DESIGN OF MAGNETRON INJECTION GUNS (MIGs) FOR GYROTRONS

Electron guns are designed by numerical simulation using computer programs known as *gun codes* [73]. These codes typically compute a self-consistent solution of Poisson's equation on a two-dimensional mesh spread over the gun's cross-sectional area. The boundary of the gun area is partially formed by electrodes at various potentials (Dirichlet boundaries) connected together by appropriate line segments with Neumann boundary conditions. The space charge of the beam is obtained by integrating a selected number of electron trajectories from the cathode through the acceleration region of the gun and calculating the resulting space charge at each mesh point along the trajectories. The solution process takes the form of iteratively solving for the potentials on the mesh, then determining the electron beam flow until a self-consistent solution is obtained. The solutions have been proved quite reliable in practice. Therefore, the design of any gun is always tested and adjusted to its final form using a gun code.

By using this technique, the design of magnetron injection guns for gyrotrons has proved to be more difficult than linear beam guns because of the requirement for a large transverse velocity compotent. The formation of a beam that meets the specifications for velocity ratio ($\alpha = v_\perp/v_z$), velocity spread ($\Delta v_z/v_z$), and the desired distribution of guiding centers, has proved to be very frustrating using computer simulation unless we start with a good initial design and have equations that tend to predict the result of design pertubations. This section provides a review of design equations that have proved to be useful for this purpose [74].

Based on the design of the RF interaction region of a gyrotron, there are five primary beam design parameters, which usually become specifications of the MIG design. These are (1) the beam energy (voltage, V_0), (2) the beam power (current, I_0), (3) the cyclotron frequency (magnetic field, B_0), (4) the average radius of the guiding centers, r_{g0}, and (5) the velocity ratio $\alpha_0 = v_{\perp 0}/v_{z0}$. The Larmor radius r_{L0} can be found from these parameters as indicated by the definitions of variables found in Table 4.1. Note that the radius, r_{L0}, is used to normalize all of the geometric variables in the design equations given in Table 4.2. Note also that the subscript 0 is used to indicate quantities pertaining to the interaction region.

Table 4.1 Definitions of Variables Used in Section 4.4

V_o	Beam voltage, volts
I_o	Beam current, amps
γ_o	$= 1 + eV_o/m_o c^2 \approx 1 + V_o(kV)/511$ = relativistic mass factor
v_o	$= c(1 - 1/\gamma_o^2)^{1/2}$ = electron velocity, m/s (c = velocity of light)
v_{zo}	$= v_o/(\alpha_o^2 + 1)^{1/2}$ = longitudinal velocity, m/s
α_o	$= v_{\perp o}/v_{zo}$ = transverse to longitudinal velocity ratio (used to get $v_{\perp o}$).
ω_c	$= 2\pi f_{co} = eB_o/m_o\gamma_o$ = relativistic cyclotron frequency, radians/s. (used to get B_o)
r_{go}	Guiding center radius at RF interaction region, m
r_{Lo}	$= v_{\perp o}/\omega_c$ = Larmor radius at RF interaction region, m
R_g	$= r_{go}/r_{Lo}$ = normalized mean guiding center radius
δR_g	$= \delta r_{go}/r_{Lo}$ = normalized full width guiding centre spread
μ	$= (R_g^2 - 1)^{-1/2}$ = cylindricity parameter
F_m	$= B_o/B_{zc}$ = magnetic field compression ratio
R_c	$= r_c/r_{Lo}$ = normalized mean cathode radius
ϕ_c	Cathode tilt angle (always positive)
L_s	$= l_s/r_{Lo}$ = normalized slant length of cathode
D_F	Cathode-anode spacing factor (select $D_F \geqslant 2$).
D_{ac}	$= d/r_{Lo}$ = normalized slant spacing between cathode and anode
Φ_a	$= eV_a/m_o c^2$ = normalized first anode voltage
E_c	Cathode electric field ($E_{max} \approx 10^7$ V/m is a conservative design value for vacuum)
J_c	Temperature limited cathode current density, A/m^2
J_L	Langmuir space charge limited current density, A/m^2

The design equations in Table 4.2 pertain to the design of a MIG of the form shown in Figure 4.27. The seven equations in the table give (in normalized form) the following design quantities in order of the equation numbers:

(1) Magnetic field compression ratio;
(2) Slant length of the cathode;
(3) Expected spread in the guiding center positions in the interaction region;
(4) Slant distance between the cathode and the first anode;
(5) Potential on the first anode;
(6) Electric field strength at the surface of the cathode;
(7) Ratio of the cathode current density to the estimated Langmuir limit.

Table 4.2 Design Trade-Off Equations for Adiabatic MIG Design (Definitions in Table 4.1)

$$F_m = \mu^2 R_c^2 \qquad (1)$$

$$\frac{L_s}{R_c} = \left(\frac{I_o}{2\pi r_{Lo}^2 J_c}\right)\left(\frac{1}{R_c^2}\right) \qquad (2)$$

$$\frac{\delta R_g}{R_g} = \left(\frac{\sin\phi_c}{\mu^2+1}\right)\left(\frac{I_o}{2\pi r_{Lo}^2 J_c}\right)\left(\frac{1}{R_c^2}\right) \qquad (3)$$

$$\frac{D_{ac}}{R_c} = \frac{D_F\,\mu}{\cos\phi_c} \qquad (4)$$

$$\Phi_a = \frac{\ln(1+D_F\,\mu)}{\ln(1+2\mu)}\left\{\left[1+\frac{4}{\mu^2}\left(\frac{1+\mu}{1+2\mu}\right)^2\left(\frac{\gamma_o^2-1}{R_c^2\cos^2\phi_c}\right)\left(\frac{\alpha_o^2}{\alpha_o^2+1}\right)\right]^{1/2}-1\right\} \qquad (5)$$

$$\frac{E_c}{E_{max}} = \left(\frac{m_o c^2/e}{E_{max}\, r_{Lo}}\right)\left(\frac{\Phi_a\cos\phi_c}{\ln(1+D_F\,\mu)}\right)\left(\frac{1}{R_c}\right) \qquad (6)$$

$$\frac{J_c}{J_L} = \left(\frac{2\pi r_{Lo}^2 J_c(1+D_F\,\mu)\beta^2}{14{\cdot}66(10^{-6})(m_o c^2/e)^{3/2}\cos^2\phi_c}\right)\left(\frac{R_c^2}{\Phi_a^{3/2}}\right) \qquad (7)$$

$$\beta = \exp(-\zeta/2)[\zeta + \tfrac{1}{10}\zeta^2 + \tfrac{5}{300}\zeta^3 + \tfrac{24}{9900}\zeta^4 + \ldots]$$

3 place accuracy up to $\zeta = 3.0$

$$\zeta = \ln(1+D_F\,\mu)$$

Each of these equations is written in terms of the five primary design specifications and four additional design variables, which we are free to manipulate to achieve an acceptable design. These variables are

r_c = the average cathode radius;
J_c = the cathode current density;
ϕ_c = the slope angle on the cathode;
D_F = the spacing factor between the cathode and the first anode of the MIG.

The equations in Table 4.2 have been reduced to a form that permits rapid design trade-offs with respect to these four parameters.

The design equations are based on adiabatic beam flow and negligible space charge in the beam. Neither assumption is always justified, and therefore cannot be ignored. One place that adiabaticity can be violated is in the region between the first and second anodes of the gun, where the electric field changes from primarily transverse in direction to longitudinal in direction. If this transistion region is short compared to the cyclotron wavelength, the adiabatic flow condition can be violated. Such a condition does not invalidate a MIG design (remember, the final validity of the design is tested by computer simulation), but it does invalidate the first-order design equations, which are then lost as an aid for adjusting the final design. In practice, it is found that a slowly varying transition region in the MIG helps facilitate convergence to a final design during the computer simulation phase.

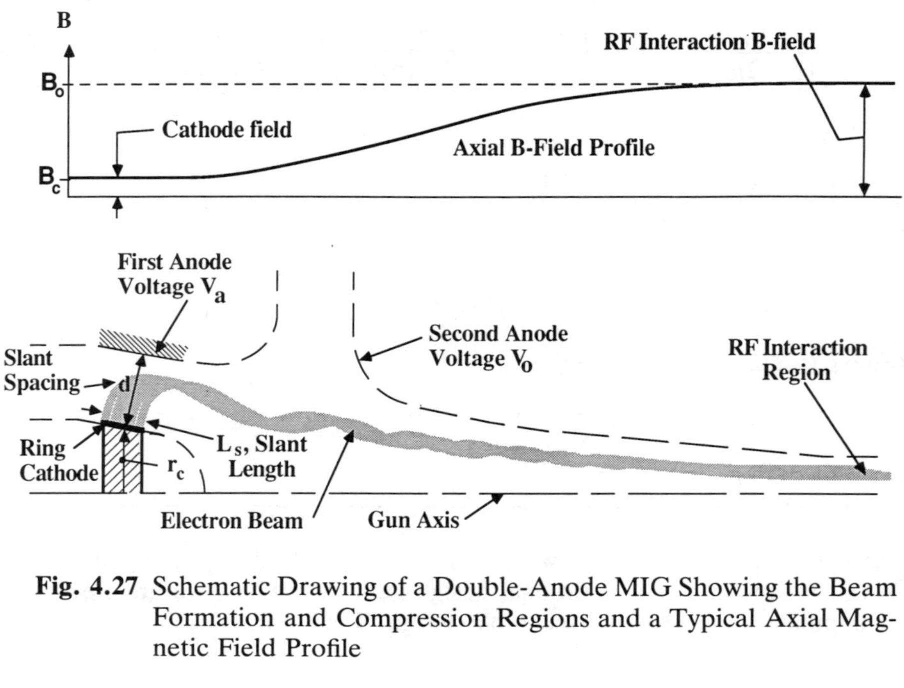

Fig. 4.27 Schematic Drawing of a Double-Anode MIG Showing the Beam Formation and Compression Regions and a Typical Axial Magnetic Field Profile

Another place in which the adiabatic beam flow assumption may not be valid is the magnetic compression region, where the transverse energy of the beam is increasing due to the magnetic mirror effect. If the magnetic field changes too rapidly over one cyclotron wavelength, the adiabatic condition can be violated. This usually only occurs when the magnetic compression ratio is too high, thus leading to long cyclotron wavelengths at the beginning of the magnetic compression region.

A third effect that can violate adiabatic beam flow in the magnetic compression region comes from the periodic space charge in the beam. At the position along the beam where the orbiting electrons are not fully mixed in phase, a periodic space-charge field is established, which can increase the velocity spread due to a pumping effect on the beam [75]. If the mesh density in the simulation is fine enough, this effect can also be simulated by the gun code.

The neglect of space charge effects in the design equations in Table 4.2, is justified by the fact that MIGs operate in the temperature limited regime well below the space-charge limit of current (Langmuir limit). This is because the strong electric field at the cathode is used in the MIG to create the transverse velocity component in the beam. When the ratio of the cathode current density to the Langmuir limit J_c/J_L reaches about 20–30%, the effects of space charge will begin to affect the design and must be taken into account. For nonrelativistic beams, there are methods for accomplishing initial design in the presence of space charge, but at greatly increased labor [76]. For relativistic beams, the design process in the presence of space charge is still an important area of research [77]. In any case, it is found in computer simulations that the effects of space charge (which reduce the cathode E-field and thus reduce α) can in most cases be effectively compensated by increasing the first anode voltage. Thus, the first-order design equations still provide a method for starting computer simulations, where the design can be optimized by parameter changes.

The cathode radius is the principal design parameter to be determined. It is usually best, therefore, to begin by making plots of $E_c/\mathrm{E}_{\max}$ and J_c/J_L as a function of the cathode radius for various selections of the other parameters. An example of such a plot is shown in Figure 4.28.

Note that $E_c/E_{\max}$ decreases as a function of cathode radius while J_c/J_L rises. This establishes a region for design over which the cathode field is safely below the desired maximum, $E_c < E_{\max}$, and the cathode current density is well below the Langmuir limit $J_c < 0.2J_L$. If an acceptable design region does not exist, then adjustments can be made by either increasing the cathode current density or the beam voltage. As noted above, a third alternative is to accept the cathode radius at the point,

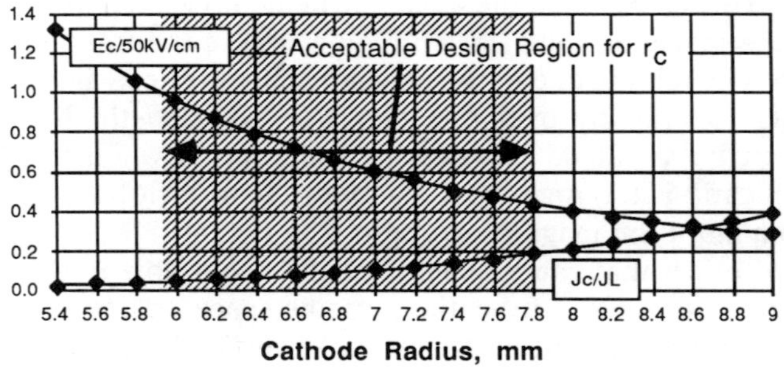

Fig. 4.28 Typical MIG Design Trade-off for a 300 GHz, 30 kV, 1 A Gun with $B_0 = 11.74T$, $\alpha_0 = 1.5$, $J_c = 2.5$ A/cm^2, $\theta_c = 20°$, and $D_F = 10$

Note: The targeted mode is $TE_{1,3}$ with the beam-guiding center located on the third radial peak.

where $E_c = E_{max}$, as long as $J_c < J_L$, proceed with the simulations, and try to make adjustments on the voltage of the first anode to compensate for the space-charge depression at the cathode.

The design equations in Table 4.2 have also been proved useful for gyrotron feasibility studies. An integral part of any question about how much power can be feasibly obtained from a given gyrotron configuration is that of how much beam power is obtainable within the imposed constraints. The equations in Table 4.2 can help answer such questions in a broad sense without resorting to specific MIG designs [78].

For further information about the MIG design process, Baird and Lawson [74] provide examples of MIG designs that were done by using the design rules in Table 4.2. The authors also provide details on the derivation of the design equations. It should be noted that, in addition to the assumptions of adiabatic beam flow and low space charge, the design equations come from applying conservation of angular momentum, and empirical rules for estimating the potential and electric field between conically shaped electrodes. Also discussed are the components of velocity spread attributable to cathode roughness and initial thermal velocities at the cathode. Each of these sources can contribute velocity spreads on the order of a few percent.

Other Beam-Forming Systems

Although the gyrotron MIG configuration illustrated in Figure 4.27 represents one of the major breakthroughs that made the gyrotron a successful device, it does not satisfy the requirements of all cyclotron interaction devices. Specifically, it cannot be used to generate a beam that is localized on the axis. Yet, there are numerous gyrotron configurations where this is required, such as a gyro-TWT in a dominant TE_{10} mode waveguide. Also, in some cases, it is even desirable to form a single beamlet with all of the electrons uniformly circling the axis, such as for the peniotron applications described in Section 4.3.2.

There are four techniques that have been tried with varying degrees of success to form an electron beamlet which encircles the axis. Like the MIG, each of these techniques can be used initially to form a beam with a small velocity ratio α, which is subsequently compressed adiabatically to increase α to the desired value in the interaction region. Unlike the MIG, the initial beamlets are formed encircling the axis.

The first technique is to simply inject a small diameter linear beam from a Pierce gun into a magnetic field at an angle. The transverse velocity in the beam then depends primarily on the angle of injection while the spreads in guiding center and velocity depend on the size of the beam. This technique was tried by Choe and Pantell [79] in the early days of verification of the CRM mechanism, and it has been used successfully in both harmonic gyrotron and peniotron experiments at the University of Utah [80]. By starting with a beam of 1 mm in diameter carrying 0.5 A at 30 kV, axial velocity spreads as low as 5% have been measured at α = 1.3 by direct beam probing. The guiding center spreads, however, were on the order of 50–70% of the Larmor radius.

A second method which has been successfully used to generate a single beamlet of rotating electrons is magnetic field reversal on a linear streaming thin hollow beam. When the radius of the initial beam is adjusted to be equal to the Larmor radius of the rotating beam after the field reversal, the result is a single rotating beamlet. The thickness of the initial beam annulus determines the velocity spread and the guiding center spread in the final beam. This approach was also used by Choe and Pantell, and in recent times it has been successfully used in peniotron experiments at Northrup [81] and in large orbit gyrotron experiments at the University of Maryland [82].

A third method which was used in early gyrotron experiments [7] and is currently being used to form the beam for a gyro-TWT [75] is the beam "wiggler," or helical pump. In this technique, a bifilar helix with dc current in opposite directions is used to produce a transverse component of B-field at the axis which rotates as a function of distance along with the helix [83]. When a thin linear electron beam is injected along the axis of the helical wiggler and the cyclotron period in the beam is adjusted to match the period of the wiggler, the result is a pumping action which creates a helical rotating beamlet. By tapering the interaction strength of the wiggler on each end, it is believed that the axial velocity spread in working devices is on the order of 5%.

The fourth method to be described is RF pumping of the cyclotron velocity in the beam. In this method, a low-frequency RF source is used to create a strong transverse electric field in a region through which a small linear flowing beam is passed. When the RF frequency is tuned to match the cyclotron frequency in the beam frame of reference, a transverse velocity component is introduced through the pumping action of the electric field. If the pumping action is strong, the beam can be effectively accelerated to high-voltage levels, in which case the RF pump becomes a major source of the beam power. This approach has been used successfully at UCLA to produce beams for large orbit gyrotron experiments [84].

In conclusion, we note that although gyrotron oscillators have been tremendously successful beyond the expectations of most researchers, the development of amplifiers continues to struggle with the problem of longitudinal velocity spread in the beam. The oscillators have a relatively short interaction region, which tends to make them insensitive to the velocity spread. The amplifiers however require an extended interaction region either for a traveling wave interaction (gyro-TWT), or for a multiple-cavity interaction (gyroklystron), and are therefore more sensitive to the velocity spread. As long as the requirement for high-power millimeter-wave amplifiers persists, there will be a need for a better beam forming technique. This better technique may be a refinement of one of those described here, or it may be some entirely different technique. This is an area where research should continue.

APPENDIX 4.A

Summary of Solymar's Coupling Coefficients [61]

Reiter's coupled mode transmission line equations [60] for a tapered waveguide are given by (4.81) in Section 4.3.4 of this chapter. They are

$$\frac{\mathrm{d}V_n}{\mathrm{d}z} = -\mathrm{j}k_{zn}Z_nI_n + \sum_k T_{kn}V_k$$

$$\frac{\mathrm{d}I_n}{\mathrm{d}z} \quad -\mathrm{j}k_{zn}Y_nV_n - \sum_k T^*_{nk}I_k$$

where n and k each represent a pair of mode indices. Solymar showed that the coupling coefficients T_{kn} in these equations can be evaluated at each z position by line integrals around the cross-sectional boundary of the waveguide by using the solutions Ψ_n and Ψ_k of the transverse wave equation.

For TE_k-to-TE_n mode coupling, when $k \neq n$ (and $k_{ck} \neq k_{cn}$), the result is

$$T_{kn} = \frac{k_{ck}^2}{k_{cn}^2 - k_{ck}^2} \int_l \tan\Theta_T \Psi_k \frac{\partial^2 \Psi_n^*}{\partial n^2} \, dl$$

and for $k = n$:

$$T_{nn} = -\frac{1}{2} \int_l \tan\Theta_T \left| \frac{\partial \Psi_n}{\partial l} \right|^2 dl$$

For TM_k-to-TM_n mode coupling, when $k \neq n$ (and $k_{ck} \neq k_{cn}$), the result is

$$T_{kn} = \frac{k_{cn}^2}{k_{ck}^2 - k_{cn}^2} \int_l \tan\Theta_T \frac{\partial \Psi_k}{\partial n} \frac{\partial \Psi_n^*}{\partial n} \, dl$$

and for $k = n$:

$$T_{nn} = -\frac{1}{2} \int_l \tan\Theta_T \left| \frac{\partial \Psi_n}{\partial n} \right|^2 dl$$

In these equations, Θ_T is the angle of the tapered waveguide wall with respect to the z-axis, and the derivatives are defined by

$$\frac{\partial}{\partial n} \equiv \hat{\mathbf{n}} \cdot \nabla \text{ and } \frac{\partial}{\partial l} \equiv \hat{\mathbf{l}} \cdot \nabla$$

The unit vectors at the wall of the waveguide form a right-handed coordinate system such that $\hat{\mathbf{n}} \times \hat{\mathbf{l}} \cdot \hat{\mathbf{z}} = 1$ with the unit vector $\hat{\mathbf{l}}$ tangential to the waveguide wall at all points along the cross-sectional boundary line.

For coupling between different mode types, the result for TE_k to TM_n is

$$T_{kn} = -\int_l \tan\Theta_T \frac{\partial \Psi_k}{\partial l} \frac{\partial \Psi_n^*}{\partial n} \, dl$$

and for TM_k to TE_n:

$$T_{kn} = 0$$

for all k and n.

In each case above, the curent coupling coefficients T^*_{nk} can be found by interchanging the mode subscripts and taking the complex conjugate.

REFERENCES

1. V.A. Flyagin, A.V. Gaponov, M.I. Petelin, and V.K. Yulpatov, *IEEE Trans.* **MTT-25**, 514–521 (1977).
2. J.L. Hirshfield and V.L. Granatstein, *IEEE Trans.* **MTT-25**, 522–527 (1977).
3. R.O. Twiss, *Australian J. Phys.* **11**, 564 (1958).
4. J. Schneider, *Phys. Rev. Lett.* **2**, 504 (1959).
5. G. Mourier, *AEU Electron. Commun.* **34**, 473–481 (1980).
6. A.V. Gaponov, *Izv. VUZ Radiofizika* **2**, 837 (1959).
7. J.L. Hirshfield and J.M. Wachtel, *Phys. Rev. Lett.* **12**, 533–536 (1964).
8. A.V. Gaponov, M.I. Petelin, and V.K. Yulpatov, *Radiophys. Quant. Electron.* **10**, 794–813 (1967).
9. N.I. Zaytsev, *et al., Radio Eng. Electron. Phys.* **19**, 103–106 (1974).
10. J.M. Baird, *Int. Electron. Dev. Meet.,* Washington, DC (1979), pp. 156–163.
11. E. Ott and W.M. Manheimer, *IEEE Trans. Plasma Sc.* **PS-3**, 1–5 (1975).
12. P. Sprangle and A.T. Drobot, *IEEE Trans.* **MTT-25**, 528–544 (1977).
13. K.R. Chu, *et al., IEEE Trans.* **MTT-27**, 178–187 (1979).
14. K.R. Chu, A.T. Drobot, H.H. Szu, and P. Sprangle, *IEEE Trans.* **MTT-28**, 313–317 (1980).
15. K.R. Chu, Y.Y. Lau, L.R. Barnett, and V.L. Granatstein, *IEEE Trans.* **ED-28**, 866–871 (1981).
16. Y.Y. Lau and L.R. Barnett, *Inte. J. Infrar. and Mill. Waves* **3**, 619–744 (1982).
17. D.C. Forster, *Proc. IEEE* **54**, 532–539 (1966).
18. K.R. Chu and J.L. Hirshfield, *Phys. Fluids* **21**, 461–466 (1978).
19. K.R. Chu, A.K. Ganguly, V.L. Granatstein, J.L. Hirshfield, S.Y. Park and J.M. Baird, *Int. J. Electron.* **51**, 493–502 (1981).
20. U.A. Shrivastava and R.W. Grow, *Int. J. Electron.* **57**, 1077–1095 (1984).
21. G. Dohler, *Int. J. Electron.* **56**, 617–627 (1984).

22. P. Vitello, *Int. J. Infrar. and Mill. Waves* (1987).
23. H. Guo, L. Chen, H. Keren, J.L. Hirshfield, *Phys. Rev. Lett.* **49**, 730–733 (1982).
24. Y.Y. Lau, *IEEE Trans.* **ED-31**, 329–337 (1984).
25. H. Uhm and R.E. Davidson, *JAP* **49**, 593–598 (1978).
26. Y.Y. Lau, J.M. Baird, L.R. Barnett, K.R. Chu, and V.L. Granatstein, *Int. J. Electron.*, **51**, 331–340 (1981).
27. P.S. Rha, L.R. Barnett, J.M. Baird, and R.W. Grow, *IEDM Tech. Digest,* 535–538 (1985).
28. V.L. Bratman, N.S. Ginzburg, G.S. Nusinovich, M.I. Petelin, and P.S. Strelkov, *Int. J. Electron.* **51**, 541–567 (1981).
29. A.W. Fliflet, *Int. J. Electron.* **61**, 1049–1080 (1986).
30. S. Ono, K. Yamanouchi, Y. Shibata, Y. Koike, *Proc. 4th Int. Cong. Microwave Tubes,* (1962), 355–363.
31. G. Dohler, *Int. J. Electron.* **56**, 617–640 (1984).
32. P. Vitello, *IEEE Trans.* **MTT-32,** 917–921 (1984).
33. L. Zhou, M. Jiang, H. Guo, and Y. He, *Int. J. Electron.,* **57**, 1065–1075 (1984).
34. M. Caplin, A. T. Lin, and K.R. Chu, *Int. J. Electron.* **53**, 659 (1982).
35. C.K. Birdsall and A.B. Langdon, *Plasma Physics via Computer Simulation* (McGraw-Hill, New York, 1985).
36. N.A. Krall and A.W. Trivlepiece, *Principles of Plasma Physics* (McGraw-Hill, New York, 1973).
37. D.E. Baldwin, I.B. Bernstein, and M.P.H. Weenink, in *Advances in Plasma Physics,* Vol. 3, A. Simon and W.B. Thompson, eds., (John Wiley and Sons, New York, 1969), 1–50.
38. F.F. Chen, *Introduction to Plasma Physics and Controlled Fusion,* Vol. 1 (Plenum Press, New York, 1984), 128–131.
39. K.R. Chu, *Phys. Fluids* **21**, 2354–2364 (1978).
40. S.V. Kolosov and A.A. Kurayev, *Radio Eng. Electron. Phys.* **19** (10) 65–73 (1974).
41. R.W. Grow and U.A. Shrivastava, *IEDM Tech. Digest,* 384–387 (1982).
42. K.R. Chu and D. Dialetis, in *Infrared and Millimeter Waves,* Chapter 3, Vol. 13, (Academic Press, New York, 1985).
43. W. Lawson and C.D. Striffler, *Phys. Fluids* **28**, 2868–2877 (1985).
44. K.R. Chu, D.S. Furuno, N.C. Luhmann, D.B. McDermott, P. Vitello, and K. Ko, *IEEE Trans.* **PS-13**, 435–443 (1985).
45. D.B. McDermott, N.C. Luhmann, Jr., D.S. Furano, and A. Kupiszenski, *Int. J. Infrar. and Mill. Waves* **4**, 639–665 (1983).
46. Y.Y. Lau, K.R. Chu, L. Barnett, and V.L. Granatstein, "Gyrotron Traveling Wave Amplifier: II. Effects of Velocity Spread and Wall

Resistivity," *Int. J. Infrar. and Mill. Waves* **2**, 395–413 (1981).
47. Y.Y. Lau, *IEEE Trans.* **ED-29**, 320–335 (1982).
48. U.A. Shrivastava and R.W. Grow, *Int. J. Electron.,* **57**, 1077–1095 (1984).
49. C.C. Johnson, *Field and Wave Electrodynamics* (McGraw-Hill, New York, 1965).
50. A.K. Ganguly and S. Ahn, "Self-Consistent Large-Signal Theory of the Gyrotron Traveling Wave Amplifier," *Int. J. Electronics* **53**, 641–658 (1982).
51. A.W. Fliflet, M.E. Read, K.R. Chu, and R. Seeley, "A Self-Consistent Field Theory for Gyrotron Oscillators," *Int. J. Electronics* **53**, 505–521 (1982).
52. P.S. Rha, "Self-Consistent Large-Signal Theory and Simulations of High Harmonic Gyrotron and Peniotron Oscillators Operating in a Magnetron-type Open Cavity," PhD Dissertation, Department of Electrical Engineering, University of Utah, Salt Lake City, UT (1986).
53. R.M. Jones and P.A. Lindsay, *Int. J. Electron.,* **61**, 937–951 (1986).
54. J.C. Slater, *Microwave Electronics* (D. Van Nostrand, New York, 1950).
55. K.E. Kreisher and R.J. Tempkin, *Int. J. Infrar. and Mill. Waves* **1**, 195–223 (1980).
56. K.E. Kreischer, B.G. Danly, J.B. Schutkeker, and R.J. Tempkin, *IEEE Trans.* **PS-13**, 364–373 (1985).
57. I.B. Bernstein, L.K. Divringi, and T.M. Smith, *Int. J. Infrar. and Mill. Waves,* **4**, 57–117 (1983).
58. L. Lewin, *Theory of Waveguides* (Newnes-Butterworths, London, 1975).
59. F. Sprleder and H-G. Unger, *IEE Electromagnetic Waves Series* **6** (1979).
60. G. Reiter, Convention on Long Distance Transmission by Waveguide, London, Jan. (1959) 54–57.
61. L. Solymar, *IRE Trans.* **MTT-7**, 379-383 (1959).
62. J.S. Levine, *Int. J. Infrar. and Mill. Waves* **5** (7), 937–952 (1984).
63. M. Thumm, *Int. J. Electron.* **57**, 1225–1246 (1984).
64. C. Moeller, *Int. J. Electron.* **53**, 587–593 (1982).
65. A.W. Fliflet, L.R. Barnett, and J.M. Baird, *IEEE Trans.* **MTT-28**, 1482–1486 (1980).
66. J.M. Baird, "Study of Input Couplers for Tapered Interaction Gyrotron Amplifiers, Including a New Generalized Theory for Tapered,

Nonideal-Wall Waveguides," Final Technical Report for Naval Research Laboratory Contract No. N00014-82-K-2053, Department of Electrical Engineering, University of Utah, Salt Lake City, UT (1985).

67. C.A. Wheeler, "Study of a High-power Elliptical Waveguide Output Coupler to Provide Beam-RF separation in Gyromonotrons," AFTER Technical Report No.3, Contract No. F30602-80-C-0215, Department of Electrical Engineering, University of Utah, Salt Lake City, UT (1985).
68. B.G. Braatz, "Computer Analysis of Arbitrarily Tapered Rectangular and Double Ridged Waveguides," AFTER Technical Report No. 21, Contract No. F30602-82-C-0161, Department of Electrical Engineering, University of Utah, Salt Lake City, UT (1986).
69. Roger F. Harrington, *Time-Harmonic Electromagnetic Fields* (McGraw-Hill, New York, 1961), 152.
70. N. Marcuvitz, *Waveguide Handbook,* MIT Radiation Laboratory Series, Vol. 10 (McGraw-Hill, New York, 1951).
71. W.R. Cowell, ed., Chapter 6, *Sources and Development of Mathematical Software,* (Prentice-Hall, Englewood Cliffts, NJ, 1984).
72. A.W. Fliflet and M.E. Read, *Int. J. Electron.,* **51**, 475–484 (1981).
73. W.B. Herrmannsfeldt, "Electron Trajectory Program," *SLAC Report No. 226,* Stanford University, Stanford, CA (1979).
74. J.M. Baird and W. Lawson, *Int. J. Electron.* **61**, 953–967 (1986).
75. J. Neilson, M. Caplan, N. Lopez, and K. Felch, *IEDM Tech. Digest,* 184–186 (1985).
76. V.N. Manuilov and Sh. E. Tsimring, *Radio Eng. Electron. Phys.* **24**, 111–119 (1979).
77. J.M. Finn, A.W. Fliflet, and W.M. Manheimer, *Int J. Electron.* **61**, 985–1003 (1986).
78. K.J. Kim, M.E. Reed, J.M. Baird, K.R. Chu, A. Brobot, J.L. Vomvoridis, A. Ganguly, D. Dialetis, and V.L. Granatstein, *Int. J. Electron.* **51**, 427–445 (1981).
79. K.K. Chow and R.H. Pantell, *Proc. IEEE* **48**, 1865–1867 (1960).
80. L.R. Barnett, J.M. Baird, U.A. Shrivastava, R.W. Grow, *IEDM Tech. Digest,* 842–844 (1984).
81. G. Dohler, D. Gallagher, C. Lowrie, R. Moats, and F. Scafuri, *IEDM Tech. Digest,* 845–848 (1984).
82. W. Lawson, W.W. Destler, and C.D. Striffler, *IEEE Trans.* **PS-13**, 444–453 (1985).
83. S.Y. Park, J.M. Baird, R.L. Smith, and J.L. Hirshfield, *J. Appl.*

Phys. **53**, 1320–1325 (1982).
84. D.S. Furuno, D.B. McDermott, and N.C. Luhmann, Jr., *IEDM Tech. Digest,* 346–348 (1986).

Chapter 5
GYROTRON EXPERIMENTAL STUDIES

V. L. Granatstein

Electrical Engineering Department
and
Laboratory for Plasma and Fusion Energy Studies
University of Maryland
College Park, Maryland

5.1 INTRODUCTION*

The status of gyrotron experiments was summarized in a previous survey chapter published in 1982 [1]. In that book, stress was primarily placed on single-cavity gyrotron oscillators (gyromonotrons) and on gyrotron traveling wave tube (gyro-TWT) amplifiers. Gyrotron oscillator development made significant advances between 1982 and 1987, and these will be discussed in the present chapter.

The gyro-TWTs were evaluated for applications involving wideband, millimeter-wave sources such as communication, radar, and electronic warfare. While this device, which uses only a simple, hollow waveguide for a circuit, holds considerable promise, relatively little work has been done recently to advance its capabilities. Instead, gyrotron amplifiers studies have shifted in large measure to examining the gyroklystron configuration. The gyroklystron does not have the bandwidth capabilities of a gyro-TWT, but it should be capable of higher gain and efficiency.

Current gyrotron research and development is driven by the requirements of microwave sources beyond current capabilities for both *controlled thermonuclear fusion research* (CTFR) [2] and *high energy physics* (HEP) [3]. In projected CTFR studies, plasma heating will require microwave sources with average power $\geqslant$ 1 MW at a wavelength $\lambda \approx$ 1 mm. In

*This chapter is based on work supported in part by the US Department of Energy.

projected HEP studies, high-gradient accelerators for use in electron-positron colliders will require phase-controlled microwave sources with peak power $\gtrsim$ 100 MW, and pulse duration $\approx$ 1 μs at 1 cm $\lesssim \lambda \lesssim$ 5 cm. This chapter reviews recent progress toward meeting these requirements. We describe advances in both gyrotron average-power capabilities and gyrotron peak-power capabilities. We review progress in phase-locking gyrotron oscillators and reducing phase jitter in gyrotron amplifiers. Finally, a design of a 10 GHz, 40 MW, gyroklystron amplifier for the collider application will be described and compared with the design parameters of a conventional klystron with similar projected capabilities.

5.2 BRIEF REVIEW OF FUNDAMENTAL GYROTRON CONCEPTS

The gyrotron is a microwave power tube that emits coherent radiation at the electron cyclotron frequency or its harmonics. Gyrotrons consist of electrons which follow helical paths around the lines of an externally applied, axial magnetic field B_0, in the presence of an electromagnetic wave with a transverse component of electric field. In order for there to be a net flow of energy from the transverse electron motion into the electromagnetic wave, the electrons must become bunched in phase within their cyclotron orbits. Such bunching can occur due to the fact that the electron cyclotron frequency ω_c is a function of electron energy such that

$$\omega_c = \frac{eB_0}{m\gamma} \tag{5.1}$$

where e and m are the electron charge and rest mass, respectively, and $\gamma = \epsilon/mc^2$, where ϵ is the total electron energy.

The phase bunching process can most easily be understood in the reference frame, where the axial electron velocity vanishes. In Figure 5.1, a large annular electron beam with radius r_b is depicted in this frame, and electrons arranged around the annulus execute circular orbits with radius r_L (the Larmor radius). Note that typically $r_L \ll r_b$. Initially, the phase of the electrons in their cyclotron orbits is random as shown. In the presence of a transverse RF electric field characteristic of a microwave-cavity TE mode, electrons either accelerate or decelerate. As a specific example, Figure 5.1 depicts an electric field with only an azimuthal component as is characteristic of TE_{0n} modes in circular waveguide. With random phasing of the electrons, there is no net energy exchange. Electron 1 will be decelerated by the azimuthal electric field and thus lose energy, while electron 2 will be accelerated and hence gain an equal amount of energy.

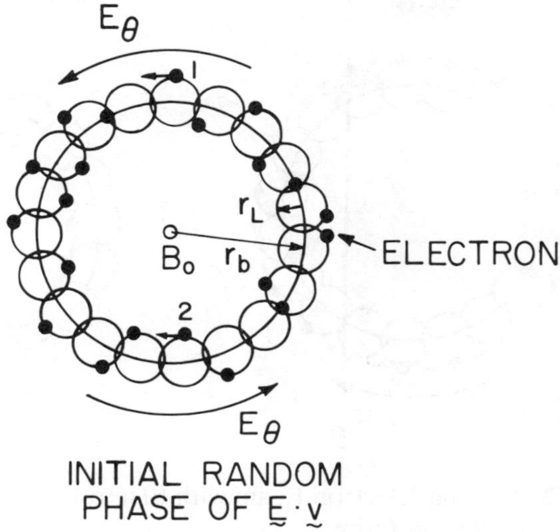

Fig. 5.1 Annular Electron Beam with Initial Random Phasing of Electrons in their Cyclotron Orbits

However, phase bunching and net energy transfer can occur if the wave frequency is somewhat larger than the initial value of the electron cyclotron frequency; that is,

$$\omega' - \frac{eB_0}{m\gamma_0'} = (\Delta\omega)' > 0 \tag{5.2}$$

where the subscript 0 denotes the initial value and the prime denotes the reference frame wherein electron axial velocity vanishes. Then, as electron 2 gains energy, its cyclotron frequency diminishes (see (5.1)), and this electron falls farther from resonance, gaining less energy on each successive cycle. On the other hand, electron 1, which initially loses energy, experiences a rising value of ω_c' and moves closer to exact resonance with the electric field, losing an increasing amount of energy on each successive cycle. Thus, an instability develops in which the wave energy grows exponentially in time and the electrons bunch in phase within their cyclotron orbit as shown in Figure 5.2. We can clearly see from Figure 5.2 that once the electrons are bunched, they may continually give up energy to an electric field which reverses its direction each half cycle of the cyclotron frequency in synchronism with the Larmor rotation of the electrons. Thus,

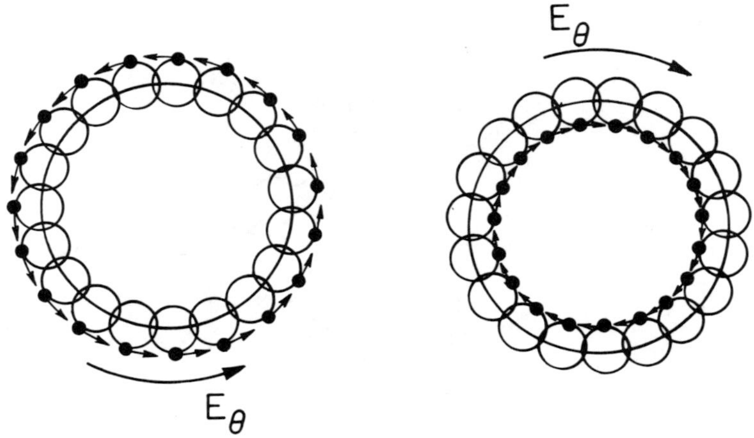

Fig. 5.2 Annular Electron Beam with Electrons Bunched in Phase in their Cyclotron Orbits

Note: Reversal of azimuthal component of electron motion on successive half-cycles shows synchronism with E_θ of electromagnetic wave at the cyclotron frequency.

an electromagnetic wave at the electron cyclotron frequency can be amplified in this process.

In a microwave cavity, the operating frequency is jointly determined by the resonant frequency of the cavity mode and the electron cyclotron frequency. Thus, the gyrotron can operate with improved stability in a higher order mode of the cavity structure as compared with other microwave tubes. Therefore, for a given operating frequency, gyrotron cavities are relatively large, and they can handle both large average power and large pulsed power.

The electric field patterns for three major mode classes that have been used in gyrotrons are sketched in Figure 5.3. The TE_{0n} modes with purely azimuthal electric field are characterized by very low wall losses, and therefore have been employed in the highest power CW gyrotrons. However, gyrotrons designed to operate in simple TE_{0n} cavities have suffered from mode competition especially with TE_{2n} modes when $n \geqslant 3$ [4, 5].

The TE_{1n} modes depicted at the bottom of Figure 5.3 allow us to split the cavity walls as shown. This has minimal perturbing effect on the TE_{1n} modes, but effectively eliminates all other mode classes and results in superior mode stability. Thus, the TE_{1n} modes may be the best choice for extremely high pulsed power operation, where high pulsed currents

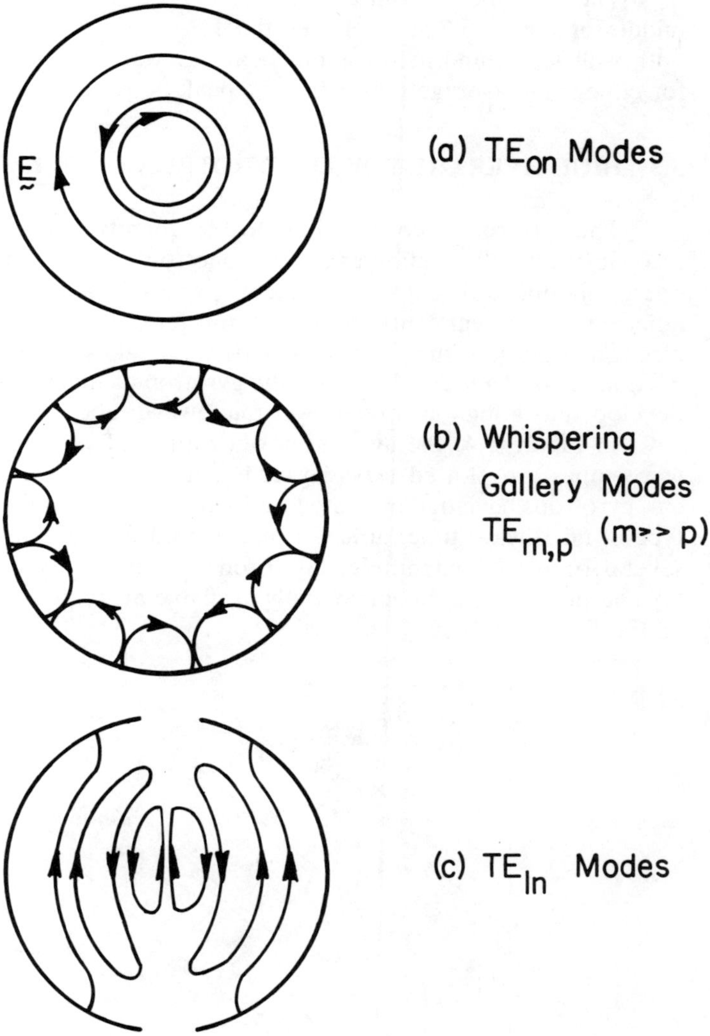

Fig. 5.3 Common Classes of Cavity Modes Used in Gyrotrons

would tend to excite many competing modes. Wall losses for the TE_{1n} modes are considerably larger than for TE_{0n} modes, but this is unimportant in pulsed operation so long as the average power requirement is modest.

The "whispering gallery" modes (TE_{mp}, $m \gg p$), depicted in the middle of Figure 5.3, lie between the TE_{0n} and TE_{1n} modes in terms of both wall losses and mode stability. Such modes are possible candidates for either high average-power or high peak-power operation.

5.3 HIGH AVERAGE-POWER GYROTRON OSCILLATORS

The gyrotron mechanism has most frequently been used with a single resonant cavity to function as a gyromonotron oscillator. In that case, the energy modulation and phase bunching occur in a portion of the cavity adjacent to the beam entry port, and transfer of energy from the bunched electron beam into the electromagnetic wave takes place in a subsequent portion of the same cavity. This is the gyrotron configuration that has been developed as a high average-power microwave tube.

In Figure 5.4, the average power ratings of various types of microwave tubes are plotted *versus* wavelength. It is apparent that for $\lambda \leq 3$ cm, gyrotrons have demonstrated far higher average power than any other type of microwave tube; at $\lambda \approx 1$ mm, the advantage in average power is several orders of magnitude. Gyrotron oscillators producing 200 kW of CW output power are commercially available at frequencies ranging from 28 GHz to 60 GHz [6].

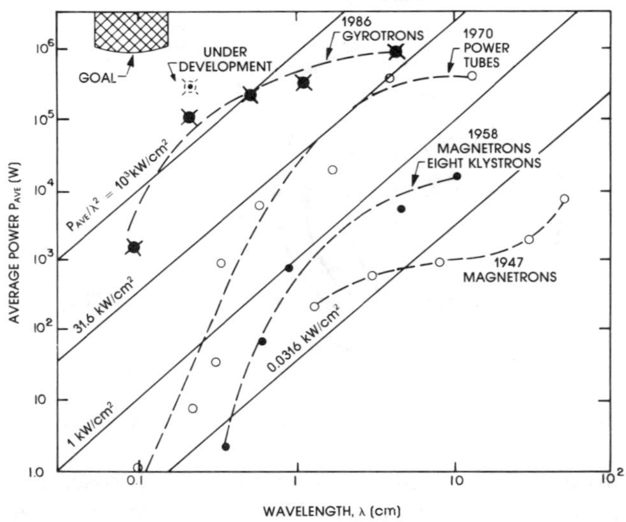

Fig. 5.4 Average Power Ratings for Various Types of Microwave Tubes Plotted *versus* Wavelength
(Based in part on data provided by R. K. Parker.)

Nevertheless, the controlled fusion research program is demanding still further advances in microwave average power capabilities. The shaded area in the upper left-hand corner of Figure 5.4 indicates the microwave source requirement for *electron cyclotron resonance heating* (ECRH) in the *compact ignition torus* (CIT), a next generation magnetic fusion experimental device planned for the 1990s. It is estimated [7] that 20 MW of ECRH power in the frequency range of 240 to 300 GHz will be required for several seconds. Of course, this power could be supplied by multiple sources, but even then, for reasons of economy, power level per tube $\geqslant$ 1 MW is desirable. Even greater challenge is presented by the recent concept of conducting ECRH at the second harmonic of the electron cyclotron frequency in the CIT, in which case the required frequency range could double (i.e., the frequency range of interest would be 480 to 600 GHz) [8].

As impressive as the CW performance of the gyrotron oscillators has been, it is still far from meeting these projected requirements. The operating gyrotron oscillator which comes closest has been developed by Varian Associates at a frequency of 140 GHz [9]. Operating with electron gun voltage of 80 kV, this gyrotron has achieved 100 kW of CW output power (with 27% efficiency), 150 kW of output power for 100 ms, and 200 kW of outpout power for 1 ms. The main resonant cavity operates in the TE_{031} mode, and is similar to that examined at MIT with 1 μs pulses [10]. Discrimination against spurious oscillation in the TE_{231} mode is achieved by prebunching the phases of the gyrating electrons in a TE_{021} cavity, which is strongly coupled to the main TE_{031} cavity. This coupled-cavity technique for achieving superior mode stability in TE_{0n1} modes was pioneered at NRL [11]. The configuration of a simple gyromonotron and a coupled cavity gyrotron oscillator are compared in Figure 5.5.

Another technique for achieving mode stability in very large gyrotron cavities is to employ whispering gallery modes ($TE_{m,n}$, $m \gg n$). These modes have their field energy concentrated close to the cavity wall, and they can be preferentially excited by a hollow electron beam of large radius. A $TE_{15,2,1}$ gyrotron is currently under development at Varian Associates, again using an 80 kV electron gun; design goals are 400 kW of output power at 140 GHz [12]. The design is similar to that used in a 1 μs pulsed experiment at MIT where output power up to 645 kW has been demonstrated with efficiency $\approx$ 20% [13]. Further upward scaling of gyrotron oscillators in average power and in frequency may require operation at harmonics of the cyclotron frequency or advanced techniques for cooling of the cavity walls.

In addition to continuing efforts to extend the capabilities of whispering gallery mode gyrotrons, other cyclotron maser oscillator configurations are also being given consideration. There is ongoing work on

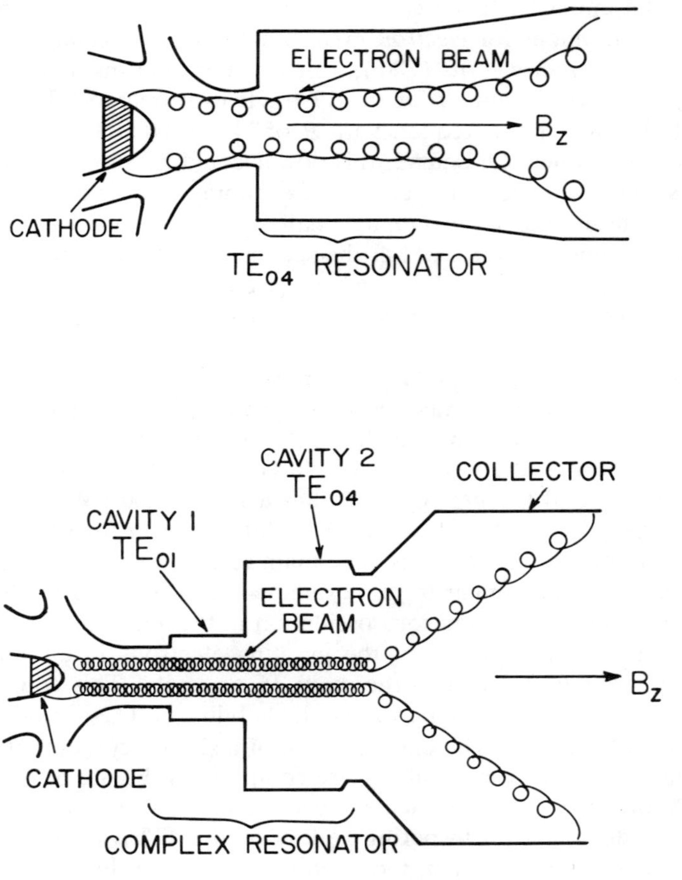

Fig. 5.5 Comparison of Simple Gyromonotron with Coupled-Cavity Gyrotron Oscillator [11]

quasioptical gyrotrons employing Fabry-Perot cavities (e.g., [14, 15]), and *cyclotron autoresonance masers* (CARMs), which exploit a large Doppler upshift to operate at frequencies far above ω_c (e.g., [16–18]). Free-electron lasers are also being considered as candidate sources for ECRH in the CIT (e.g., [19, 20]).

5.4 HIGH PEAK-POWER GYROTRON OSCILLATORS DRIVEN BY PULSE-LINE ACCELERATORS

Linear accelerators require high peak power microwave sources with relatively short pulse duration, rather than high average power, as required in the plasma heating application. The suitability of the gyrotron mechanism for generating high peak power has been initially studied in *pulse-line accelerator* experiments. In an early experiment, 1 GW was generated at 8 GHz in a super-radiant oscillator experiment that was relatively inefficient [21].

Pulse-line gyrotron experiments in the USSR have studied the gyrotron mechanism in a TE_{1n} oscillator cavity configuration. The parameters and results of some of these experiments are listed in Table 5.1

Whispering gallery mode gyrotron oscillators with high peak-power output have been recently studied at NRL [24]. The pulsed power for this study was provided by a "single-shot" 775 kV Febetron with a 55 ns pulse duration, where the electron beam was drawn from a cold cathode by plasma-induced field emission. The electrons are subsequently given a substantial increase in their transverse momentum by passing through a nonadiabatic dip in magnetic field produced by a *pump* magnet coil. Transverse momentum is then further increased between the pump magnet and the gyrotron cavity by adiabatic compression. The experimental configuration is sketched in Figure 5.6

At 35 GHz, approximately 100 MW of output power was produced in the TE_{621} mode with magnetic field in the cavity set at 24.4 kG. This power level corresponds to an efficiency of $\approx 8\%$ based on a current of 1.6 kA flowing through the gyrotron cavity. By tuning the cavity magnetic field, it proved possible to achieve a stepwise frequency tuning ability with the gyrotron operating at high power in a family of TE_{m21} modes. This tuning ability is shown in Figure 5.7, where the observed operating frequency in modes with azimuthal eigenvalue varying from 4 to 10, is plotted *versus* the axial magnetic field applied to the cavity. Stepwise tuning ability was achieved over a frequency range of 28 to 49 GHz. The reader may note that stepwise tuning ability is of interest in the plasma heating application.

For application to RF accelerators, tuning ability is of limited interest, but phase control is essential. Hundreds or even thousands of microwave sources will be required to drive a large accelerator, and these sources must produce radiation with the proper phase to accelerate the electron

Table 5.1 Gyromonotron Oscillator Experiments Employing Pulse-Line Accelerators and Field Emission Cathodes (Single Pulse)

Frequency (GHz)	*Power* (MW)	*Pulse Length* (ns)	*Efficiency*	*Electron Mode*	*Energy* (keV)	*Reference*
10	60	30	15%	TE_{13}	350	[22]
40	40	20	10%	TE_{13}	350	[23]

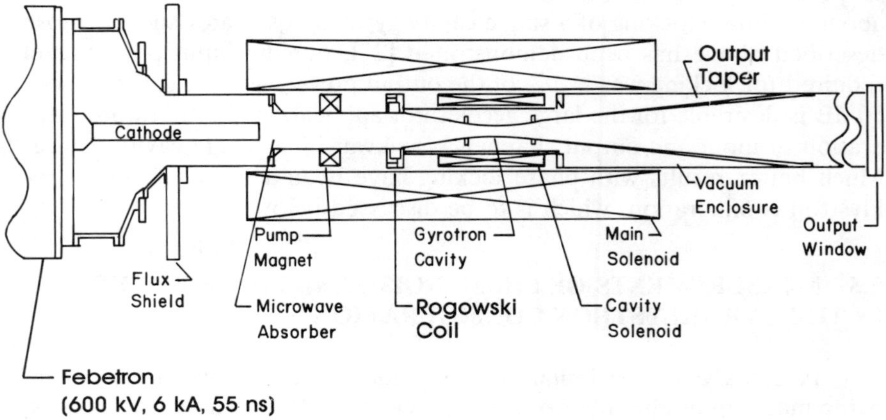

Fig. 5.6 Experimental Configuration for 100 MW K_a-Band Gyrotron Oscillator Study [24]

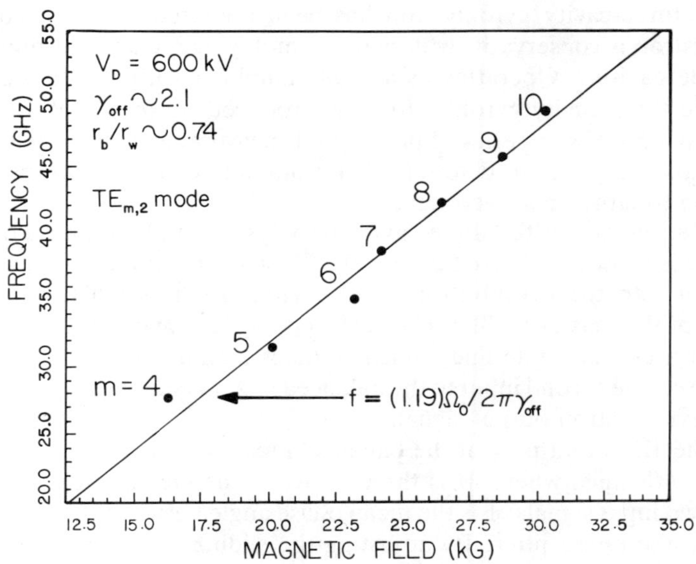

Fig. 5.7 Stepwise Frequency Tunability with the Gyrotron Operating at High Power in Family of TE_{m2} Modes

Note: This tuning ability is shown where the observed operating frequency in modes with azimuthal eigennumber varying from four to ten is plotted *versus* the axial magnetic field applied to the cavity [24].

or positron bunch as it passes. The required phase control accuracy is about 1°. Phase locking of a single cavity gyrotron oscillator such as those described above has been demonstrated [25], but the input power level required for locking was ≥ 1% of the output power, while a *gain* of about 50 dB is desirable for the large accelerator application. Furthermore, separation of input and output channels is awkward in a single-cavity device. Much better results with phase locking have been achieved in the gyroklystron configuration, which is to be discussed below.

5.5 MEASUREMENTS OF PHASE NOISE AND PHASE LOCKING IN THE GYROKLYSTRON CONFIGURATION

A gyroklystron is similar in construction to a conventional klystron, being made up of a number of resonant cavities separated by drift spaces, which are free of electromagnetic waves. However, in a gyroklystron, bunching takes place in the phases of the electrons within their cyclotron orbits, rather than in the axial position of the electrons, and hence no small-gap re-entrant cavities are required.

A three-cavity gyroklystron has been operated at NRL [26]. This gyroklystron is conservatively designed, employing rectangular fundamental-mode cavities. Operating as a stable amplifier with a 30 kV electron gun, the three-cavity gyroklystron has produced up to 54 kW of output power at 4.5 GHz. A series of phase jitter measurements has been carried out on this amplifier to determine the suitability of gyroklystrons for application to large linear accelerators.

A schematic of the three-cavity gyroklystron and the phase detector circuitry is shown in Figure 5.8. A 1 W CW signal is coupled from a master oscillator into the gyroklystron input cavity. At the same time, a small sample of the master oscillator signal is fed to one arm of a balanced mixer through a coaxial cable line, which contains a sliding short for adjusting the phase. The second input to the balanced mixer is supplied with a portion of the gyroklystron output signal.

The IF output port of the balanced mixer yields a signal $S(\phi) = M \cos\phi + M\delta \sin\phi$, where M is the product of the electric field amplitude of the two input signals, ϕ is the mean phase angle between the two signals, and δ is the phase jitter. By adjusting the sliding short, the phase jitter can be found as $\delta = S(\pi/2)/S(0)$. Early measurements [27] recorded phase jitter of 20° during the pulse with a jitter frequency of about 6 MHz. This jitter was accompanied by a 1% ripple in beam current at about the same frequency. Installation of a capacitive filter between the cathode and final anode of the magnetron injection gun reduced the current ripple to 0.3% and the phase jitter to 0.75° [28].

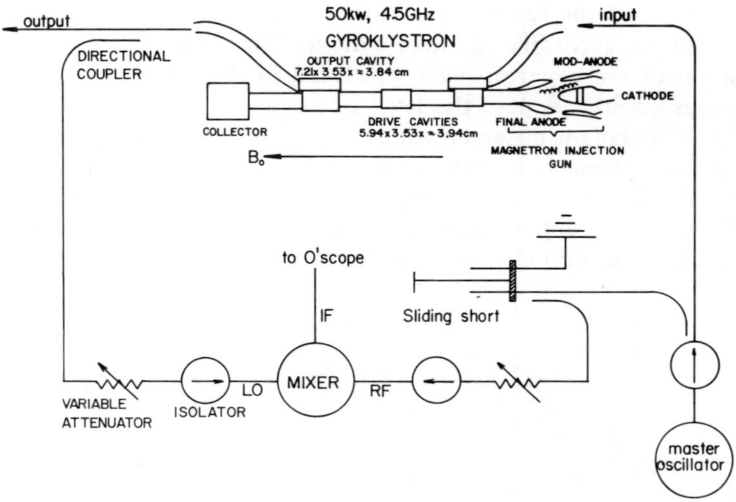

Fig. 5.8 Schematic of 54 kW, 4.5 GHz, Three-Cavity Gyroklystron and Phase Detector Circuitry [27, 28]

Pulse-to-pulse jitter was also measured because of its relevance to starting many amplifiers with synchronized phase. This measurement was made by observing the mixer output for many pulses at a fixed time interval after the start of each pulse. Pulse-to-pulse phase jitter was measured as <0.25°.

While these measurements of phase jitter <1° are encouraging, it should be noted that they were made in a relatively low power gyroklystron. However, because phase jitter can be reduced by feedback circuitry if required, there is promise that even much higher power gyroklystrons can be operated with phase jitter sufficiently small for the large linear accelerator application.

The 4.5 GHz gyroklystron could also be made to operate with the output cavity freely oscillating. This was accomplished by mechanically tuning the resonant frequency of the output cavity. There has been study of the phase locking of the free oscillations in the output cavity by a small signal introduced into the input cavity [29]. Results are shown in Figure 5.9, where the ratio of input drive power to output power P_d/P_0 to achieve locking is plotted *versus* the frequency difference between the two signals; magnetic field was slightly tapered along the axis so that the input cavity absorbed radiation effectively at the drive frequency. The solid line in

Figure 5.9 shows the level of P_d/P_0 that would be expected in a single-cavity oscillator according to Adler's theory [30]. It is clear that the pre-modulation of the electron beam produced in the buncher cavities and drift spaces greatly reduces the level of drive power required for phase lock. Phase locking is observed with drive power almost 40 dB below the output power. Because microwave tubes that drive RF accelerators must be controlled in phase, but not necessarily in amplitude, it would be possible to use either stable gyroklystron amplifiers or phase-locked gyroklystrons with their output cavities in free oscillation.

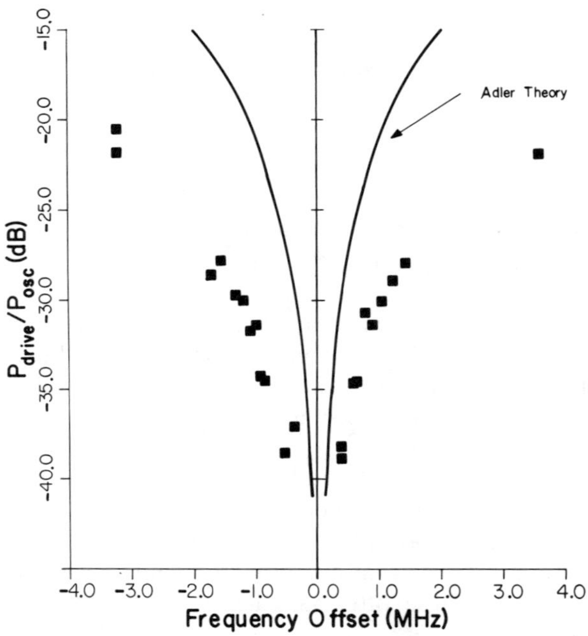

Fig. 5.9 Phase Locking of Free Oscillations in Output Cavity by Small Signal Introduced into the Input Cavity

Note: Results are shown where the ratio of input drive power to output power, P_d/P_0, is plotted *versus* the frequency difference between the two signals [29].

5.6 DEVELOPMENT OF A GYROKLYSTRON AMPLIFIER FOR DRIVING LINEAR SUPERCOLLIDERS

Development of high peak-power gyroklystron amplifiers is underway at the University of Maryland. The purpose of this developmental project is specifically to evaluate the applicability of gyroklystrons to driving linear supercolliders. The first amplifier (*circa* 1988) has been designed to operate at a frequency of 10 GHz with peak power in the range 36 MW to 48 MW.

Due to the high power level and because only a fraction of the electron velocity is directed along the tube axis, a relatively large electron energy (500 keV) was chosen to avoid excessive potential depression of the beam with respect to the circuit walls. The ratio of transverse to axial velocity must be greater than unity to make possible high device efficiency, and a value of $v_\perp / v_\parallel = 1.5$ was chosen. This choice led to a Larmor radius so large that the beam cannot easily fit into a drift tube, which is cut off to the fundamental mode. Then, the TE_{011} circular mode was chosen for the gyroklystron cavities. The drift tubes can be made sufficiently small to cut off the TE_{01} mode. Lower order modes (namely, TE_{11}, TM_{01}, and TE_{21}) are suppressed by slotting the drift tubes to prevent the flow of axial wall current. A photograph of a slotted drift tube is shown in Figure 5.10. This design concept was presented at the Particle Accelerator Conference in 1985 [31].

Fig. 5.10 Slotted Drift Tube

The complete gyroklystron is sketched in Figure 5.11, which shows the gyroklystron circuit to be composed of four cavities. The rotating

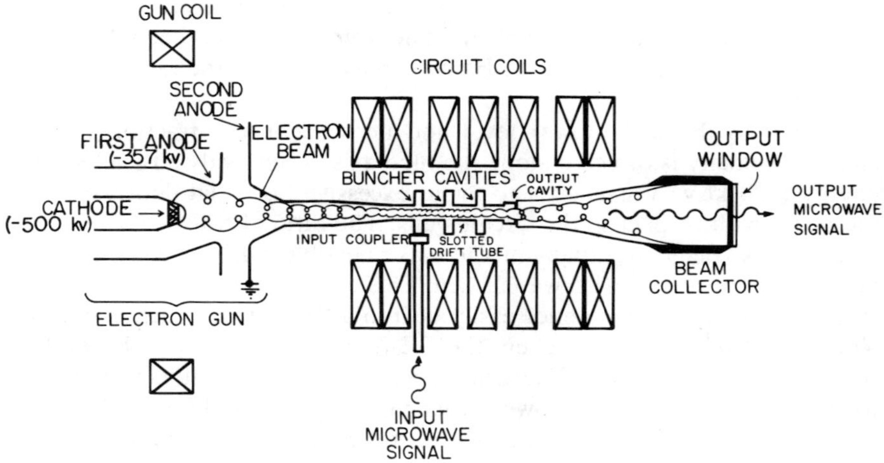

Fig. 5.11 Sketch of 40 MW, 10 GHz Gyroklystron [34]

annular electron beam is generated by a magnetron injection gun, the design optimization of which is described in the literature [32]. The peak dc electric field on any electrode in the gun is ⩽91 kV/cm. Axial velocity spread in the circuit region is a minimum of ≈6% when beam current is 160 A. For a beam current of 240 A, axial velocity spread rises to ≈8%.

On the basis of this electron beam, the gyroklystron circuit design was optimized by using a partially self-consistent code developed by K. R. Chu [33]. The dimensions and Q of each cavity as well as the length of each drift space are optimized for maximum device efficiency at a current of 160 A, while ensuring that operation is below the threshold current for any mode in the bunching cavities. The output cavity will self-oscillate when there is no input signal, but will operate stably when the incoming electron beam is prebunched. The circuit dimensions and other design details are described in the literature [34]. A calculated gain curve is plotted in Figure 5.12 for the case of beam current I = 160 A. Peak output power of 36 MW is predicted at a saturated gain of 63 dB.

The design parameters of this X-band gyroklystron are summarized in Table 5.2, where the parameters are compared with those of a conventional klystron being developed at the Stanford Linear Accelerator Center [35]. Interestingly, although the gyroklystron has somewhat higher operating frequency, gain, and output power, the maximum gradient at the electron gun electrodes is much lower. The maximum gradient for the

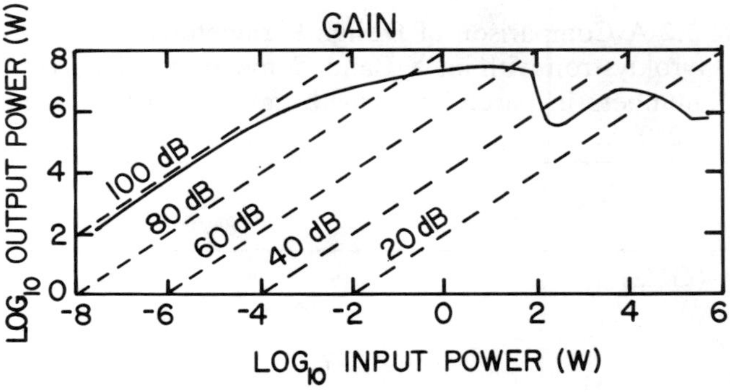

Fig. 5.12 Calculated Gain Curve Plotted for the Case of Beam Current I = 160 A (Curve provided by P. Latham.)

gyroklystron is 91 kV/cm compared with 300 kV/cm for the conventional klystron. The reason for this large difference is somewhat difficult to enunciate because the electron gun configurations are so different: a double-anode, temperature-limited, MIG producing a hollow beam of spiralling electrons in the case of the gyroklystron; a Pierce gun producing a solid beam of streaming electrons in the case of the conventional klystron. Nevertheless, it is worthwhile for the reader to note that the gyroklystron has larger cathode area and smaller electron beam perveance, and these are usually factors leading to lower gradients at the gun electrodes.

The lower gradients in the gyroklystron gun imply that we could scale the gyroklystron to operate at higher voltage and higher output power. For example, we could produce a gyroklystron design by using a TE_{01} circuit similar to that described above, while increasing both the voltage and current by 60% to maintain a constant relative electrostatic potential depression in the circuit region. This might produce an increase in output power to a level of ≈100 MW. The electron gun for such a gyroklystron would have gradients at the electrons larger than 91 kV/cm, but still well below the 300 kV/cm value that characterizes the conventional klystron design in Table 5.2.

It might also be possible to develop higher power gyroklystrons that would operate with a higher order cavity mode (e.g., TE_{02}) and drift spaces of larger cross section. Such devices could have the same 500 kV voltage and approximately the same current density as the gyroklystron in Table 5.2. However, cathode area and total current would be much larger. Scaling in this fashion would require a more complex circuit design to supress spurious modes especially in the drift spaces [31].

Table 5.2 A Comparison of Design Parameters for an X-Band Gyroklystron with an X-Band Conventional Klystron
Note: The numbers in parentheses in the gyroklystron column refer to operation at 240 A.

	Gyroklystron	*Conventional Klystron*
Frequency (GHz)	10	8.568
Voltage (kV)	500	330
Current (A)	160 (240)	152
Saturated Gain (dB)	$\geqslant$60	50
Peak Output Power (MW)	36 (48)	30
Pulse Duration (μs)	1.5	1.0
Efficiency (%)	45.5 (40)	60
Number of Cavities	4	4
Current Density in Interaction Circuit (A/cm^2)	67.2 (101)	330
Beam Area Convergence Ratio	12:1	36:1
Cathode Area (cm^2)	28.6	16.6
Current Density at Cathode (A/cm^2)	5.6 (8.4)	9.2
Maximum Gradient at Electrodes (kV/cm)	91	300
Perveance (μp)	0.45 (0.68)	0.8

Finally, it should be noted that the pulse duration in the gyroklystron of Table 5.2 is considerably larger than the ≈110 ns pulse length that would be required for the collider application at X band, and much larger peak power is therefore attainable in principle by use of a pulse compression circuit. This power multiplication technique is, in fact, being invoked up to a factor of 16 in conceptual designs of linear colliders [36]. A pulse compressor design that is well matched to the gyroklystron, because the compressor functions best with the TE_{01} mode, has been described by Farkas [37].

5.7 SUMMARY

High-power gyrotron technology is in a state of rapid development. During the year 1987, a 400 kW CW gyrotron oscillator at 140 GHz is expected to become operational, representing a fourfold advance in average power capability. However, even this demonstration falls far short of the perceived requirements for high average-power, millimeter-wave sources for plasma heating in advanced magnetic fusion experiments (i.e., power $\geqslant$ 1 MW and frequency in the range of 240 to 600 GHz). Some promising concepts for further research and development in high average power and high-frequency sources include whispering-gallery-mode gyrotron oscillators operating at cyclotron harmonics, quasioptical gyrotrons using Fabry-Perot cavities, and free-electron lasers.

Also imminent is the first demonstration of a high peak-power, X-band gyroklystron amplifier, which would be relevant to driving linear supercolliders. Peak power in the range of 36 to 48 MW is predicted with pulse duration of 1.5 μs. By using pulse compression, this gyroklystron might deliver peak power on the order of 600 MW. The design parameters of the gyroklystron promise the feasibility of further upward scaling in peak power or frequency. However, we should remember that the 36 MW gyroklystron represents a thousandfold advance in peak power over existing gyroklystron capability. Thus, the proof of the pudding will be in the results of the upcoming experimental demonstration.

ACKNOWLEDGMENTS

The author is indebted to the following lists of his colleagues for very useful discussions and criticisms: A. Drobot, D. Farkas, K. Felch, A. Fliflet, T. V. George, S. Gold, P. E. Latham, W. Lawson, J. McAdoo, A. McCurdy, A. Mondelli, R. K. Parker, C. D. Striffler, P. Vitello, and P. Wilson.

REFERENCES

1. V.L. Granatstein, M.E. Read, and L.R. Barnett, in K. Button, ed., *Infrared and Millimeter Waves* (Academic Press, New York, 1982), Vol. 5, 287–304.
2. V.L. Granatstein and P.L. Colestock, eds., *Wave Heating and Current Drive in Plasmas* (Gordon and Breach Science Publishers, New York, 1985).
3. V.L. Granatstein and A. Mondelli, in M. Month and M. Dienes, eds., *Physics of High Energy Accelerators, Conf. Proc. No. 153,* (AIP, New York, 1987), 1501–1571.
4. R.J. Temkin, *et al., Int. J. Infrar. and Mill. Waves* **3**, 427 (1982).
5. B. Arfin, *et al., IEEE Trans. Electron Dev.* **ED-29**, 1911 (1982).
6. H. Jory, *Proc. Int. Symp. on Heating in Toroidal Plasmas* **2**, 1424 (1984).
7. US Department of Energy, Division of Magnetic Fusion, Meeting of the Ignition Technical Oversight Committee (ITOC), San Diego, CA, June 1986.

8. T.V. George, private communication.
9. K. Felch, R. Bier, L.J. Craig, H. Huey, L. Ives, H. Jory, N. Lopez, and S. Spang, *Int. J. Electr.* **61**, 701 (1986); see also, *Proc. 11th Int. Conf. Infrared and Millimeter Waves,* 43 (1986).
10. K.E. Kreischer, J.B. Schutkeker, B.G. Danly, W.J. Mulligan, and R.J. Temkin, *Int. J. Electr.* **57**, 835 (1984).
11. Y. Carmel, K.R. Chu, M. Read, A.K. Ganguly, D. Dialetes, R. Seeley, J.S. Levine, and V.L. Granatstein, *Phys. Rev. Lett.* **50**, 112 (1983).
12. K. Felch, private communication.
13. K.E. Kreischer, S. Spira, and R.J. Temkin, *Proc. 11th Int. Conf. Infrared and Millimeter Waves,* 229 (1986).
14. T.A. Hargreaves, K.J. Kim, J.H. McAdoo, S.Y. Park, R.D. Seeley, and M.E. Read, *Int. J. Electr.* **57**, 977 (1984).
15. M.Q. Tran, A. Bondeson, A. Perrenoud, S. Alberti, B. Isaak, and P. Muggli, *Int. J. Electr.* **61**, 1029 (1986).
16. I.E. Botvinnik, V.L. Bratman, A.B. Volkov, C.G. Denisov, B.D. Kol′chugin, and M.M. Ofitserov, *Sov. Tech. Phys. Lett.* **8**, 596 (1982).
17. A.T. Lin and C.C. Lin, *Int. J. Infrar. and Mill. Waves* **6**, 41 (1985).
18. A. Fliflet, *Int. J. Electr.* **61**, 1049 (1986).
19. T.J. Orzechowski, E.T. Scharlemann, B. Andereson, V.K. Niel, W.M. Fawley, D. Prosnitz, S.M. Yarema, D.B. Hopkins, A.C. Paul, A.M. Sessler, and J.S. Wurtele, *Phys. Rev. Lett.* **57**, 2172 (1986).
20. W.W. Destler, V.L. Granatstein, I.D. Mayergoyz, and Z. Segalov, *J. Appl. Phys.* **60**, 521 (1986).
21. V.L. Granatstein, *et al., Plasma Phys.* **17**, 23 (1975).
22. V.I. Krementsov, *et al., Sov. Phys. JETP* **48** (6), 1084 (1978).
23. S.N. Voronkov, *et al., Sov. Phys. Tech. Phys.* **27**, 61 (1983).
24. S.H. Gold, A.W. Fliflet, W.M. Manheimer, W.M. Black, R.B. McCowan, V.L. Granatstein, A.K. Kinkead, D.L. Hardesty, and M. Sucy, *Proc. 11th Int. Conf. Infrared and Millimeter Waves,* 351 (1986); see also, *Phys. Fluids* (1987).
25. M.E. Read, R. Seeley, and W.M. Manheimer, *IEEE Trans. Plasma Sc.* **PS-13**, 398 (1985).
26. W.M. Bollen, A.H. McCurdy, J.H. McAdoo, A.K. Ganguly, V.L. Granatstein, and R.K. Parker, *IEEE Trans. Nucl. Sc.* **NS-32**, 2879 (1985).
27. J. McAdoo, W.M. Bollen, V.L. Granatstein, R. Parker, R. Smith, G. Thomas, and A. McCurdy, *IEEE Trans. Nucl. Sc.* **NS-32**, 2963 (1985).
28. J. McAdoo, W.M. Bollen, A. McCurdy, V.L. Granatstein, and R.K. Parker, *Int. J. Electr.* **61**, 1025 (1986).

29. A.H. McCurdy, C.M. Armstrong, W.M. Bollen, R.K. Parker, and V.L. Granatstein, *Phys. Rev. Lett.* **57**, 2379 (1986).
30. R. Adler, *Proc. IRE* **34**, 351 (1946).
31. V.L. Granatstein, P. Vitello, K.R. Chu, K. Ko, P.E. Latham, W. Lawson, C.D. Striffler, and A. Drobot, *IEEE Trans. Nucl. Sc.* **NS-32**, 2957 (1985).
32. W. Lawson, J. Calame, V.L. Granatstein, G.S. Park, C.D. Striffler, and J. Neilson, *Int. J. Electr.* **61**, 969 (1986).
33. K.R. Chu, V.L. Granatstein, P.E. Latham, W. Lawson, and C.D. Striffler, *IEEE Trans. Plasma Sc.* **PS-13**, 417 (1985).
34. W. Lawson, J. Calame, V.L. Granatstein, P.E. Latham, J. McAdoo, M. Reiser, C.D. Striffler, D. Welsh, and F.J. Williams, *Proc. Particle Accelerator Conf.* (1987).
35. T.G. Lee, oral presentation at the SLAC-KeK Workshop, March 1987.
36. P.B. Wilson, SLAC-PUB-3985, May 1986.
37. Z.D. Farkas, *IEEE Trans. Microwave Theory Tech.* **MTT-34**, 1036 (1986).

Chapter 6
FREE-ELECTRON LASERS

John A. Pasour

Mission Research Corporation
Alexandria, Virginia

6.1 INTRODUCTION

It would be difficult to identify a technology that has matured more rapidly than the *free-electron laser* or FEL. During a period of about 10 years, great strides have been made in both the theoretical understanding of the FEL interaction and the experimental verification of the many promising predictions of the theory [1, 2]. Experimental research has advanced from the first demonstrations of the physical mechanism to the operation of numerous successful devices at laboratories around the world. These FELs operate at wavelengths from 10^{-5} cm to 1 cm, and some have power levels in excess of any other source in their spectral regime. Now FELs are being touted as everything from "Star Wars" weapons of the future to medical miracle workers.

This large amount of research on FELs has been stimulated by the unique capabilities of the device and the many applications that require these features. In addition, the FEL is conceptually quite simple: it consists only of an energetic electron beam together with some means of forcing the electrons to oscillate in a regular fashion. The most important FEL attributes are broad frequency-tuning ability, high efficiency, high power, and wide bandwidth. These features are desirable for a wide range of applications. For example, plasma heating at the electron cyclotron resonance in high-field fusion devices requires efficient millimeter to submillimeter sources at powers $\geqslant$ 10 MW. High-resolution, long-range radar also needs powerful millimeter to submillimeter sources. At shorter wavelengths, there is a need for efficient tuned sources for biomedical and photochemical applications, laser isotope separation, materials processing, and physics research, in addition to the obvious and highly publicized military applications. FELs are also being examined for various advanced particle accelerator concepts, such as high-frequency, high-gradient RF accelerators and the inverse FEL, in which the phase of the interaction is adjusted so that the electrons gain energy from the radiation field, rather than supply it with energy.

One of the most attractive features of the FEL is its tuning ability. In typical lasers, the operating frequency is fixed and determined by the quantized energy levels of the electrons in the excited medium. The FEL, however, can be described in purely classical terms. The operating frequency is determined by continuously variable experimental parameters, such as the kinetic energy of the electrons. Consequently, the FEL may be continuously tuned by varying these parameters.

The other important features of the FEL are its high-power capabilities and its high efficiency. Because there is no physical lasing medium that must support the radiation field, problems of heating or breakdown, which plague conventional solid or gaseous lasers, are absent. Enormous powers can be deposited in relativistic electron beams propagating in vacuum. Efficiencies of transferring this electron kinetic energy to FEL radiation have reached 40% at longer wavelengths.

To give a picture of the current technology of FEL devices as compared to other radiation sources, the peak-power levels attained as a function of wavelength are shown in Figure 6.1. At longer wavelengths, FELs already offer the highest power available from any source. At shorter wavelengths, the peak FEL power is still less than that from conventional lasers, but FELs can operate at frequencies unattainable by other coherent sources. The power levels of short-wavelength FELs are also rapidly increasing as more experimental research reaches completion.

In this chapter, we will first present a simple theoretical overview of the FEL mechanism, and discuss various operating regimes and efficiency-enhancement schemes. We will then give a brief experimental history of the FEL. We will outline modifications of FEL theory due to the space-charge effects and finite waveguide geometry, which characterize the longer wavelength FELs, detail the major components of the FEL, and summarize the results of some contemporary millimeter-wave FEL experiments. A brief discussion of the ongoing research and the technological challenges remaining to be solved will conclude the chapter.

6.2 FEL THEORY

The FEL is conceptually quite simple, consisting only of an electron beam, a periodic pump field, and the radiation field. The electron beam passes through the pump field and begins to oscillate. (The pump field is

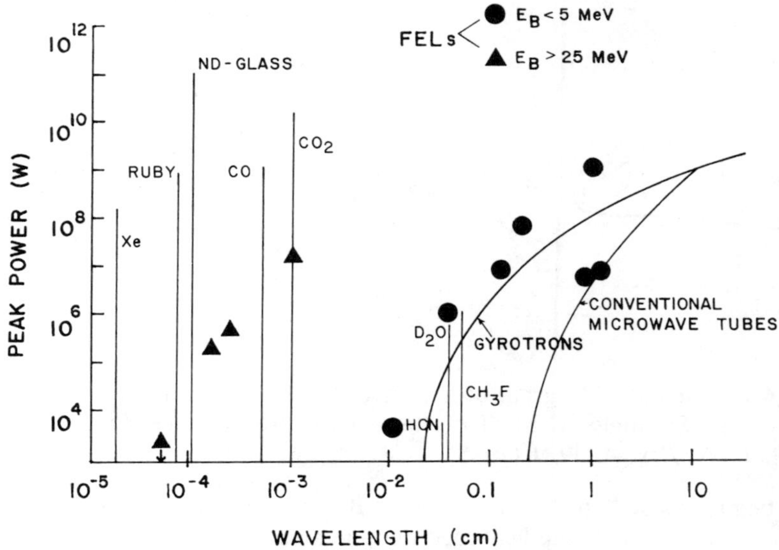

Fig. 6.1 Peak-Power Output of FELs and Other Representative Sources

usually a static periodic magnetic field called a "wiggler," but in principle any field can be used if it is capable of producing a transverse electron oscillation.) The oscillating electrons radiate, and the combination of the radiation field and the wiggler field produces a beat wave, which tends to bunch the electrons in the axial direction. The bunched electrons radiate more coherently, thereby strengthening the radiation, which causes more bunching, then produces stronger and more coherent radiation, and so forth. The electrons in the beam are not bound to a nucleus, hence the term "free-electron."

This discussion of FEL theory consists of a summary, rather than a derivation of the important results. Our objective is to provide a physical understanding of the interaction without a lot of mathematical detail. Such an understanding is necessary for full appreciation of the special capabilities and problems of the FEL. The additional details can be found in the literature [3, 4].

6.2.1 Physical Mechanism

A typical configuration of the FEL is illustrated in the simple schematic diagram of Figure 6.2. The periodic magnetic wiggler field is perpendicular to the FEL axis. Electrons injected into the wiggler along the

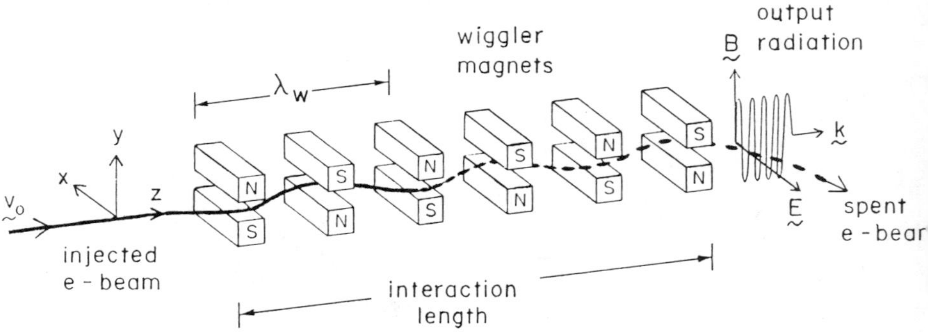

Fig. 6.2 Schematic Diagram of a Typical FEL Configuration
Source: P. Sprangle, C.M. Tang, and C.W. Roberson, *Nucl. Instrum. Meth. Phys. Res. A239,* 1 (1985).

axis begin to oscillate due to the $\mathbf{v} \times \mathbf{B}$ force. The oscillating electrons radiate, as any accelerated charged particle must do. The radiation electric field is in the same direction as the transverse particle velocity, so that once the radiation field is established, the electrons can either lose or gain energy, depending on the phase of their oscillatory velocity with respect to the field. Because the electrons are initially distributed randomly in phase, a bunching mechanism is required in order to achieve coherent radiation. This bunching force is provided by the beating of the radiation and wiggler fields, resulting in a so-called *ponderomotive wave*, which is characterized by a periodic axial force.

To show explicitly how the ponderomotive force arises, we assume a particular form for the magnetic wiggler field: $\mathbf{B}_w = B_w \cos(k_w z)\ \mathbf{e}_y$. Here, the wiggler period is $\lambda_w = 2\pi/k_w$, and $\mathbf{e}_y$ is the unit vector in the y direction. The radiation fields are assumed to have amplitudes E_R and B_R, and to vary as $\cos(kz - \omega t)$, where the frequency and wavelength are related by the usual vacuum dispersion relation: $\omega = ck = 2\pi c/\lambda$. Electrons injected into the wiggler with axial velocity v_0 will acquire a transverse oscillatory velocity (or *wiggle velocity*). The amplitude of the wiggle velocity can be easily shown from the force equation to be $\mathbf{v}_w = eB_w/(\gamma_0 m k_w)$ $\sin(k_w z)\ \mathbf{e}_x$, where $-e$ is the charge of an electron, $\gamma_0 = (1 - v_0^2/c^2)^{-1/2}$ is the relativistic mass factor, c is the speed of light, and m is the rest mass of the electron. The $\mathbf{v}_w \times \mathbf{B}_R$ term in the force equation then produces an axial force, which can be shown to vary as $\sin[(k + k_w)z - \omega t]$. This ponderomotive force hence has a phase velocity $v_{\text{ph}} = \omega/(k + k_w)$, which must be approximately equal to the electron axial velocity for the electrons to remain in phase with the wave. Thus, setting $\omega/(k + k_w) = v_z$ and using $k = c/\omega$, in order to have synchronism between the ponderomotive wave and the electrons (so that bunching may occur), we find that the frequency must satisfy $\omega = (1 - v_z/c)^{-1} v_z k_w$. In terms of the wavelength, this expression can be written $\lambda = \lambda_w/[\beta_z(1 + \beta_z)\gamma_z^2]$, where we have defined $\beta_z = v_z/c$. For highly relativistic electron beams ($v_0 \approx c$), the relation simplifies to $\lambda \approx \lambda_w/2\gamma_z^2$, from which it is obvious that the radiation wavelength can be much smaller than the wiggler wavelength. By using the relation $(\gamma_0/\gamma_z)^2 = 1 + (\gamma_0\beta_\perp)^2$, assuming that $\beta_\perp = \beta_w$, and averaging over a wiggler period, the resonance condition can be written as

$$\lambda = (\lambda_w/2\gamma_0^2)\,(1 + K_w^2/2) \tag{6.1}$$

where $K_w = eB_w/mck_w$ is the wiggler strength parameters. This is the well known FEL resonance condition, from which it is obvious that the FEL can be tuned by varying either the electron beam energy or the wiggler field.

Free-electron lasers, like other lasers, can operate as either amplifiers or oscillators. In the amplifier configuration, an input signal injected into the wiggler along with the electron beam is amplified during a single pass through the wiggler. To be practical, amplifiers require high growth rates, so that large gains in the radiation field can occur in a reasonable distance. We shall see that this requirement typically limits amplifiers to the infrared-to-millimeter wavelength regime. In an oscillator, the radiation is confined between two mirrors. No input signal is required, but instead the fraction of the spontaneously emitted radiation that has the resonant frequency is amplified during repeated passes through the interaction region. A small fraction of the radiation can be coupled out of the resonator through a partially transmitting mirror, just as in conventional lasers. Because the radiation can grow over many transits, a high gain per pass is not necessary (assuming that the electron beam pulse is of sufficient duration). Consequently, oscillators are appropriate for short wavelength operation, where, as we shall see, the gain is often relatively low. One other advantage of oscillators as compared with amplifiers is that they can operate at frequencies where other sources do not exist because an input signal is not necessary.

6.2.2 FEL Operating Regimes

There are three major operating regimes of FELs, depending on the electron beam and wiggler parameters, from which we can obtain relatively simple expressions for growth rates and efficiencies. These regimes are referred to as the *Compton* or *single-particle regime,* the *Raman* or *collective regime,* and the *strong pump* or *high-gain Compton regime.* Free-electron lasers powered by high energy, low current, and high quality beams, such as are produced by RF linacs, microtrons, or storage rings, can be expected to operate in the Compton regime. In the Compton regime, space charge effects are negligible. The radiation growth rate is usually low due to the single-particle nature of the interaction, so oscillator operation is preferable. Another limitation of the Compton regime is that the intrinsic efficiency is low, typically 1% or less. However, because the beam energy is high and the beam quality is good, Compton FELs can operate at short wavelengths in the visible or UV region.

The Raman regime is characterized by high current (kA) and relatively low energy (MeV) electron beams in which space-charge effects are very important. These beams are generally produced by pulse-line accelerators, high-current modulators, or induction linacs. The relatively low beam energy limits operation to long wavelengths (millimeter to submillimeter), but impressively large gains on the order of 40–50 dB per pass and intrinsic efficiencies of about 10% have been demonstrated in several experiments. Operation in either the amplifier or the oscillator mode is possible.

The third operating regime is similar in some respects to both the Compton and Raman regimes. The beam current is usually high so that the gain is large, but the wiggler field is so strong that the ponderomotive force dominates the space-charge forces, hence the name "strong pump" or "high-gain" Compton.

A more detailed discussion of these various regimes and a derivation of the growth rates and efficiencies in each are given by Sprangle *et al.* [3]. Here, we simply present the results of that work in Table 6.1. Note that the scaling of the growth rate and efficiency with various parameters is different in the various regimes. Clearly, not all FELs can fall into one of these simple categories because transition regions must exist. Nevertheless, the simple expressions can be quite useful for "back-of-the-envelope" calculations to predict FEL performance in particular situations.

6.2.3 Efficiency Enhancement

As discussed above, the intrinsic efficiency of the FEL can be quite small, especially in the Compton regime. Physically, the intrinsic efficiency is determined by particle trapping in the ponderomotive potential wells, as shown in Figure 6.3. Ideally, monoenergetic electrons are injected into the wiggler with a velocity slightly larger than the phase velocity of the ponderomotive wave. As the electrons interact with the wave, some are

Table 6.1 FEL Operating Regimes

	Gain or Growth Rate	*Intrinsic Power Efficiency*
Compton (single particle, low gain)	$\pi F \frac{\nu}{\gamma}(\beta_w)^2 \frac{L^3}{r_b^2 \lambda_w} f(\theta)$	$\left(\frac{1}{2}\right) \frac{\lambda_w}{L}$
Raman (collective, high gain)	$(\pi \gamma_z F)^{1/2}(\nu/\gamma)^{1/4} \frac{\beta_w}{\sqrt{r_b \lambda_w}}$	$\frac{1}{\pi \gamma_z}\left(\frac{\nu}{\gamma}\right)^{1/2} \frac{\lambda_w}{r_b}$
High-Gain Compton (single particle)	$2F^{1/3}\left(\frac{\nu}{\gamma}\right)^{1/3}\left(\frac{r_b}{\lambda_w}\right)^{1/3} \frac{\beta_w^{2/3}}{r_b}$	$0.18\left(\frac{\nu}{\gamma}\right)^{1/3}\left(\frac{\lambda_w \beta_w}{r_b}\right)^{2/3}$

Note: The quantities used in this table have the following definitions: $\nu = I/17$ is Budker's parameter, I is the beam current in kiloamperes, L is the wiggler length, r_b is the beam radius, $F = r_b^2/r_{rad}^2$ is the filling factor, $f(\theta) = \partial(\sin\theta/\theta)^2/\partial\theta$, $\theta = (1 - v_0/v_{ph})\tau\omega/2$ and $\tau = L/v_0$ is the electron's transit time.

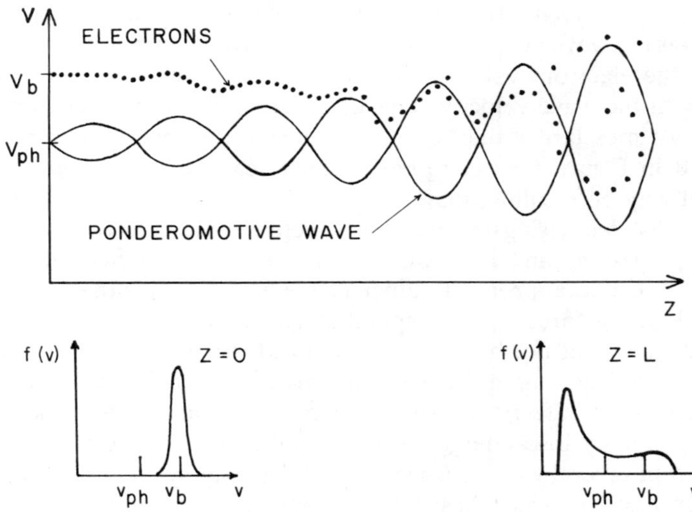

Fig. 6.3 Idealized Phase-Space Diagram (top) Showing Electrons Losing Energy to and Becoming Trapped in the Ponderomotive Potential Wells

Note: The FEL interaction results in a net decrease of electron energy, as shown in the initial and final electron distribution functions (bottom).

accelerated and others are decelerated, depending on the relative phase of the electrons with respect to the wave. If the electrons initially have a velocity above the resonance value, then those that lose energy will move closer to resonance and interact more strongly with the wave, resulting in a net decrease of electron kinetic energy and a corresponding increase in the radiation amplitude. Eventually, the electrons are decelerated to a velocity less than the phase velocity of the ponderomotive wave. The electrons then begin to be reaccelerated by the wave, resulting in a loss of radiation amplitude and the saturation of the interaction. The electrons are essentially trapped in the ponderomotive potential wells (often called "buckets"). The fractional electron energy loss that occurs before saturation is the intrinsic FEL efficiency.

In order to enhance the efficiency of the FEL, it is necessary either to find a way to continue to extract energy from the electrons after they are trapped in the buckets, or to recover the kinetic energy of the electrons exiting the FEL and recycle the energy to the accelerator. It turns out that both options are possible and have been demonstrated in experiments.

The first method of extending the interaction is to decrease the phase velocity of the ponderomotive wave after the electrons are trapped [5, 6]. If the phase velocity (i.e., the bucket velocity) is decreased gradually enough as the wave and the trapped electrons travel through the interaction region, the electrons will remain trapped and their velocity will decrease along with the wave velocity. This decrease in electron kinetic energy can be many times larger than can occur during the initial trapping process depicted in Figure 6.3. The phase velocity can be changed in a straightforward way be recalling that $v_{\text{ph}} = \omega/(k + k_w) \approx c(1 - \lambda/\lambda_w)$. Hence, by spatially decreasing the wiggler wavelength λ_w, the phase velocity of the trapping wave can be decreased to produce the efficiency enhancement.

The second approach to efficiency enhancement is the application of an accelerating force to the trapped electrons. Because the electrons are trapped, they will not be accelerated out of the buckets unless the accelerating force is too large. Instead, the force produces a relative phase shift of the electrons, effectively holding them in the tops of the buckets, where they continually lose energy to the wave. The accelerating force can be provided in several ways, perhaps most obviously by an axial electric field. Another method which has the same effect is gradually decreasing the amplitude of the wiggler field, thereby decreasing the electron wiggle velocity. By conservation of energy, if the energy corresponding to the wiggle velocity decreases at the same rate that the radiation field energy increases, the axial electron velocity will remain constant. Thus, the energy that would ordinarily appear as an axial acceleration is instead transferred to the radiation field.

Therefore, enhanced growth of the radiation field can be achieved by spatially decreasing the wiggler field amplitude, the period, or both. These schemes can be applied to either the Compton or Raman regimes of the FEL. One potential drawback to these schemes is that in the trapped particle mode of operation, it is possible for the trapped electrons to become unstable, resulting in the generation of sideband radiation at frequencies slightly different from the desired values [7]. This instability is still being investigated, but the results of recent experiments at Lawrence Livermore National Laboratory (LLNL) [8] indicate that the instability does not necessarily preclude effective efficiency enhancement. The LLNL tapered wiggler experiments, which will be described below, resulted in an efficiency enhancement by a factor of 5 (to about 40%) over the untapered value.

Finally, the overall system efficiency can obviously be increased by reusing the electron beam energy which remains after the interaction. This approach is practical for both *electrostatic accelerators,* in which depressed collectors can be used, and *RF accelerators,* in which the spent electrons are passed through a decelerating RF structure (see [1, Vol. 9]). In either case, the recovered energy can be directly applied to the acceleration of additional electrons. A practical problem with this approach is that the energy spread induced by the FEL interaction itself makes beam recovery more difficult.

It is clear that, in most applications, we will desire the highest possible overall efficiency. Consequently, we cannot stress too strongly the importance of the efficiency enhancement capability of the FEL. It represents a major advantage with respect to conventional lasers.

6.3 FEL EXPERIMENTAL HISTORY

The first operation of a device employing what we now call the "FEL mechanism" was reported in 1960 by Phillips [9]. He called the device a *ubitron* and built several microwave-tube versions that behaved impressively at wavelengths of about 10 cm. It was 15 years later before the experiments were performed that led to the present-day FELs. These experiments evolved along two separate paths, depending on the type of accelerator used and the operating regime. In the Compton regime, the work was pioneered by J. Madey and his colleagues at Stanford University using an RF linac operating at 25–50 MeV [10, 11]. The Raman experiments were pioneered at the Naval Research Laboratory (NRL) and Columbia University using pulse-line accelerators, which produced 1 MeV electron beams at currents of tens of kiloamperes [12, 13]. The first Raman experiments were referred to as *stimulated scattering,* but after the Stanford group popularized the term FEL, the Raman devices were also referred to by that name.

6.3.1 Compton FELs

The early Compton regime experiments were primarily designed to verify some of the important theoretical predictions, such as growth rate and efficiency. The Stanford device was first operated as an amplifier at 10.6 μm and later as an oscillator at 3.4 μm. Both experiments used a super-conducting RF linac having a peak current of about 2 A. The electron beam consisted of a train of pulses 1-mm in length, separated by about 25 m. Although the fractional energy spread was very small (about 5×10^{-4}), the efficiency was less than 10^{-4} and the gain was so small (a few percent per pass) that it was very difficult for the device to oscillate. One of the technical problems characteristic of RF linac FEL oscillators is the extremely tight tolerance on mirror placement. Because the electron bunches are so short and are spaced so far apart, the mirror spacing must be held to within a micron over distances of 10–15 m in order to cause the radiation pulses to continue to overlap the electron pulses during repeated passes through the resonator.

In the early 1980s, a series of FEL amplifier experiments at 10.6 μm was carried out at Los Alamos, TRW, and Boeing/Spectra Technologies to demonstrate ponderomotive trapping and efficiency enhancement [1]. In these experiments, powerful CO_2 lasers were used as the input source to establish the ponderomotive wave. These experiments also pioneered the use of permanent-magnet wigglers, which were much more amenable to tapering than Stanford's helical electromagnet.

More recent Compton regime experiments underway at Los Alamos, Boeing, and Stanford are aimed at shorter wavelength operation, higher efficiency, and higher power [2]. A joint TRW-Stanford experiment has already demonstrated FEL operation in the visible spectrum. At Los Alamos, emphasis is being placed on efficiency enhancement with a tapered wiggler and RF electron beam recovery. Peak powers of 10 MW have been achieved from Los Alamos' oscillator at 10 μm. At Boeing, a visible FEL oscillator is under development. A key advancement has been the development of much higher peak electron currents in the RF linacs, both by improving the electron guns and by more strongly bunching the electrons before accelerating them. One of the major technical problems thus far has been maintaining the high degree of accelerator stability required for this high-current operation. However, researchers are optimistic that efficiencies up to 20% can be achieved in forthcoming experiments.

Other infrared FEL experiments are being performed using cyclic RF accelerators called *microtrons* (see [1, Vol. 9]). These devices are very compact because the electron beam follows a spiral orbit, which brings it through the same RF accelerating cavity many times. Microtrons for FELs have energies of about 20 MeV and peak currents of a few amperes. Their compact size makes them suitable for applications in which enormous power levels are not required and space is at a premium.

Far-infrared FEL experiments are being performed at the University of California at Santa Barbara, using an electrostatic accelerator (Van de Graaf generator). These accelerators typically have very low currents, but by recollecting the current (as previously discussed), pulses of several amperes can be generated for tens of microseconds. These accelerators are characterized by extremely good voltage stability and high beam quality. The oscillator experiment has produced >10 kW of infrared radiation, and further experiments should significantly exceed that level [14].

A number of Compton FEL experiments using electron storage rings are also in progress or have been completed as of this writing [15]. These high-energy (hundreds of MeV) beams hold the promise of UV operation, although the efficiencies are expected to be quite low. Although the electrons make many passes through the wiggler, the thermalization due to the FEL interaction and the *synchrotron radiation* emitted because of the high electron energy limit the FEL efficiency to about 1%. For some applications requiring tuned, coherent, UV radiation, that 1% efficiency is acceptable.

6.3.2 Raman FELs

The origins of the Raman FELs can be traced to experiments at Cornell University and NRL in the early 1970s [16, 17]. These experiments were performed by passing intense electron beams (1 MeV, >10 kA, 50 ns single pulses) through periodic structures or rippled magnetic fields. Subsequent experiments at NRL demonstrated the use of an *electromagnetic radiation wiggler* [18], and, in a collaboration with Columbia University, explored various magnetic wigglers and FEL configurations, including an oscillator [19].

The Raman FELs initially produced MW power levels at millimeter to centimeter wavelengths, but at efficiencies of less than 1%. This low efficiency was primarily due to the relatively poor beam quality (large energy spread), which was typical of the pulse-line accelerators. Subsequent efforts at NRL to improve the beam quality by redesigning the electron-accelerating diodes and beam-transport systems eventually led to the generation of 75 MW pulses of 4 mm radiation at an efficiency of 6% [20, 21]. In the amplifier mode, a power gain of 50 dB with >3% efficiency was achieved [22].

The gain in most Raman experiments is so high that many of them have been operated in the super-radiant mode, in which the spontaneously emitted radiation is amplified during a single pass through the wiggler.

Most of these experiments have employed an axial magnetic field superimposed on the wiggler field to confine the high-current beams. The axial field can enhance the FEL interaction by resonantly increasing the electron wiggler velocity, but this field can also complicate the interaction by causing cyclotron emission [23].

A typical experimental arrangement [20] is shown schematically in Figure 6.4. As is true of practically all the FELs in this regime, this device operated in the single-pulse mode. The FEL was powered by a Marx generator connected by a pulse-forming line to a field-emission diode. A 1–2 MeV, 1–2 kA electron beam was extracted through an anode aperture,

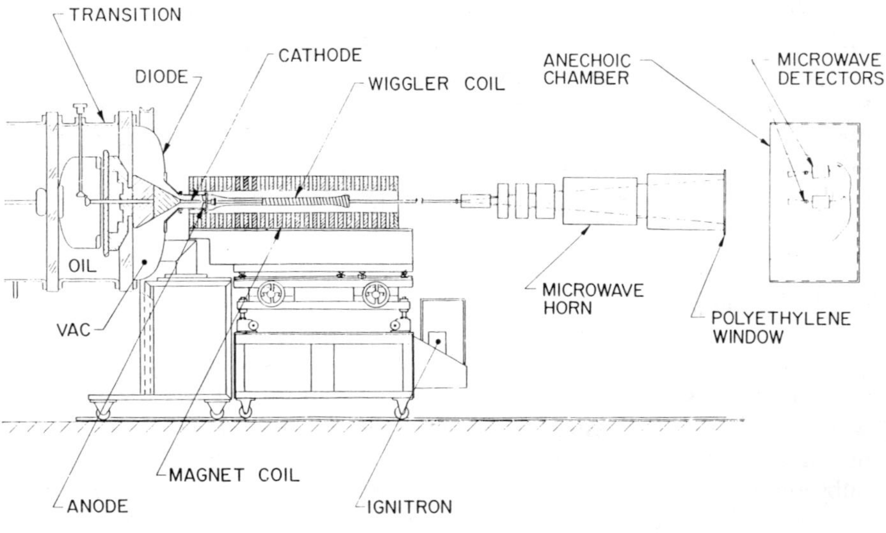

Fig. 6.4 Schematic of the VEBA Super-Radiant FEL Experiment [20] at NRL
(Courtesy of S.H. Gold.)

which collected >90% of the diode current. However inefficient, the extracted beam was of quite high quality, having an axial velocity variation of ≈0.1%. The helical wiggler and the solenoidal magnets were driven by pulsed capacitor banks. The low duty cycle permitted the use of high-current windings with no provisions for cooling.

Experiments at a number of laboratories around the world have been very important in the development of long-wavelength FELs. The details of all these experiments cannot be described here, but we will highlight some of the important contributions. Many experiments have been performed over the years at MIT, with much of the work aimed at understanding the complications that arise due to the axial magnetic field, which is usually used in the Raman experiments. Recently, researchers at MIT achieved efficiencies of about 12% without efficiency enhancement [24]. Experiments at Hughes [25] and TRW [26] are aimed at developing more practical millimeter-wave FELs, with such features as compactness, high system efficiency, and higher repetition rate. The group at Columbia has performed pioneering amplifier and oscillator [27] experiments at millimeter wavelengths and has developed a Thomson scattering diagnostic for directly measuring the axial energy spread of electrons in the FEL [28]. Recently, experiments have begun at the University of Maryland to study *electromagnetic-wave-driven FELs* and *rotating-beam FELs* [29]. Groups at Hebrew University and Tel Aviv University are studying electrostatic and longitudinal wiggler FELs [30]. Studies of super-radiant, long-wavelength FELs have also been carried out at Ecole Polytechnique in France [31].

The FEL is now the highest power source available at millimeter wavelengths (see Figure 6.1). Ongoing experiments at LLNL, NRL, Columbia, and MIT are pushing toward high powers at shorter wavelengths. Recently, high-power millimeter-wave FEL oscillators have been reported at NRL and Columbia [27, 32], with the NRL experiment using a unique *long-pulse induction linac* to produce a steady 1 μs pulse at powers up to 10 MW. High-power amplifier research is being carried out at LLNL using *short-pulse induction linacs* [33]. The major accomplishment of this experiment is the demonstration of efficiency enhancement to 40% at a power level of 1 GW [8]. In this experiment, operated at a wavelength of 8 mm, the wiggler amplitude was tapered. The present experiment is performed using the *electron test accelerator* (4 MeV, 1 kA), but the next series of experiments will use the *advanced test accelerator* (50 MeV, 10 kA) to amplify 10.6 μm radiation.

6.4 AXIAL MAGNETIC FIELD EFFECTS

As we noted above, the large majority of millimeter-wave FELs have employed an axial magnetic field, which significantly affects the FEL interaction in several fundamental ways. First, the electron trajectories are strongly modified by the axial field, particularly when the wiggler-induced motion is near resonance with the cyclotron motion due to the axial field. This gyroresonant effect strengthens the FEL interaction by increasing the amplitude of the wiggler-driven oscillations. Second, the axial field directly affects the strength of the FEL coupling through its effect on the ponderomotive potential, and the field can change the nature of the FEL instability. Third, as we mentioned previously, the axial field makes possible an additional instability, the *cyclotron maser,* which can play a role in these devices.

6.4.1 Raman Operation

To understand the effect of an axial field on the FEL interaction, it is necessary to understand the nature of the collective (Raman) FEL. The fundamental FEL relations presented in Section 6.2 must be modified because of the space-charge forces of the electrons and the finite waveguide geometry in which these devices operate. The emission from a Raman FEL results from the interaction between the slow space-charge wave on the electron beam:

$$\omega = \beta_z c(k + k_w) - p_1\omega_p\Phi^{1/2}/\gamma^{1/2}\gamma_z \tag{6.2a}$$

and the electromagnetic waveguide mode:

$$\omega^2 = c^2k^2 + \omega_c^2 \tag{6.2b}$$

Here, $\omega_p = (Ne^2/\epsilon_0 m)^{1/2}$ is the nonrelativistic plasma frequency, $\omega_c = (\omega_{co}^2 + p_2\omega_p^2/\gamma)^{1/2}$ is the effective waveguide cut-off frequency, adjusted for the presence of the electron beam, ω_{co} is the empty waveguide cut-off frequency, p_1 and p_2 are frequency-dependent numerical factors less than unity, which characterize the waveguide geometry, and Φ is a correction to the space-charge dispersion relation due to the combined presence of the axial and magnetic fields, defined as [34]:

$$\Phi = 1 - \{\Omega_0\gamma_z^2\beta_w^2/[(\beta_z^2 + \beta_w^2)\Omega_0 - \gamma\beta_z^2k_w v_z]\} \tag{6.3}$$

where $\Omega_0 = eB_z/m$ is the nonrelativistic cyclotron frequency in the axial field. The value of Φ can be positive, negative, or zero, depending on the parameters of the experiment.

Maximum gain of the FEL instability occurs near the frequency corresponding to the intersections of the space charge and electromagnetic modes. The solution of (6.2a) and (6.2b) leads to the radiation frequency:

$$\omega = \beta_z c k_w'\gamma_z^2\{1 \pm \beta_z[1 - (\omega_c/k_w'\gamma_z v_z)^2]^{1/2}\} \tag{6.4}$$

where the effective wiggler wave number is

$$k_w' = k_w - p_1\omega_p\Phi^{1/2}/\gamma^{1/2}\gamma_z v_z$$

A plot of these curves is shown in Figure 6.5. Note that there are upper and lower intersections corresponding to the ± sign of (6.4). Also shown is the cyclotron line $\omega = \Omega_0/\gamma + kv_z$, with its intersections, from which we can see that the generation of cyclotron emission can interfere with the operation and interpretation of the FEL.

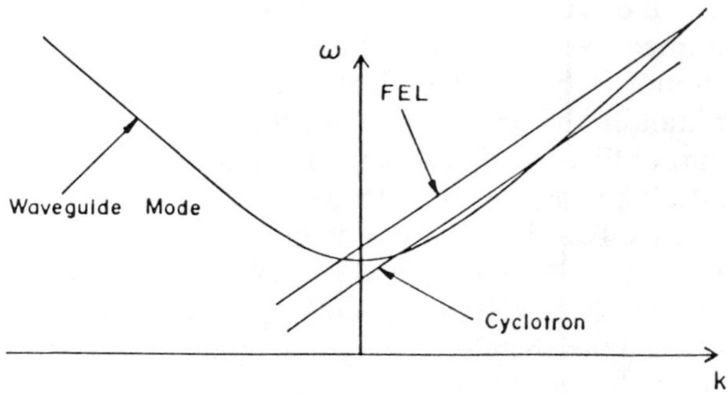

Fig. 6.5 Dispersion Diagrams for a Waveguide FEL
Note: Interaction occurs near the intersection of the waveguide mode and the line resulting from the FEL resonance condition that $v_{ph} = v_z$. Also shown is the cyclotron beam line for the case when an axial magnetic field is present. Cyclotron emission can occur at the intersection of this line with the waveguide mode. Usually, multiple modes can exist, resulting in a more complicated diagram.

6.4.2 Electron Motion

From (6.4), we may see that the output frequency of the FEL depends on the axial velocity component. If we assume an idealized helical wiggler with a purely sinusoidal axial variation and no radial variation, we may show from the equations of motion that a constant radius helical trajectory can result [35]. The velocity components of these trajectories are

$$\beta_z = (1 - \gamma^{-2} - \beta_w^2)^{1/2} \tag{6.5a}$$

$$\beta_w = \beta_z\Omega_w/(\Omega_0 - \gamma k_w v_z) \tag{6.5b}$$

These idealized orbits can be nearly achieved in practice, provided that the wiggler has an adiabatic entrance. Clearly, a resonance occurs when the denominator of (6.5b) vanishes. However, as resonance is approached, the radial excursions of the electrons become large and the assumptions made in the derivation of (6.5) are no longer valid.

The solutions of the fourth-order polynomial formed by (6.5) are plotted in Figure 6.6, using the parameters from the NRL experiment [20]. The solutions fall into two categories: the so-called Group I orbits apply when $\Omega_0 < \gamma k_w v_z$ (below resonance), and the Group II orbits result above resonance. Also indicated are the corrections that result when a nonideal wiggler is assumed. As resonance is approached from either above or below, axial motion is transformed into perpendicular motion.

The variation in the parameter Φ for the two types of orbits is shown in the lower part of Figure 6.6. The sign of Φ significantly affects the nature of the FEL instability. For $\Phi > 0$, we recover the usual FEL regime, in which the collective FEL instability couples a stable electromagnetic wave with a stable beam-space-charge mode. However, the regime above gyroresonance, in which $\Phi < 0$, is particularly interesting because the FEL interaction then couples the electromagnetic mode to a beam-space-charge wave, which is itself unstably driven by the presence of the wiggler field [36]. This regime is predicted to offer broader bandwidth as well as higher efficiency, and is the regime in which the NRL's VEBA experiments operate best [20, 21]. For these reasons, an axial magnetic field is an attractive option that may be incorporated in devices which would otherwise not require such a feature for beam formation and space-charge control.

6.4.3 Experimental Verification

The most definitive and detailed experimental study of these axial field effects was carried out at MIT by Fajans [37]. He used a 175 kV, 10–50 A electron beam to measure frequency, gain, and efficiency as a function of the axial and wiggler field amplitudes, and studied both amplifier and oscillator configurations. The narrowband (<2%) output radiation was tuned from 7 to 21 GHz by varying the electron energy, as shown in Figure 6.7. Output powers of 100 kW at an interaction efficiency of 12% and of 1 MW at 8% efficiency were achieved. The results of the experiments agreed very well with three-dimensional theory and simulations [36], as evidenced by the data shown in Figure 6.7.

6.5 FEL COMPONENTS

Any FEL consists of four major components, each of which is critical to the success of the device. First, the electron accelerator controls the beam energy, current, pulse duration and pulse repetition rate, and beam

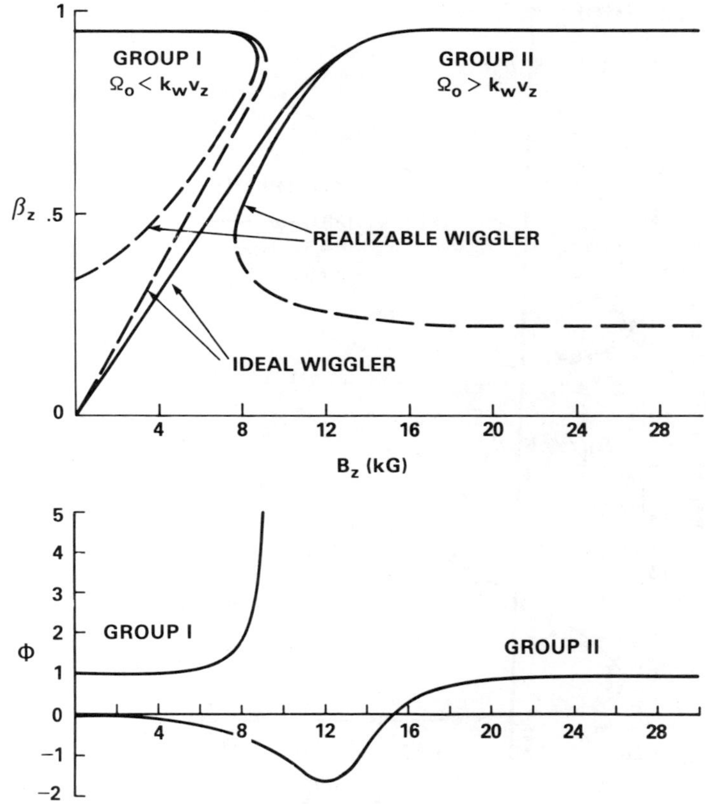

Fig. 6.6 Plots of Axial Electron Velocity (top) and the Parameter Φ Defined in Eq. (6.3) (bottom) as a Function of Axial Magnetic Field *Note*: $\gamma = 3.4$ and $B_w = 630$ G, typical of NRL's VEBA FEL [20, 21]. The dashed lines represent unstable electron orbits. (Courtesy of S.H. Gold.)

quality. Second, the electron optics or focusing fields determine the electron beam size and, if not carefully designed, the fields can cause a deterioration in beam quality. Third, the wiggler, which must be designed to be compatible with the electron beam, is the major contributor to the ponderomotive potential, and the wiggler controls efficiency enhancement. Fourth, the radiation optics determine the optical beam quality, and they must be able to tolerate the desired power levels. We will briefly address some important issues affecting the choice or design of these major components.

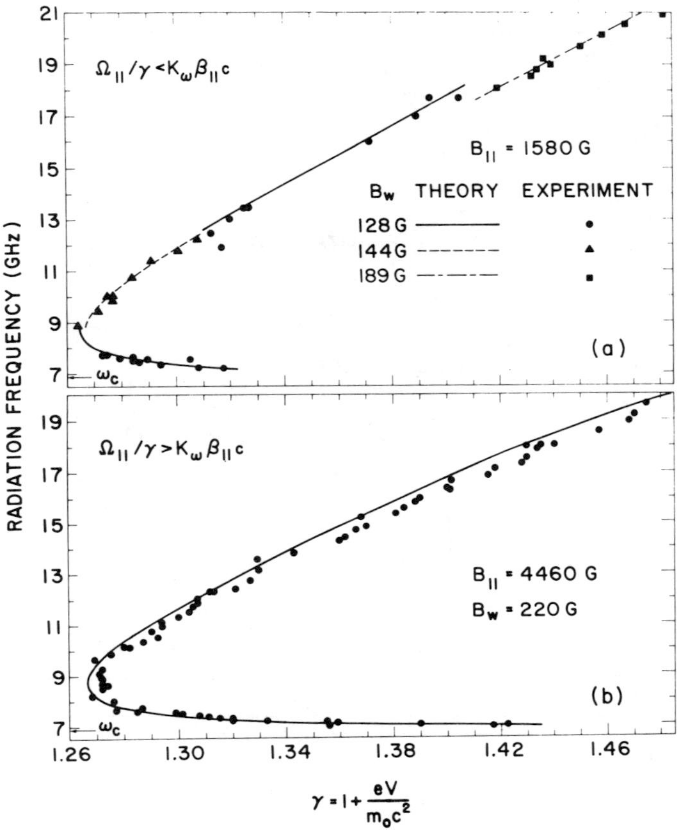

Fig. 6.7 Radiation Frequency as a Function of Electron Energy in the MIT FEL Amplifier Experiment [37]

Note: Results are shown both below resonance (top) and above resonance (bottom). (Courtesy of J. Fajans.)

6.5.1 Electron Acceleration and Transport

The choice of an accelerator is largely determined by the desired operating energy, but the design must be compatible with the other FEL requirements. A listing of the various types of accelerators appropriate for FEL experiments at longer wavelengths, and the advantages and disadvantages of each, are shown in Table 6.2. Each type of accelerator has its own particular features, which must be carefully considered when designing an experiment. For example, the induction linac is capable of high-

Table 6.2 Accelerators for FELs in the 0.1–10 mm Range

Type	*Advantages*	*Disadvantages*
Modulators	Moderately high current Long-duration, repetitive pulses Relatively inexpensive	Voltage limited to ≈500 kV
Pulse Lines	Very high current Relatively inexpensive Feasible up to a few MV	Limited pulse duration (≈100 ns) Difficult to pulse repetitively
Induction Linacs	High current Scalable to high voltage	Relatively expensive Long pulse duration requires very large cores
Van de Graafs	Excellent voltage flatness Long pulse duration (with beam recovery) Relatively high voltage	Low current (tens of amperes with beam recovery)

current operation and the electron energy can be scaled to high values by adding more induction modules (e.g., 50 MeV at 10 kA from the ATA at Livermore). However, the system is very long (ATA is more than 100 m in length) and complex. Because the versatility of the FEL allows operation over such a large parametric range, it is unreasonable to expect any single type of accelerator to be generally preferable. The accelerator must be chosen on a case-by-case basis.

Electron beam focusing is usually provided by electromagnets, either solenoids or quadrupoles. Some beam focusing or steering is always required to transport the beam from the accelerator to the wiggler. Most millimeter-wave FELs have also employed magnetic focusing through the wiggler because the combination of high current and relatively low energy would otherwise tend to cause the beam to defocus due to its own space charge.

6.5.2 Wiggler Types

Nearly all FEL experiments have been performed with either a *helical wiggler* or a *linear wiggler* [1, 2]. Helical wigglers usually consist of a pulsed, bifilar winding, which produces a magnetic field perpendicular to the axis, but with a direction that rotates about the axis. Such a wiggler provides radial focusing for the beam so that an external focusing field can be avoided if so desired [38]. However, the helical wiggler is compatible with an external axial magnetic field to enhance the FEL interaction, as described in the previous section. The major disadvantage of the helical wiggler is the difficulty of tapering it for efficiency enhancement, although a permanent-magnet helical wiggler has been recently proposed, which could be more easily tapered [39]. The reader should also note that helical wigglers produce circularly polarized radiation. Schematic diagrams of helical wigglers, together with the field formulas, are shown in Figure 6.8.

Linear wigglers can be constructed from electromagnets or permanent magnets [40]. In either case, tapering is straightforward. The linear wiggler produces a field in a single transverse direction with a sinusoidally varying amplitude. It therefore produces linearly polarized radiation. However, the linear wiggler provides focusing in only one plane (along the direction of the wiggler field), and it is not compatible with an axial magnetic field, which induces an outward drift of off-axis electrons [41]. Focusing can be provided by quadrupoles, by shaping the wiggler-field pole pieces to provide a small degree of field curvature, or by using a "square" wiggler [41, 25]. Typical linear wiggler configurations and field formulae are shown in Figure 6.9.

6.5.3 Optical System

The radiation optics required depend on the wavelength of the FEL. Infrared and shorter wavelength FELs require optical mirrors and lenses. A major concern for high-power operation is the damage threshold for these components. At 10 μm, for example, the maximum single-pulse power density that can be tolerated on a copper mirror is about $150/\tau^{1/2}$ MW/cm^2, where τ is the pulse duration in microseconds. Resonator designs capable of handling ultra-high-power levels are being actively studied for short-wavelength FELs.

Because of the large optical diffraction at longer wavelengths, most millimeter-wave FELs are operated in overmoded waveguide, so microwave techniques are appropriate. For amplifier or super-radiant operation, the procedures are straightforward. The major concern is to prevent the excitation of unwanted modes. In practice, unwanted modes have been almost totally eliminated by carefully controlling the electron beam and judiciously choosing the FEL parameters.

Long-wavelength oscillators pose an additional problem: the design of mirrors that can be used in the waveguide without perturbing the electron beam. The simplest approach is to place an aperture in the waveguide

Helical Wigglers

$$B_x = B_0 \cos k_w z, \quad B_y = B_0 \sin k_w z, \quad B_z = 0 \quad (r = 0)$$

$$K_w = \frac{eB_0\lambda_w}{2\pi mc} = 0.0934\ B_0(kG)\ \lambda_w(cm)$$

$$\lambda = \frac{\lambda_w}{2\gamma^2}(1 + K_w^2)$$

- Bifilar Helix

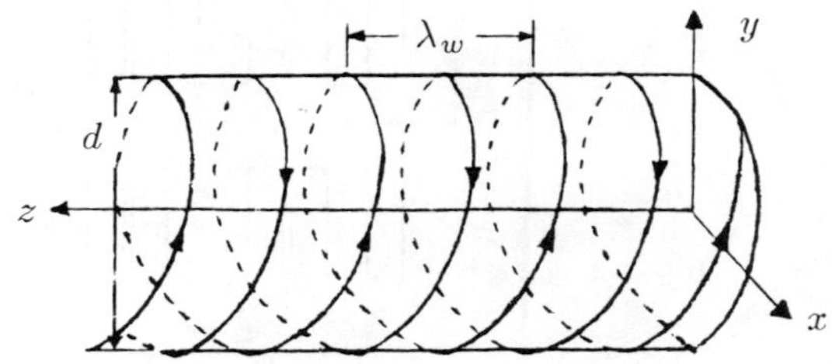

$$B_0 = 8\pi \times 10^{-7} I\left[\xi K_0(\xi) + K_1(\xi)\right], \quad \xi = \pi d/\lambda_w$$

If $\pi d > \lambda_w$, then $K \approx 3.8 \times 10^{-4} I e^{-0.9\xi}$.

- Permanent Magnets

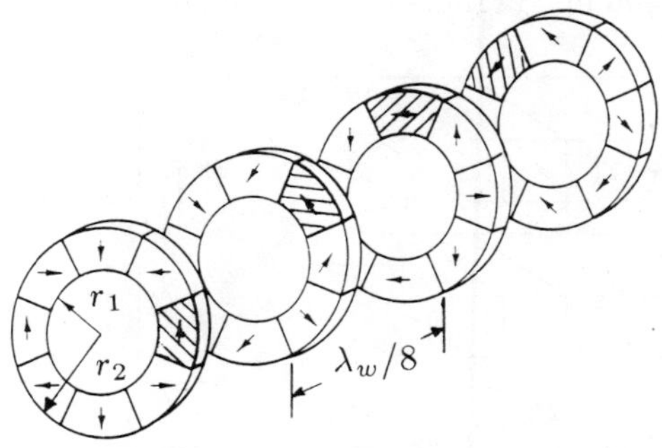

$$B_0 = 0.8\ B_r\ \left[T(2\pi r_1/\lambda_w) - T(2\pi r_2/\lambda_w)\right], \quad \text{where}$$

$$T(x) = K_0(x) + (x/2)K_1(x), \quad B_r = \text{remanent field.}$$

Fig. 6.8 Schematic Diagrams of Helical Wigglers with the Corresponding Spatial Variations of Magnetic Field

Planar Magnets

$$B_x = 0, \quad B_y = B_0 \sin k_w z, \quad B_z = 0 \qquad (r = 0)$$

$$\lambda = \frac{\lambda_w}{2\gamma^2}\left(1 + K_w^2/2\right)$$

- Electromagnet

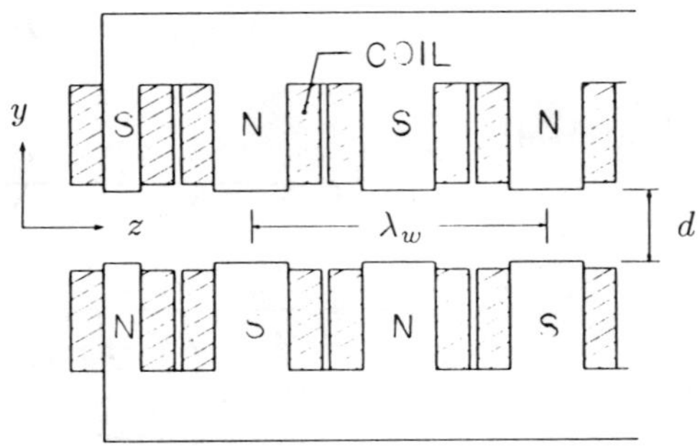

$$B_0 = \frac{32\mu_0 NI}{\sqrt{2}\pi\lambda_w}\left[\frac{1}{\sinh(\xi)} - \frac{1}{3\,\sinh(3\xi)}\right], \qquad \xi = \pi d/\lambda_w .$$

- Permanent Magnets (Pure or Hybrid)

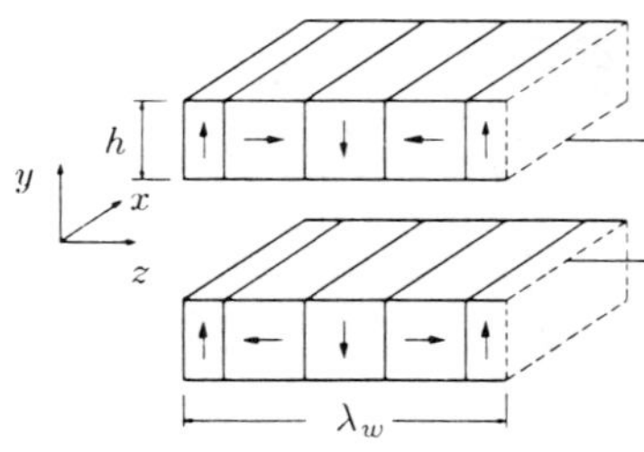

$$B_0 = 1.8\, B_r\,\left(1 - e^{-k_w h}\right) e^{-k_w d/2}$$

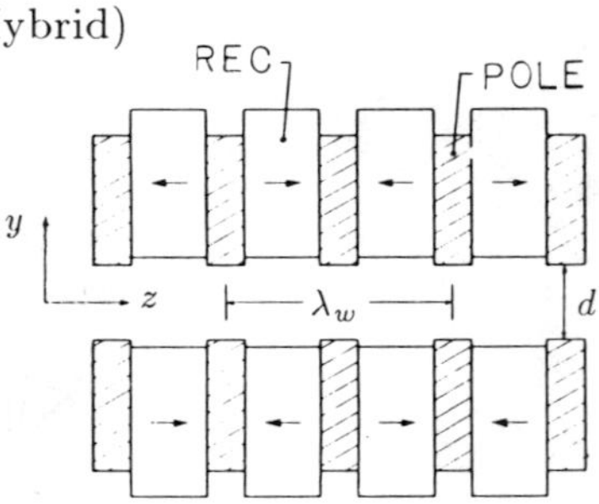

$$B_0 = \frac{4\, B_r}{\sqrt{2}\pi}\left[\frac{1}{\sinh(\xi)} - \frac{1}{3\,\sinh(3\xi)}\right]$$

B_r = remanent field $\approx 9\ kG$ for REC

Fig. 6.9 Schematic Diagrams of Linear (or Planar) Wigglers with the Corresponding Spatial Variations of Magnetic Field through which the electron beam can pass. The discontinuity created by the aperture can reflect 50% or more of the FEL radiation, which is often enough given the high gain of these devices. A more sophisticated technique is the use of a Bragg reflector, which consists of a length of corrugated waveguide [42]. By properly choosing the shape and number of corrugations, it has been possible to obtain reflectivities in excess of 99% [43]. The Bragg reflector works much like a grating: the wave is reflected when the periodicity of the corrugations is half the wavelength of the guided radiation. Consequently, a Bragg reflector operates only at a single wavelength or its harmonics (for a single waveguide mode). The fixed wavelength can be desirable in some applications, but it is also a drawback in that it precludes the tuning of the FEL.

6.5.4 Electron Beam Quality

The most important and often most difficult task in any FEL is the generation and transport of a high-quality electron beam. A variation in axial beam energy can greatly reduce the electron bunching process, thereby lowering the growth rate and efficiency. A rule of thumb is that the fractional energy spread must be small compared with the intrinsic interaction efficiency, which tends to place a more stringent requirement on beam quality at short wavelengths. Electron energy variations can be produced by the beam *emittance* (the intrinsic, random, perpendicular velocity components of the electrons), gradients in the wiggler field, the beam's own space charge, and any voltage variations that occur during acceleration.

The normalized electron-beam emittance $\epsilon_n = \beta\gamma r_b\theta$, where r_b is the beam radius and θ is the angular spread of the electron velocities, is an important quantity in the characterization of beam quality for FELs. Even if all of the electrons have identical total energies, the emittance produces a spread in the axial energy component which is given by $\Delta\gamma_z/\gamma_z = \epsilon_n^2/2r_b^2$. Some of the most important contributors to beam emittance are nonuniform cathode emission, cathode surface roughness, nonlinear accelerating electric fields or focusing fields, magnetic field aberrations or misalignment, nonlinear space-charge forces, and mismatches between focusing elements. Only by very careful design and fabrication can the beam emittance be kept small enough for efficient FEL operation at short wavelenghts.

Another major contributor to axial energy spread in FELs is the wiggler field gradient. The wiggler gradient contribution arises because in any physically realizable wiggler, the magnetic field amplitude must increase away from the axis. In simple terms, the higher off-axis field induces higher wiggle velocities than the lower field on axis. The higher perpendicular velocity results in a lower axial velocity for the off-axis particles. Quantitatively, $\Delta\gamma_z/\gamma_z = (eB_w r_b/2mc)^2$. Consequently, this effect is more pronounced at higher wiggler fields and larger beam radii. Because the FEL gain and efficiency typically increase with wiggler field and do so inversely with beam radius, it is usually much better to reduce the beam radius than to reduce the wiggler field amplitude. However, the electron beam radius should not be substantially smaller than that of the radiation beam.

The electron space charge also contributes to the energy spread because of the electrostatic potential drop across the beam. This potential energy variation produces a variation in the kinetic energy if the total energy of all the electrons is the same. This effect reduces the kinetic energy of on-axis particles relative to the off-axis particles. Note that the general effect is opposite that of the wiggler gradient, so that to some extent the two contributions cancel. However, it is the total energy and not just the axial component that is affected by the space charge. The total energy spread due to space charge is $\Delta\gamma/\gamma = \nu/\gamma$, where ν is the normalized current $I/17$ kA. It is possible to design the accelerator and focusing fields so that most of this total energy spread appears in the transverse direction, rather than the axial direction. In the ideal case of Brillouin flow, all of the space-charge-induced energy spread is in the transverse component.

We stress here that the whole issue of electron beam quality is an integral part of the total FEL research effort. Experiments and numerical simulations to improve cathodes, electron guns, and focusing geometry are being actively pursued at many laboratories. Since the mid-1980s, there has been improvement by more than an order of magnitude in beam brightness, which has been achieved at several energy and current levels, and the prospects are good for further improvement.

6.6 EXAMPLES

To give a better picture of actual FEL devices, we present a brief description of two particular experiments. The first is a high-power amplifier at Lawrence Livermore National Laboratory [8, 33]. The LLNL experiment is important because it has demonstrated the high efficiencies that can be achieved with the previously described enhancement techniques. The second experiment to be discussed is a high-gain, long-pulse oscillator experiment at NRL [32]. The NRL experiment has demonstrated essentially steady-state operation for 1 μs, corresponding to nearly 200 transits of the radiation within the resonator. The results of this experiment offer encouragement about the prospects for a two-stage FEL oscillator, which will be discussed briefly in the next section.

6.6.1 LLNL Amplifier

The Electron Laser Facility (ELF) at LLNL is shown schematically in Figure 6.10. The device is powered by the *experimental test accelerator* (ETA), which is an induction linac capable of producing 5 MeV, 10 kA, 50 ns electron pulses at a 1 Hz continuous rate. After emerging from the accelerator, the beam is passed through a series of magnetic transport, focusing, and conditioning elements so that the beam reaching the FEL is of sufficient quality. The emittance selector shown in the schematic is a long cylindrical pipe in a solenoid with adjustable field. At higher fields, more beam current is transmitted, but there is higher emittance. A crucial element of a FEL is the proper matching of the electron beam into the wiggler. The additional focusing magnets shown serve this purpose.

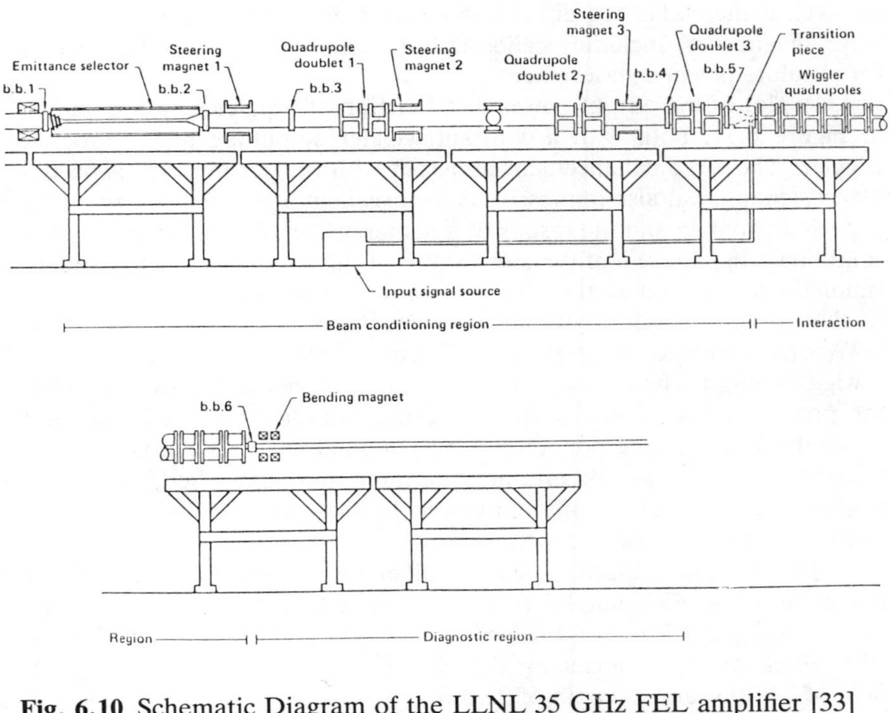

Fig. 6.10 Schematic Diagram of the LLNL 35 GHz FEL amplifier [33] *Note*: Overall length of the portion illustrated is about 18 m. (Courtesy of T. Orzechowski.)

The ELF wiggler is a planar electromagnet having a 9.8 cm period. Each two periods are separately controlled for arbitrary tapering of the wiggler amplitude. Because planar magnets provide focusing in only the direction parallel to the field, quadrupoles are superimposed on the wiggler to provide focusing in the opposite direction. The maximum available wiggler length is 3 m, but any shorter value can be selected by simply turning off the power supplies to the unwanted portion. To maintain alignment of the beam inside the wiggler, the first period has a profile of 0.3 B_w, -0.8 B_w.

The input power to be amplified is supplied by a 60 kW magnetron at a wavelength of 8.6 mm. The power is coupled into the 3 cm × 10 cm rectangular waveguide which passes through the wiggler in the TE_{01} mode, so that the electric field is parallel to the broad wall and the electron wiggle velocity. Up to 30 kW can be injected in the proper mode into the FEL.

The FEL output is radiated into a large chamber, which provides 41 dB of attenuation through diffraction. Without this attenuation, the high-power output of the device breaks down the air in the laboratory. A receiver at the end of the diffraction chamber leads to a variety of microwave diagnostics, including calibrated attenuators, filters, and detectors for absolute power measurement.

A plot of the output power as a function of wiggler length is shown in Figure 6.11, both with a constant wiggler amplitude and a tapered wiggler. The large enhancement in output with wiggler tapering is particularly striking, but also impressive is the excellent agreement between the experimental data and the results of a numerical simulation (shown by the solid line). In fact, one of the motivations of this experiment is to validate simulations, which may then be used to determine scaling of the device to shorter wavelengths. Without tapering, the peak output of about 200 MW corresponds to an interaction efficiency of about 8% and occurs with a wiggler length of just under 1.5 m. As the wiggler length is increased, the power decreases as a result of the trapped electrons oscillating and thermalizing inside the ponderomotive potential wells. When the field is tapered, beginning just before the saturation point is reached, the power continues to increase rapidly until over 35% of the electron energy has been converted into radiation.

The success of LLNL's FEL amplifier provides optimism that high-power, infrared FEL amplifiers can also be demonstrated. Experiments are beginning at LLNL's ATA facility (50 MeV, 10 kA) to test a 10.6 μm FEL amplifier. The wiggler for that device has a period of 8 cm, a peak field of 2.3 kG, and a length of 25 m.

6.6.2 NRL Oscillator

The FEL oscillator [32] is shown schematically in Figure 6.12, along with sample oscillograms of the electron beam current injected into the

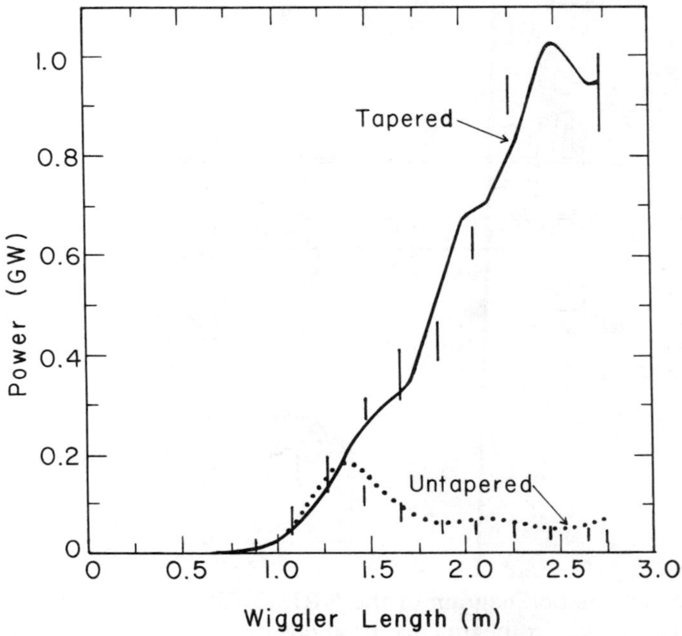

Fig. 6.11 Output Power *versus* Wiggler Length for the LLNL 35 GHz FEL Amplifier [8, 33]
Note: Vertical lines represent the range of experimentally measured powers, while the solid and dotted lines result from numerical simulations for the tapered and untapered wigglers, respectively. $I \approx 850$ A, $V \approx 3$ MeV, and $B_w \approx 3.6$ kG (before tapering).

wiggler and the output microwave pulse. This device is driven by a 750 kV, 750 A induction linac, which has a 2 μs pulse duration. The electron beam is injected into the FEL resonator through an aperture and a microwave reflector, positioned just outside the wiggler. The beam current entering the wiggler is varied between 50 and 300 A by varying the size of the aperture. The wiggler is a bifilar helix having a 4 cm period with a six-period-long adiabatic entry and exit. The constant amplitude section of the wiggler is 22 periods long, and the field can be varied up to 1 kG. The electrons are focused solely by the wiggler field. The radiation is coupled out of the resonator through a partially transmitting output mirror, followed by a microwave horn. This oscillator operates at a wavelength of 8–12 mm, depending on the electron energy and wiggler field strength.

Various mirror types have been employed, including the Bragg reflector shown in the diagram. This Bragg reflector had a reflectivity of 65%

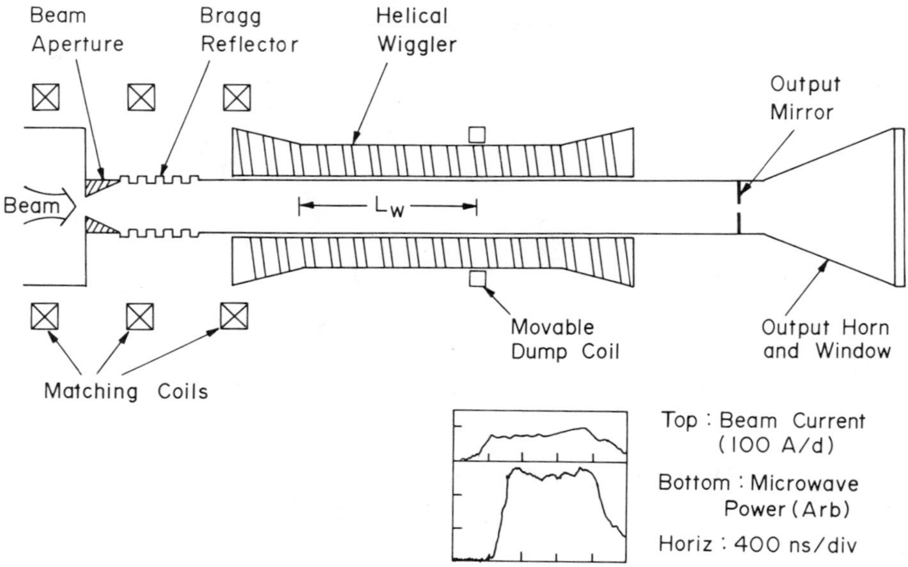

Fig. 6.12 Schematic Diagram of the NRL Oscillator Experiment [32]
Note: Also shown are beam current and output radiation oscillograms.

at 32 GHz, with a bandwidth of about 1%. The only operating parameter significantly affected by the type of output mirror is the pulse duration. A wire-mesh reflector has been found to break down due to the microwave field, resulting in a significantly shortened output pulse.

The oscillator start-up is studied by varying the FEL interaction length. The dump coil shown in the schematic is moved along the wiggler to dump the beam to the waveguide wall at the desired location. Because this is a single-shot experiment (one pulse every few minutes), no special beam dump is necessary. The start-up time (defined as the delay between beam injection and the growth of microwaves to a measurable level) is plotted *versus* interaction length in Figure 6.13, along with the peak output power. There is a sharp threshold interaction length L_{th}, below which the loss per pass exceeds the gain, so that oscillation never occurs. For lengths greater than L_{th}, the power is relatively constant, but the turn-on time drops dramatically with increasing interaction length. Qualitatively, this behavior is typical of an oscillator, in which the saturation level is essentially independent of the gain, but a larger gain per pass reduces the start-up time.

In a high-gain FEL oscillator, the cavity power grows with time as [4, 44]:

$$P = P_0 \exp\{2[\Gamma L_w - \ln(m/R^{1/2})]t/T_b\} \quad (6.6)$$

where P_0 is the spontaneous radiation power level, Γ is the linear electric field growth rate, R is the net power reflectivity of the mirrors, L_w is the interaction length, and $T_b = 2L/c$ is the round-trip bounce time in the resonator of length L. The factor m represents the coupling loss for the interaction. For a Raman interaction $m = 2$, but in the Compton regime $m = 3$.

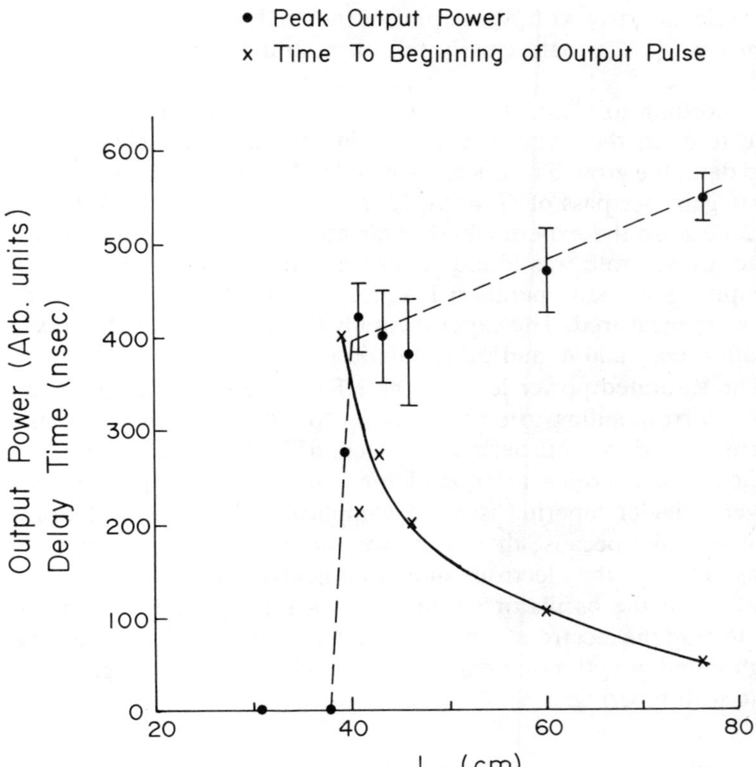

Fig. 6.13 Output Power and Oscillator Turn-on Time *versus* Effective Wiggler Length from the NRL Experiment
Note: $I \approx 100$ A, $V \approx 500$ keV, and $B_w \approx 400$ G.

From (6.6), it is clear that a threshold interaction length exists for net gain per pass, such that

$$\Gamma = \ln(m/R^{1/2})/L_{\text{th}} \tag{6.7}$$

Consequently, the growth rate can be determined by measuring L_{th} if we assume exponential growth. A more rigorous test of the theory and a verification of exponential growth can be achieved by rewriting (6.6) as

$$0.5 \ln(P_t/P_0) = L_w N_b \Gamma - N_b \ln(m/R^{1/2}) \tag{6.8}$$

where P_t is the power at the turn-on time T and $N_b = T/T_b$. Note that $L_w N_b$ varies linearly with N_b. A plot of the NRL oscillator data is shown in Figure 6.14. The data clearly fall on a straight line, as predicted by (6.8).

According to (6.8), the growth rate is simply the ratio of the logarithmic term to the slope of the straight line in Figure 6.14. From the plotted data, the growth rate is determined to be 0.045 cm^{-1}, which implies a power gain per pass of $G = \exp(2\Gamma L_w) = 3000$. (Here, we have used $m = 2$ because the experimental parameters indicate Raman operation, and the growth rate was found experimentally to vary linearly with B_w, also implying Raman operation.) At larger values of B_w, gains per pass of 10^4 were measured. The experimentally determined growth rates agree well with theory and numerical simulation.

The saturated power level in the NRL experiment range from 1 to 10 MW, corresponding to efficiencies up to 6%. Simulations of this experiment have shown efficiencies in excess of 20% with a tapered wiggler [44]. Clearly, a tapered wiggler oscillator could be a very impressive device. However, wiggler tapering is more complicated with an oscillator than with an amplifier because there is no well defined point at which to begin tapering. Ideally, the electrons should be nearly trapped before tapering begins, but in the oscillator multiple passes through the wiggler are required to trap the electrons. Therefore, a trade-off must be made between the high initial growth rate of an untapered wiggler and the high efficiency possible with tapering.

6.7 FUTURE DIRECTIONS

Ongoing experimental FEL work in the visible to infrared portions of the spectrum is directed toward increasing the efficiency and power. Higher current and better quality electron beams, high-power optical systems that incorporate sophisticated components such as grazing incidence mirrors, and electron energy recovery are all being actively pursued in the laboratory. Studies of harmonic generation and sideband instabilities are

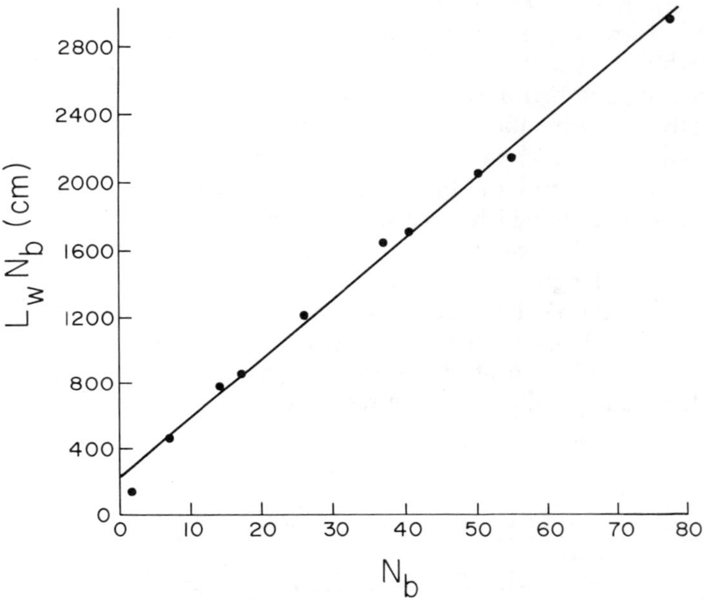

Fig. 6.14 Results of the NRL Experiment Plotted for Comparison with Eq. (6.8)

Note: N_b is the number of round-trip bounces of the radiation in the resonator before power is detected, and L_w is the effective wiggler length.

also an important part of the effort. At some laboratories, notably Stanford, an effort to develop practical, user-facility FELs for research and medical applications is underway. In several of these experiments, interaction lengths are proposed to be many times longer than the Rayleigh range of the optical beam. In order for these devices to work, the radiation beam must be guided and focused by the electron beam. Consequently, optical guiding is a subject of intense theoretical and experimental interest as of this writing [45, 46].

In the far-infrared to microwave portion of the spectrum, there is also an effort to increase power and efficiency, especially at the short-wavelength end of the region. Perhaps more importantly, experiments are beginning to emphasize more practical systems with high overall efficiency. Such devices are typically powered by repetitively pulsed modulators with carefully designed electron guns employing thermionic cathodes. Oscillator experiments are underway at Hughes [24, 43] as is an amplifier experiment at NRL [47]. A system employing beam recovery has been proposed by TRW [25]. These experiments could lead to compact, high-power, tube devices, which would be useful for a variety of applications, including plasma heating and high-resolution radar systems.

Finally, experimental work is beginning on a two-stage FEL [48, 49]. In this device, shown schematically in Figure 6.15, the radiation produced by a conventional FEL (the first stage) is reflected back on the electron beam. This radiation, which has a wavelength $\lambda_1 = \lambda_w/2\gamma_z^2$, acts as a wiggler for the second stage interaction. Radiation produced in the second stage has a wavelength $\lambda_2 = \lambda_1/4\gamma_z^2 = \lambda_w/8\gamma_z^4$. Consequently, very short wavelength radiation can be produced with moderate electron energy. For example, a 5 MeV electron beam together with a 3 cm period wiggler would produce radiation at about 150 μm in a conventional FEL. The same system could produce 1.5 μm radiation in a two-stage FEL.

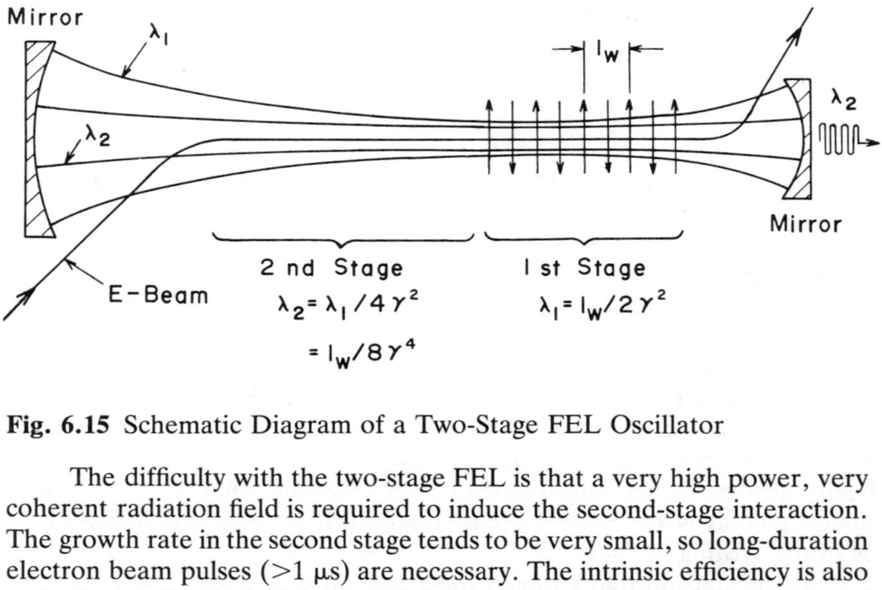

Fig. 6.15 Schematic Diagram of a Two-Stage FEL Oscillator

The difficulty with the two-stage FEL is that a very high power, very coherent radiation field is required to induce the second-stage interaction. The growth rate in the second stage tends to be very small, so long-duration electron beam pulses (>1 μs) are necessary. The intrinsic efficiency is also typically small, but efficiency enhancement is possible if the first- and second-stage radiation fields are strong enough to produce trapping in the ponderomotive wave. The electron beam quality must be very high for successful operation of an efficient two-stage FEL at short wavelengths. Experiments are planned at the University of California at Santa Barbara, using an electrostatic generator, and at NRL, using a long-pulse induction linac.

6.8 CONCLUSIONS

The FEL has clearly become a major new radiation source. In recent years, the interest in FELs has grown dramatically as experimental devices have verified the exciting predictions of theory. In the millimeter-wavelength regime, the FEL has the highest power of any available source. When coupled with the inherent tuning ability and reasonable efficiency of FEL sources, the uniquely high-power capability is even more impressive. Not surprisingly, what has until now been an experimental laboratory device is being seriously considered for various specialized applications. Although difficult technical issues remain to be solved before the FEL can be used for some of the more exotic applications, there has been rapid progress.

REFERENCES

1. "S.F. Jacobs *et al.*, eds., *Free-Electron Generators of Coherent Radiation,*" *Physics of Quantum Electronics,* Vols. 7,8,9 (Addison-Wesley, Reading, MA, 1980–1982).
2. *IEEE J. Quant. Electr.* **QE-21** (7) Special Issue on Free-Electron Lasers (1985).
3. P. Sprangle, R.A. Smith, and V.L. Granatstein, in K.J. Button, ed. *Infrared and Millimeter Waves,* Vol. 1 (Academic Press, New York, 1979), 279–327.
4. T.C. Marshall, *Free-Electron Lasers* (Macmillan, New York, 1985).
5. P. Sprangle, C.M. Tang, and W. Manheimer, *Phys. Rev. A* **21**, 302 (1980).
6. N.M. Kroll, P.L. Morton, and M.N. Rosenbluth, *IEEE J. Quant. Electr.* **QE-17**, 1436 (1981); See also, Ref. [1], Vol. 7, 89.
7. R.W. Warren, B.E. Newnam, and J.C. Goldstein, Ref. [2], 882.
8. T.J. Orzechowski, *et al., Phys. Rev. Lett.* **57**, 2172 (1986).
9. R.M. Phillips, *IRE Trans. Electron Dev.* **7**, 231 (1960).
10. L.R. Elias, *et al., Phys. Rev. Lett.* **36**, 717 (1976).
11. D.A.G. Deacon, *et al., Phys. Rev. Lett.* **38**, 892 (1977).
12. V.L. Granatstein, *et al., IEEE Trans. Microwave Theory Tech.* **MTT-22**, 1000 (1974).
13. P.C. Efthimion and S.P. Schlesinger, *Phys. Rev. A* **16**, 633 (1977).
14. A. Amir *et al., Appl. Phys. Lett.* **47**, 1251 (1985).
15. M. Billardon *et al.*, Ref. [2], 805.
16. J.A. Nation, *Appl. Phys. Lett.* **17**, 491 (1970).

17. M. Friedman and M. Herndon, *Phys. Rev. Lett.* **28**, 210 (1972).
18. V.L. Granatstein, *et al., Appl. Phys. Lett.* **30**, 384 (1977).
19. D.B. McDermott, *et al., Phys. Rev. Lett.* **41**, 1368 (1978).
20. R.K. Parker, *et al., Phys. Rev. Lett.* **48**, 238 (1982).
21. S.H. Gold, *et al., Phys. Fluids* **27**, 746 (1984).
22. S.H. Gold, *et al., Phys. Rev. Lett.* **52**, 1218 (1984).
23. R.E. Shefer and G. Bekefi, in Ref. [1], Vol. 9, 703.
24. J. Fajans, *et al., Phys. Rev. Lett.* **53**, 246 (1984).
25. R.J. Harvey and F.A. Dolezal, *Nucl. Instrum. Meth. in Phys. Res.* **A250**, 274 (1986).
26. M.Z. Caponi, *et al.,* in Ref. [1], Vol. 7, 523; H. Boehmer, *et al., Bull. APS* **31**, 1482 (1986).
27. J. Masud, *et al., Phys. Rev. Lett.* **56**, 1567 (1986).
28. S.C. Chen and T.C. Marshall, *Phys. Rev. Lett.* **52**, 425 (1984).
29. W.W. Destler, *et al., Phys. Fluids* **28**, 1962 (1985).
30. P. Avivi, *et al., Bull. APS* **30**, 1539 (1985).
31. K. Felch, *et al., Bull APS* **26**, 945 (1981).
32. J. Mathew and J.A. Pasour, *Phys. Rev. Lett.* **56**, 1805 (1986).
33. T.J. Orzechowski, *et al., Phys. Rev. Lett.* **54**, 889 (1985).
34. H.P. Freund and P. Sprangle, Phys. Rev. A **28**, 1835 (1983).
35. L. Friedland, *Phys. Fluids* **23**, 2376 (1980).
36. H.P. Freund, *Phys. Rev. A* **27**, 1977 (1983); H.P. Freund and A.K. Ganguly, *Phys. Rev. A* **28**, 3438 (1983).
37. J. Fajans, *et al., Phys. Fluids* **28**, 1995 (1983).
38. J.A. Pasour and S.H. Gold, in Ref [2], 845.
39. P. Diament, in Ref. [2], 1094.
40. K. Halbach, *Nucl. Instr. and Meth.* **169**, 1 (1980); *IEEE Trans. Nucl. Sc.* **NS-28**, 3136 (1981).
41. J.A. Pasour, *et al., J. Appl. Phys.* **53**, 7174 (1982).
42. V.L. Bratman, *et al., IEEE J. Quant. Electr.* **19**, 282 (1983).
43. R.J. Harvey, *et al., Proc. 8th Int. Free-Electron Laser Conf.,* Glasgow, Scotland, Sept. 1986; see also *Nucl. Instrum. Meth. Phys. Res.* (1987).
44. W.P. Marable, P. Sprangle, and C.M. Tang, *Proc. 7th Int. Free-Electron Laser Conf.*, Sept. 1985; see also *Nucl. Instrum. Meth. Phys. Res.* A250, 159 (1986).
45. C.M. Tang and P. Sprangle, in Ref. [1], Vol. 9, 627 (1982).
46. E.T. Scharlemann, A.M. Sessler, and J.S. Wurtele, *Phys. Rev. Lett.* **54**, 1925 (1985).
47. R.H. Jackson, D.E. Pershing, C.M. Armstrong, and R.K. Parker, *Proc. of 8th Int. Free-Electron Laser Conf.,* Glasgow, Scotland, Sept. 1986; see also *Nucl. Instrum. Meth. Phys. Res.* (1987).

48. L.R. Elias, *Phys. Rev. Lett.* **42**, 977 (1979).
49. P. Sprangle and R.A. Smith, *Phys. Rev. A* **21**, 293 (1980).

Chapter 7

THE AXIAL INJECTION ORBITRON MASER

John M. Burke

National Research Council Associate
and
Plasma Physics Division
Naval Research Laboratory
Washington, DC

Wallace M. Manheimer

Plasma Physics Division
Naval Research Laboratory
Washington, DC

Edward Ott

Science Applications International Corporation
McLean, Virginia
and
Departments of Electrical Engineering and
Physics and Astronomy
University of Maryland
College Park, Maryland

7.1 INTRODUCTION

Many devices have been developed over the past 50 years to generate centimeter- and millimeter-wavelength electromagnetic waves. The most highly developed of these radiation sources are slow-wave devices such as magnetrons, klystrons, traveling wave tubes, and backward wave oscillators [1]. The output power capabilities of slow-wave devices is limited because progressively smaller dimensions of the electron beam and interaction circuit are required with shorter wavelengths. Hence, the generation of millimeter-wavelength electromagnetic waves with slow-wave radiation sources becomes difficult due to fabrication difficulties and heat dissipation.

The development of fast-wave devices, principally the *electron cyclotron maser* (ECM) or gyrotron [2, 3] has made possible the efficient generation of high-power millimeter-wavelength electromagnetic waves. One drawback to these devices with respect to size and cost, however, is that they require large magnetic fields. Thus, there is current interest in developing a high-power millimeter-wave source that does not require the use of a large magnetic field. In this chapter, we propose and analyze a device configuration that appears to hold great promise in fulfilling this goal. The device is called the *axial injection orbitron maser* or AXIOM [4], and it is a combination of the basic orbitron maser concept of Alexeff and Dyer [5] and the standard ECM configuration. The AXIOM employs coaxial cylindrical geometry with the center conductor held at a positive potential with respect to the outer conductor, as shown in Figure 7.1. Electrons are radially confined by the positive potential of the center conductor and orbit around it in low angular momentum, noncircular orbits as shown in Figure 7.2.

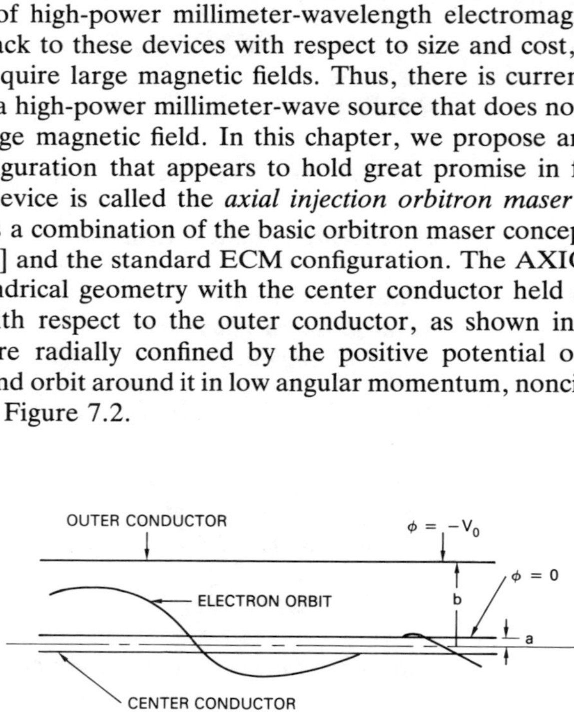

Fig. 7.1 The Orbitron Configuration

We will demonstrate that a high-frequency, high-efficiency maser that operates at high-power levels can be realized by using the AXIOM concept. Electrons would be injected with a large value of axial momentum in a region with a low radial electric field. They would then be adiabatically compressed into the interaction region where the radial electric field is

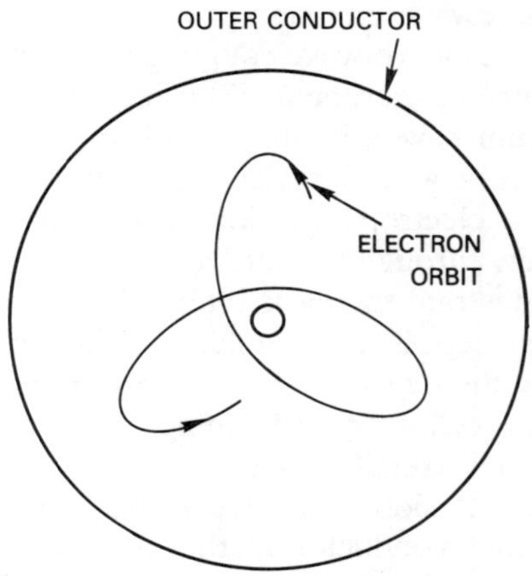

Fig. 7.2 An Axial View of an Electron Orbit

large. The adiabatic compression process has the advantage that it increases the electron density as well as the ratio of their perpendicular to parallel energy. The electrons would not be axially trapped in the interaction region as in the early orbitron experiments but would rapidly transit the interaction region where they would resonantly transfer their energy to the radiation fields. This injection scheme is very similar to the one used in most ECM devices which have been demonstrated to operate at high powers with high efficiencies. The AXIOM would also operate in high vacuum, as is the case with most ECM devices. Most importantly, axial injection allows one to maximize the efficiency of the maser interaction by providing a means to produce an electron beam whose distribution function is optimized for the power level requirements of the device. It also provides a means of optimally positioning the electron beam to interact with a given waveguide or cavity mode.

The AXIOM concept differs in a number of ways from previous radiation sources. The AXIOM is a fast-wave device that exploits the rich harmonic content and high-power density of an axially injected and compressed electron beam with low angular momentum and noncircular electron orbits. In the first orbitron experiments of Alexeff and Dyer [5], electrons were introduced into the interaction region by creating a glow discharge between the center and outer conductor. Broadband microwave emission was observed in a series of short bursts of radiation. It is now believed that the microwave emission process in the glow discharge orbitron is due to nonlinear wave-wave coupling of counterstreaming plasma waves [6]. Later orbitron experiments [7] using a radial electron injection scheme in high vacuum have demonstrated low-power centimeter-wavelength emission, which is almost certainly due to the process detailed in this work. The radial electron injection scheme provides no means to control the power flow through the interaction region, and can therefore only be used to demonstrate the basic interaction mechanism.

The AXIOM concept is also similar to other previously developed devices, most notably the *heliotron* [8, 9], the *spiratron* [10] and the *strophotron* [11, 12]. The heliotron and the spiratron are devices in which axially injected electrons execute circular orbits about a center conductor in a coaxial waveguide. A slow-wave structure is placed in the waveguide (an axially slotted inner conductor in the heliotron and a helix in the spiratron) to interact with the electron beam. These devices demonstrated the ability to generate centimeter wavelength emission but suffer from the low output power capabilities of slow-wave devices. Also, high voltages on the center conductor are required in a heliotron for high-frequency operation. By use of the notation of Figure 7.1, the frequency of a nonrelativistic electron orbiting the center conductor in a circular orbit of radius r is

$$f = \frac{2.111}{r(\text{mm})}\sqrt{\frac{V_0(\text{kV})}{\ln\left(\frac{b}{a}\right)}}\ \text{GHz} \tag{7.1}$$

Due to these problems, the development of the heliotron and the spiratron ceased with the introduction of solid-state devices in the early 1960s. The strophotron is actually very similar to the AXIOM, except that it requires a strong magnetic field to confine and drift the electrons through the interaction region.

For high harmonic operation, we wish to take advantage of the structure of an electron orbit with a small minimum radius, so the natural inclination is to examine modes with the largest fields near the center conductor. For this reason, we restrict our attention to TEM modes. In these, E_r and B_θ are both proportional to $1/r$ between the center and outer conductor. We have succeeded in analyzing the interaction of a TEM waveguide or cavity mode with low angular momentum electrons transiting the interaction region of an orbitron. This allows us to quantify the spatial growth rate for an infinite-length traveling wave tube amplifier, as well as the threshold values of the cavity Q multiplied by the electron beam power required for self-oscillation in a finite-length oscillator configuration. We have also developed the basic design equations necessary for the axial electron-injection scheme.

The chapter is organized as follows. Section 7.2 presents the linear theory of the TEM orbitron maser. The dispersion relation for the axially uniform orbitron is derived by solving the linearized Vlasov equation by the method of characteristics. The integration over unperturbed orbits is performed by expressing the radial velocity of the electron multiplied by the radial part of the electric field's eigenfunction in a Fourier series. The radial part of the eigenfunction for the TEM mode has a particularly simple form, $1/r$. It will be shown that the interaction strength at a given harmonic of the radial oscillation frequency is proportional to the square of the Fourier coefficient for that frequency in the Fourier series expansion of the electron's radial velocity divided by its radius. The radial oscillation period for a nonrelativistic electron and other quantities having to do with the electron orbits in an orbitron which are necessary to evaluate the linear theory of the orbitron interaction are calculated in Section 7.3. We will show that σ, which is defined as the ratio of the maximum to the minimum radius of the electron's orbit, is an important parameter. Specifically, the maximum harmonic number for which there can be a strong interaction is roughly $\sigma/2$. In Section 7.4, we use the technique developed in Section 7.2 to calculate the power transferred from the electron beam to the radiation fields as the beam traverses a finite length cavity. The results of this calculation are combined with the definition of the cavity Q and an expression for the energy stored in the cavity in terms of the radiation electric field amplitude to find the beam power required to make the cavity self-oscillate. A cold beam distribution function in which each electron is assumed to have a single value of energy, angular momentum, and axial momentum is presented in Section 7.5. This distribution function allows easy evaluation of the results of Sections 7.2 and 7.4, which are given in terms of a general distribution function. Section 7.6 presents the results for the infinite-length traveling wave tube amplifier and the finite-length oscillator configurations. The spatial growth rate is given *versus* frequency and beam parameters for the traveling wave tube amplifier. For the finite-length oscillator, we plot the cavity Q multiplied by the threshold beam power required for self-oscillation *versus* the voltage between the center and outer conductor and beam parameters. Finally, in Section 7.7, we discuss the consistency of the approximations used in the theory, some mode-selection schemes, and present a conceptual design of the beam-injection system. We also use the constancy of the angular momentum of an electron in a TEM orbitron maser with an azimuthally symmetric electron beam to calculate the maximum possible efficiency of the device.

7.2 PHYSICAL MODEL: THE AXIALLY UNIFORM ORBITRON

The configuration of the orbitron maser that we will describe is shown in Figures 7.1 and 7.2. It consists of an electron beam propagating axially through a coaxial waveguide of circular cross section with inner radius a and outer radius b. The axis of the electron beam coincides with that of the waveguide. The electrons are radially confined by the electric field between the center and outer conductors of the waveguide, and move in axis encircling orbits with a substantial fraction of their kinetic energy transverse to the axis.

The following simplifying assumptions are made:

1. The electron beam distribution function and the cavity fields are both independent of the azimuthal angle θ;
2. The electron beam is of low enough density so that its self-electrostatic field can be neglected compared to the static confining electric field due to the applied voltage difference between the inner and outer surfaces of the waveguide;
3. The cavity fields are a first order perturbation of the static confining electric field;
4. The perturbed distribution function f^1 is of first order with respect to the initial distribution function f^0.

The TEM mode is the lowest order mode for a coaxial waveguide and we shall restrict the analysis of the maser action in the orbitron to this mode. The TEM mode has a particularly simple structure which greatly simplifies the analysis compared with that of the TE mode which also should give rise to a maser interaction.

In this section, we examine the orbitron interaction by calculating the dispersion relation for an axially uniform system. This will set the stage for the calculations in Section 7.4 of start oscillation conditions for a finite-length cavity, as well as show how to handle the effect of the complex orbit.

If we use the convention shown in Figure 7.1 for the potentials on the conductors, the electrostatic potential energy of an electron is

$$q\phi(r) = P_0 \ln\left(\frac{r}{a}\right) \tag{7.2}$$

where

$$P_0 = \frac{eV_0}{\ln\left(\dfrac{b}{a}\right)} \tag{7.3}$$

The Hamiltonian can then be written

$$H = \frac{p_r^2}{2m} + \frac{P_\theta^2}{2mr^2} + \frac{p_z^2}{2m} + P_0 \ln\left(\frac{r}{a}\right) \tag{7.4}$$

where

$$\mathbf{p} = m\mathbf{v} \tag{7.5}$$

and the electron is assumed to be nonrelativistic. Now, since the Hamiltonian is independent of time, θ and z we have three constants of the motion H, P_θ, and p_z, where

$$P_\theta = rp_\theta = mr^2\dot{\theta} \tag{7.6}$$

To find the motion of an electron, we first determine the radius as a function of time by combining (7.4) and (7.5),

$$\mathrm{d}t = \frac{\mathrm{d}r}{\sqrt{\dfrac{2H}{m} - \dfrac{P_\theta^2}{m^2r^2} - \dfrac{p_z^2}{m^2} - \dfrac{2P_0}{m}\ln\left(\dfrac{r}{a}\right)}} \tag{7.7}$$

Once $r(t)$ is known, θ as a function of time can be computed by integrating (7.6) in time.

The angular momentum term in (7.7) provides an outward radial force which dominates at small r, the electrostatic potential provides an inward force which dominates at large r. Thus, the orbit has an inner and outer radial turning point, r_i and r_o. At each turning point, the radial velocity is, of course, zero.

The orbit is characterized by a shape, which is invariant given the constants of motion, and an initial value. For the more conventional case of a sinusoidal or circular orbit, the former corresponds to the amplitude and frequency, while the latter corresponds to the phase. To specify the shape of the orbit, we will consider an orbit such that $t = 0$ is the time at which $r = r_o$ and define this orbit as $r = R(t)$ and $v_r = V(t)$. The reader may note that $R(t)$ is an even function of t, while $V(t)$ is an odd function of t. Schematics of $R(t)$ and $V(t)$ are shown in Figure 7.3. The orbit has a full period $T(H, P_\theta, p_z)$, which depends only on the constants of the motion.

To calculate the dispersion relation for the orbitron interaction in an axially uniform system we use Maxwell's equations to find an equation that relates the perturbed electric field in the waveguide to the perturbed current. We then find the perturbed current in terms of the perturbed electric field by integrating Vlasov's equation over unperturbed orbits. Combining the two relations yields the dispersion relation.

Now, starting with Maxwell's equations:

$$\nabla \cdot \mathbf{E} = \frac{\rho}{\epsilon_0} \tag{7.8}$$

$$\nabla \cdot \mathbf{B} = 0 \tag{7.9}$$

$$\nabla \times \mathbf{E} + \frac{\partial \mathbf{B}}{\partial t} = 0 \tag{7.10}$$

$$\nabla \times \mathbf{B} - \frac{1}{c^2}\frac{\partial \mathbf{E}}{\partial t} = \mu_0 \mathbf{J} \tag{7.11}$$

we combine (7.10) and (7.11) with an assumed time dependence of $e^{-i\omega t}$ to obtain an equation for the electric field:

$$\nabla \times \nabla \times \mathbf{E} - \frac{\omega^2}{c^2}\mathbf{E} = i\omega\mu_0 \mathbf{J} \tag{7.12}$$

To proceed, we expand **E** as a linear superposition of orthogonal vacuum waveguide modes. Here, we focus specifically on the TEM mode, because this mode has maximum electric field near the center conductor, where, as we will see, the orbit has the most harmonic structure. The waveguide fields of the TEM mode are given by

$$\mathbf{E}^1(r, \theta, z, t) = E_1 \frac{a}{r} e^{i(kz - \omega t)} \hat{r} \tag{7.13}$$

$$\mathbf{B}^1(r, \theta, z, t) = E_1 \frac{a}{cr} e^{i(kz - \omega t)} \hat{\theta} \tag{7.14}$$

and we assume that the beam current is small so that the mode is nearly a vacuum mode. This is analogous to assumptions made in calculating the linear theory of the ECM. Then, as discussed by Ott and Manheimer [13], once we determine the form of $J_r^1(r, \theta, z, t)$ driven by the waveguide fields given by (7.13) and (7.14), we take the dot product of (7.12) with the complex conjugate of (7.13) and integrate over the waveguide cross section to find the dispersion relation:

$$\left[k^2 - \frac{\omega^2}{c^2}\right] \int |\mathbf{E}^1|^2 \, r dr d\theta = i\omega\mu_0 \int \mathbf{E}^{1*} \cdot \mathbf{J}^1 \, r dr d\theta \tag{7.15}$$

Finding the perturbed current means integrating over the unperturbed orbits which are unspecified as yet, and quite complicated. However, we will see shortly that the effect of this complex orbit can be characterized by a single parameter.

The perturbed current density, in terms of the perturbed distribution function, is given by

$$J_r^1(r, \theta, z, t) = -\mathrm{e} \int v_r f^1(\mathbf{r}, \mathbf{p}, t) \, \mathrm{d}^3 p \tag{7.16}$$

We use the Vlasov equation:

$$\frac{\partial f(\mathbf{r}, \mathbf{p}, t)}{\partial t} + \mathbf{v} \cdot \frac{\partial f(\mathbf{r}, \mathbf{p}, t)}{\partial \mathbf{r}} + q(\mathbf{E}(\mathbf{r}, t) + \mathbf{v} \times \mathbf{B}(\mathbf{r}, t)) \cdot \frac{\partial f(\mathbf{r}, \mathbf{p}, t)}{\partial \mathbf{p}} = 0 \tag{7.17}$$

to find the perturbed distribution function, where

$$f(\mathbf{r}, \mathbf{p}, t) = f^0(\mathbf{r}, \mathbf{p}) + f^1(\mathbf{r}, \mathbf{p}, t) \tag{7.18}$$

Then, to first order,

$$f^1(\mathbf{r}, \mathbf{p}, t) = \mathrm{e} \int_{-\infty}^{t} \mathrm{d}t' \, [\mathbf{E}^1(\mathbf{r}', t') + v'(\mathbf{r}', t') \times \mathbf{B}^1(\mathbf{r}', t')] \frac{\partial f^0(\mathbf{r}', \mathbf{p}')}{\partial \mathbf{p}'} \tag{7.19}$$

The quantities r and v_r implicit in (7.19) are evaluated along the unperturbed orbit, where the orbit is initialized by specifying that the velocity $V(t)$ is v_r and the position $R(t)$ is r at time $t = t'$. If $\hat{t}(r, v_r)$ is the time between 0 and $T(H, P_\theta, p_z)$ such that $V(\hat{t}) = v_r$ and $R(\hat{t}) = r$, the unperturbed orbit is specified by $R(y)$ and $V(y)$, where

$$y = t' - t + \hat{t}(r, v_r) \tag{7.20}$$

By inserting (7.13) and (7.14) into (7.19), and using the fact that

$$\frac{\partial f^0(H, P_\theta, p_z)}{\partial p_r} = v_r \frac{\partial f^0(H, P_\theta, p_z)}{\partial H} \tag{7.21}$$

we find an expression for the perturbed distribution function:

$$f^1(\mathbf{r}, \mathbf{p}, t) = eaE_1 \left[\left(1 - \frac{v_z}{c}\right) \frac{\partial f^0}{\partial H} + \frac{1}{c} \frac{\partial f^0}{\partial p_z} \right] \times \int_{-\infty}^{t} dt' \, e^{i(kz'(t') - \omega t')} \frac{V(t' - t + \hat{t}(r, v_r))}{R(t' - t + \hat{t}(r, v_r))} \tag{7.22}$$

Because $V(y)/R(y)$ is an odd function of y, it can be expanded as a sine series:

$$\frac{V(y)}{R(y)} = \sum_{s=1}^{\infty} a_s(H, P_\theta, p_z) \sin(s\Omega(H, P_\theta, p_z)y) \tag{7.23}$$

where $\Omega(H, P_\theta, p_z) = 2\pi/T(H, P_\theta, p_z)$ and $a_s(H, P_\theta, p_z)$ is the Fourier coefficient, which is a function of the constants of the motion.
Now, if we let

$$x = t' - t \tag{7.24}$$

$$v_z' = v_z \tag{7.25}$$

then,

$$kz'(t') = kz + kv_z x \tag{7.26}$$

$$t' = x + t \tag{7.27}$$

Inserting (7.23), (7.26), and (7.27) into (7.22) yields the perturbed distribution function:

$$f^1(\mathbf{r}, \mathbf{p}, t) = eaE_1 \left[\left(1 - \frac{v_z}{c}\right) \frac{\partial f^0}{\partial H} + \frac{1}{c} \frac{\partial f^0}{\partial p_z} \right] \times \sum_{s=1}^{\infty} a_s(H, P_\theta, p_z) \int_{-\infty}^{0} dx \, e^{i[kz - \omega t + (kv_z - \omega)x]} \times \sin\{s\Omega(H, P_\theta, p_z)[x + \hat{t}(r, v_r)]\}$$

$$= -\frac{eaE_1}{2}\left[\left(1 - \frac{v_z}{c}\right)\frac{\partial f^0}{\partial H} + \frac{1}{c}\frac{\partial f^0}{\partial p_z}\right]$$

$$\times \sum_{s=1}^{\infty} a_s(H, P_\theta, p_z)\left\{\frac{e^{i[kz - \omega t + s\Omega(H,P_\theta,p_z)\hat{t}(r,v_r)]}}{[s\Omega(H, P_\theta, p_z) - \omega + kv_z]}\right.$$

$$\left. + \frac{e^{i[kz - \omega t - s\Omega(H,P_\theta,p_z)\hat{t}(r,v_r)]}}{[s\Omega(H, P_\theta, p_z) + \omega - kv_z]}\right\} \tag{7.28}$$

and combining (7.16) and (7.28), we find that

$$i\omega\mu_0 J_r^1 = i\mu_0\omega\frac{e^2aE_1}{2}\sum_{s=1}^{\infty}\int d^3p a_s(H, P_\theta, p_z)v_r$$

$$\times \left[\left(1 - \frac{v_z}{c}\right)\frac{\partial f^0}{\partial H} + \frac{1}{c}\frac{\partial f^0}{\partial p_z}\right]$$

$$\times \left\{\frac{e^{i[kz - \omega t + s\Omega(H,P_\theta,p_z)\hat{t}(r,v_r)]}}{[s\Omega(H, P_\theta, p_z) - \omega + kv_z]}\right.$$

$$\left. + \frac{e^{i[kz - \omega t - s\Omega(H,P_\theta,p_z)\hat{t}(r,v_r)]}}{[s\Omega(H, P_\theta, p_z) + \omega - kv_z]}\right\} \tag{7.29}$$

Equations (7.13), (7.15), and (7.29) then yield the dispersion relation:

$$\left(k^2 - \frac{\omega^2}{c^2}\right)a^2\ln\left(\frac{b}{a}\right) = i\mu_0\omega\frac{e^2a^2}{2}$$

$$\times \sum_{s=1}^{\infty}\int dr\int d^3p v_r a_s(H, P_\theta, p_z)$$

$$\times \left[\left(1 - \frac{v_z}{c}\right)\frac{\partial f^0}{\partial H} + \frac{1}{c}\frac{\partial f^0}{\partial p_z}\right]$$

$$\times \left\{\frac{e^{i[s\Omega(H,P_\theta,p_z)\hat{t}(r,v_r)]}}{[s\Omega(H, P_\theta, p_z) - \omega + kv_z]}\right.$$

$$\left. + \frac{e^{-i[s\Omega(H,P_\theta,p_z)\hat{t}(r,v_r)]}}{[s\Omega(H, P_\theta, p_z) + \omega - kv_z]}\right\} \tag{7.30}$$

Notice that at this point there is still a fairly complex dependence on r and v_r through the exponential of $\Omega(H, P_\theta, p_z)\hat{t}(r, v_r)$. Utilizing the Jacobian to transform the integration over d^3p to an integration over $dH dP_\theta dp_z$,

$$\left(k^2 - \frac{\omega^2}{c^2}\right) \ln\left(\frac{b}{a}\right) = i\mu_0\omega \frac{e^2}{2} \sum_{s=1}^{\infty} \int_a^b dr \int dH\, dP_\theta dp_z \frac{1}{r} \frac{v_r}{|v_r|} \sum_{\pm}$$
$$\times a_s(H, P_\theta, p_z) \left[\left(1 - \frac{v_z}{c}\right) \frac{\partial f^0}{\partial H} + \frac{1}{c} \frac{\partial f^0}{\partial p_z}\right]$$
$$\times \left\{ \frac{e^{i[s\Omega(H,P_\theta,p_z)\hat{t}_\pm(r,v_r)]}}{[s\Omega(H, P_\theta, p_z) - \omega + kv_z]} + \frac{e^{-i[s\Omega(H,P_\theta,p_z)\hat{t}_\pm(r,v_r)]}}{[s\Omega(H, P_\theta, p_z) + \omega - kv_z]} \right\} \tag{7.31}$$

where the summation over $\pm$ is a summation over $\hat{t}_+(r, v_r)$ and $\hat{t}_-(r, v_r)$. Here $\hat{t}_+(r, v_r)$ is the time for the particle to reach r with positive v_r. Using the time origin of Figure 7.3, we see that $T/2\ (H, P_\theta, p_z) < \hat{t}_+(r, v_r) < T(H, P_\theta, p_z)$. Similarly, $\hat{t}_-(r, v_r)$ is the time required for the particle to reach r with negative radial velocity and $0 < \hat{t}_-(r, v_r) < T/2\ (H, P_\theta, p_z)$. We now bring the integration over radius inside the integration over momentum space and perform the integration over radius by changing the variable of integration to $\hat{t}(r, v_r)$. Then,

$$\frac{dr}{r} = \frac{V(\hat{t})}{R(\hat{t})} d\hat{t} \tag{7.32}$$

and (7.31) becomes

$$\left(k^2 - \frac{\omega^2}{c^2}\right) \ln\left(\frac{b}{a}\right) = i\mu_0\omega \frac{e^2}{2} \sum_{s=1}^{\infty} \int dH dP_\theta dp_z\, a_s(H, P_\theta, p_z) \left[\left(1 - \frac{v_z}{c}\right) \frac{\partial f^0}{\partial H} + \frac{1}{c} \frac{\partial f^0}{\partial p_z}\right]$$
$$\times \left\{ \int_{\frac{T}{2}(H,P_\theta,p_z)}^{T(H,P_\theta,p_z)} \frac{V(\hat{t})}{R(\hat{t})} d\hat{t} \left[\frac{e^{i[s\Omega(H,P_\theta,p_z)\hat{t}]}}{[s\Omega(H, P_\theta, p_z) - \omega + kv_z]} + \frac{e^{-i[s\Omega(H,P_\theta,p_z)\hat{t}]}}{[s\Omega(H, P_\theta, p_z) + \omega - kv_z]} \right] \right.$$
$$\left. - \int_{\frac{T}{2}(H,P_\theta,p_z)}^{0} \frac{V(\hat{t})}{R(\hat{t})} d\hat{t} \left[\frac{e^{i[s\Omega(H,P_\theta,p_z)\hat{t}]}}{[s\Omega(H, P_\theta, p_z) - \omega + kv_z]} + \frac{e^{-i[s\Omega(H,P_\theta,p_z)\hat{t}]}}{[s\Omega(H, P_\theta, p_z) + \omega - kv_z]} \right] \right\} \tag{7.33}$$

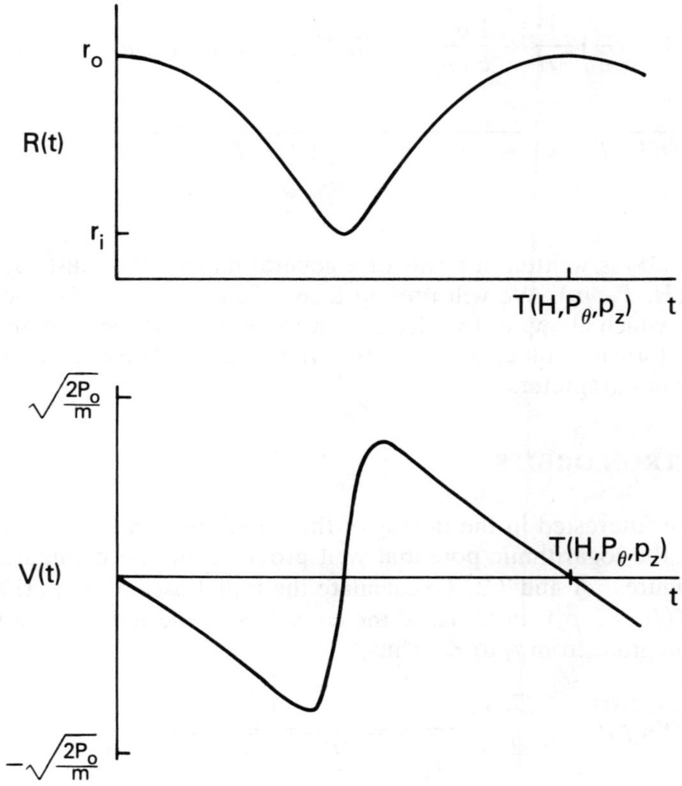

Fig. 7.3 The Radius and Radial Velocity of an Electron Orbiting the Center Conductor *versus* Time

Now, we have

$$\frac{V(\hat{t})}{R(\hat{t})} = \sum_{s'=1}^{\infty} a_{s'}(H, P_\theta, p_z) \sin[s'\Omega(H, P_\theta, p_z)\hat{t}] \tag{7.34}$$

Inserting (7.34) into (7.33) and integrating over $\hat{t}$, we obtain the dispersion relation for a forward traveling wave in an axially uniform orbitron in its final form:

$$\left(k^2 - \frac{\omega^2}{c^2}\right) \ln\left(\frac{b}{a}\right) = -\frac{\mu_0 e^2 \omega}{4} \sum_{s=1}^{\infty} \int \mathrm{d}H \mathrm{d}P_\theta \mathrm{d}p_z$$
$$\times \left\{ \left[\left(1 - \frac{v_z}{c}\right) \frac{\partial f^0}{\partial H} + \frac{1}{c} \frac{\partial f^0}{\partial p_z} \right] T(H, P_\theta, p_z)\, a_s^2(H, P_\theta, p_z) \right.$$
$$\left. \times \left[\frac{1}{[s\Omega(H, P_\theta, p_z) - \omega + kv_z]} - \frac{1}{[s\Omega(H, P_\theta, p_z) + \omega - kv_z]} \right] \right\} \tag{7.35}$$

Equation (7.35) is written in terms of a general unperturbed distribution function $f^0(H, P_\theta, p_z)$. We will present a specific distribution function in Section 7.5, which allows us to calculate, in Section 7.6, the spatial growth rate of the forward traveling wave as a function of frequency and the electron beam parameters.

7.3 ELECTRON ORBITS

We are interested in the details of the radial motion of an electron confined in the logarithmic potential well provided by the configuration shown in Figures 7.1 and 7.2. To calculate the radial oscillation period of the orbit, $T(H, P_\theta, p_z)$, in terms of the constants of the motion, we take (7.7) and integrate from r_i to r_o. Thus,

$$\frac{T}{2}(H, P_\theta, p_z) = \sqrt{\frac{m}{2}} \int_{r_i}^{r_o} \frac{\mathrm{d}r}{\sqrt{H - \dfrac{P_\theta^2}{2mr^2} - \dfrac{p_z^2}{2m} - P_0 \ln\left(\dfrac{r}{a}\right)}} \tag{7.36}$$

where r_o and r_i are the roots of the equation:

$$v_r = \sqrt{\frac{2}{m}} \sqrt{H - \frac{P_\theta^2}{2mr^2} - \frac{p_z^2}{2m} - P_0 \ln\left(\frac{r}{a}\right)} = 0 \tag{7.37}$$

To put (7.36) into a simpler form, we transform variables as follows [14]. Let

$$r = \frac{P_\theta}{\sqrt{2mP_0}} \frac{1}{\xi} \tag{7.38}$$

and

$$\beta(H, P_\theta, p_z) = \frac{H}{P_0} - \frac{p_z^2}{2mP_0} - \ln\left(\frac{P_\theta}{\sqrt{2mP_0a}}\right) \tag{7.39}$$

Equation (7.36) can then be written

$$T(H, P_\theta, p_z) = \frac{P_\theta}{P_0}\,\psi(\beta) \tag{7.40}$$

where

$$\psi(\beta) = \int_{\xi_o}^{\xi_i} \frac{d\xi}{\xi^2\sqrt{\beta + \ln\xi - \xi^2}} \tag{7.41}$$

and ξ_o and ξ_i are the two roots of the equation:

$$\beta(H, P_\theta, p_z) + \ln(\xi) - \xi^2 = 0 \tag{7.42}$$

We can use the particularly simple form of (7.40) to compute the derivatives $\partial T/\partial H$ and $\partial T/\partial p_z$, which are necessary to evaluate the general result given in (7.35) for the specific distribution function that will be described in Section 7.5. Indeed,

$$\begin{aligned}\frac{1}{T}\frac{\partial T(H, P_\theta, p_z)}{\partial H} &= \frac{1}{P_0}\frac{d}{d\beta}\ln[\psi(\beta)] \\ &= \frac{1}{P_0}\,\chi(\beta)\end{aligned} \tag{7.43}$$

and, similarly,

$$\frac{1}{cT}\frac{\partial T(H, P_\theta, p_z)}{\partial p_z} = -\frac{v_z}{c}\frac{1}{P_0}\,\chi(\beta) \tag{7.44}$$

Now, when we evaluate the theory given in Section 7.2, we choose to specify the constants of the motion P_θ and p_z in terms of the more physically interpretable parameters σ and α, where σ is the ratio of the maximum to the minimum radius of the electron's orbit and α is the square root of the ratio of the electron's perpendicular energy to the parallel energy; that is,

$$\sigma = \frac{r_o}{r_i} = \frac{\xi_i}{\xi_o} \tag{7.45}$$

and

$$\alpha = \sqrt{\frac{H_\perp}{H_\parallel}} \tag{7.46}$$

We will now show that $\beta(H, P_\theta, p_z)$ is uniquely specified by the single quantity σ. Equating the left-hand side of (7.42) for the root ξ_i to the left-hand side of (7.42) for the root ξ_o and using (7.45), we find the relation:

$$\xi_o^2 = \frac{\ln(\sigma)}{\sigma^2 - 1} \tag{7.47}$$

Inserting (7.47) into (7.42) then yields an equation for $\beta(H, P_\theta, p_z)$ in terms of σ. Thus,

$$\beta(\sigma) = \frac{\ln(\sigma)}{\sigma^2 - 1} - \frac{1}{2}\ln\left(\frac{\ln(\sigma)}{\sigma^2 - 1}\right) \tag{7.48}$$

The parameter $\chi(\beta)$ is therefore completely determined by the parameter σ and we define the function $\eta(\sigma)$ in terms of $\chi(\beta)$:

$$\eta(\sigma) = \chi[\beta(\sigma)] \tag{7.49}$$

A graph of $\eta(\sigma)$ *versus* σ is shown in Figure 7.4.

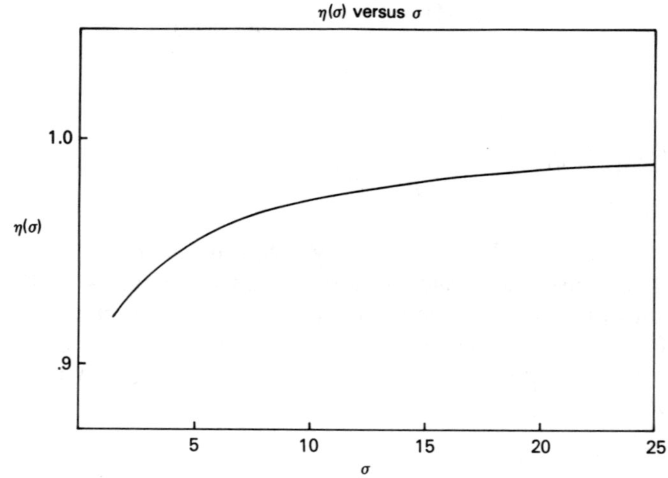

Fig. 7.4 The Parameter $\eta(\sigma)$ *versus* σ

The radial oscillation period in terms of H, σ, and α is given by

$$T = a\sqrt{\frac{2m}{P_0}}\exp\left[\frac{\alpha^2 H}{(1+\alpha^2)P_0} - \frac{\ln(\sigma)}{\sigma^2-1}\right]\zeta(\sigma) \tag{7.50}$$

where

$$\zeta(x) = \int_1^x \frac{d\xi}{\xi^2\sqrt{\dfrac{\ln(\sigma)}{\sigma^2-1}(1-\xi^2)+\ln(\xi)}} \tag{7.51}$$

The function:

$$t\left(\frac{r_0}{r}\right) = a\sqrt{\frac{m}{2P_0}}\exp\left[\frac{\alpha^2 H}{(1+\alpha^2)P_0} - \frac{\ln(\sigma)}{\sigma^2-1}\right]\zeta\left(\frac{r_0}{r}\right) \tag{7.52}$$

is the time it takes the electron to go from r_0 to r. To compute the radial oscillation period in the two limiting cases of a nearly circular orbit, $\sigma \to 1$, and in the zero angular momentum case, where $\sigma \to \infty$, we note that

$$\zeta(1) = \pi \tag{7.53}$$

and

$$\zeta(\infty) = \sqrt{\pi} \tag{7.54}$$

The function $\hat{T}(\sigma)$, where

$$\hat{T}(\sigma) = \frac{T}{a\sqrt{\dfrac{2m}{P_0}}\exp\left[\dfrac{\alpha^2 H}{(\alpha^2+1)P_0}\right]} \tag{7.55}$$

is displayed in Figure 7.5.

To evaluate the expression given in (7.35) for the dispersion relation for a forward traveling wave tube amplifier and the result of the oscillator calculation in the section that follows, we must also calculate the coefficient $a_s(H, P_\theta, p_z)$ in the Fourier series for $V(t)/R(t)$ given in (7.23) and its derivatives with respect to H and p_z. That is,

$$\frac{V(t)}{R(t)} = \sum_{s=1}^{\infty} a_s(H, P_\theta, p_z)\sin\left[\frac{2\pi s}{T(H, P_\theta, p_z)}t\right] \tag{7.56}$$

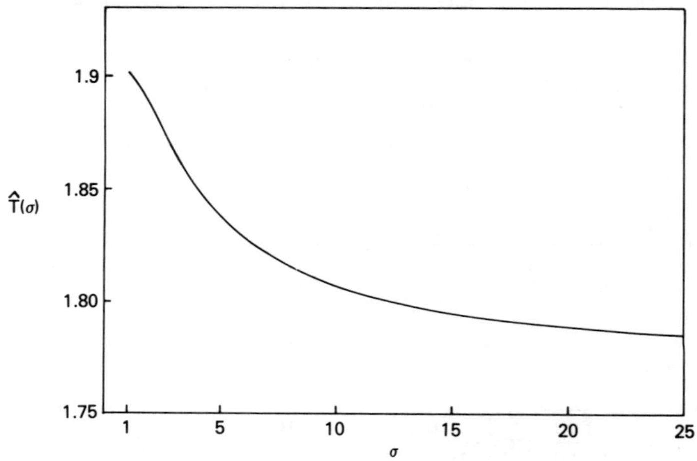

Fig. 7.5 The Normalized Radial Oscillation Period *versus* σ

and, therefore,

$$a_s(H, P_\theta, p_z) = \frac{4}{T(H, P_\theta, p_z)} \int_0^{\frac{T}{2}(H, P_\theta, p_z)} \frac{V(t)}{R(t)} \sin\left[\frac{2\pi s}{T(H, P_\theta, p_z)}\, t\right] \mathrm{d}t \tag{7.57}$$

Now, if we define

$$\kappa_s \equiv \frac{T(H, P_\theta, p_z)}{2\pi}\, a_s(H, P_\theta, p_z) \tag{7.58}$$

and change variables to $r_0/r(t)$, we find that

$$\kappa_s = -\frac{2}{\pi} \int_1^{\sigma} \frac{\mathrm{d}x}{x} \sin\left[s\pi\, \frac{\zeta(x)}{\zeta(\sigma)}\right] \tag{7.59}$$

We see that κ_s is a function of only σ and $\kappa_s^2(\sigma)$ is plotted *versus* s/σ for various values of σ and s in Figure 7.6. To compute the value of s, where $\kappa_s(\sigma)$ begins to decrease in amplitude, we combine (7.57) and (7.58) and integrate by parts, to obtain

$$\kappa_s(\sigma) = \frac{4s}{T}\int_0^{\frac{T}{2}} \ln\left[\frac{r_0}{r(t)}\right] \cos(s\Omega t)\, dt \tag{7.60}$$

The harmonic content of the radial oscillation period is principally determined by the motion of the electron near the minimum radius turning point, so we expand the ln $[r_0/r(t)]$ term in the integrand around $r(t) = r_i$ and examine its form. Using (7.52), we find

$$\begin{aligned}\ln\left[\frac{r_0}{r(t)}\right] &= \ln(\sigma)\left\{1 - \frac{\zeta^2(\sigma)\sigma^2}{\ln(\sigma)T^2}\left[\frac{2\sigma^2\ln(\sigma)}{\sigma^2 - 1} - 1\right]\left(t - \frac{T}{2}\right)^2 + \ldots\right\} \\ &\approx \ln(\sigma)\cos\left[\tilde{\omega}\left(t - \frac{T}{2}\right)\right]\end{aligned} \tag{7.61}$$

for $t \approx T/2$, where

$$\tilde{\omega} = \frac{\sigma\zeta(\sigma)\Omega}{\pi\sqrt{2\ln(\sigma)}}\sqrt{\frac{2\sigma^2\ln(\sigma)}{\sigma^2 - 1} - 1} \tag{7.62}$$

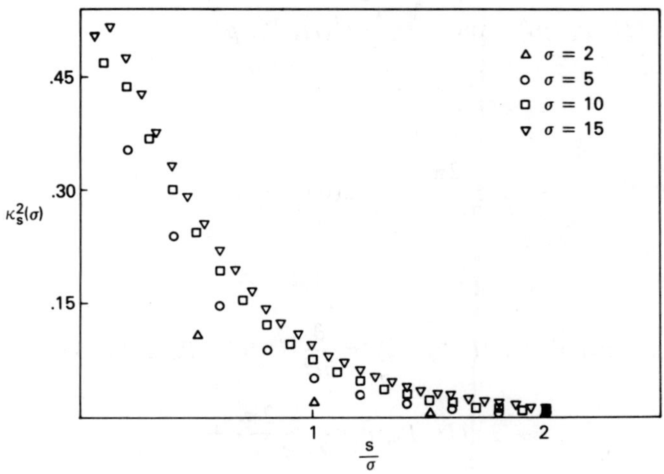

Fig. 7.6 The Coupling Coefficient $\kappa_s^2(\sigma)$ *versus* Harmonic Number for Various Values of σ

The radial oscillation contains frequencies up to $\sim\tilde{\omega}$, and therefore $\kappa_s(\sigma)$ should begin to decrease in amplitude for

$$s\Omega > \tilde{\omega} \tag{7.63}$$

or

$$\frac{s}{\sigma} \geqslant \frac{\zeta(\sigma)}{\pi\sqrt{2\ln(\sigma)}}\sqrt{\frac{2\sigma^2\ln(\sigma)}{\sigma^2 - 1} - 1} \tag{7.64}$$

Evaluating the expression on the right-hand side of (7.64), we find that $\kappa_s(\sigma)$ decreases for

$$\frac{s}{\sigma} \sim \frac{1}{2} \tag{7.65}$$

more or less independently of σ, as can be seen from Figure 7.6.

Last, to compute

$$\frac{\partial}{\partial H}\, a_s^2(H,\, P_\theta,\, p_z) \quad \text{and} \quad \frac{1}{c}\frac{\partial}{\partial p_z}\, a^2(H,\, P_\theta,\, p_z)$$

we have

$$a_s(H,\, P_\theta,\, p_z) = \frac{2\pi}{T(H,\, P_\theta,\, p_z)}\,\kappa_s(\sigma) \tag{7.66}$$

so,

$$\begin{aligned}\frac{\partial}{\partial H}\, a_s^2(H,\, P_\theta,\, p_z) &= 2a_s(H,\, P_\theta,\, p_z)2\pi\,\frac{\partial}{\partial H}\,T^{-1}(H,\, P_\theta,\, p_z)\kappa_s(\sigma) \\ &\quad + 2a_s(H,\, P_\theta,\, p_z)\,\frac{2\pi}{T(H,\, P_\theta,\, p_z)}\,\frac{\partial\kappa_s(\sigma)}{\partial H} \\ &= -2\Omega^2\kappa_s^2(\sigma)\,\frac{1}{P_0}\,[\eta(\sigma) - \kappa_s'(\sigma)]\end{aligned} \tag{7.67}$$

where

$$\kappa_s'(\sigma) = \frac{\dfrac{\partial \kappa_s(\sigma)}{\partial \sigma}}{\kappa_s(\sigma)} \frac{\partial \sigma}{\partial \beta} \tag{7.68}$$

Similarly,

$$\frac{1}{c} \frac{\partial}{\partial p_z} a_s^2(H, P_\theta, p_z) = 2\Omega^2 \kappa_s^2(\sigma) \frac{v_z}{c} \frac{1}{P_0} [\eta(\sigma) - \kappa_s'(\sigma)] \tag{7.69}$$

7.4 FINITE-LENGTH CAVITY

We will now calculate the power transferred from an electron beam injected axially through the cavity that is shown schematically in Figure 7.7 to the fields of a TEM cavity mode. The assumptions 1 to 4 used in Section 7.2 will again be applied here. The result of the transferred power calculation will then be used in Section 7.6 to compute the threshold beam power required to make the cavity self-oscillate.

The cavity fields for a TEM mode in the cavity of Figure 7.7 are given by

$$\mathbf{E}^1(r, \theta, z, t) = E_1 \frac{a}{r} \sin(kz) \sin(\omega t)\hat{r} \tag{7.70}$$

and

$$\mathbf{B}^1(r, \theta, z, t) = E_1 \frac{a}{cr} \cos(kz) \cos(\omega t)\hat{\theta} \tag{7.71}$$

where

$$k = \frac{l\pi}{L} \tag{7.72}$$

and

$$\omega = ck = l\frac{c\pi}{L} \tag{7.73}$$

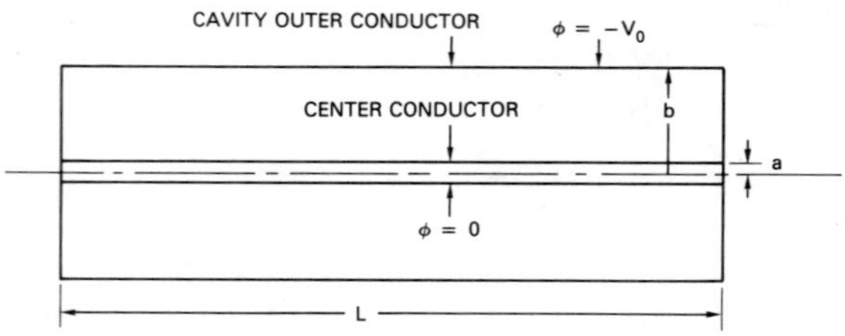

Fig. 7.7 An Idealized Orbitron Cavity

Using (7.17), (7.18), (7.70), and (7.71) to obtain the perturbed distribution function by integrating over unperturbed orbits, we find that

$$f^1(\mathbf{r}, \mathbf{p}, t) = eaE_1 \int_{t-z/v_z}^{t} \mathrm{d}t' \left[\sin(kz'(t')) \sin(\omega t') \frac{1}{r'(t')} \frac{\partial f^0}{\partial p'_r} - \frac{v_z}{c} \cos(kz'(t')) \cos(\omega t') \frac{1}{r'(t')} \frac{\partial f^0}{\partial p'_r} + \frac{1}{c} \cos(kz'(t')) \cos(\omega t') \frac{v'_r(t')}{r'(t')} \frac{\partial f^0}{\partial p'_z} \right] \tag{7.74}$$

The integration is taken from the time the electron enters the cavity to time t. Proceeding in the same manner as we did between (7.19) and (7.28) of Section 7.2, we find the perturbed distribution function:

$$\begin{aligned} f^1(\mathbf{r}, \mathbf{p}, t) = \frac{eaE_1}{4} \sum_{s=1}^{\infty} \Bigg\{ & \frac{\partial f^0}{\partial H} a_s(H, P_\theta, p_z) \\ & \times [L_s(\omega, k) - L_s(\omega, -k) - L_s(-\omega, k) + L_s(-\omega, -k)] \\ & + \left[\frac{1}{c} \frac{\partial f^0}{\partial p_z} - \frac{v_z}{c} \frac{\partial f^0}{\partial H} \right] a_s(H, P_\theta, p_z)[L_s(\omega, k) \\ & + L_s(\omega, -k) + L_s(-\omega, k) + L_s(-\omega, -k)] \Bigg\} \end{aligned} \tag{7.75}$$

where

$$L_s(\omega, k) = \frac{\cos\{s\Omega(H, P_\theta, p_z)\hat{t}(r, v_r) - \omega t - [s\Omega(H, P_\theta, p_z) - \omega]z/v_z\}}{[s\Omega(H, P_\theta, p_z) - \omega + kv_z]} - \frac{\cos[s\Omega(H, P_\theta, p_z)\hat{t}(r, v_r) - \omega t + kz]}{[s\Omega(H, P_\theta, p_z) - \omega + kv_z]} \tag{7.76}$$

We are interested in calculating the power transferred from the electron beam to the cavity fields. Equation (7.70) for the TEM mode cavity electric field shows that the electric field has no azimuthal or axial components, so we only have to compute the radial component of the perturbed current in order to calculate the transferred power. Then,

$$\mathbf{J}^1(r, \theta, z, t) = -e\int d^3p v_r f^1(\mathbf{r}, \mathbf{p}, t)\hat{r} \tag{7.77}$$

and the time-averaged power loss of all of the electrons in the cavity, P, is therefore

$$P = -a\omega E_1 \int_0^{2\pi/\omega} dt \int_0^L dz \int_a^b r\, dr \frac{J_r^1(r, \theta, z, t)\sin(kz)\sin(\omega t)}{r} \tag{7.78}$$

Combining (7.75), (7.77), and (7.78), we have

$$\begin{aligned} P = {} & \frac{e^2a^2E_1^2}{4}\omega \sum_{s=1}^{\infty} \int_0^{2\pi/\omega} dt \sin(\omega t) \int_0^L dz \sin(kz) \int_a^b dr \int d^3p v_r \\ & \times \left\{ \frac{\partial f^0(H, P_\theta, p_z)}{\partial H} a_s(H, P_\theta, p_z) \right. \\ & \times [L_s(\omega, k) - L_s(\omega, -k) - L_s(-\omega, k) + L_s(-\omega, -k)] \\ & + \left[\frac{1}{c}\frac{\partial f^0(H, P_\theta, p_z)}{\partial p_z} - \frac{v_z}{c}\frac{\partial f^0(H, P_\theta, p_z)}{\partial H}\right] a_s(H, P_\theta, p_z) \\ & \left. \times [L_s(\omega, k) + L_s(\omega, -k) + L_s(-\omega, k) + L_s(-\omega, -k)] \right\} \end{aligned} \tag{7.79}$$

We now transform the integration over d^3p to an integration over $dH\, dP_\theta dp_z$ and perform the integration over r by the technique used between (7.31) and (7.35) of Section 7.2. This yields

$$P = \frac{e^2 a^2 E_1^2 \omega}{8} \sum_{s=1}^{\infty} \int_0^{2\pi/\omega} \mathrm{d}t \sin(\omega t) \int_0^L \mathrm{d}z \sin(kz) \int \mathrm{d}H \mathrm{d}P_\theta \mathrm{d}p_z$$

$$\times \left\{ \frac{\partial f^0(H, P_\theta, p_z)}{\partial H} T(H, P_\theta, p_z) a_s^2(H, P_\theta, p_z)[G_s(\omega, k) - G_s(\omega, -k)\right.$$

$$- G_s(-\omega, k) + G_s(-\omega, -k)]$$

$$+ \left[\frac{1}{c}\frac{\partial f^0(H, P_\theta, p_z)}{\partial p_z} - \frac{v_z}{c}\frac{\partial f^0(H, P_\theta, p_z)}{\partial H}\right] T(H, P_\theta, p_z) a_s^2(H, P_\theta, p_z)$$

$$\left. \times [G_s(\omega, k) + G_s(\omega, -k) + G_s(-\omega, k) + G_s(-\omega, -k)] \right\} \tag{7.80}$$

where

$$G_s(\omega, k) = \frac{\sin[\omega t + (s\Omega(H, P_\theta, p_z) - \omega) z/v_z] - \sin(\omega t - kz)}{(s\Omega(H, P_\theta, p_z) - \omega + kv_z)} \tag{7.81}$$

Integrating over t and z, we find the general expression for the power transferred from the electron beam to the cavity fields. Thus,

$$P = \frac{e^2 a^2 E_1^2 L}{8l} \sum_{s=1}^{\infty} \int \mathrm{d}H \mathrm{d}P_\theta \mathrm{d}p_z$$

$$\times \left\{ \frac{\partial f^0(H, P_\theta, p_z)}{\partial H} T(H, P_\theta, p_z) a_s^2(H, P_\theta, p_z)\right.$$

$$\times [F_s(\omega, k) - F_s(\omega, -k) + F_s(-\omega, k) - F_s(-\omega, -k)]$$

$$+ \left[\frac{1}{c}\frac{\partial f^0(H, P_\theta, p_z)}{\partial p_z} - \frac{v_z}{c}\frac{\partial f^0(H, P_\theta, p_z)}{\partial H}\right] T(H, P_\theta, p_z) a_s^2(H, P_\theta, p_z)$$

$$\left. \times [F_s(\omega, k) + F_s(\omega, -k) - F_s(-\omega, k) - F_s(-\omega, -k)] \right\} \tag{7.82}$$

where

$$F_s(\omega, k) = \frac{k^2 L^2 \{1 - \cos(kL) \cos[(s\Omega(H, P_\theta, p_z) - \omega) L/v_z]\}}{[s\Omega(H, P_\theta, p_z) - \omega + kv_z] \left\{ k^2 L^2 - [s\Omega(H, P_\theta, p_z) - \omega]^2 \dfrac{L^2}{v_z^2} \right\}} \tag{7.83}$$

Equation (7.82) is a general expression for the beam power loss. In the next section we will specialize to an idealized distribution function which will allow easy calculation of the threshold beam power required for the cavity to self-oscillate.

7.5 THE DISTRIBUTION FUNCTION OF THE ELECTRON BEAM

Before (7.35) and (7.82) can be evaluated, we must specify an equilibrium distribution function which is constructed from the constants of the motion, H, P_θ, and p_z. We choose the simplest distribution function possible which is that of a monoenergetic beam with one value of angular and axial momentum:

$$f^0(\mathbf{r}, \mathbf{p}, t) = A\delta(H - H_0)\delta(P_\theta - P_{\theta 0})\delta(p_z - mv_0) \tag{7.84}$$

where A is determined from the condition:

$$\frac{2\pi}{m} e \int_a^b r\mathrm{d}r \int \mathrm{d}^3p \, p_z f^0(\mathbf{r}, \mathbf{p}, t) = I \tag{7.85}$$

and I is the negative of the current of electrons through the cavity. To evaluate the constant A in (7.84) we insert (7.84) into (7.85) and transform the integration over d^3p into an integration over the constants of the motion. Then,

$$I = \frac{2\pi Ae}{\sqrt{2m}} \int_a^b r\mathrm{d}r \int \frac{\mathrm{d}H \, \mathrm{d}P_\theta \, \mathrm{d}p_z \, p_z \delta(H - H_0)\delta(P_\theta - P_{\theta 0})\delta(p_z - mv_0)}{r\sqrt{H - \dfrac{P_\theta^2}{2mr^2} - \dfrac{p_z^2}{2m} - P_0 \ln\left(\dfrac{r}{a}\right)}}$$

$$= \sqrt{2m}\,\pi Aev_0 \int_{r_i}^{r_o} \mathrm{d}r \frac{1}{\sqrt{H_0 - \dfrac{P_{\theta 0}^2}{2mr^2} - \dfrac{mv_0^2}{2} - P_0 \ln\left(\dfrac{r}{a}\right)}} \tag{7.86}$$

Using (7.36) for $T(H, P_\theta, p_z)$, the radial oscillation period of an electron with total energy H, angular momentum P_θ, and axial momentum p_z, we see that (7.86) can be written as

$$I = \pi T(H_0, P_{\theta 0}, mv_0)ev_0 A \tag{7.87}$$

Therefore, for an electron beam with energy H_0, current I, angular momentum $P_{\theta 0}$, and axial momentum mv_0, the distribution function is

$$f^0(H, P_\theta, p_z) = \frac{I}{\pi e v_0 T(H_0, P_{\theta 0}, m v_0)} \times \delta(H - H_0)\delta(P_\theta - P_{\theta 0})\delta(p_z - m v_0) \quad (7.88)$$

The distribution function given in (7.88) is an idealized representation of the distribution function that would be present in an actual device. However, such a distribution function has many advantages from an analytical point of view. The principal advantage is that it allows for simple evaluation of the theory and yields results that are easily interpreted. The distribution function can also be used as the basis for superposing many solutions together to model the situation where the beam has arbitrarily spread in energy, angular momentum, and axial momentum. The superposition principle is valid here because the self-fields of the beam have been neglected and the theory is otherwise linear.

7.6 RESULTS

The parameters we will use to describe the results of the linear theory of the orbitron traveling wave tube amplifier and oscillator are similar to the parameters used to describe the ECM. Indeed, the voltage between the center and outer conductor in an orbitron, V_0, plays the role of the magnetic field in an ECM. The parameters H_0, the total electron energy, α the square root of the ratio of the electron's perpendicular to parallel energy, and b the radius of the inner surface of the waveguide or cavity outer conductor are also used for the ECM. Two parameters, the radius of the center conductor, a, and the ratio of the maximum to the minimum radius of the electron's orbit σ, have no ECM analogy. The remaining eight quantities all have direct ECM counterparts. For the traveling wave tube amplifier we have the harmonic number s, the electron beam current I, the frequency ω, and the spatial growth rate k_i. The quality factor of the cavity Q, the length of the cavity L, the mode number l, and the threshold beam power required for the cavity to self-oscillate P_b^{th}, are used for the oscillator. We hope this comparison of parameters will help to clarify the results that we are about to present for those readers who are familiar with the ECM.

7.6.1 Traveling Wave Tube Amplifier

Dispersion Relation

Combining (7.35) for the dispersion relation with (7.88) for the idealized beam distribution function, we obtain an explicit expression for the dispersion relation for a forward traveling wave in an axially uniform configuration:

$$\left(k^2 - \frac{\omega^2}{c^2}\right)\ln\left(\frac{b}{a}\right) = \frac{\mu_0 e^2 \omega}{4} \frac{I}{\pi e v_0 T(H_0, P_{\theta 0}, m v_0)} \sum_{s=1}^{\infty}$$
$$\times \left(1 - \frac{v_0}{c}\right) \frac{\partial}{\partial H} \left\{ T(H, P_{\theta 0}, m v_0) a_s^2(H, P_{\theta 0}, m v_0) \right.$$
$$\left.\left. \times \left[\frac{1}{(s\Omega(H, P_{\theta 0}, m v_0) - \omega + k v_0)} - \frac{1}{(s\Omega(H, P_{\theta 0}, m v_0) + \omega - k v_0)} \right] \right\} \right|_{H=H_0}$$
$$+ \frac{1}{c} \frac{\partial}{\partial p_z} \left[T(H_0, P_{\theta 0}, p_z) a_s^2(H_0, P_{\theta 0}, p_z) \left\{ \frac{1}{(s\Omega(H_0, P_{\theta 0}, p_z) - \omega + k v_z)} \right. \right.$$
$$\left. \left. - \frac{1}{(s\Omega(H_0, P_{\theta 0}, p_z) + \omega - k v_z)} \right\} \right]_{p_z = m v_0} \tag{7.89}$$

Using (7.43), (7.44), (7.67), and (7.69) for the derivatives of $T(H, P_\theta, p_z)$ and $a_s^2(H, P_\theta, p_z)$ developed in Section 7.3, and expressing the answer in terms of H_0, σ, and α, we obtain the final expression for the dispersion relation:

$$\left(k^2 - \frac{\omega^2}{c^2}\right) = \frac{\mu_0 I \omega \Omega^2}{4\pi v_0 V_0} \sum_{s=1}^{\infty} \kappa_s^2(\sigma) \left\{ \left[\left(1 - \frac{2v_0}{c}\right) \frac{(2\kappa_s'(\sigma) - \eta(\sigma))}{(s\Omega - \omega + k v_0)} \right. \right.$$
$$\left. + \left(1 - \frac{2v_0}{c}\right) \frac{\eta(\sigma) s\Omega}{(s\Omega - \omega + k v_0)^2} - \frac{k v_0 P_0}{m v_0 c (s\Omega - \omega + k v_0)^2} \right]$$
$$- \left[\left(1 - \frac{2v_0}{c}\right) \frac{(2\kappa_s'(\sigma) - \eta(\sigma))}{(s\Omega + \omega - k v_0)} \right.$$
$$+ \left(1 - \frac{2v_0}{c}\right) \frac{\eta(\sigma) s\Omega}{(s\Omega + \omega - k v_0)^2}$$
$$\left. \left. + \frac{k v_0 P_0}{m v_0 c (s\Omega + \omega - k v_0)^2} \right] \right\} \tag{7.90}$$

Spatial Growth Rate

The model we used to find the dispersion relation for the infinite-length orbitron assumed that the electron beam interacted only with a forward traveling wave. In a real device, reflections at the output can cause a backward wave to be present in the interaction region. If the amplitude of the reflected signal reaches a level large enough for the loop gain of the system to exceed unity, the instability that gives rise to the maser interaction changes from convective to absolute, and the amplifier thus self-oscillates. Internal feedback through coupling to the backward wave can also cause an absolute instability [15]. It is important to know the parametric dependences of the transition from convective to absolute instability when designing an actual device [16], and various schemes can be used to avoid the onset of an absolute instability such as lining the inner surface of the cavity with an absorbing material. Here, we choose to ignore the problem of absolute instability and simply solve (7.90) to find the spatial growth rate for a forward traveling wave in an infinite-length orbitron amplifier as a function of the frequency and beam parameters. The reader should understand, however, that the resultant growth rates may not be realized in an actual device due to the onset of an absolute instability.

Now, to solve (7.90) for the spatial growth rate, we assume that $\omega \approx s\Omega$, and only the sth term is resonant in the infinite sum. Then, (7.90) can be written in a more tractable form:

$$\left(k^2 - \frac{\omega^2}{c^2}\right)(s\Omega - \omega + kv_0)^2 = \frac{\mu_0 \omega I \Omega^2 \kappa_s^2(\sigma)}{4\pi v_0 V_0}\left\{\left(1 - \frac{2v_0}{c}\right)[2\kappa_s'(\sigma) - \eta(\sigma)] \times (s\Omega - \omega + kv_0) + \left(1 - \frac{2v_0}{c}\right)\eta(\sigma)s\Omega - \frac{kv_0 P_0}{mv_0 c}\right\} \tag{7.91}$$

Equation (7.91) is of fourth order in k and has nine free parameters a, b, α, σ, s, V_0, H_0, I, and ω. In the numerical results that follow, we reduce the number of free parameters by normalizing the radius of the center conductor a and the spatial growth rate k_i to the waveguide outer conductor radius b. We also normalize the frequency ω to the frequency c/b. We fix the value of α, which is the square root of the ratio of the perpendicular to parallel energy, at 1.5, a value typical of most ECM experiments [17]. The parameter b/a has been chosen to be 100; the harmonic number s is taken to be 1, the electron energy H_0 is fixed at 10,000 eV, and the ratio of the maximum to the minimum radius of the electron's orbit σ is taken to be 5. The three remaining parameters: V_0, the voltage between the center and outer conductor; I, the electron beam current; the normalized frequency, $\omega c/b$, will be varied. Figure 7.8 plots the spatial growth rate *versus* frequency for $V_0 = 13{,}500$ V and $I = 100$ mA. This is plotted over the lowest frequency interval that gives growth. The voltage chosen gives a maximum electron radius of about $0.1b$. Thus, the electron (with $\sigma = 5$) passes close to the center conductor yielding a large growth rate. Figure 7.9 plots the spatial growth rate *versus* the beam current for the value of

V_0 used in Figure 7.8. The growth rate is seen to scale as $\sim I^{1/3}$. Last, the spatial growth rate is plotted *versus* V_0 in Figure 7.10 for the voltages that are consistent with radial confinement of the electrons. The beam current is taken to be 100 mA and the frequency was varied here and in Figure 7.9 to maximize the spatial growth rate.

7.6.2 Oscillator

Beam Power Transfer to the Cavity Fields

If we combine (7.82) and (7.88), then we obtain an explicit expression for the power transferred from the electron beam to the cavity fields:

$$
\begin{aligned}
P = &-\frac{e^2a^2E_1^2L}{8l}\cdot\frac{I}{\pi e v_0 T(H_0, P_{\theta 0}, mv_0)}\sum_{s=1}^{\infty} \\
&\times\left\{\frac{\partial}{\partial H}[T(H, P_{\theta 0}, mv_0)a_s^2(H, P_{\theta 0}, mv_0)\right. \\
&\times (F_s(\omega, k) - F_s(\omega, -k) + F_s(-\omega, k) - F_s(-\omega, -k))]|_{H=H_0} \\
&+\frac{1}{c}\frac{\partial}{\partial p_z}[T(H_0, P_{\theta 0}, p_z)a_s^2(H_0, P_{\theta 0}, p_z) \\
&\times (F_s(\omega, k) + F_s(\omega, -k) - F_s(-\omega, k) - F_s(-\omega, -k))]|_{p_z=mv_0} \\
&-\frac{v_0}{c}\frac{\partial}{\partial H}[T(H, P_{\theta 0}, mv_0)a_s^2(H, P_{\theta 0}, mv_0) \\
&\left.\times (F_s(\omega, k) + F_s(\omega, -k) - F_s(-\omega, k) - F_s(-\omega, -k))]|_{H=H_0}\right\}
\end{aligned}
\tag{7.92}
$$

where $F_s(\omega, k)$ is given by (7.83). By using (7.43), (7.44), (7.67), and (7.69) for the derivatives of $T(H, P_\theta, p_z)$ and $a_s^2(H, P_\theta, p_z)$ developed in Section 7.3 and expressing the answer in terms of H_0, σ, and α, we obtain the final expression for the beam power loss:

$$
P = \frac{\pi l a^2 \ln\left(\frac{b}{a}\right)E_1^2 I\Omega^2\tau^2}{4V_0}\sum_{s=1}^{\infty}\kappa_s^2(\sigma)\{P_s^1 + P_s^2 + P_s^3 + P_s^4\} \tag{7.93}
$$

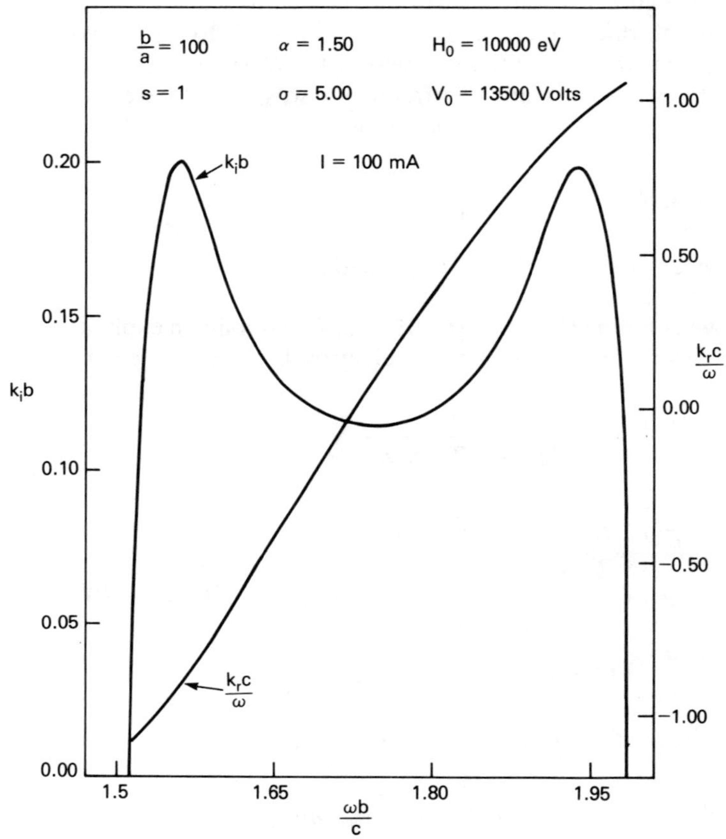

Fig. 7.8 The Spatial Growth Rate and Wavenumber *versus* Frequency for an Orbitron Traveling Wave Tube Amplifier

$$P_s^1 = \left(1 - \frac{2v_0}{c}\right)[\eta(\sigma) - 2\kappa_s'(\sigma)]F(\Delta) + \eta(\sigma)\left(1 - \frac{2v_0}{c}\right)s\Omega\tau G(\Delta) - \frac{P_0}{mv_0c}H(\Delta) \tag{7.94}$$

$$P_s^2 = -\left(1 + \frac{2v_0}{c}\right)[\eta(\sigma) - 2\kappa_s'(\sigma)]L(\Delta) - \eta(\sigma)\left(1 + \frac{2v_0}{c}\right)s\Omega\tau M(\Delta) - \frac{P_0}{mv_0c}N(\Delta) \tag{7.95}$$

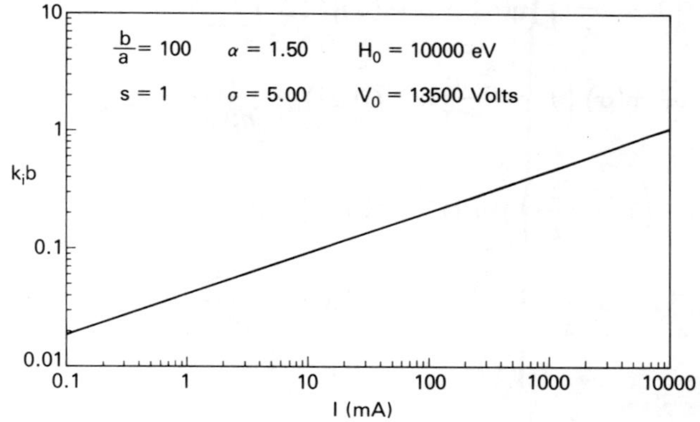

Fig. 7.9 The Spatial Growth Rate *versus* Electron Beam Current for an Orbitron Traveling Wave Tube Amplifier

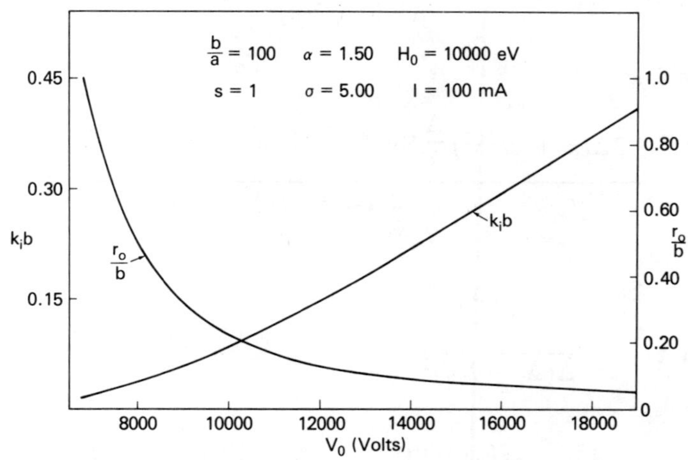

Fig. 7.10 The Spatial Growth Rate and Maximum Electron Radius *versus* Voltage between Center and Outer Conductor for an Orbitron Traveling Wave Tube Amplifier

$$P_s^3 = \left(1 + \frac{2v_0}{c}\right)[\eta(\sigma) - 2\kappa_s'(\sigma)]F(\Delta')$$

$$+ \eta(\sigma)\left(1 + \frac{2v_0}{c}\right)s\Omega\tau G(\Delta') + \frac{P_0}{mv_0c}H(\Delta') \tag{7.96}$$

$$P_s^4 = -\left(1 - \frac{2v_0}{c}\right)[\eta(\sigma) - 2\kappa_s'(\sigma)]L(\Delta')$$

$$- \eta(\sigma)\left(1 - \frac{2v_0}{c}\right)s\Omega\tau M(\Delta') + \frac{P_0}{mv_0c}N(\Delta') \tag{7.97}$$

$$F(\Delta) = -\frac{\sin^2\left(\dfrac{\Delta}{2}\right)}{\Delta^2(\Delta - 2l\pi)} \tag{7.98}$$

$$G(\Delta) = \frac{(3\Delta - 4l\pi)\sin^2\left(\dfrac{\Delta}{2}\right) - (\Delta - 2l\pi)\dfrac{\Delta}{2}\sin(\Delta)}{\Delta^3(\Delta - 2l\pi)^2} \tag{7.99}$$

$$H(\Delta) =$$

$$\frac{(\Delta^2 - 3\Delta l\pi + 2l^2\pi^2)\dfrac{\Delta}{2}\sin(\Delta) - (2\Delta^2 - 5l\pi\Delta + 4l^2\pi^2)\sin^2\left(\dfrac{\Delta}{2}\right)}{\Delta^3(\Delta - 2l\pi)^2} \tag{7.100}$$

$$L(\Delta) = -\frac{\sin^2\left(\dfrac{\Delta}{2}\right)}{\Delta(\Delta - 2l\pi)^2} \tag{7.101}$$

$$M(\Delta) = \frac{(3\Delta - 2l\pi)\sin^2\left(\dfrac{\Delta}{2}\right) - (\Delta - 2l\pi)\dfrac{\Delta}{2}\sin(\Delta)}{\Delta^2(\Delta - 2l\pi)^3} \tag{7.102}$$

$$N(\Delta) =$$

$$\frac{(\Delta - l\pi)(\Delta - 2l\pi)\dfrac{\Delta}{2}\sin(\Delta) - (2\Delta^2 - 3l\pi\Delta + 2l^2\pi^2)\sin^2\left(\dfrac{\Delta}{2}\right)}{\Delta^2(\Delta - 2l\pi)^3} \tag{7.103}$$

$$\Delta = [s\Omega(H_0, P_{\theta 0}, mv_0) - \omega + kv_0]\tau \tag{7.104}$$

$$\Delta' = [s\Omega(H_0, P_{\theta 0}, mv_0) + \omega + kv_0]\tau \tag{7.105}$$

and

$$\tau = \frac{L}{v_0} \tag{7.106}$$

Threshold Beam Power Required for Self-Oscillation

The model that we used to calculate the beam energy loss to the cavity fields assumed an idealized cavity with discrete eigenfrequencies. The cavity has a finite Q, however, and this implies that the cavity resonates in a narrow band of eigenfrequencies with a bandwidth of approximately $2\omega/Q$. We will show in Section 7.7 that the frequency width for beam energy loss to the cavity fields is approximately $2\pi/\tau$, and the finite Q cavity can therefore be treated as idealized, if

$$\frac{\pi}{\tau} \gg \frac{\omega}{Q} \tag{7.107}$$

This condition is usually easily satisfied for any reasonable value of Q.

From the definition of Q, we have

$$Q = \frac{\omega E_s}{P_L} \tag{7.108}$$

where E_s is the electromagnetic energy stored in the cavity, and P_L is the electromagnetic power lost. Thus, we see that the threshold condition for oscillation occurs when the power lost from the cavity in the form of electromagnetic radiation is just balanced by the power input from the electron beam, or

$$P \geqslant P_L \tag{7.109}$$

which implies that

$$P > \frac{\omega E_s}{Q} \tag{7.110}$$

Now, the beam power P_b is related to the beam current by

$$P_b = H_0 I/e \tag{7.111}$$

and the stored energy for a TEM mode in the cavity is given by

$$E_s = \frac{\pi}{2}\,\epsilon_0 a^2 E_1^2 L \ln\left(\frac{b}{a}\right) \tag{7.112}$$

If we combine (7.73), (7.93), (7.110), (7.111), and (7.112), we obtain the relation:

$$P_b^{\mathrm{th}} = \frac{2\pi\epsilon_0 c V_0 (H_0/e)}{Q\Omega^2\tau^2 \sum_{s=1}^{\infty} \kappa_s^2(\sigma)(P_s^1 + P_s^2 + P_s^3 + P_s^4)} \tag{7.113}$$

Equation (7.113) is a function of nine free parameters V_0, H_0, α, σ, b, a, L, l, and Q. In the numerical examples that follow, we reduce the number of free parameters by normalizing the cavity length L and the radius of the center conductor a to the cavity outer conductor radius b. We take α to be 1.5, as we did in the numerical examples for the traveling wave tube amplifier. The parameter b/a has been chosen to be 100, and L/b is taken to be 10. Finally, from (7.113), we see that Q and P_b^{th} can be combined to form a single quantity, QP_b^{th}. The four remaining parameters will be varied: V_0, the voltage between the center and outer conductors; H_0, the electron energy; σ, the ratio of the maximum to the minimum radius of the electron's orbit; l, the mode number. Figure 7.11 plots QP_b^{th} *versus* V_0 for $l = 1$, 2, and 5, $H_0 = 10{,}000$ eV, and $\sigma = 5$. This is plotted in the voltage range where the $s = 1$ interaction of the $l = 1$ mode occurs. The $l = 1$ interaction has the widest voltage tuning range and the largest QP_b^{th} for a given harmonic number s. Figure 7.12 shows QP_b^{th} *versus* H_0 for $s = l = 1$, 2, and 5 and $\sigma = 5$. V_0 is varied here and in Figure 7.13 to minimize QP_b^{th}. The start power requirements for the lowest order mode are seen to increase rapidly as a function of the injection energy. Last, Figure 7.13 displays QP_b^{th} *versus* σ for $H_0 = 10{,}000$ eV, $l = s = 1$, 2, 5, and $l = s/2 = 1$, 2. QP_b^{th} is seen to decrease as σ increases with saturation occurring when $\sigma \sim 5\, s/l$.

7.7 DISCUSSION

7.7.1 Consistency of Approximations

We wish to verify that the assumptions we made in Section 7.2 are consistent with the results of the calculations performed in Sections 7.2

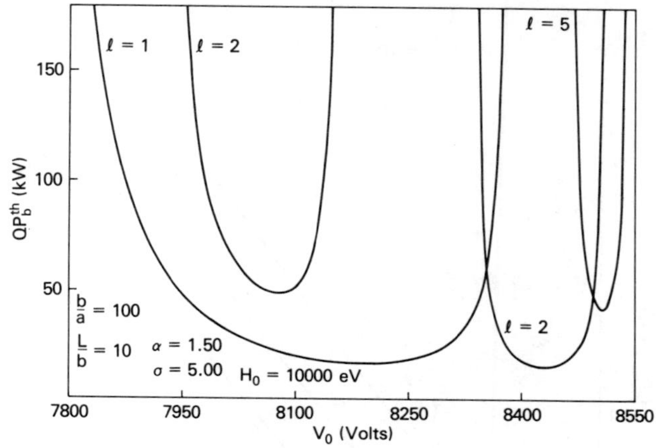

Fig. 7.11 The Cavity *Q* Multiplied by the Threshold Beam Power Required for Self-Oscillation *versus* Voltage between Center and Outer Conductor for Various Modenumbers in an Orbitron Oscillator

and 7.4. First, we consider assumption 2, the requirement that the electric field due to the electron beam is small compared with the confining electric field due to the voltage between the inner and outer surfaces of the waveguide. Now, the electric field of the azimuthally symmetric electron beam is largest compared to the confining electric field for radii greater than the maximum radius of the electron beam, r_o. Therefore, we have the relation:

$$\frac{I}{2\pi\epsilon_0 v_0 r_o} \ll \frac{V_0}{\ln\left(\dfrac{b}{a}\right) r_o} \tag{7.114}$$

For the examples presented in Section 7.6, $v_0 \approx 0.1c$, $V_0 \approx 10{,}000$ V, and $b/a = 100$, for which (7.114) yields

$$I \ll 3.6 \text{ A} \tag{7.115}$$

which is easily satisfied in such cases.

Next, to verify the validity of assumption 3 of Section 7.2, we must show that the amplitude of the radiation electric field in the interaction region, E_1, is small compared with the confining electric field; that is,

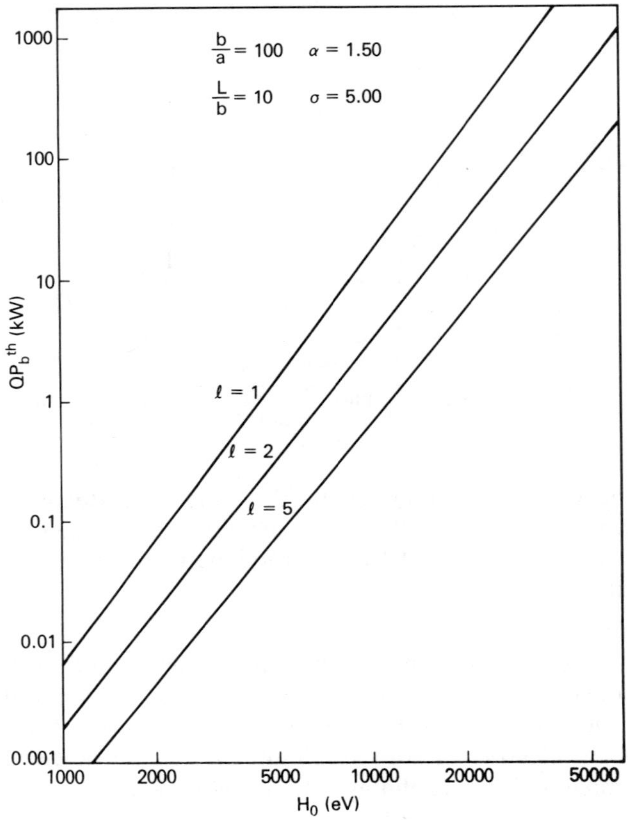

Fig. 7.12 The Cavity Q Multiplied by the Threshold Beam Power Required for Self-Oscillation *versus* Electron Beam Energy for Various Modenumbers in an Orbitron Oscillator

$$\frac{a}{r} E_1 \ll \frac{V_0}{\ln\left(\frac{b}{a}\right) r} \tag{7.116}$$

We choose to verify that (7.116) is valid for the oscillator configuration having a larger radiation electric field amplitude in the interaction region

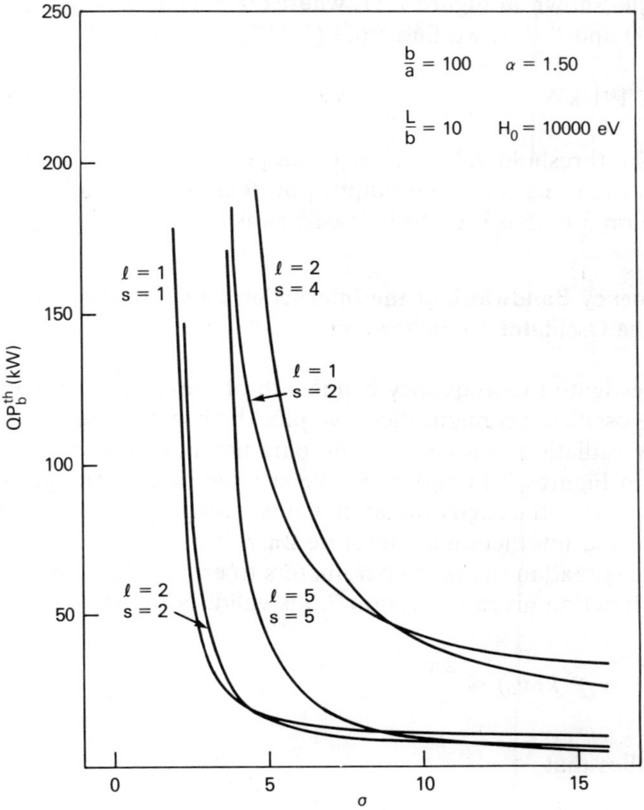

Fig. 7.13 The Cavity Q Multiplied by the Threshold Beam Power Required for Self-Oscillation *versus* Parameter σ for Various Mode and Harmonic Numbers in an Orbitron Oscillator

as compared with the amplifier configuration for a given output power. If we write (7.116) in terms of the output power, using (7.108) and (7.112), we have

$$QP_L \ll l\,\frac{\pi^2 c\epsilon_0}{2}\,\frac{V_0^2}{\ln\left(\dfrac{b}{a}\right)} \tag{7.117}$$

In the example shown in Figure 7.11, where $QP_b^{\text{th}} = 17$ kW for $V_0 \approx 8200$ V, $b/a = 100$ and $l = 1$, we find from (7.117) that

$$QP_L \ll 191 \text{ kW} \tag{7.118}$$

Therefore, the threshold value of the beam power required for self-oscillation is less than one-tenth the output power needed to violate (7.116), and assumption 3 is thus seen to be easily satisfied.

7.7.2 Frequency Bandwidth of the Interaction and Allowable Beam Spread for the Oscillator Configuration

To investigate the frequency bandwidth of the orbitron maser interaction in the oscillator configuration, we plot the beam energy transferred to the cavity radiation fields *versus* the parameter Δ for the $l = 1$ and $l = 2$ modes in Figures 7.14 and 7.15. We can see that the range of Δ for beam energy loss to the cavity radiation fields is roughly 2π. The frequency bandwidth of the interaction is therefore $2\pi/\tau$. This allows us to compute the allowable spread in the beam parameters to ensure that the cold beam distribution function given in Section 7.5 is valid. Now, we have

$$(s\delta\Omega - \omega + k\delta v_0) \ll \frac{2\pi}{\tau} \tag{7.119}$$

and this implies that

$$\frac{\delta\Omega}{\Omega} \ll \frac{2\pi}{s\Omega\tau} \ll \frac{1}{sN} \tag{7.120}$$

where N is the number of turns the electron makes around the central conductor while transiting the cavity. We also have a condition on the spread in v_z, which is

$$\frac{\delta v_z}{v_z} \ll \frac{2\pi}{kv_0\tau} \tag{7.121}$$

and, using (7.72) and (7.106), we find that

$$\frac{\delta v_z}{v_z} \ll \frac{2}{l} \tag{7.122}$$

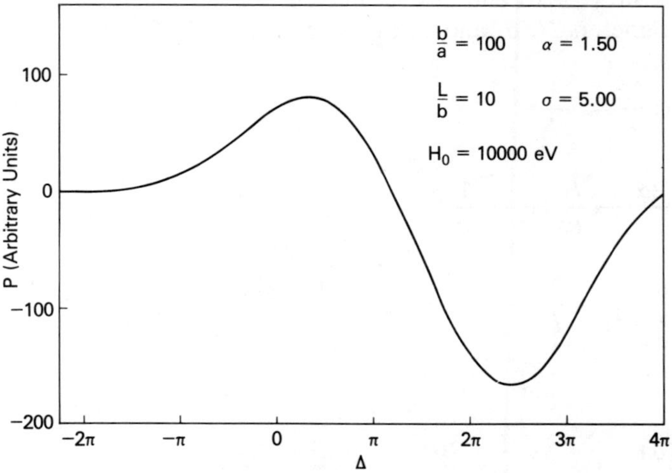

Fig. 7.14 The Electron Beam Power Transferred to the Radiation Fields *versus* Parameter Δ for $l = 1$, $s = 1$

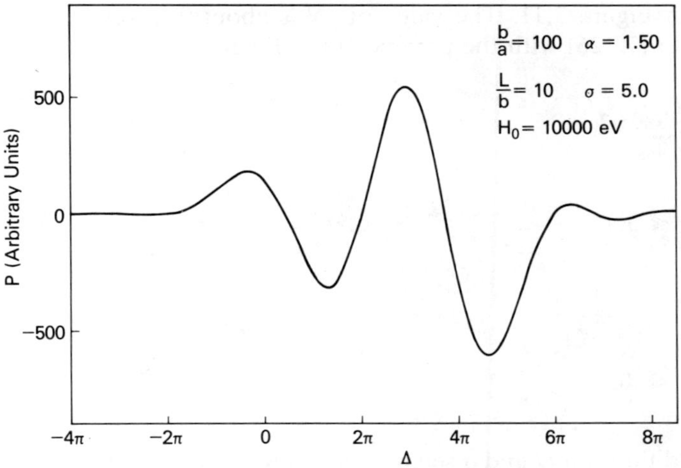

Fig. 7.15 The Electron Beam Power Transferred to the Radiation Fields *versus* Parameter Δ for $l = 2$, $s = 1$

We can write (7.120) and (7.122) in terms of the allowable spread in the beam parameters H, σ, and α, by using (7.43), (7.46), and (7.50):

$$\eta(\sigma) \frac{\delta H}{P_0} \ll \frac{1}{sN} \tag{7.123}$$

$$\frac{2\alpha}{(1 + \alpha^2)^2} \frac{H}{P_0} \delta\alpha \ll \frac{1}{sN} \tag{7.124}$$

$$\frac{\alpha\delta\alpha}{(1 + \alpha^2)} \ll \frac{2}{l} \tag{7.125}$$

and

$$\frac{\frac{\partial \hat{T}(\sigma)}{\partial\sigma}}{\hat{T}(\sigma)} \delta\sigma \ll \frac{1}{sN} \tag{7.126}$$

where $\partial\hat{T}/\partial\sigma$ is the derivative of the curve shown in Figure 7.5.

For a numerical example of the allowable spread in the beam parameters, we use the $l = 2$ mode with the lowest threshold beam power shown in Figure 7.11. The value of sN is about 9 in this case. By using (7.123) to (7.126) with the parameters of Figure 7.11, we find that

$$\frac{\delta H}{H} \ll .02 \tag{7.127}$$

$$\frac{\delta\sigma}{\sigma} \ll 3.7 \tag{7.128}$$

and

$$\frac{\delta\alpha}{\alpha} \ll .05 \tag{7.129}$$

The conditions on H and σ seem to be easily achievable, but the allowable spread in the parameter α could pose a problem. Any effort to design an electron gun for a TEM orbitron oscillator should pay close attention to the spread in the beam parameters.

7.7.3 Nonlinear Efficiency Limits

Here, we consider the maximum energy that can be transferred from the electron beam to a TEM waveguide mode in an orbitron. Because the electric field of a TEM mode has only a radial component, the angular momentum of an electron that interacts with this mode is conserved. Also, the electron has given the maximum amount of energy to the TEM mode if it exits the interaction region with no radial velocity. Therefore, to compute an upper bound on the power transfer from the electron beam to the TEM mode, we calculate the change in energy of electrons that enter the interaction region with energy H_0 having nonzero radial velocity and exit the interaction region with energy H^* having no radial velocity, consistent with the axial velocity change of the electrons due to the interaction with the TEM mode.

We consider the case of a forward traveling wave tube amplifier. The analysis is also valid for the oscillator configuration if the electron beam primarily interacts with the forward wave component of the standing wave in the cavity. (If the electron beam principally interacts with the backward wave in the cavity, the analysis is performed with the right-hand side of (7.130) having the opposite sign.) Now, from (7.13) and (7.14), we see that the only nonzero components of the TEM waveguide mode are E_r and B_θ, and

$$E_r = cB_\theta \tag{7.130}$$

Then, the equations of motion are

$$\frac{\mathrm{d}H}{\mathrm{d}t} = -ev_rE_r \tag{7.131}$$

and

$$\frac{\mathrm{d}p_z}{\mathrm{d}t} = -ev_rB_\theta \tag{7.132}$$

Hence, combining (7.130), (7.131), and (7.132), we find that

$$\frac{\mathrm{d}H}{\mathrm{d}t} = c\frac{\mathrm{d}}{\mathrm{d}t}p_z \tag{7.133}$$

This implies that $H - cp_z$ is a constant of the motion for electrons that transit the interaction region.

Now, the Hamiltonian of an electron entering the interaction is given by

$$H_0 = \frac{p_r^2}{2m} + \frac{p_z^2}{2m} + \Psi(r) \tag{7.134}$$

where $\Psi(r)$ is an equivalent potential:

$$\Psi(r) = \frac{P_\theta^2}{2mr^2} + P_0 \ln\left(\frac{r}{a}\right) \tag{7.135}$$

The function $\Psi(r)$ is a monotonically decreasing function of r for $a \leqslant r < r^*$, and a monotonically increasing function of r for $b \geqslant r > r^*$. A schematic plot of $\Psi(r)$ is given in Figure 7.16. In this figure $H_\perp$, where

$$H_\perp = H_0 - \frac{p_z^2}{2m} \tag{7.136}$$

denotes the perpendicular energy of an electron entering the interaction region, and r_i and r_o are the respective minimum and maximum radial position of an electron with perpendicular energy $H_\perp$. For the electron to exit the interaction region with minimum perpendicular energy, the electron must have radial position r^* and no radial velocity. To compute the value of r^*, we find the radial position of the minimum of $\Psi(r)$, using the relation:

$$\left.\frac{\partial \Psi(r)}{\partial r}\right|_{r=r^*} = 0 \tag{7.137}$$

This yields

$$r^* = \frac{P_\theta}{\sqrt{mP_0}} \tag{7.138}$$

Now, writing the angular momentum, parallel energy, and the Hamiltonian in terms of the parameters r_o, σ, and α defined in Section 7.3, we have

$$P_\theta = r_o \sqrt{\frac{2mP_0 \ln(\sigma)}{\sigma^2 - 1}} \tag{7.139}$$

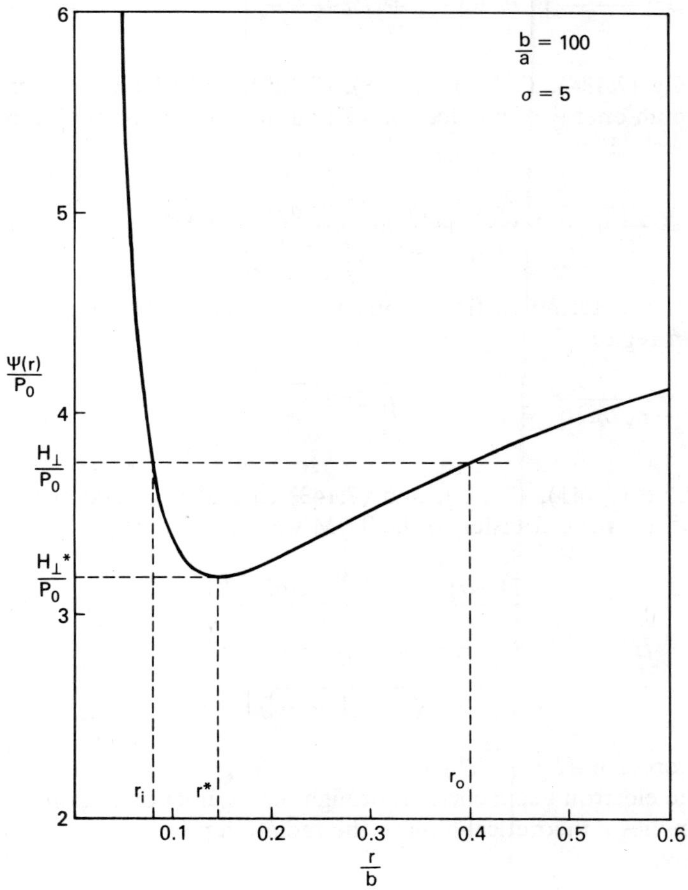

Fig. 7.16 The Equivalent Potential of an Electron Orbiting the Center Conductor of an Orbitron

$$H_{\parallel} = \frac{p_z^2}{2m}$$

$$= \frac{H_0}{1 + \alpha^2} \tag{7.140}$$

and

$$H_0 = \left(\frac{1 + \alpha^2}{\alpha^2}\right)\left[P_0 \ln\left(\frac{r_o}{a}\right) + P_0 \frac{\ln(\sigma)}{\sigma^2 - 1}\right] \tag{7.141}$$

Then, using (7.134), (7.135), (7.138), (7.139), and (7.140), we find that the minimum energy of an electron exiting the interaction region is given by

$$H^* = \frac{P_0}{2} + P_0 \ln\left(\frac{r_o}{a}\right) + \frac{P_0}{2} \ln\left(\frac{2 \ln(\sigma)}{\sigma^2 - 1}\right) + H_\parallel^* \tag{7.142}$$

where $H_\parallel^*$ is determined from using the constant of the motion in the interaction region, $H - cp_z$, and (7.140):

$$H^* - c\sqrt{2mH_\parallel^*} = H_0 - c\sqrt{\frac{2mH_0}{(1 + \alpha^2)}} \tag{7.143}$$

We combine (7.141), (7.142), and (7.143) to find the maximum energy that an electron can transfer to the TEM waveguide mode:

$$H_0 - H^* = \frac{P_0}{2} \frac{\left[\dfrac{2\ln(\sigma)}{\sigma^2 - 1} - 1 - \ln\left(\dfrac{2\ln(\sigma)}{\sigma^2 - 1}\right)\right]}{\left[1 - \sqrt{\dfrac{2H_0}{mc^2(1 + \alpha^2)}}\right]} \tag{7.144}$$

to lowest order in $(H_0 - H^*)/mc^2$.

If the electron beam current through the interaction region is I, then (7.144) implies a theoretical limit to the radiation power output of a TEM orbitron maser:

$$P_{\max} = \frac{IV_0}{2\ln\left(\dfrac{b}{a}\right)} \frac{\left[\dfrac{2\ln(\sigma)}{\sigma^2 - 1} - 1 - \ln\left(\dfrac{2\ln(\sigma)}{\sigma^2 - 1}\right)\right]}{\left[1 - \sqrt{\dfrac{2H_0}{mc^2(1 + \alpha^2)}}\right]} \tag{7.145}$$

Similarly, the theoretical limit to the radiation efficiency, $E_{\max}$, is given by

$$E_{\max} = \frac{P_0}{2H_0} \frac{\left[\dfrac{2\ln(\sigma)}{\sigma^2 - 1} - 1 - \ln\left(\dfrac{2\ln(\sigma)}{\sigma^2 - 1}\right)\right]}{\left[1 - \sqrt{\dfrac{2H_0}{mc^2(1 + \alpha^2)}}\right]} \tag{7.146}$$

7.7.4 Mode Selection

The TEM mode is the lowest order mode of a coaxial cavity. The TEM mode has large electric fields near the center conductor, which is an advantage for operation at high harmonics of the radial oscillation frequency. However, unlike the TE modes, the TEM mode has the disadvantage that there is no cut-off frequency. The cavity outer wall radius cannot be reduced below the cut-off radius to provide a strong reflection of the forward or backward wave that makes up the standing wave in the cavity.

We can use two methods to provide reflection at the entrance or exit regions of the cavity at the desired operating frequency. One method is to modulate the outer cavity wall with a periodicity λ_w. Fortunately, we see from the results of the discussion of the maximum possible efficiency of the TEM orbitron in Section 7.7.3 that the efficiency of the device is maximized when the voltage V_0 is maximized. This implies that the maximum radius of the electrons orbit, r_o, is minimized for high efficiency operation. We can therefore have large ripples in the outer wall of the cavity without intercepting the electron beam while operating in the high-efficiency regime of the device. The other scheme involves periodically placing azimuthally symmetric conducting breaks at the radius of the cavity outer conductor with radial waveguides extending into the insulator. The radial waveguide sections would be tuned to provide an optimal cavity Q at the desired operating frequency and minimal cavity Q at unwanted frequencies. This scheme has the advantage that it will not strongly cut off TE modes to produce a large cavity Q for an unwanted mode. The scheme also has the feature that conducting breaks are required in any case to increase the potential of the outer conductor from its large negative value in the interaction region to a small negative or ground potential in the collector region. Last, we should mention that all TE modes produce azimuthal currents in the cavity outer conductor and TEM modes have no azimuthal wall currents. Lossy material can therefore be incorporated into the cavity design to dissipate azimuthal currents, and thus to lower the value of Q for a TE mode, while having a small effect on cavity Q for a TEM mode.

A conceptual design of a TEM orbitron cavity is shown in Figure 7.17. Further analysis is required to find the shape and number of ripples in the cavity outer wall, and the spacing and length of the radial waveguides to provide optimal reflection at a given operating frequency.

7.7.5 Beam Formation

We envision forming an azimuthally symmetric electron beam in accordance with assumption 1 of Section 7.2. The electrons would be

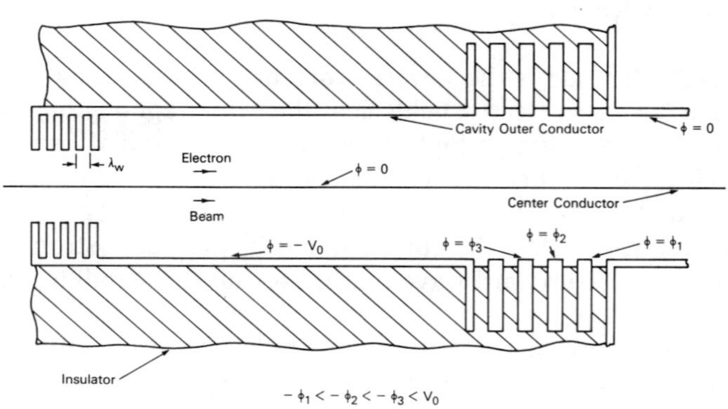

Fig. 7.17 A Conceptual Design of a TEM Orbitron Cavity

emitted from an azimuthally symmetric tungsten dispenser cathode through a grid into the beam-injection region, as shown in Figure 7.18. The electrons would be injected with a large value of axial momentum, a small value of radial momentum, and, of course, no angular momentum due to the azimuthal symmetry. The angular momentum would be created and controlled by the presence of a weak azimuthally symmetric magnet field. As an illustration, we assume that the magnetic vector potential of the applied field is given by

$$\mathbf{A}(r, z) = \frac{rB_0}{2}\,\hat{\theta} \tag{7.147}$$

in the beam-formation region. The canonical angular momentum of the injected electrons would then be

$$P_\theta = \frac{-er_e^2}{2}\,B_0 \tag{7.148}$$

where r_e is the radius of the grid. The magnetic field would be shielded out of the beam-compression region and the interaction region by the presence of an iron shield, which surrounds the downstream portion of the device to remove the injection magnetic fields and any other stray magnetic field that are present. The electrons then enter the beam-compression region with the value of angular momentum given by (7.148),

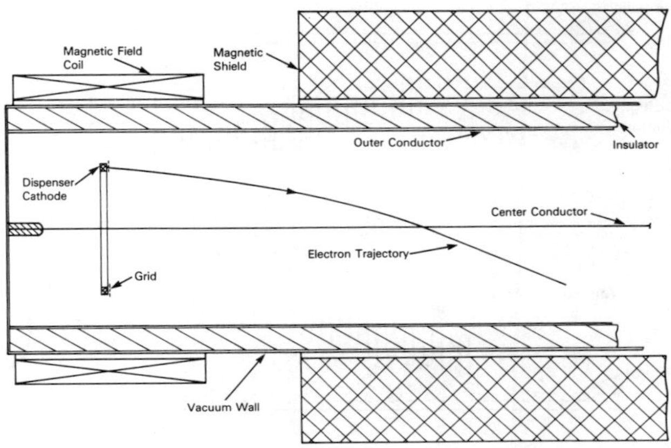

Fig. 7.18 A Conceptual Design of the Beam-Injection Region of an Axial-Injection Orbitron

although the field has been removed. To find the magnitude of the magnetic field required in the beam-injection region, we use (7.139) and (7.148):

$$B_0 = -\frac{2r_o}{er_e^2}\sqrt{\frac{2mP_0\ln(\sigma)}{\sigma^2 - 1}} \tag{7.149}$$

and for $V_0 = 10{,}000$ V, $b/a = 100$, $\sigma = 5$, $r_o = 1$ mm, and $r_e = 1$ cm, we find that $B_0 = 8.1$ G.

An alternative to the above injection scheme is to use a nonazimuthally symmetric cathode and grid structure, shaped similarly to a turbine, to inject the beam. This scheme has the advantage that it does not require a magnetic field to give the beam angular momentum. The beam would not be completely azimuthally symmetric, however, and the ability to control independently the two beam parameters σ and α would be lost. It might be advantageous to use this type of injection scheme in a production device, once the optimal beam parameters become known.

The beam exits the injection region and enters the compression region as shown in Figure 7.19. Here, the beam is adiabatically compressed so that it enters the interaction region with a small axial momentum and a small maximum radius. If the beam is adiabatically compressed, the value of the radial action S is invariant, where

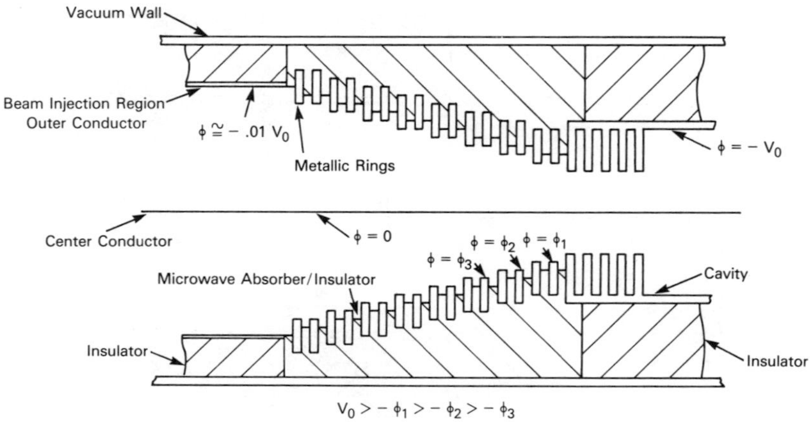

Fig. 7.19 A Conceptual Design of the Beam-Compression Region in an Axial-Injection Orbitron

$$S = \oint p_r \mathrm{d}r \tag{7.150}$$

Evaluating (7.150) using (7.5), (7.37), (7.38), (7.39), and (7.42), we find that

$$S = P_\theta \int_{\xi_o}^{\xi_i} \frac{\mathrm{d}\xi}{\xi^2} \sqrt{\beta + \ln(\xi) - \xi^2} \tag{7.151}$$

P_θ is conserved because the compression region is azimuthally symmetric. Therefore, the parameter β is an adiabatic invariant because the integral is solely a function of β. We see from (7.48) that σ is therefore also an adiabatic invariant. Now, using the fact that the parameters H_0, σ, β, and P_θ are invariant in the beam-compression region, we can use (7.3), (7.39), (7.139), and (7.140) to find two equations that relate the parameters a, b, r_o, V_0, and α in the interaction region to their values as they exit the injection region a', b', r_o', V_0', and α',

$$r_o \sqrt{P_0} = r_o' \sqrt{P_0'} \tag{7.152}$$

and

$$\frac{\alpha^2}{(1+\alpha^2)}\frac{P_0'}{P_0} + \frac{P_0'}{H_0}\ln\left(\frac{\sqrt{P_0}\,a}{\sqrt{P_0'}\,a'}\right) = \frac{\alpha'^2}{(1+\alpha'^2)} \tag{7.153}$$

where

$$P_0' = \frac{eV_0'}{\ln\left(\frac{b'}{a'}\right)} \tag{7.154}$$

For a numerical example of the design equations for an adiabatically compressed beam, we will calculate P_0' and α' for the interaction region parameters of Figure 7.8, and $a' = a$, $b' = 1.5b$, and a compression ratio of 10. Now, using (7.3) and (7.152), we find that

$$V_0' = 146.9 \text{ V} \tag{7.155}$$

and inserting this value of V_0' into (7.153) yields the value of α required at the entrance to the beam-compression region:

$$\alpha' = .118 \tag{7.156}$$

Finally, a conceptual design of a TEM orbitron oscillator is shown in Figure 7.20.

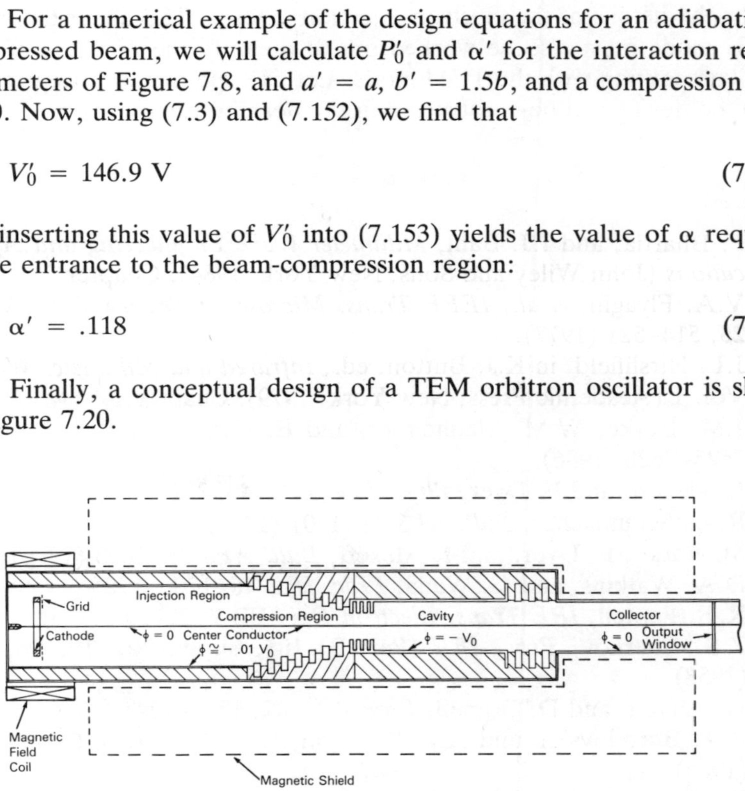

Fig. 7.20 A Conceptual Design of the Axial-Injection TEM Orbitron

7.8 CONCLUSIONS

In this chapter, we have presented the linear theory of the TEM orbitron maser and gained an understanding of the basic operation of the device in both the oscillator and traveling wave tube amplifier configurations. There appears to be great promise in developing the AXIOM concept as an inexpensive high-efficiency source of high-power millimeter-wavelength electromagnetic waves.

Further work is required for the area of mode selectivity in the cavity and for the details of the electron-injection scheme. The linear theory of the TE mode must also be worked out to quantify the problem of TE-mode competition with the TEM mode. Let us hope this will occur and a prototype device will be constructed in the near future.

REFERENCES

1. P. Bhartia, and I.J. Bahl, *Millimeter Wave Engineering and Applications* (John Wiley and Sons, New York, 1984), Chapter 2.
2. V.A. Flyagin, *et al., IEEE Trans. Microwave Theory Tech.* **MTT-25**, 514–521 (1977).
3. J.L. Hirshfield, in K.J. Button, ed., *Infrared and Millimeter Waves,* Vol. 1 (Academic Press, New York, 1979), Chapter 1.
4. J.M. Burke, W.M. Manheimer, and E. Ott, *Phys. Rev. Lett.* **56**, 2625–2628 (1986).
5. I. Alexeff, and F. Dyer, *Phys. Rev. Lett.* **45**, 351–354 (1980).
6. R.W. Schumacher, *Bull. APS* **31**, 1601 (1986).
7. M. Rader, F. Dyer, and I. Alexeff, *Bull. APS* **31**, 1603 (1986).
8. D.A. Watkins, and G. Wada, *Proc. IRE* **46**, 1700–1705 (1958).
9. R.H. Pantell, *IRE Trans. Electron Dev.* **ED-7**, 22–29 (1960).
10. Z.S. Chernov, *Proc. IEE* (Part B) **105**, Suppl. No. 10, 594–596 (1958).
11. H. Alfven, and D. Romell, *Proc. IRE* **42**, 1239–1241 (1954).
12. P.A. Borodovskii, and A.F. Buldygin, *Soc. Radiophys.* **8**, 683–689 (1965).
13. E. Ott, and W.M. Manheimer, *IEEE Trans. Plasma Sc.* **PS-3**, 1–5 (1975).
14. R.H. Hooverman, *J. Appl. Phys.* **34**, 3505–3508 (1963).
15. R.J. Briggs, *Electron Stream Interaction with Plasma* (MIT Press, Cambridge, MA, 1964), Chapter 2.
16. Y.Y. Lau, *et al., Int. J. Infrar. Mill. Waves* **2**, 373–393 (1981).
17. K.R. Chu, *Phys. Fluids* **21**, 2354–2364 (1978).

Chapter 8
THE ORBITRON MICROWAVE MASER

Igor Alexeff

University of Tennessee
Knoxville, Tennessee

8.1 INTRODUCTION*

The *orbitron microwave maser* is an extremely simple and compact device that produces power in the difficult submillimeter and millimeter wavelength regions. In its most simple form, the orbitron maser comprises a central positive wire placed inside a microwave cavity. Electrons, which are confined electrostatically in orbits around the wire, couple to cavity modes and radiate. Because the wire can be made quite thin, electrons close to its surface are in a high electrostatic field, orbit at a high frequency, and emit very-high-frequency radiation. The instability driving the emission is quite strong because electrons trapped in a logarithmic potential well around a wire are negative-mass unstable [1]. Two typical models, along with a pen for size, are shown in Figure 8.1.

Two advantages of the orbitron maser over other advanced microwave tubes, such as the gyrotron and free-electron laser, are that the orbitron does not require a high magnetic field to produce submillimeter microwave radiation. The orbitron maser does not use any magnetic field and it does not require relativistic electrons to cause the negative-mass instability as in magnetic devices, such as the gyrotron.

The orbitron maser has caused much interest in microwave laboratories around the world. In addition to the original work at the University of Tennessee, orbitrons have been studied at the Hughes Research Laboratories, NRL, Grinnell College, in Israel, and in laboratories in the

*This chapter is based on work supported by the US AFOSR under Grant No. AFOSR 86-0100.

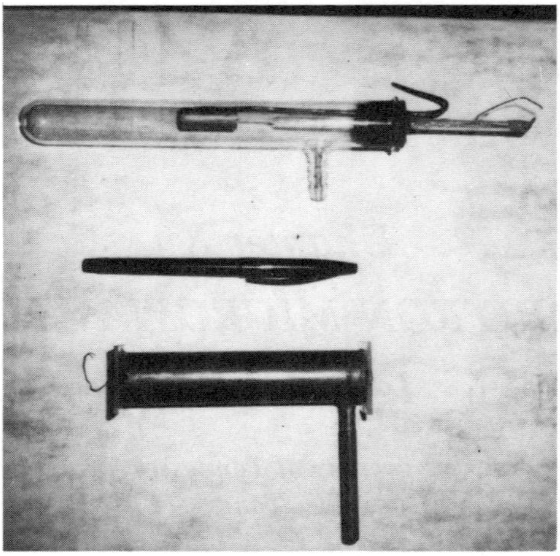

Fig. 8.1 Typical Orbitron Maser Designs
Note: The upper device is the one with which the radiation at 1 THz was generated. The lower device has plastic windows at both ends, and was used for plasma density measurements with an externally generated probe signal. Both are gas-filled, pulsed devices. A fountain pen is used for size information.

People's Republic of China. The simplicity of the device, and the extremely high frequencies produced (1 THz) suggest that orbitron masers will have a niche in the microwave world of the future.

8.2 BASIC PHILOSOPHY

The basic concept in the design of the orbitron maser was to create a very-high-frequency microwave device by combining the most appropriate features of lasers with those of conventional microwave vacuum tubes. In this respect, the orbitron maser is intended to bridge the gap between microwave tubes and lasers. Thus, instead of electron beams, the concept was created to use electrons orbiting around positive wires, in analogy to electrons orbiting positive nuclei, to produce a lasing medium. These orbits can be so small that very high frequencies can be obtained without such heroic measures as frequency upshifts by relativistic electron beams. To obtain more power, many wires could be placed in the cavity to form a lasing medium locked in phase by a common microwave field.

Another laser concept is to use cavity resonators instead of discrete traveling wave structures [2]. The problem with discrete structures is that, for very high frequencies, the structures become very small, difficult to fabricate, and hard to cool. Cavity resonators can be overmoded, and hence quite large and easy to cool. However, frequency selection can still be obtained by properly treating the cavity wall. For example, the wall could be composed of an interference filter, in analogy with the multilayer dielectric filters used at the ends of laser tubes.

One advantage of using a medium of electrons orbiting positively charged wires is that the electron orbits in the logarithmic potential well are negative-mass unstable. In other words, the electron cloud tends spontaneously to bunch in azimuth, and consequently to radiate collectively. In fact, in the orbitron pressure gauge tube [3] of Professor R. G. Herb, the instability is apparently spontaneous in the absence of a cavity resonator and limits the low-pressure performance of the gauge [4].

To improve the performance of our orbitron masers at high frequency, we have etched our wires to produce alternate thick and thin sections. Computer simulation demonstrates that the electrons are trapped around the thicker wire sections, in analogy to electrons being trapped around positive nuclei in the atoms used in lasers [5]. In this case, we consider the device to most closely resemble a radio frequency laser, or maser.

Very high potentials and electric fields can be used in the device, because the central wire is positive, and electron field-emission cannot occur. In addition, due to the logarithmic nature of the potential well, the high field region is limited in volume.

8.3 EARLY EXPERIMENTS AND THEORY

The concepts used in the orbitron microwave maser grew out of the author's early experiences in the nuclear research laboratory of Professor R. G. Herb at the University of Wisconsin. There, I worked on the *evaporion vacuum pump,* in which chemically active gases were gettered by evaporated titanium. Chemically inert gases, such as argon, were ionized by an electron beam, then electrostatically driven into the titanium-coated wall [6]. One model of the pump gave remarkably good performance with argon. Subsequently, the good performance was found to be due to intense argon ionization caused by electrons orbiting an anode of fine, parallel wires. The success of this pump led Professor Herb to extend the principle to a new form of ultra-high-vacuum gauge, which was subsequently marketed [3].

The basic concept of the trapping of electrons around a positive wire remained with the author, and I subsequently applied it to a fast-opening plasma switch, which I patented [7]. I also noted that the orbiting frequencies of electrons confined about thin wires at moderate potentials could be very high, so an orbitron microwave maser could be very useful. A simple theoretical model, using a resistive wall, instead of a reactive cavity resonator, was worked out while on an airplane trip from Knoxville to Santa Barbara, CA, and the experimental program thus started [1].

Early experiments succeeded with the excellent experimental help of Fred Dyer of the University of Tennessee. The experimental observation of intense microwave emission with wavelengths between 3 cm and 5.5 mm from a pulsed, gas-filled tube, along with the primitive resistive-wall theory, were published in *Physical Review Letters* [1]. In addition, a US patent application on the principle was filed, and subsequently granted [8].

The basic configuration used in these early experiments is shown in Figure 8.1. A low-pressure gas environment (about 20 μm) was provided by a bell-jar equipped vacuum system, as shown in Figure 8.2. A thin tungsten wire was the anode (typically 20 mils in diameter). An empty beer can served the three functions of cavity resonator, discharge cathode, and axial confinement of electrons by means of the negative ends. A high-voltage pulsed power supply (600–30,000 V) was provided by a capacitor in series with a spark gap. The sequence of operations was as follows. First, the spark gap fired. Then, a discharge built up in the orbitron maser. After a delay time that was inversely proportional to pressure (typically 10 μs), a microwave pulse was emitted. The power levels observed were typically several watts. Finally, the device changed to a low-voltage, nonemitting mode. Because the microwave emission occurs after the spark gap fires, the microwave emission does not come from the spark gap. This conclusion is verified by localizing the emission with microwave horns. Also, the spark gap does not emit substantial power at frequencies above about 10 GHz.

The basic equation giving the highest possible fundamental frequency for circular orbits is given below:

$$f = \frac{1}{2\pi r}\left[\frac{-ZeV}{m\ln(R/r')}\right]^{1/2} \tag{8.1}$$

where f is the emitted frequency (Hz), Z and e are the sign and charge of an electron (-1.60^{-19} C), V is the potential applied to the central wire (V), r is the radius of the electron orbit (m), m is the mass of an electron (0.91^{-30} kg), ln is the natural logarithm, R is the radius of the outer electrode or cavity resonator (m), and r' is the radius of the wire (m). For the highest emitted frequency, the electrons must have the smallest possible

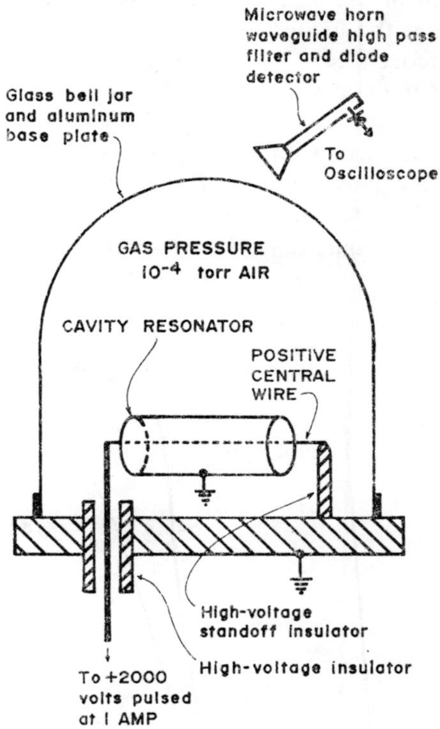

Fig. 8.2 The Original Orbitron Maser Configuration Inside a Bell Jar

orbit, or $r = r'$. For a wire of one mil (25 μm) in diameter inside a cavity resonator of one inch (2.54 cm) in diameter and an applied voltage of 30 kV, the predicted maximum frequency is 3.5^{11} Hz, corresponding to a wavelength of 0.85 mm. These values are typical for operating devices that have been constructed, and the predicted frequency limit is approached and often attained, thus supporting the model. Harmonic production can exceed this limit, as may electron space-charge effects (by, in effect, reducing the outer radius R).

Note that (8.1) predicts that the frequency f increases as energy is removed from an electron and it sinks deeper into the potential well around the wire (r decreases). Such an increase of frequency as energy is decreased is referred to as a *negative-mass* effect. It cannot occur in linear systems, only in circular ones. Such a system can be shown to be inherently unstable. (See the references at the end of this chapter.)

A verification of the orbitron theory is discussed in the cited paper [1]. The idea is that the highest emitted fundamental frequency is given by simple orbit theory (8.1). The highest fundamental is proportional to the square root of the applied voltage. The results presented in this early paper are shown in Figure 8.3, and show how higher frequencies require higher voltages. One further interesting observation was that a very weak magnetic field (a few gauss) turned off the microwave emission, although the field had no other observed effect on the discharge. Figure 8.4 shows the effect of an axial magnetic field on microwave emmission.

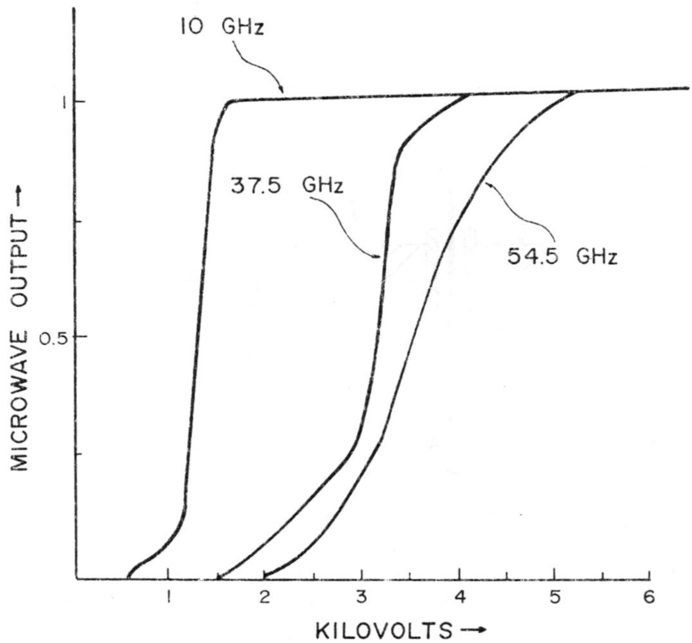

Fig. 8.3 Observed Upper Frequency Increase with Increasing Voltage

An attempt to demonstrate maser amplification with this same apparatus appeared to be successful. A 3 cm microwave signal was introduced into the side of the cavity. When the orbitron power pulse was reduced below some well defined threshold, microwave emission did not appear, unless the 3 cm signal was present. The defect in the experiment was that it was done with microwave diodes and horns, however, and the narrow-band nature of the apparently amplified signal could not be verified.

The decision was made to progress as rapidly as possible to higher frequencies because the orbitron maser could outperform other devices at

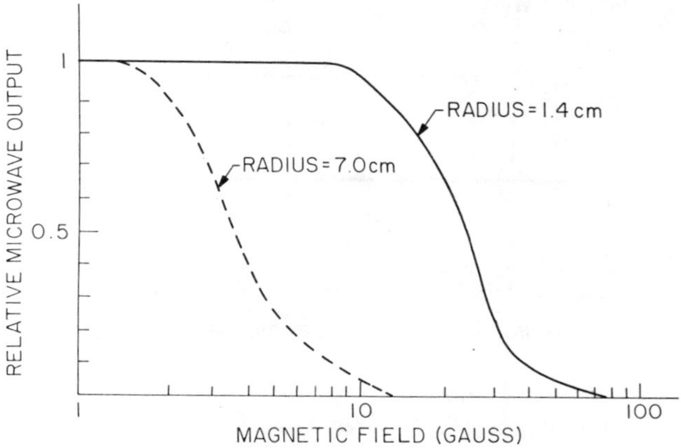

Fig. 8.4 Observed Reduction of Microwave Emission with Increasing Axial Magnetic Field
Note: The two curves are for different radii of the cavity resonator.

that range. The first barrier was the inability of the device to produce wavelengths shorter than 5.5 mm. One possible limitation was that the very high frequencies were being produced by electron orbits too close to the wire. Such orbits are not affected by the confining electric field of the end plates, escape axially through the holes in the end plates, and do not participate in the maser action. A way of overcoming this limitation was to etch grooves in the central wire, which produced an axial electrostatic confining field, as shown in Figure 8.5. In effect, the wire has been converted into a chain of positive nuclei. Computer simulation demonstrated the effectiveness of this axial confinement [5], and in experiments the orbitron maser produced wavelengths down to 2 mm.

For wavelengths shorter than 2 mm, the limitations were in the diagnostic devices, rather than in the orbitron maser. Eventually, an indium antimonide crystal detector, operating at liquid helium temperatures, was obtained and debugged. The radiation was focused by plastic lenses, and the frequencies were determined by transmitting the radiation through many layers of fine wire mesh. By using these techniques, orbitrons with very fine wires (less than $\frac{1}{2}$ mil or $\frac{1}{80}$ mm in diameter), as shown in Figure 8.1 (top), and very high voltages (30,000 V), radiation was observed at $\frac{1}{3}$ mm wavelength, or 1 THz [9]. The power at 1 THz was about 1 W. This emission is considerably higher in frequency and in power than that produced by devices of comparable simplicity.

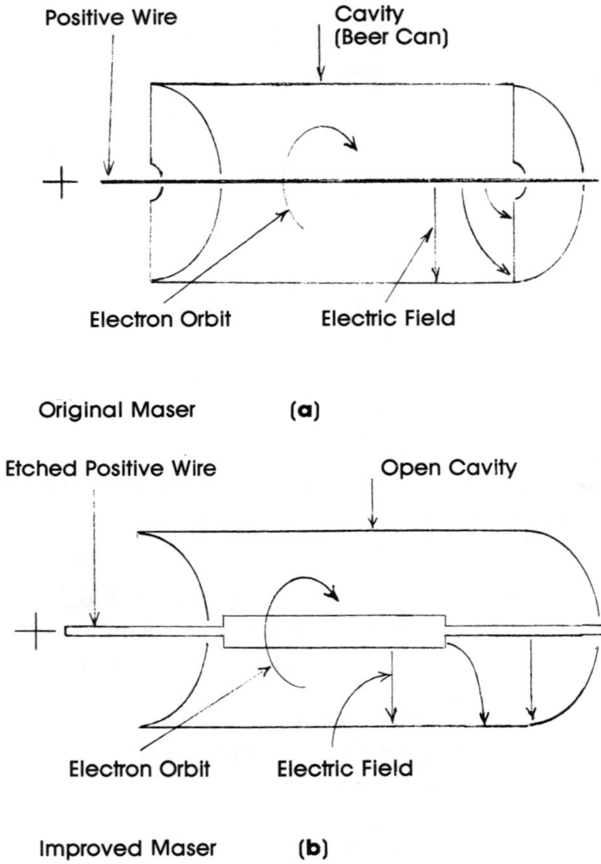

Fig. 8.5 Modifications in the Electric Field Around a Wire Caused by Etching

One problem with using wire meshes for frequency determination is that wire meshes do not have a sharp transmission cut-off with decreasing frequency. These difficulties have led us to develop some new and superior devices, which, unfortunately, are proprietary at the time of this book's publication (*circa* 1987).

8.4 RECENT EXPERIMENTS AND THEORY

Recent orbitron maser experiments have concentrated on two areas: first, the production of high-vacuum, steady-state devices; second, a demonstration that the emission of microwave radiation from the gas-filled, pulsed devices is due to orbiting electrons, rather than plasma oscillations.

The first operation of steady-state, high-vacuum orbitron masers was accomplished with the able help of Mark Rader. The devices were very simple systems to which an oxide-coated cathode was added in the form of a wire parallel to the central wire. The success of the experiment was due to having a very sensitive Hewlett-Packard panoramic receiver on hand, so that the low level radiation from an unoptimized orbitron could be detected and increased by appropriate adjustments. Line radiation was observed at applied voltages of about 1 kV and at currents of a few milliamperes. The highest frequency emitted was close to that predicted by simple theory for circular orbits just grazing the wire anode (differing by a factor of between $\frac{1}{4}$ and 1). Because the device suffered from space-charge limitation of electron flow, multiple wires were tried in order to obtain more current. The orbiting electron clouds around each wire were expected to phase-lock to the common radiation field in the cavity. These experiments appeared to be successful, as the power output and highest frequency emitted increased as the number of wires was increased from one to seven. The highest frequency emitted in the steady state so far is about 10 GHz, the limitation being the overheating of the device when the voltage is increased to produce higher frequency. The use of many anode wires to share the heat load is anticipated to remove the frequency limitation.

The gas-filled, pulsed orbitron maser tubes are useful because high voltages can be applied for short times to produce extremely high frequencies without overheating. The gas is necessary for the production of ions to reduce electron space charge. This permits the higher electron currents, which appear necessary to obtain very-high-frequency emission. Some authors have stated that the operation of the gas-filled devices has nothing to do with orbiting electrons, but is caused by the emission by plasma oscillations (at the second harmonic) from the plasma produced by the gas [10]. For the observed emission at 1 THz, this explanation appears unlikely because the plasma frequency in the device would be considerably lower, even if the gas were 100% ionized. However, the experiments described below were done at lower frequencies to check the plasma hypothesis.

The most obvious way of checking an orbitron maser to observe the occurrence of plasma emission is to check directly the radio-frequency properties of the plasma itself [11]. Consequently, several devices were constructed in which the cavity resonator was also a piece of waveguide. A steady-state, externally produced radio-frequency signal was sent through the waveguide, while the waveguide was also operated as an orbitron maser, as shown by the bottom device in Figure 8.1. The object was to determine the plasma frequency in the orbitron maser during and after orbitron emission by observing the cut-off of the external signal by the plasma filling the device. The results were quite clear: under appropriate operating conditions, the externally generated frequency penetrated the plasma, while the orbitron was emitting a frequency over four times higher. Thus, plasma oscillations were not responsible for microwave emission in these cases. A cross-check with a Langmuir probe, which is used to measure plasma density, verified that it was too low to cause the observed emission. However, under other conditions (higher gas pressure or lower orbitron frequency), the externally transmitted signal was indeed cut off, and, in this case, the result was ambiguous.

A surprising discovery was that more resonant systems existed other than the empty cavity resonator. Line emission was found corresponding to resonances of the coaxial system formed by the cavity resonator with the central wire. Low-frequency lines were observed, which corresponded to the external circuitry supplying power to the device. A low-frequency line was found in steady-state experiments with gas filled tubes, which corresponded to the excitation of background plasma in the device. This low-frequency line was tuned by changing the background plasma density. Thus, the powerful negative-mass instability of the orbiting electrons couples to whatever resonant system is available.

Our theoretical advances clearly show the negative-mass instability to be present. The advances include a treatment of the orbitron maser for circular orbits that predicts enhanced emission at harmonics of the orbit frequency [12], a Vlasov treatment of an electron cloud with a distribution in energy that predicts emission slightly above the frequency of the interacting, orbiting electrons [13], a treatment of noncircular orbits that shows enhanced emission at harmonics [14], an electrostatic analog of the loss cone for particles confined in a magnetic mirror [15], and a magnetically coupled system in addition to the original, electrostatically coupled one [16]. Computer simulation has also been used successfully, in particular, in demonstrating the enhanced confinement of orbiting electrons around a multiply etched wire [5].

8.5 SUMMARY OF RECENT ORBITRON WORK AT OTHER LABORATORIES

The first group outside of ours to become interested in the orbitron was the Hughes Research Laboratory at Malibu, CA. Under the direction of Jay Hyman, work was initiated by Jay Palmer. The attempts to produce operating orbitrons were unsuccessful, and so the group purchased two units from the University of Tennessee. The work was extended by Robert Schumacher and Robin Harvey, who came to the conclusion that the observed emission in the gas-filled orbitron was due solely to plasma oscillations [10]. The Hughes researchers were also unable to obtain emission from a hot-cathode-equipped vacuum orbitron.

Next, Bill Case from Grinnell College also became interested. With help from our group at Tennessee, he produced an operating device and studied the mode structure inside the cavity resonator [17].

The Naval Research Laboratory in Washington, DC became very interested, and has done a comprehensive study in analytic theory and computer simulation. First, Y. Y. Lau observed from work originating in analytic and computer studies of relativistic betatrons that the orbitron configuration (radially outward electric field, no magnetic field) was the most unstable configuration of those under study [18]. Therefore, this configuration was very good for the production of microwave radiation. A comprehensive analytical study was carried on by Manheimer, who investigated highly eccentric orbits around the wire, and showed that strong emission occurred at harmonics of the periodic orbit frequency [19]. Ganguli led a comprehensive computer study of the saturation mechanism and efficiency of power production for a beam-driven orbitron. The conclusion was that the possible efficiency was low compared with other devices such as gyrotrons [20].

Yacob Ben-Aryeh in Israel was interested in the possibility that orbitron emission was related to an electron-gas atom collisional process that he had been studying [21]. He visited the University of Tennessee, conducted experiments there, and took an orbitron back to Israel. Subsequently, the device was reported to be operational.

Liu Shenggang, the President of the Chengdu Institute of Radio Engineering, the People's Republic of China, spent several weeks visiting the University of Tennessee, experimenting with orbitrons, and giving lectures. He reports that there is a large program for constructing and studying orbitrons at Chengdu, and that the basic concepts have been extended to new types of hybrid devices [22].

8.6 FUTURE PROSPECTS

The basic experimental work on orbitron microwave masers at the University of Tennessee, plus the generally favorable reception elsewhere, suggest that there is a real need for the device. The reason is that it can produce reasonable amounts of microwave radiation in the difficult gap in the electromagnetic spectrum between microwaves and light. In addition, the device is quite simple, small in size, easy to fabricate, and reliable.

Applications for such a device are numerous, such as high resolution radar, compact radar (small antenna structures), secure radar (not detectable at long ranges due to atmospheric absorption), broadband communication systems (due to the high frequency), and basic research (easy access to a new frequency range).

The goals of our program at the University of Tennessee include producing still higher frequencies and constructing a steady-state orbitron maser of moderate power. We believe that multiple-anode systems, analogous to a laser with a large lasing volume, will prove to be very useful. The progress has been slower than desired because we not only have to develop the microwave emitter, but also the frequency measuring and detecting systems at the same time.

REFERENCES

1. I. Alexeff and F. Dyer, *Phys. Rev. Lett.* **45**, 351 (1980).
2. D.A. Watkins and G. Wada, *Proc. IRE* **46**, 1700 (1958).
3. W.G. Mourad, T. Pauley, and R.G. Herb, *Rev. Sci. Instrum.* **35**, 661 (1964). The name "Orbitron" has been applied to the microwave maser with the permission of Professor R.G. Herb; see also G.W. McClure, *Appl. Phys. Lett.* **2**, 233 (1963).
4. R.G. Herb, private communication. The glow discharge goes out at low pressures, and an external filament is required, unlike the case of the magnetic (Penning) gauge. Professor Herb also noted that his ion gauges "communicated" with each other; in retrospect, this was probably by microwaves through the vacuum system.
5. I. Alexeff and F. Dyer, *Proc. IEEE Int. Conf. Plasma Sc.* (IEEE Pub. No. 85–CH–2199–8), Paper 3q2 (1985), p. 47.
6. I. Alexeff and E.C. Peterson, *1955 Vacuum Symp. Trans.* (Committee on Vacuum Techniques, Boston, 1956), Vol. 1 (2nd mtg), p. 87. R.H. Davis and A.S. Divatia, *Rev. Sci. Instrum.* **25**, 1193 (1954).
7. I. Alexeff and F. Dyer, *IEEE Trans. Plasma Sc.* **PS-8**, 163 (1980). I. Alexeff, US Patent No. 4,291,255, Sept. 22, 1981.
8. I. Alexeff, US Patent No. 4,459,511, July 10, 1984.
9. I. Alexeff, F. Dyer, and W. Nakonieczny, *Int. J. Infrar. and Mill. Waves* **6**, 481 (1985).
10. R.W. Schumacher and R.J. Harvey, *Bull. APS* **29**, 1179, (1984); R.W. Schumacher and R.J. Harvey, *Proc. IEEE Int. Conf. Plasma Sc.* (IEEE Pub. No. 84–CH–1958–8) (1984), p. 109.

11. M. Rader, F. Dyer, and I. Alexeff, *IEEE Trans. Plasma Sc.* **PS-15**, 56 (1987).
12. I. Alexeff, *IEEE Trans. Plasma Sc.* **PS-12**, 280 (1984).
13. I. Alexeff, *Phys. Fluids* **28**, 1990 (1985).
14. I. Alexeff and F. Dyer, *Proc. IEEE Int. Conf. Plasma Sc.* (IEEE Pub. No. 84–CH–1958–8) (1984), p. 109.
15. W. Nakonieczny and I. Alexeff, *Proc. IEEE Int. Conf. Plasma Sc.* (IEEE Pub. No. 84–CH–1958–8) (1984), p. 23.
16. I. Alexeff and W. Nakonieczny, *Proc. IEEE Int. Conf. Plasma Sc.* (IEEE Pub. No. 85–CH–2199–8), Paper 3q1 (1985), p. 47.
17. W. Case, private communication.
18. Y.Y. Lau and D. Chernin, *Phys. Rev. Lett.* **52**, 1425 (1984); D. Chernin and Y.Y. Lau, *Phys. Fluids* **27**, 2319 (1984).
19. J.M. Burke, W.M. Manheimer, and E. Ott, *Phys. Rev. Lett.* **56**, 2652 (1986).
20. A.K. Ganguly, H.P. Freund, and S. Ahn, "Nonlinear Theory of the Orbitron Maser in Three Dimensions," (to be published).
21. A. Rosenberg, Y. Ben-Aryeh, J. Feldsteiner, and J. Politch, *Phys. Rev. Lett.* **49**, 1917 (1982).
22. Liu Shenggang, private communication; Liu Shenggang, 10th International Conference on Infrared and Millimeter Waves, Lake Buena Vista, FL, Dec. 1985 (IEEE Pub. No. 85–CH–2204–6), paper f4.7.

20. A. [illegible] Soliton Maser in Three Dimensions" (to be published).
21. A. [illegible], Y. Ben-Aryeh, J. Feldsteiner, and [illegible], Phys. Rev. Lett. 48, [illegible] (1982).
22. Liu [illegible], private communication [illegible]

Part III
SLOW-WAVE DEVICES AND VIRTUAL CATHODE OSCILLATORS

Chapter 9

THEORY OF CROSSED-FIELD DEVICES AND A COMPARATIVE STUDY OF OTHER RADIATION SOURCES

Y. Y. Lau

Plasma Physics Division
Naval Research Laboratory
Washington, DC

This chapter gives an elementary exposition of the theory of crossed-field microwave devices. It provides a re-examination of the interaction mechanism in light of the more modern radiation sources, such as the orbitron, gyrotron, peniotron, and gyromagnetron. The treatment emphasizes simplicity and accessibility to workers who are not familiar with the field. The operating conditions of magnetrons will be described first, together with a qualitative description of the physical processes by which the energy of the electrons is transferred to the RF. These processes involve an interplay of the streaming motion and transverse migration of the electrons in dc electric and magnetic fields, and phase focusing provided by the RF fields of the corrugated walls. The more modern devices mentioned above are based on certain combinations of these processes, depending on the presence, or absence, of the dc electric field, the dc magnetic field, and the wall corrugation. The essential features of the interaction mechanism can be quantified by the use of a sheet-beam model and the various radiation sources are then characterized by a single dimensionless parameter (h), which is proportional to the equilibrium electric field experienced by the sheet beam. The wall corrugation enters only in the admittance in the theory of beam-wave interaction. A small-signal theory is developed from first principle for general values of h. The decisive dependence of the stability on the equilibrium type (h) is described in detail. While the validity of the simple theory has been ascertained by various rigorous tests, several important features remain to be tested in laboratory experiments. These features, together with other outstanding issues, will be addressed.

9.1 INTRODUCTION*

The magnetron is a widely used radiation source (e.g., radar, microwave oven, *et cetera*), which was developed very early but is perhaps the most difficult to treat theoretically [1–3]. Design studies have largely relied on empiracal experiences and the two formulas, the Hull condition and the Buneman-Hartree condition, which were derived almost half a century ago. Modern advances in electronic computing still have not provided sufficiently accurate models to predict confidently or explain the multiple facets usually observed in experiments. Such a state of imperfect understanding of the magnetrons remains despite the strong efforts of the many leaders in the field. Resurgence of interest in crossed-field devices is in part triggered by the more recent developments of high-power diodes [4] and relativistic magnetrons [5, 6].

Much of the difficulty in a theoretical study of magnetrons lies with their complex geometries. Controversy already emerged at the very first step. What should be taken as the basic motion of the electrons in a *smooth bore* magnetron, in the *absence* of RF fields? Should we take the highly idealized model of laminar flow, or should we really need to use the more complicated "multistream" model, in which the electrons follow cycloidal orbits that commence and finish on the cathode surface? To this date, a definite answer still has not emerged, but general opinion seems to be more in favor of the laminar flow model, especially prior to full oscillation. Note that the italicized words above already indicate considerable simplification. In fact, in virtually all existing analytical theories, the presence of a corrugated structure is assumed to have no effect on the unperturbed motion of the electrons [7]. The wall corrugation enters only in the RF modes that are excited (Figure 9.1(a)). It is possible that such an assumption, made for convenience of analysis, may have already eliminated a significant amount of interesting physics. (We need not go very far to see this. For example, it is the periodicity introduced in the streaming motion of the electrons in the equilibrium state (by a magnetic wiggler) that is responsible for free-electron laser action as is explained elsewhere in this book.)

*This chapter is based on work supported by the Office of Naval Research.

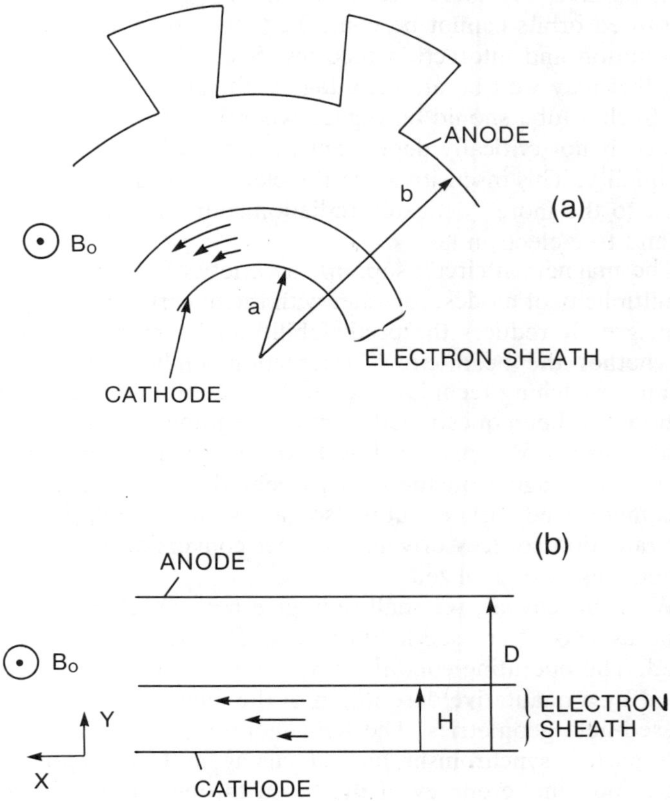

Fig. 9.1 (a) Laminar Flow Model of Magnetron; (b) Further Idealization

Another feature contributing to the complexity of magnetrons is the fact that the electron-emitting cathode is part of the interaction circuit. This differs from many other microwave devices, such as gyrotron [8], klystron [3], free-electron laser [9], gyromagnetron [10], and peniotron [11], where the interaction region is separated from that of beam generation. The control of gun noises and "beam quality" becomes problematical. This, of course, is the reason why the beam cannot be well characterized as described in the preceding paragraphs. This problem of equilibrium modeling appears to be an intrinsic difficulty for theorists when the RF draws directly upon the electrostatic energy of the system. (Another case in point is the more recent invention of orbitron [12], where the unperturbed orbits cannot be controlled, thereby leading to a variety of interpretation and interaction theories. See below for more detail.) Actually, this may well be an advantage with regard to the utility of these tubes. Such a tube should be rugged when in operation because the performance is not critically dependent on whether the beam has a superb beam quality. This insensitivity to the electron beam quality is in marked contrast to the more "delicate" radiation sources, such as traveling wave tubes and free-electron lasers.

The magnetron circuit supports a variety of electromagnetic modes. The multiplicity of modes, together with the uncertainties in the electronic motion, greatly reduces the predictability of beam-circuit interaction. In fact, whether the beam-circuit interaction analysis using the standard impedance-matching technique is adequate at all in describing magnetron operation, has been questioned from the beginning. This concern was not without basis and was reviewed by J. M. Osepchuk [2, p. 275]. It perhaps explains why magnetron theory lags behind those of the more modern devices mentioned earlier, but it also means that the full potential of high-power radiation sources driven by direct conversion (from potential energy) remains to be realized.

With this caveat, we shall only give here an elementary exposition of some aspects of crossed-field devices. The laminar flow model will be adopted. The operating conditions will be given in the next section, to be followed by a qualitative description of the mechanism of energy transfer in crossed-field geometries. The RF generation is a result of an interplay of wave-particle synchronism, phase focusing, and conversion of both electrostatic and kinetic energy of the beam through its transverse and longitudinal motion in the RF fields supported by the corrugated walls. Some (but not all) of these features are shared by other, more contemporary microwave sources, such as the orbitron, gyrotron, peniotron, and gyromagnetron, and will be discussed in Section 9.3. The comparison of these various devices forms an integral part of this chapter. This can be achieved by the use of highly idealized models, which will be described in Sections 9.4 and 9.5. Despite its scope, the exposition is self-contained, requiring a minimal amount of mathematics. Detailed justification and extension of the theory will be referenced as the occasion arises. Some of the topics discussed are still under active investigation.

9.2 BRILLOUIN FLOWS, HULL CUT-OFF CONDITION, AND BUNEMAN-HARTREE OSCILLATION CONDITION

The laminar flow model that has been studied extensively in crossed-field devices is the Brillouin flow. It is basically an $E \times B$ drift, ignoring the cycloidal motion of the electrons (Figure 9.1). It is the mean flow field of the electron cloud with which the RF interacts over a long time scale. Although the Brillouin flow is difficult to achieve experimentally, recent particle simulation of magnetrons show that the space-charge distribution and the electronic motion contain a large component of the Brillouin flow, especially during the start-up stage. In the absence of curvature effects, consideration of the planar Brillouin flow immediately leads to the Hull cut-off condition and the Buneman-Hartree condition. These conditions determine the range of voltage (at a given magnetic field) in which a magnetron operates. When curvature and relativistic effects are included, the derivation would be more involved, but the main features remain unchanged for these conditions.

The following assumptions are used in the derivation of the Brillouin flow:

(a) the flow is laminar and parallel to the sole;
(b) there is no variation in the direction of the flow, and spatial inhomogeneity associated with the corrugated structure is thus ignored;
(c) all electrons are emitted from the cathode with zero velocity and zero energy.

In the derivation of the cut-off condition and the oscillation condition, the RF fields, when present, are assumed to have infinitesimally small amplitude, and thus all fields are essentially static. (The last two assumptions are also used when these conditions are derived from single particle motion as in the case of a cylindrical model.) The generalization to cylindrical models with relativistic effects will be indicated.

For simplicity, we shall first consider the planar, nonrelativistic model. The governing equations for the system depicted in Figure 9.1(b) read

$$\vec{v} = \hat{x} v(y) \tag{9.1}$$

$$-|e|V + \frac{1}{2} m_0 v^2 = 0 \tag{9.2}$$

$$v(y) = \frac{1}{B_0} \frac{\partial V}{\partial y} = -E_0/B_0 \tag{9.3}$$

Equation (9.1) follows from assumptions (a) and (b), and (9.2) from assumption (c) of the previous paragraph. Equation (9.3) accounts for the force balance in the flow field. Here, $v(y)$ is the flow speed and $V = V(y)$ is the voltage distribution with $V(0) = 0$ at the cathode, e is the electronic charge, E_0 is the electric field consistent with the space charges, and B_0 is the external constant magnetic field. The diamagnetic field is ignored.

Upon differentiating (9.2) with respect to y and eliminating $\partial V/\partial y$ according to (9.3), we obtain a solution for $v(y)$:

$$v(y) = \omega_c y \tag{9.4}$$

where $\omega_c = |eB_0/m_0|$ is the cyclotron frequency. In obtaining (9.4), we have used the boundary condition $v(0) = 0$. The voltage distribution reads

$$V(y) = \frac{1}{2}\frac{m_0 v^2}{|e|} = \frac{1}{2}\frac{m_0}{|e|}\,\omega_c^2 y^2 \tag{9.5}$$

and the electric field is

$$E_0(y) = -\frac{\partial V}{\partial y} = -\frac{m_0}{|e|}\,\omega_c^2 y \tag{9.6}$$

The charge density is

$$\rho_0 = \epsilon_0 \frac{\partial E_0}{\partial y} = -\frac{m_0}{|e|}\,\omega_c^2 \epsilon_0$$

which may be written as

$$\omega_p^2 = \omega_c^2$$

where $\omega_p = (\rho_0 e/m_0\epsilon_0)^{1/2}$ is the plasma frequency. Note that the laminar flow condition demands that $\omega_p = \omega_c$ be satisfied.

Given separation D, operating voltage V, and magnetic field B_0, the thickness H of the space-charge layer (Figure 9.1(b)) may be calculated. Note first that the electric field equals $m_0\omega_c^2 H/|e|$ between the region $y = H$ and $y = D$ (9.5). Then, the total voltage drop becomes

$$V = V(H) + \frac{m_0\omega_c^2 H}{|e|}(D - H) = \omega_c B_0 H\left(D - \frac{H}{2}\right) \tag{9.7}$$

Solving for H, we obtain

$$H = D[1 - (1 - V/V_c)^{1/2}] \tag{9.8}$$

where

$$V_c = \frac{1}{2}\frac{|e|}{m_0} B_0^2 D^2 \tag{9.9}$$

is the Hull cut-off voltage.

Thus, the Brillouin flow is characterized by a velocity field of constant shear (9.4) and a constant charge density, extending from the cathode to a distance H, which is given by (9.8) in terms of the voltage V, magnetic field B_0, and anode-cathode separation D. The electron density satisfies $\omega_p = \omega_c$.

The Hull cut-off voltage V_c has the following interpretation. At a given magnetic field V_c is the maximum voltage allowed for magnetic insulation, preventing the electrons from reaching the anode. This can be seen by setting $H = D$ in (9.7). The parabolic dependence of V_c on B_0 (9.9) is sketched in Figure 9.2. For parameters below the Hull parabola, the magnetron is insulated. The same condition (9.9) is obtained from single-particle orbital considerations. An electron emitted from rest at the cathode will not be able to reach the anode if $V < V_c$, if the fields are time-independent and azimuthally symmetric.

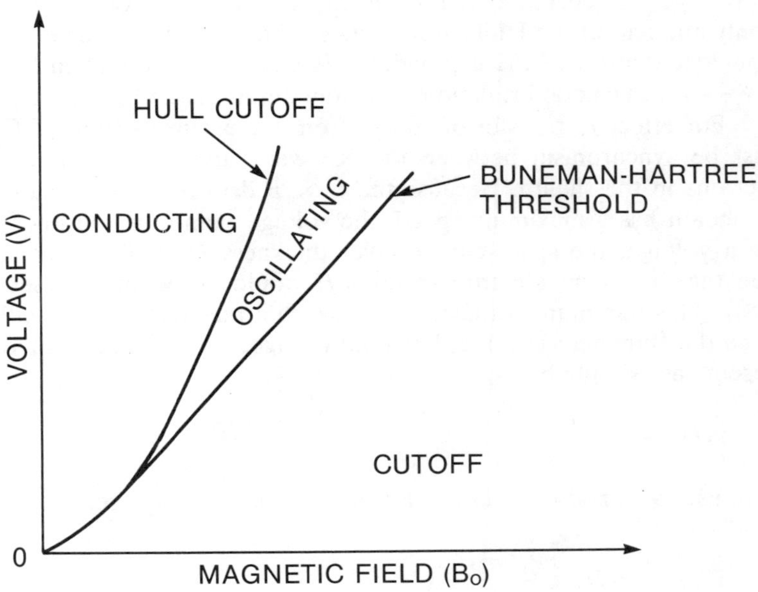

Fig. 9.2 Hull Cut-Off Voltage and Buneman-Hartree Threshold Voltage
Note: The latter curve is tangent to the former.

An alternative form of (9.8) is

$$H = D\{1 - [1 - (B_H/B_0)^2]^{1/2}\}$$

where

$$B_H = \left(\frac{2m_0V}{|e|D^2}\right)^{1/2} \tag{9.10}$$

is known as the Hull cut-off magnetic field. It is the minimum magnetic field required for magnetic insulation at a given voltage. When the operating voltage V is small compared with V_c, we may approximate (9.8) as

$$H \approx \frac{1}{2}D\left(\frac{V}{V_c}\right) = \frac{D}{2}\left(\frac{B_H}{B_0}\right)^2$$

which shows that the space-charge layer thickness increases linearly with the voltage, but decreases as $1/B_0^2$ as B_0 increases. If the operating voltage is only one-half of the Hull cutoff voltage, the space-charge layer occupies about one-quarter of the gap width. We can easily show from (9.8) that as $V \to V_c$, magnetic insulation is rapidly diminished.

For effective transfer of energy from the electrons to the RF, there must be synchronism between the RF wave and at least some of the electrons in the sheath. Because the $E \times B$ drift of the electrons within the sheath has a maximum speed, the voltage cannot be too low. At too low a voltage, the space-charge layer thickness H would be small, and even the maximum electron speed $\omega_c H$ would be below v_{ph} (see (9.4), (9.8)). This minimum voltage, above which a magnetron can oscillate, is called the Buneman-Hartree threshold voltage, and it is determined in the present case simply by

$$v(H) = v_{\text{ph}} \tag{9.11}$$

If we use (9.8) and (9.4) in (9.11), the threshold voltage reads

$$V_{BH} = B_0 D v_{\text{ph}} - \frac{m_0}{2|e|}v_{\text{ph}}^2 \tag{9.12}$$

which depends on the phase velocity of the mode under consideration. This threshold voltage is also shown in Figure 9.2. We can easily show that, regardless of the phase velocity, the straight line representing the threshold voltage in Figure 9.2 is always tangent to the Hull parabola.

When both relativistic and cylindrical effects are included, the Hull cut-off magnetic field and the threshold voltage are modified. If we use D^* to represent the equivalent gap width $(b^2 - a^2)/2a$, (Figure 9.1(a)), the Hull cut-off magnetic field becomes

$$B_H = \left(\frac{m_0 c}{|e|D^*}\right)\left[\frac{2|e|V}{m_0 c^2} + \left(\frac{|e|V}{m_0 c^2}\right)^2\right]^{1/2} \tag{9.10a}$$

and the threshold voltage (9.12) is modified to read

$$V_{BH} = B_0 D^* v_{ph} - \frac{m_0 c^2}{|e|}\left\{1 - \left[1 - \left(\frac{v_{ph}}{c}\right)^2\right]^{1/2}\right\} \tag{9.12a}$$

This relativistically correct threshold voltage was obtained by Walker [13] using a Hamiltonian description of orbits. When this threshold voltage is maximized with respect to the phase speed, we obtain the cut-off voltage (9.10a). Again, the threshold voltage curve is tangent to the cut-off voltage curve, as in Figure 9.2. Magnetron oscillation takes place when V and B_0 lie between these curves, but usually closer to the Buneman-Hartree threshold voltage [6]. Note that (9.10) and (9.12) are recovered from (9.10a) and (9.12a) in the nonrelativistic limit $c \to \infty$. There are recent modifications of these voltages, including the effect of diamagnetic current [14] and axial current [15]. All of these derivations made three key assumptions: (a) the external field is azimuthally symmetric; (b) the amplitude of RF is so low that the energetics of orbital motion is not affected; (c) the fields are time-independent. In practice, none of these assumptions are valid. We refer to Chapter 10 by J. Benford [16] for a detailed discussion of the experiments and these conditions.

The validity of the old concepts outlined above has been tested in computer simulations. A. Palevsky [5] recently developed a two-dimensional, time-dependent, fully electromagnetic, and fully relativistic particle-simulation code to model pulsed relativistic magnetrons [17]. Some of the simulation results are shown in Figures 9.3 and 9.4. Figure 9.3 shows the contour levels of charge densities at an early stage, before oscillation starts, and the arrows indicate the theoretical Brillouin layer. The corresponding velocity profile (not shown) follows the Brillouin flow to within a few percent. The contour plot at a much later stage (when oscillation is fully developed) is shown in Figure 9.4. The spoke formation is evident. Note that the spatial extent of the sheath is approximately given by the

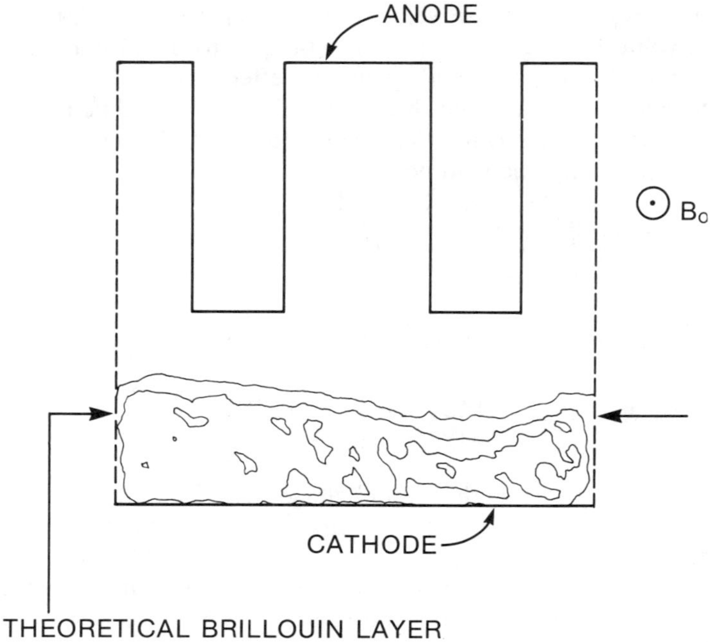

Fig. 9.3 Density Contours Obtained from Palevsky's Particle Simulation [5] During the Early Stage
Note: The arrows indicate the theoretical Brillouin layer.

Brillouin layer, even at this late stage, when the beam-circuit interaction is highly nonlinear. The velocity profile corresponding to Figure 9.4 contains considerable spread about the theoretical Brillouin flow, and confirms the significant back-bombardment of electrons at the cathode [18]. These simulation results show that the electron flow is a mixture of Brillouin flow and a somewhat weaker multistream component. The results also corroborate the suggestion made by Buneman [19], a long time ago, that the instabilities associated with such shear flows may play a key role in starting magnetron oscillation. We shall return later to the stability of the laminar flow.

9.3 GAIN MECHANISM IN CROSSED-FIELD DEVICES AND COMPARISON WITH OTHER RADIATION SOURCES

In this section, we describe in a qualitative manner the transfer of electron energy (kinetic and potential) to the RF. We shall indicate certain

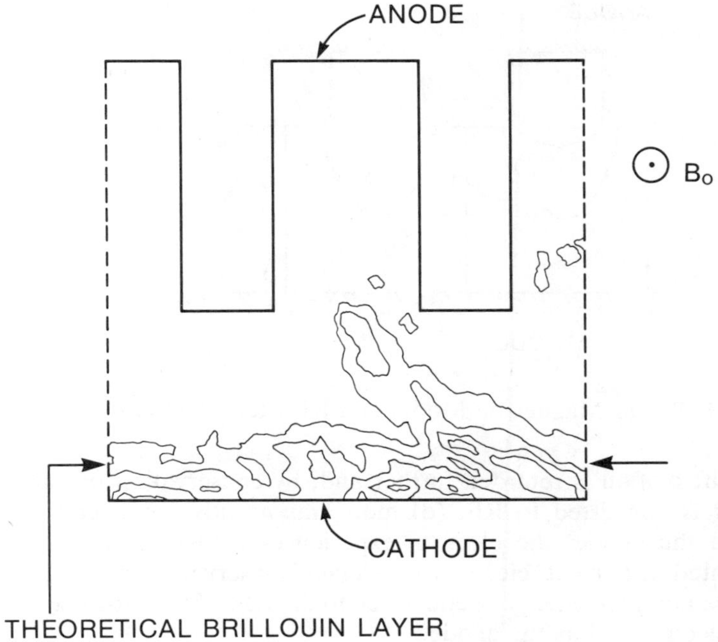

Fig. 9.4 Density Contours at a Later Stage of Fully Developed Oscillation
Note: The magnetron voltage is reduced, leading to a narrower Brillouin sheath, as shown here.

features that are unique to the magnetron configuration. We shall also contrast several other microwave sources, which share some, but not all, of these features. A simple theory will be given in Section 9.4.

If the externally imposed electric field and axial magnetic field are sufficiently strong, the average motion of the electrons is approximately the $E \times B$ drift. The curvature effect of the magnetrons is secondary in importance, and we may consider a planar model, as shown in Figure 9.5. Let us consider the action on electron A due to the combined dc fields E_0 and B_0, and an RF electric field E_1. The RF electric field is assumed to have a phase velocity approximately equal to the unperturbed drift velocity v_0 of the electron and a phase shift of π from one cavity to the next, as shown in Figure 9.5. The RF electric field E_1 affects the motion of electron A in four ways: (a) it retards the drift motion of the electron because E_1 and v_0 are in the same direction (A carries a negative charge); (b) the $E_1 \times B_0$ drift brings test electron A upward to a location closer to the wall corrugation (anode), where the RF field is stronger; (c) as a result,

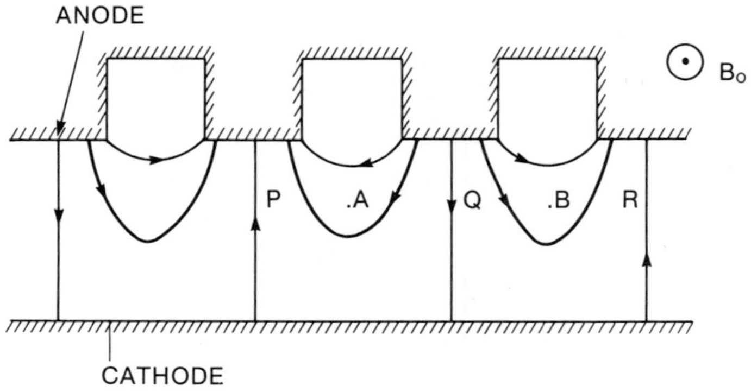

Fig. 9.5 Planar Magnetron Model and RF Electric Fields of the π-Mode

the drift motion is retarded further and, by conservation of energy, this energy is converted to RF; (d) more importantly, as electron A drifts toward the anode, the electron experiences a loss of potential energy associated with the dC electric field. Again, conservation of energy requires that the potential energy be converted to the RF. These processes continue until electron A hits the anode.

Consider now the action of the RF electric field on test electron B. (Figure 9.5). Test electron B is accelerated by the field, but this acceleration is gradually weakened because the $E_1 \times B_0$ drift brings electron B downward, toward the cathode, where the RF electric field is weaker. While electron B gains energy from the RF, on the whole, the amount of energy this electron gains is less than the amount of energy electron A would lose to the waves, and the RF signal thus grows. Note that this RF growth is at the expense of both the kinetic and potential energy of the electrons, the potential energy giving the dominant contribution.

The configuration shown in Figure 9.5 also demonstrates the interesting property of phase focusing in magnetron. Suppose now that the phase speed v_{ph} of the RF is slightly less than the drift speed v_0 of the electrons. At a later time, electron A will advance ahead of the wave, reaching a position near P relative to the wave. Near P, E_1 opposes E_0. Therefore, the electron drift velocity $(E_0 + E_1)/B_0$ is lowered, thereby reducing the difference between v_{ph} and the drift velocity. This effectively brings the electron from position P back to the original position A. For electron B, insofar as $v_0 > v_{ph}$, at a later stage, it reaches position near Q with respect to the wave. Near Q, E_1 is in the same direction as E_0, and the combined drift velocity $(E_1 + E_0)/B_0$ is therefore increased, enlarging the difference between the drift velocity and the phase speed of the wave. In other words, electrons such as B, which gain energy from the RF, are quickly put out of synchronism.

The same argument applies when $v_0 < v_{ph}$. For example, if $v_0 < v_{ph}$, electron A will lag behind the wave, moving to a position near Q with respect to the RF. At Q, the drift speed $(E_1 + E_0)/B_0$ increases as E_1 reinforces E_0 there. This effectively brings A closer to synchronism with the wave than it was originally.

From the description of the two preceding paragraphs, we can see that there is a built-in phase focusing of the "favorable" electrons (like A, which gives its energy to the RF), and a phase defocusing of the "unfavorable" electrons (like B, which gains energy from the RF). This built-in focusing property is largely responsible for the high efficiency of magnetrons. This property is also the reason charge spokes are formed (Figure 9.4), and perhaps is a physical basis for the construction of non-linear soliton solutions [20], which are shape-preserving.

The rather fascinating interplay between dc and ac electric fields, drift motion, and the presence of a corrugated structure in a crossed-field configuration can be contrasted with other microwave sources that share some of these features. We have in mind the peniotron, orbitron, gyrotron, and gyromagnetron. All of these devices are currently under active study, and were invented more than 20 years after magnetrons became operational. In a sense, they are much "simpler" devices.

A model of peniotron [11] is shown in Figure 9.6(a). It consists of an annular electron layer rotating inside a ridged waveguide. The beam is guided only by an external magnetic field. This system differs from a magnetron in that the RF energy derives from the kinetic energy of the electron beam, the potential energy is absent. However, the peniotron is similar to the magnetron in one important aspect. The spatial inhomogeneity of the RF fields produced by the ridged waveguides may cause the guiding center of an electron to drift sideways and, as a result, to turn virtually all of its kinetic energy into radiation, implying an extremely high efficiency.

To see the action of the RF field inhomogeneity on the energy transfer, consider two electrons A and B in Figure 9.6(b, c). Assume that the electrons are *nonrelativistic*, (~10 keV, as typical in peniotron) and that the peniotron's cyclotron frequency is one-half of the RF frequency (i.e., operation at second cyclotron harmonics). Electron A is in an acceleration phase, its rotational energy increases and, as a result, its radius enlarges. As electron A's radius enlarges, it moves half a cyclotron period later to a position where the RF field is stronger, but at which time the RF field decelerates electron A. Thus, electron A gives more energy to the RF during the second half of the cyclotron period than it gains during the first

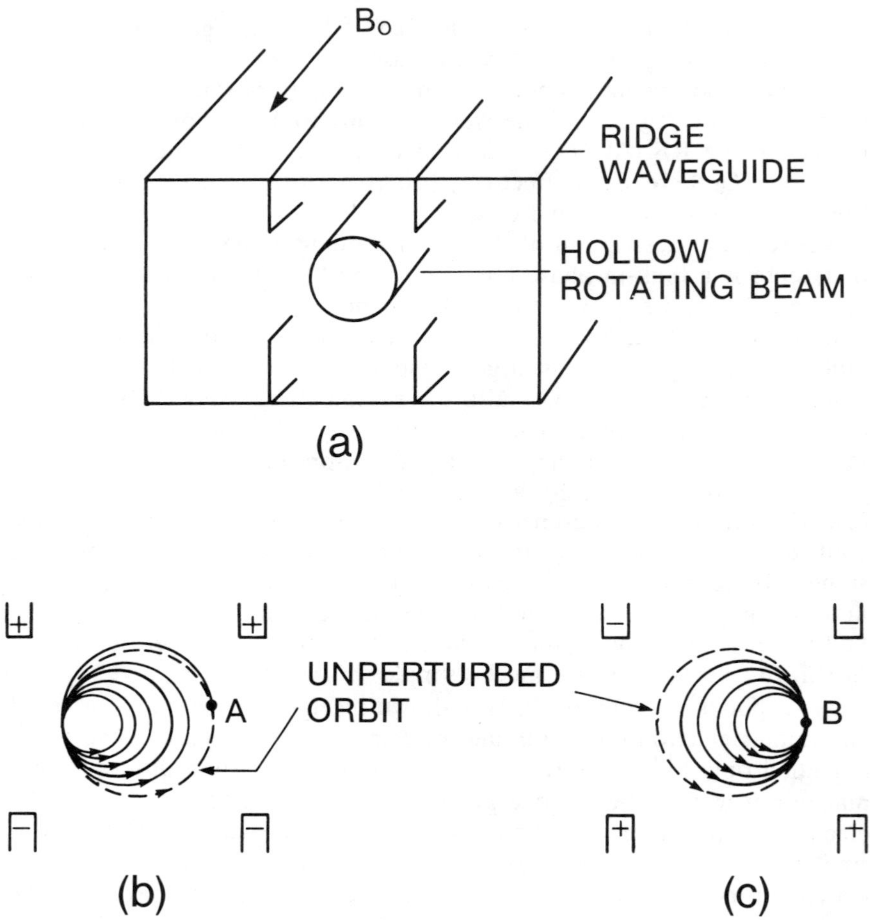

Fig. 9.6 (a) Peniotron Model; (b), (c) Action of RF Electric Field on Electron A (Initial Accelerating Phase) and on Electron B (Initial Decelerating Phase)
Note: The dotted circles denote the unperturbed orbits.

half. On the whole, the energy of electron A decreases and its guiding center shifts to the left (Figure 9.2(b)). The electron energy is transferred to the RF. Electron B (Figure 9.6(c)) is initially decelerated by the RF field and, as a result, its radius decreases. Half a cyclotron period later, electron B experiences a weaker accelerating electric field. Thus, because of the spatial inhomogeneity of the RF electric field created by the ridges, electron B loses energy to the RF field during a cyclotron period and the electron's guiding center drifts to the right (Figure 9.2(c)). Note that both electrons A and B yield their kinetic energy to the RF. Computer simulation shows that the decelerated electrons can be virtually stopped, leading to a theoretical efficiency close to 100% [22, 11]. (Experimental results may be more difficult to interpret, however [21].)

The above description indicates that the mechanism of energy transfer in peniotron need not involve beam bunching along the unperturbed orbit of an electron beam. As in a magnetron, the transverse migration of electrons into regions of stronger RF field created by the ridges leads to the high efficiency. For these and other reasons, the modes of operation of peniotron and crossed-field devices are sometimes known as the transverse modes. A more precise definition will be given in Section 9.4. We should also point out that the above description is not the only possible mechanism responsible for the radiation generated in a peniotron [21]. We shall say more about this at the end of this section, when we treat the gyromagnetron [10, 23]. The gyromagnetron is topologically identical to the peniotron, but is considered to operate with an entirely different mechanism.

Orbitron [12] is similar to a magnetron in that an electrostatic field is present and the RF derives most of its power from the electrostatic field. A model of the orbitron, which has been the subject of some theoretical study, is shown in Figure 9.7. In this figure, a cylindrical layer of electrons rotates about a center wire, which is positively charged with respect to the outer cylinder. The radial electrostatic field provides the centripedal acceleration and no external magnetic field is present [24]. In some sense, the orbitron may be considered as diametrically opposite to the peniotron. The orbitron uses only an electrostatic field for electronic motion, whereas the peniotron uses only a magnetic field. Moreover, the peniotron relies on complicated wall structures, whereas a smooth wall suffices for the orbitron, in which case the importance of RF field inhomogeneity diminishes.

Because the role of RF field inhomogeneity is minimal in an orbitron, the way for the beam to yield its energy (kinetic and potential) is by a bunching mechanism along the particle orbit for the model shown in Figure 9.7. This is indeed found to be the case. Suppose that some charge perturbation is set up on the beam. A test electron moving ahead of this charge bunch is accelerated, it moves outward to a larger radius. This leads to a decrease of the rotational frequency because the angular frequency decreases as $1/r$ for an electrostatically focused beam (Figure 9.7). Azimuthally, this test electron moves toward the bunch and behaves as if

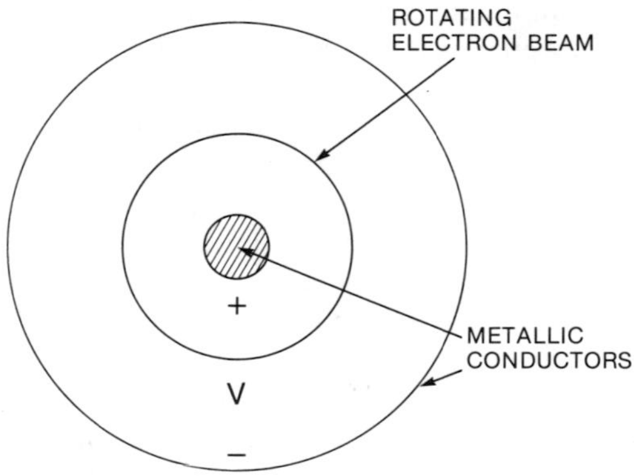

Fig. 9.7 Orbitron Model
Note: This model is used in [12, 27, 28].

having a *negative inertia*—it falls back toward the bunch upon being pushed [12, 27, 28]. Similarly, a test electron that is initially behind the charge perturbation is decelerated. The test electron moves to a smaller radius, but its angular frequency increases, effectively catching up with the bunch in azimuth. Thus, there is strong tendency for beam bunching in the orbitron configuration shown in Figure 9.7. We shall show in the next section that this negative mass behavior is indeed very pronounced—far more pronounced than that exhibited in a large orbit gyrotron at low beam energy (<10 keV) (see (9.27) below).

A highly bunched beam can readily yield radiation. In so doing, the beam contracts radially in an orbitron, and falls to a region of lower potential energy. We can easily estimate that the radiation draws energy primary upon the electrostatic energy, as in a magnetron. However, as the beam falls toward the center wire, its angular frequency may be considerably detuned from the mode of operation, thereby limiting the operating efficiency. Unless some form of phase focusing is introduced, orbitrons cannot be expected to yield efficiencies comparable to those of magnetrons. Thus, the model shown in Figure 9.7 has the peculiar property that it yields tremendous small-signal gain, but the efficiency may be modest. The nonlinear theory recently developed by Ganguly *et al.* [25] showed that this is indeed the case.

The electron beam in a large-orbit gyrotron (Figure 9.8(a)) requires an external axial magnetic field to provide the rotation. There is no externally imposed dc electric field in the circuit. In the simplest geometry, the wall corrugation is absent and the transfer of energy is through charge bunching along the beam orbit Figure 9.8(a, b). Similar to the orbitron,

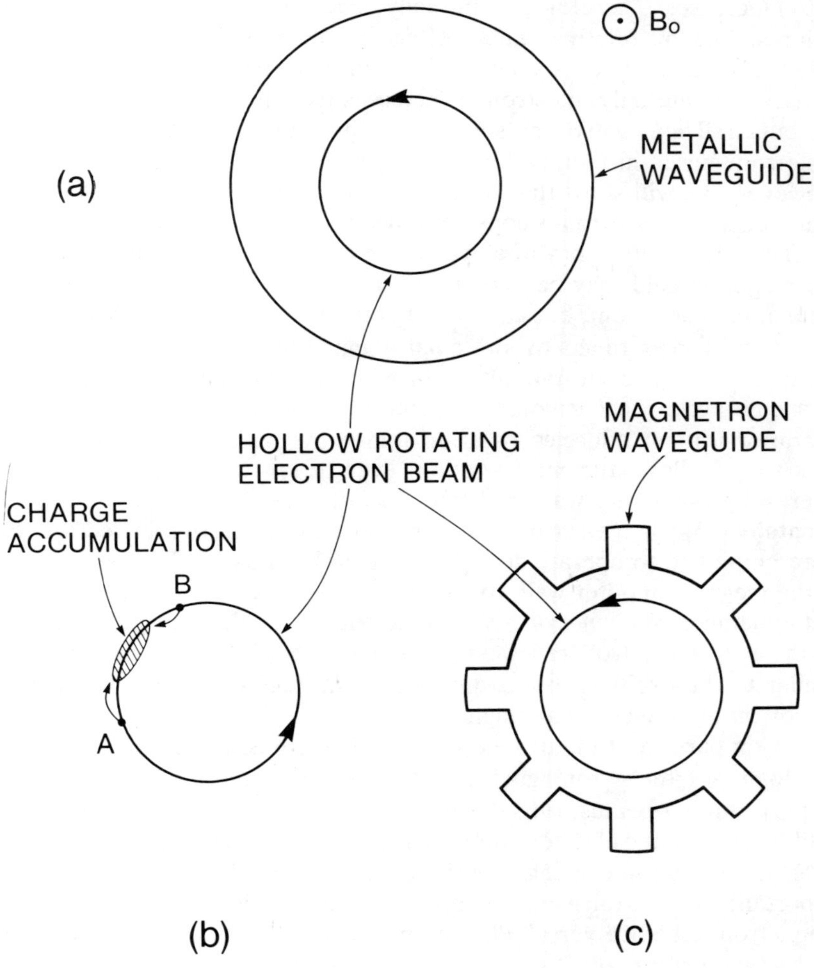

Fig. 9.8 (a) Gyrotron Model; (b) Negative Mass Effect; (c) Gyromagnetron Model

Note: A magnetron waveguide is used to encourage harmonic generation. See Destler *et al.* [23] and interpretation given in [10].

the bunching is caused by a negative-mass effect. Here, the negative-mass effect is relativistic in origin, but already noticeable when the beam energy exceeds 5 keV. If there is a charge perturbation in the electron beam, then test electron A, in front of the bunch, accelerates. The rotational frequency $\omega_0 = |e|B_0/m_0\gamma_0(A)$ of electron A decreases as its relativistic mass factor

$\gamma_0(A)$ increases. Therefore, azimuthally, electron A falls back to the bunch as a result of this relativistic mass effect. Similarly, electron B decelerates, $\gamma_0(B)$ decreases, but its angular frequency $\omega_0(B) = |e|B_0/m_0\gamma_0(B)$ increases. Azimuthally, electron B increases its speed and catches up with the bunch. The negative-mass instability thus developed is identical to the cyclotron maser instability [36], which leads to radiation in gyrotrons. However, we will show that, at low beam energy (<10 keV), the small-signal gain in a gyrotron is considerably smaller than the orbitron. On the other hand, the efficiency of a gyrotron may be noticeably higher because the magnetic field may be detuned so that $\omega - \omega_0$ is sufficiently small to yield appreciable gain, but large enough to yield substantial efficiency [8].

Gyrotron is tuned by the axial magnetic field. The magnetic field requirement can be substantially reduced if the tube can operate efficiently at a high cyclotron harmonic. A promising way to encourage harmonic operation is to introduce corrugation at the waveguide wall (Figure 9.8(c)). This would allow efficient operation, even at low beam energy. Such an interesting possibility was predicted theoretically [10] and verified experimentally [23] when electron beams with energy at tens of kiloelectronvolts were employed to operate at high cyclotron harmonics. The resemblance of the magnetron outer wall, together with the gyrotron mechanism, leads to the name *gyromagnetron.* We should emphasize that this interpretation of the gyromagnetron relies on the longitudinal bunching of the beam, similar to the orbitron, but is quite different from that normally given to peniotron or conventional magnetron.

Comparison of Figures 9.8(c) and 9.6(a) immediately shows that the topologies of the gyromagnetron and the peniotron are identical! Both employ annular beams, ridged waveguides, and only an external magnetic field to provide the electron rotation. Even the beam energy and frequency range are similar in existing experiments. The immediate question is: How important is the transverse peniotron mode in the operation of gyromagnetron, and vice versa? Phrased more bluntly, the question becomes: Is the peniotron really a cyclotron maser in disguise, or does a gyromagnetron really operate according to what is envisioned to happen in a peniotron? These are interesting questions, which have generated considerable controversy, and a resolution is not immediate [29]. Thus, given the magnetic field and the beam and circuit parameters, we cannot even predict with confidence in which mode the tube should operate [21, 29]!

If the simple peniotron-gyromagnetron configuration can cause uncertainties in our understanding of their operation at this late stage, simply due to the corrugation of the waveguide circuit, we should hardly be surprised that the situation for magnetron would be much worse because there is not even a consensus about the choice of a suitable equilibrium model. Thus, it would seem appropriate to present a tractable theory here, which can cover the effects of corrugated structure, curvature, and relativity, for an electron flow under a general combination of dc electric field and magnetic field, including the effects of the ac and dc self-fields. Such a theory will be presented in the following sections by the use of a simple model (Figure 9.9). We shall isolate the circuit effects and the intrinsic beam stability, and examine the coupling.

9.4 STABILITY OF LAMINAR FLOWS

We shall now give a simple exposition of the various types of instabilities encountered in crossed-field devices. These instabilities have been invoked to explain the intense noise observed and the initial stages of spoke formation [2]. For simplicity of mathematics and ready comparison with other radiation sources, only the electron sheet model shown in Figure 9.9 will be covered in some detail. Although highly simplified, this sheet model retains the essential features of the physical mechanism described in the previous sections. The sheet model also demonstrates the profound effects of the equilibrium type on the stability of an electron beam. These effects were only noticed in recent years [27, 28]. In fact, the extent to which they would enter into the case of a thick beam remains an unanswered question. It is highly probable that the effects are large because different waves may be excited across a thick shear layer and they may interact. We shall address some of these issues in the next section.

We shall begin with the *longitudinal mode*. The real part of the frequency ω of the longitudinal modes satisfies

$$\omega \approx l\omega_0 \tag{9.13a}$$

where l is the azimuthal mode number and ω_0 is the angular frequency of rotation of the electron. When a longitudinal mode is present, an electron experiences the RF field at an almost constant phase since $\omega - l\omega_0 \approx 0$, and there would be a strong interaction between the electron beam and the RF. This is particularly true if ω is also close to the natural mode of the circuit. The longitudinal modes are important in gyrotron, orbitron, and crossed-field devices.

The frequencies of the *transverse modes* satisfy

$$\omega - l\omega_0 \approx \pm\kappa \tag{9.13b}$$

where κ is a natural frequency of oscillation about the guiding center when an electron orbit is perturbed (e.g., $\kappa = eB_0/mc$ for gyrotron, peniotron, and the planar magnetron, but $\kappa = \sqrt{2}\omega_0$ for the orbitron model (Figure 9.7), where an electron circulates under a central force field, generated by a positively charged thin wire, with $1/r$ dependence). The peniotron is thought to operate with the transverse mode [11, 22, 29]. The transverse mode is also important in crossed-field configurations [19].

Both transverse and longitudinal modes have been formulated exactly [27, 28] for a cylindrical laminar flow of arbitrary density profile, including dc and ac space-charge effects for a general combination of radial electric field $E_0(r)$ and axial field $B_0(r)$. A simple dispersion relation has been obtained when specialized to a thin sheet (Figure 9.9). The theory presented below will highlight the remarkable features.

9.4.1 Longitudinal Instability

The free energy that drives the longitudinal modes unstable may be either the potential energy or the kinetic energy. The potential energy may arise from the dc space charge of the electron beam, in which case the *diocotron instability* is excited [30–36], or from the externally imposed potential, as in the orbitron model. In the orbitron case, the negative-mass instability is dominant. On the other hand, it is the kinetic energy of the electrons that is converted to the RF field in gyrotrons. The transition of these various instabilities, and their relative importance in different regimes, will be covered in the following subsections.

Diocotron Instability—An Instability Caused by dc Space Charges

The diocotron instability [19, 30–33] was discovered during the initial phase of magnetron research, and was also one of the first plasma instabilities ever discovered. The free energy that drives this instability resides solely with the electrostatic energy of the self-field of the electron sheet. There are several (equivalent) ways to view this instability. Two are given below for the model shown in Figure 9.10, which is a further simplification of the model in Figure 9.9.

The original papers attributed the diocotron instability to a hydrodynamic instability of shear flow [30]. The velocity shear is due to the self-field of the electron sheet, which is assumed to have a small, but finite thickness. It was found that the diocotron instability has the same growth rate as the Kelvin-Helmholtz instability of a vortex sheet in classical hydrodynamics [37]. The evolution of the instabilities, such as the curling up of the sheet, is also similar.

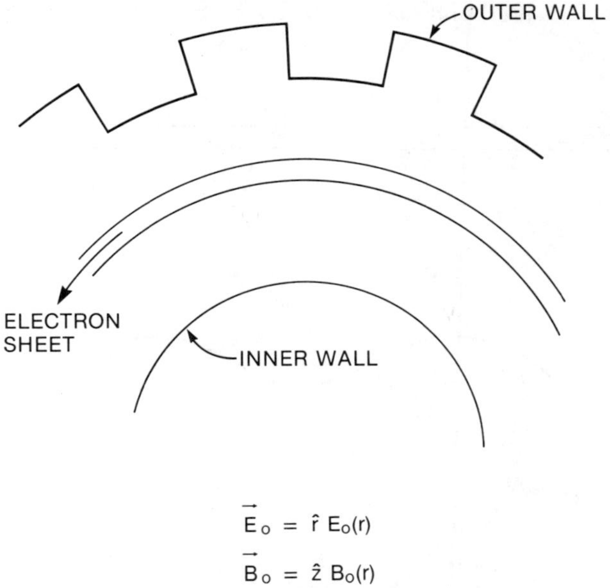

Fig. 9.9 Model for which a Simple Dispersion Relation Is Obtained
Note: This model is representative of crossed field devices, gyrotron, orbitron, peniotron, gyromagnetron, depending on the circuits as well as the magnitude and polarity of E_0 and B_0.

Associated with the dc self-field of the electron sheet is a velocity shear (Figure 9.10(a)) in the equilibrium $E \times B$ drift. The velocity differential due to this self-field is

$$\Delta v = \frac{E_s}{B_0} = \frac{\rho_0 \tau}{\epsilon_0 B_0} = \frac{\sigma_0}{\epsilon_0 B_0}$$

where E_s is the total change of the self electric field across the electron layer of thickness τ and constant electron charge density ρ_0. Note that Δv depends only on the surface charge density σ_0, but is independent of τ, and this is a good approximation for a general density profile if τ is sufficiently small. The growth rate of the Kelvin-Helmholtz instability in a thin vortex sheet is [37]:

$$\omega_i = \frac{1}{2} k \Delta v = \frac{1}{2} (k\tau) \omega_p^2 / \omega_c \tag{9.14}$$

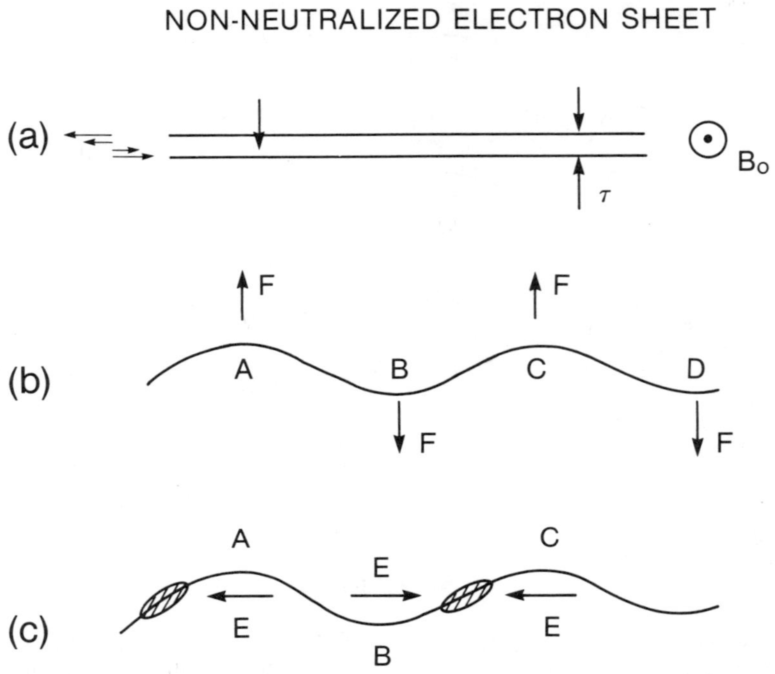

Fig. 9.10 Physical Origin of the Diocotron Instabilities as Viewed in the Rest Frame of the Mean Flow: (a) Velocity Shear Generated by the Electron Self-Field; (b), (c) Ripple on the Sheet and its Reinforcement

where k is the wavenumber in the direction of flow. Expression (9.14) gives the diocotron growth rate in the long wavelength limit $k\tau \ll 1$. note that the growth rate is proportional to the beam current.

An equally transparent way to see the physical origin of the diocotron growth is through the process depicted in Figure 9.10(b, c). Suppose that a sinusoidal ripple is introduced on the electron sheet (Figure 9.10(b)). Electrons in positions A and C experience an electrostatic force in the upward direction, due to the electrons in their immediate neighborhood. Electrons in positions B and D experience a similar force, but in the downward direction. These forces produce an $E \times B$ drift to the left for electrons at A and C, but to the right for electrons at B and D. This leads to an accumulation of charges between B-C and a deficiency of charges between A-B (Figure 9.10(c)). The electric field associated with these charge perturbations, as shown in Figure 9.10(c), produces an $E \times B$ drift of electrons, which reinforces the original ripple, and the instability grows as a result.

The diocotron effect was considered to be a major cause for the intense noise and fluctuation observed in crossed-field devices and high-power diodes. This effect was also thought to contribute to a significant increase of the gain in crossed-field amplifiers [2, 3]. There are many refinements of the diocotron instabilities, including the effects of finite beams, geometries, and finite Larmor radius, [30–36]. Instead of going into these topics, we shall stay with the thin-layer model (Figure 9.9) so as to bring out the crucial dependence of stability on the equilibrium type. For example, we shall show below that the diocotron instability may be stabilized if the curvature effects are taken into account.

General Longitudinal Instability without dc Space-Charge Effects (Negative Mass, Cyclotron Maser, Orbitron Modes, *et cetera)*

The curvature and relativistic effects have been ignored in the above description of the diocotron instabilities. In practice, the curvature effects of the electron orbits are crucial in many radiation sources. The relativistic effect may already be important for an electron beam with energy as low as 5 keV, as shown in the original experiments on cyclotron maser [38]. To examine all of these, we now pretend here that the electron layer is neutralized to rule out the diocotron instability. The questions become: Does it make a difference in the stability if the equilibrium rotation is supported by a radial electric field alone (like an orbitron), by an axial magnetic field alone (like a large orbit gyrotron), or by a combination of both (like a smooth-bore magnetron)? How and in what way would the relativistic effects enter? All of these questions can be answered with the following simple study.

Consider a thin, neutralized electron layer concentrically rotating at velocity $\vec{v}_0 = \hat{\theta} v_0(r) = \hat{\theta} r \omega_0(r)$ inside some waveguide structure (Figure 9.9). The E-layer has a small thickness τ, and uniform charge density extending from $r = r_1$ to $r = r_2$ with mean radius R. For the time being, we shall leave unspecified the relative strength of the radial electric field E_0 and the vertical magnetic field B_0, which are needed to provide the circular motion of the beam. Thus, v_0 is governed by

$$\gamma_0 \frac{v_0^2}{r} = -\frac{e}{m_0}(E_0 + v_0 B_0) \tag{9.15}$$

It is convenient to introduce a dimensionless quantity h to characterize the strength of the electric field of the equilibrium:

$$h = \frac{-erE_0}{m_0 \gamma_0^3 v_0^2} \tag{9.16}$$

Note that $\gamma_0^2 h$ is equal to the ratio of the electric force to the centrifugal force in equilibrium [39]. By using this definition and the equilibrium condition (9.15), we may characterize the equilibrium type according to the value of h (Figure 9.11):

(1) $h = 0$ corresponds to the large-orbit gyrotron (Figure 9.8), where the equilibrium rotation is supported by an axial magnetic field alone ($E_0 = 0$).

(2) $h = 1/\gamma_0^2$ corresponds to the orbitron model (Figure 9.7), in which the rotation is supported solely by a radial electric field ($B_0 = 0$, see (9.15), (9.16)).

(3) $h \gg 1/\gamma_0^2$ corresponds to an *inverted* magnetron, with the cathode at the outer conductor and the anode at the inner conductor. The rotation is approximately given by the $E \times B$ drift. (Centrifugal force is small.)

(4) $h \ll -1/\gamma_0^2$ corresponds to a conventional magnetron, with the cathode at the inner conductor and the anode at the outer conductor. Again, the rotation is approximately given by the $E \times B$ drift, and the centrifugal force is small compared with either the electric or Lorentz force in equilibrium.

(5) The planar limit is formally recovered as $r \to \infty$ (fixing E_0, v_0). That is, $|h| \to \infty$ corresponds to the planar limit (see (9.15), (9.16)).

These various special values are labelled on the h axis in Figure 9.11, and a simple dispersion relation is derived below for general values of h.

For the longitudinal modes, the beam interacts with an RF field mainly through the azimuthal component of the electric field ($E_{1\theta}$), which the thin beam experiences. Conservation of energy gives

$$ev_0 E_{1\theta} = d\epsilon/dt \tag{9.17}$$

where ϵ is the total energy (kinetic and potential) of the electron beam. Using the chain rule, we express

$$d\epsilon/dt = (d\dot{\theta}/dt)/(d\dot{\theta}/d\epsilon) = (\ddot{\eta}/R)/(d\dot{\theta}/d\epsilon)$$

in terms of the azimuthal displacement η of an electron from its unperturbed position. Thus, upon linearization, (9.17) becomes

$$\ddot{\eta} = eRv_0E_{1\theta}\left(\frac{d\omega_0}{d\epsilon}\right) \equiv eE_{1\theta}/M_{eff} \tag{9.18}$$

where the equilibrium value $\dot{\theta} = \omega_0$ is expressed as a function of the particle energy. In analogy with the force law "$F = ma$," we define in (9.18) an effective mass $M_{eff} = (Rv_0 d\omega_0/d\epsilon)^{-1}$. It is not difficult to show from (9.15) and (9.16) that

$$M_{eff} = -m_0\gamma_0\left(\frac{1 + \gamma_0^2h^2}{\beta_0^2 + 2h}\right) \tag{9.19}$$

where $\beta_0 = v_0/c$.

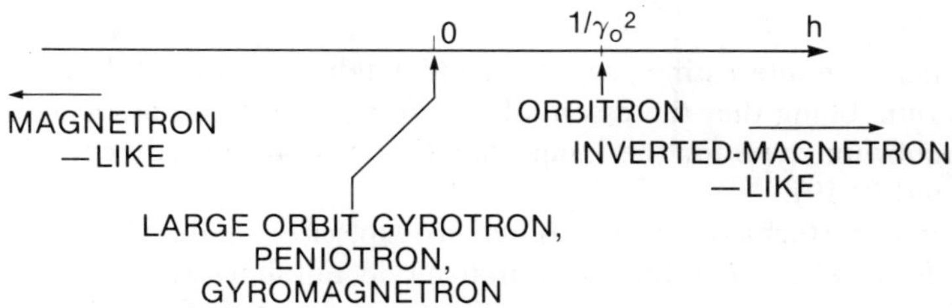

Fig. 9.11 Correspondence between the Values of the Normalized Electric Field h and the Various Types of Equilibria

Equations (9.18) and (9.19) are very interesting. The relationships govern the longitudinal (azimuthal) dynamics of the electron beam. The crucial properties of gyrotron, orbitron, gyromagnetron, smooth-bore magnetron and its inverted configuration, or even cyclic particle accelerators, such as betatrons, are contained in this effective mass. Relativistic effects, curvature effects, and the magnitude and sign of h determine this effective inertia, which, in turn, determines the response of an electron to an accelerating or decelerating RF electric field. For example, M_{eff} is negative for $h > -\beta_0^2/2$, it is positive when $h < -\beta_0^2/2$, it is infinite at $h = -\beta_0^2/2$, and it has a local maximum when $h = 1/\gamma_0^2$ (Figure 9.12).

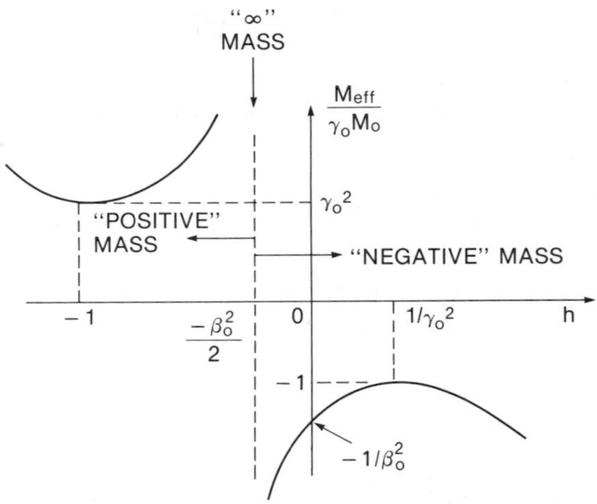

Fig. 9.12 Normalized Effective Mass $M_{\text{eff}}/\gamma_0 m_0$ as a function of h
Note: Compare with Figure 9.11.

All of these are interesting properties and will be considered shortly. Note for the time being that the essential dynamical properties associated with the equilibrium type for the longitudinal modes are already contained in (9.18) and (9.19).

For a perturbation proportional to $\exp[j\omega t - jl\theta]$, d/dt stands for $j(\omega - l\omega_0)$, where l is the azimuthal mode number. A surface charge density perturbation σ_1, associated with the azimuthal displacement η, is given by

$$\sigma_1 = -\frac{\sigma_0}{R}\frac{\partial\eta}{\partial\theta} = \frac{il\sigma_0}{R}\eta$$

$$= \frac{-il\sigma_0}{R}\frac{eE_{1\theta}}{(\omega - l\omega_0)^2 M_{\text{eff}}} \tag{9.20}$$

in which $E_{1\theta}$ is understood to be the value determined at $r = R$. In writing the last expression, we have used (9.18). Equation (9.20) expresses the charge density perturbation in the beam in response to some imposed azimuthal electric field. Thus far, the surrounding electromagnetic structure has not entered into consideration.

To complete the analysis, we calculate what kind of electromagnetic waves would be excited if there were an RF surface charge perturbation

σ_1 located at $r = R$. The presence of a surface charge σ_1, by Gauss law, produces a discontinuity in the radial component of the RF electric field E_{1r}. Thus,

$$E_{1r}^{+} - E_{1r}^{-} = \sigma_1/\epsilon_0 \tag{9.21}$$

where, and in the following, the superscript plus (or minus) is used to designate the position just outside (or inside) the radius under consideration. We next express $E_{1r}^{\pm}$ in terms of $E_{1\theta}^{\pm}$ by introducing the wave admittances b_+ and b_- at the outer and inner edges of the beam. Specifically, b_+ and b_- are defined by

$$b_+ = iE_{1r}^{+}/E_{1\theta}^{+} \tag{9.22}$$

$$b_- = -iE_{1r}^{-}/E_{1\theta}^{-} \tag{9.23}$$

Because the tangential electric field ($E_{1\theta}$) is approximately continuous across the electron layer, we have $E_{1\theta}^{+} \approx E_{1\theta}^{-} \approx E_{1\theta}$, and hence (9.21) to (9.23) give

$$\frac{\sigma_1}{\epsilon_0} = -i(b_+ + b_-)E_{1\theta} \tag{9.24}$$

Whereas (9.20) expresses the dynamic response of the beam to an RF electric field without reference to the circuit, (9.24) gives the circuit response to some beam excitation (σ_1), irrespective of how σ_1 is produced dynamically. For self-consistency, we obtain from (9.20) and (9.24) the dispersion relationship:

$$\begin{aligned}(\omega - l\omega_0)^2 &= \frac{l\omega_a^2}{(b_+ + b_-)(M_{\text{eff}}/m_0\gamma_0)} \\ &= -\omega_p^2 \frac{l(\tau/R)}{(b_+ + b_-)}\left(\frac{\beta_0^2 + 2h}{1 + \gamma_0^2 h^2}\right)\end{aligned} \tag{9.25}$$

where

$$\omega_a^2 = \frac{2(c^2/R^2)\nu}{\gamma_0} = \frac{Ne^2}{2\pi\epsilon_0 m_0\gamma_0 R^2} \tag{9.26}$$

In (9.26), $N = 2\pi R\sigma_0/e$ is the number density per unit axial length, ν is the dimensionless *Budker parameter* measuring the beam density (current). (Note that $\nu = Ne^2/4\pi\epsilon_0 m_0 c^2$ in MKS units, and it equals e^2N/m_0c^2 in CGS units.)

The dispersion relationship (9.25) governs the longitudinal stability of a model electron sheet in orbitrons, gyrotrons, gyromagnetrons, and (smooth-bore [7]) magnetrons. The effects of the waveguide wall enter only in the normalized wave admittance $b_+ + b_-$ in (9.25), and the dynamics (equilibrium type) enter in the factor $(\beta_0^2 + 2h)/(1 + \gamma_0^2 h^2)$, which is also the mass factor in (9.19). The wave admittance b_+ and b_- is a function of geometry and frequency ω, and for the longitudinal modes that we are studying, $\omega \approx l\omega_0$. The mathematical expressions of b_+ and b_- for various cases are given in Appendix B of Chernin and Lau [28]. The major points are as follows. When the waveguide walls are lossless, $b_+ + b_-$ are real. The sum can either be positive or negative. When it is positive (negative), the beam is said to experience a capacitive (inductive) structure [40]. For frequency ω close to cut-off frequencies (ω_g) of the vacuum waveguide mode, $b_+ + b_- \propto (\omega - \omega_g)$, and the growth rate becomes large, *regardless of the sign of* M_{eff}. Under this condition (the *synchronous condition*), (9.25) becomes a cubic polynomial of $(\omega - l\omega_0)$, leading to a rate of growth (gain) proportional to $I^{1/3}$, which is characteristic of gain in longitudinal interactions, as in gyrotrons [8, 36, 40, 41]. These statements on the wave admittances may be extended to cases where the waveguide walls become corrugated, as in a gyromagnetron [10].

For convenience in discussing the beam dynamics, we shall assume that $b_+ + b_-$ is real and positive (except when stated otherwise). We may draw the following conclusions regarding the dynamics of the electron beam.

(1) When $h = 0$, an instability exists as a result of the relativistic effect. This is the well known negative-mass instability [42, 40], the growth rate of which is given by $\omega_i^2 = \omega_p^2(l\tau/R)\beta_0^2/(b_+ + b_-)$, which vanishes in the nonrelativistic limit $\beta_0 \to 0$. This instability was predicted, and it places a limit on the beam current in cyclic accelerators. The instability turns out to be identical to that of the cyclotron maser [36], which causes gyrotrons to radiate.

(2) The negative-mass growth persists if $h > -\beta_0^2/2$. The growth rate is maximized with respect to h when $h = 1/\gamma_0^2$, as is readily demonstrated from (9.25). This case corresponds to the orbitron configuration (Figures 9.11, 9.12). In other words, for a given rotational energy and geometry, the negative-mass effect is most pronounced when the equilibrium is supported by a radial electric field alone, as in an orbitron model [27, 28]. By setting $h = 0$ and $h = 1/\gamma_0^2$ in

(9.25), we obtain the comparison of the small-signal growth between a large-orbit gyrotron and an orbitron:

$$\frac{\omega_i(\text{orbitron})}{\omega_i(\text{gyrotron})} = \frac{1}{\beta_0} \tag{9.27}$$

As β_0 equals 0.044, 0.14, 0.37, 0.55, and 0.78, when the rotational energy is 1 keV, 5 keV, 40 keV, 100 keV, and 300 keV, respectively, we see that the small-signal gain for an electron layer in an orbitron configuration, in principle, can be rather high in comparison with the other radiation sources, especially at low beam energy. A potential use of this enhanced charge bunching in klystrons has been proposed by the author [43].

(3) The dispersion relation (9.25) suggests that the negative-mass instability is suppressed if

$$h < -\beta_0^2/2 \tag{9.28}$$

Thus, the negative-mass instability may be stabilized by a negative radial dc electric field of a suitable magnitude. In terms of an external potential V imposed between the inner conductor at $r = a$ and the outer conductor at $r = b$ (see Figure 9.16), the stability condition (9.28) reads

$$|eV| > \left(\frac{m_0c^2}{2}\right) \beta_0^4\gamma_0^3 \ln(b/a) \tag{9.29}$$

Note that this stabilization mechanism is independent of the beam velocity spread, and is insensitive to the beam current, container geometry, or mode number [27]. While it is not practical to stabilize a high-energy electron beam against the negative-mass instability by this method, due to the γ_0^3 dependence in (9.29). This method becomes attractive for cyclic acceleration of high-energy ions (~500 MeV) of intermediate atomic mass (atomic number of order 20).

Points (1), (2), and (3) are illustrated (Figure 9.13) for two cases, with nonrelativistic and relativistic beam kinetic energies (1 keV and 300 keV), respectively. The growth rates shown in this figure already include the (weaker) diocotron effects, which will be described more fully later.

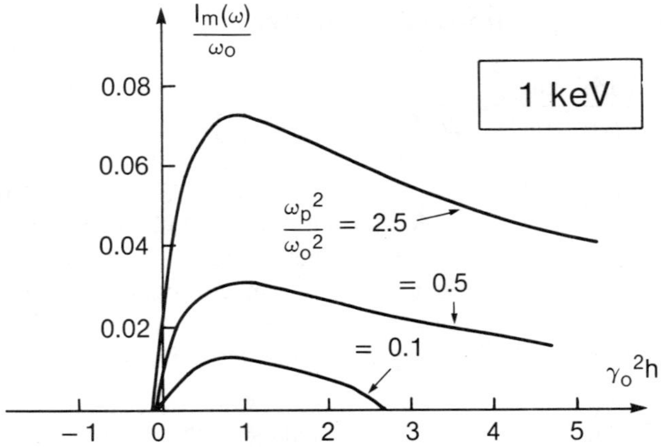

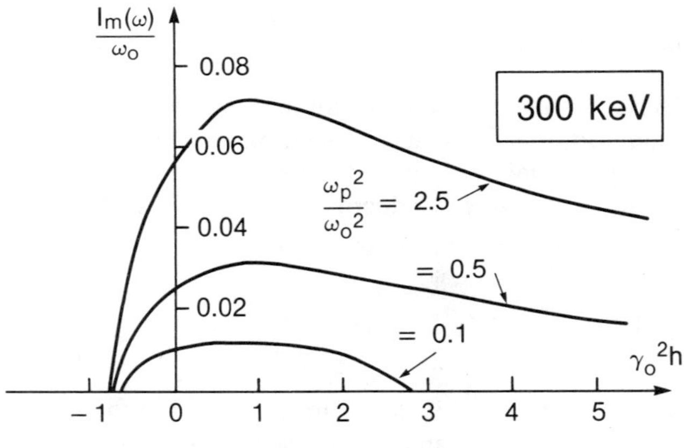

Fig. 9.13 Growth Rates of Longitudinal Modes in a Thin Electron Layer as a Function of the Equilibrium Type (h): (a) at Nonrelativistic Energy; (b) at Relativistic Energy

Note: The growth rates are calculated from (9.31) with $l = 1$, $\tau/R = 0.016$, $b_+ = 5$, $b_- = 3$, and the beam kinetic energy and density are specified in the figure.

(4) If the wall is lossy, $b_+ + b_-$ becomes complex, and the resistive instabilities would result [44], whether the effective mass is positive or negative [45]. However, even this resistive instability can be stabilized if $h = -\beta_0^2/2$, as is evident in the dispersion relation (9.25). Physically, when $h = -\beta_0^2/2$, the effective mass of a rotating

electron is infinite (see Figure 9.12). The beam is very rigid azimuthally and incapable of transferring its rotational energy to the resistive wall, which is the physical mechanism for the excitation of the resistive instability [46].

(5) The negative mass instability disappears in the planar geometry limit, as we would intuitively expect. This is also reflected in the dispersion relation (9.25). In this planar limit $R \to \infty$, $l \to \infty$, but l/R, τ, and E_0 remain finite. Then, $h \to \infty$ from (9.16) and the right-hand member of (9.25), tends to zero. What remains is the diocotron instability, which is not included in (9.25) because the self-field has been explicitly ignored in its derivation.

If viewed somewhat differently, relation (9.25) is the lowest order dispersion when we expand the growth rate in terms of the small parameter (τ/R). The diocotron growth given by (9.14) is the residual instability in the absence of the curvature effect [39]. This was found to be the case in an exact formulation of the eigenmodes.

An Exact Formulation and the Verification of the Simplified Dispersion Relation

The implications outlined above were based on the highly simplified derivation given there. This derivation was motivated by the analytical and numerical solution of an exactly formulated eigenvalue equation for cylindrical laminar flows supported by a general combination of radial electric field and axial magnetic field with self-consistent density and velocity profiles. The exact formulation is fully relativistic and fully electromagnetic, including both dc and ac space-charge effects. For TE modes with dependence $\exp(j\omega t - jl\theta)$, the eigenvalue equation for $\phi = rE_{1\theta}$ is of the form [27, 28]:

$$\frac{d}{dr}\left(A\,\frac{d\phi}{dr}\right) + B\phi = 0 \tag{9.30}$$

where A and B are rather complicated functions of ω, l, and r. This eigenvalue equation governs *both* the longitudinal and transverse modes. These modes manifest themselves as singularities of the equation. The complex eigenfrequencies $\omega = \omega(l)$ for both types of modes are obtained by matching the solutions of ϕ at the beam edges to those of the vacuum solution exterior to the beam (or $\phi = 0$ at the conducting boundary with which the electron sheath happens to be in contact).

The eigenvalue equation (9.30) was solved both analytically and numerically for a thin beam. In the analytic procedure, we expand all quantities in powers of (τ/R). To two orders in (τ/R), the growth rate is given by [27, 28]

$$\omega_i^2 = \left(\frac{1}{b_+ + b_-}\right)\omega_p^2\left(\frac{l\tau}{R}\right)\frac{(\beta_0^2 + 2h)}{(1 + \gamma_0^2 h^2)} + \frac{l^2\tau^2}{R^2}\Lambda\omega_0^2 \tag{9.31}$$

where

$$\begin{aligned}\Lambda = {} & \frac{1}{4}\frac{1}{\gamma_0^6(1 + h)^2(1 + \gamma_0^2 h^2)} \\ & \times \left\{\left(\frac{\omega_p^2}{\omega_0^2}\right)^2[\beta_0^2 + 2h + (1 + h)^2] + 2\left(\frac{\omega_p^2}{\omega_0^2}\right)\gamma_0^4(\beta_0^2 + 2h)^2\right. \\ & \left. - \gamma_0^6(1 + \gamma_0^2 h^2)(\beta_0^2 + h)^2\right\} - \left[\frac{1}{2}\frac{(1 + h)\omega_p^2/\omega_0^2}{\gamma_0^2(1 + \gamma_0^2 h^2)}\left(\frac{b_+ - b_-}{b_+ + b_-}\right)\right]^2\end{aligned} \tag{9.32}$$

All quantities in this expression were defined in the previous section [39]. Note that the first term of (9.31) is the same as the right-hand side of (9.25), except for the generalization [39] of h in (9.31). In the planar limit $R \to \infty$, $l \to \infty$, $l/R \to k$ is finite, then $h \to \infty$, and the first term of (9.31) disappears. By using the last term of (9.31) and noting that only the term involving ω_p^4/ω_0^4 remains in this planar limit, we obtain, for $b_+ = b_-$,

$$\omega_i^2 = (k^2\tau^2/4\gamma_0^4)\omega_p^4/\omega_c^2$$

This equation is merely the relativistic version of the diocotron instability for a sheet beam (see (9.14); see also [33]). Thus, the diocotron instability is, in fact, recovered and the negative-mass instability is removed in the planar limit. In other words, (9.31) includes both the dc and ac space charge effect.

For a thin beam, the second term in (9.31) is usually negligible as compared with the preceding term. Thus, stabilization of the negative-mass instability by the use of the criterion (9.28) would also imply stabilization of the diocotron instability for a thin beam. Figure 9.13 shows the growth rates according to the dispersion relation (9.31) in two examples.

Three interesting features are noted in the figure: (a) maximum small-signal growth for the orbitron configuration ($h = 1/\gamma_0^2$); (b) stabilization of negative-mass *and* diocotron instability when $h \leq -\beta_0^2/2$; (c) persistence of small-signal growth at low beam energy for the gyrotron configuration ($h = 0$). This growth is due more to the diocotron effect than to the negative-mass effect. (See [36] for more detail.)

The validity of the analytic theory has been confirmed [28] by comparing the growth rates with those obtained from a direct numerical integration of the governing differential (9.30). Some examples of this comparison are shown in Figures 9.14 and 9.15. The solid curves represent the eigenvalues according to numerical integration of (9.30), and the dashed curves are according to the analytic formula (9.31). Note that the stability condition and the peak growth rate predicted for orbitrons, according to (9.25), are confirmed in Figure 9.14. There is remarkable agreement in the dependence on beam current (Figure 9.15). Similar checks have been performed on the dependence of the growth rates on geometry,

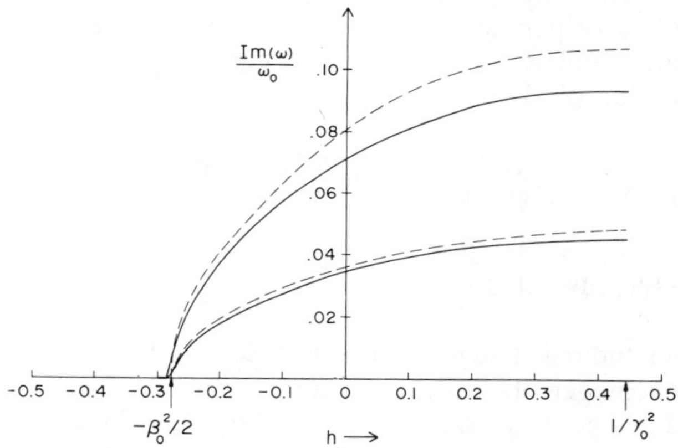

Fig. 9.14 Normalized Growth Rate for Longitudinal Modes as a Function of h for a Test Case with $a = 0.6$ m, $b = 2.6$ m, $R = 1.0$ m, $\tau/R = 0.02$, $\gamma_0 = 1.5$, $l = 1$

Note: A solid curve indicates data obtained from a numerical solution of (9.30); the dashed line is obtained from (9.31). The upper pair of curves is for $\nu/\gamma_0 = 7.88 \times 10^{-3}$, $\omega_p^2/\omega_0^2 = 1.42$; the lower pair is for $\nu/\gamma_0 = 1.57 \times 10^{-3}$, $\omega_p^2/\omega_0^2 = 0.28$.

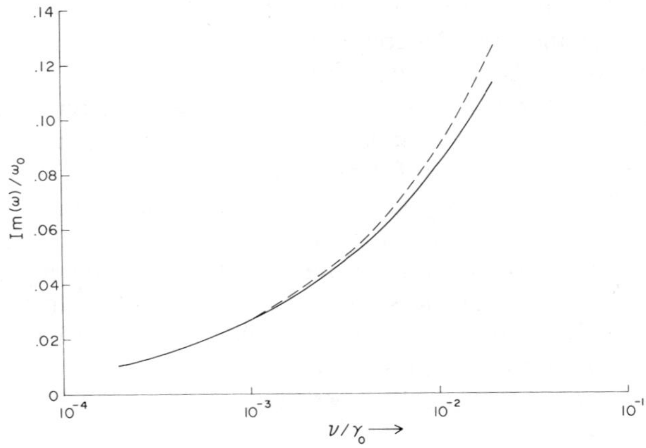

Fig. 9.15 Normalized Growth Rate for the Longitudinal Modes *versus* ν/γ_0 for a Test Case with $a = 0.5$ m, $b = 2.2$ m, $r_1 = 0.99$ m, $r_2 = 1.01$ m, $\gamma_0 = 3$, $l = 1$

Note: A solid curve denotes data obtained from a numerical solution of the eigenvalue problem (9.30); a dashed curve denotes data from the dispersion relation (9.31).

mode number, γ_0, τ/R, *et cetera*, [28]. All such checks confirm the validity of the analytic dispersion relation (9.31).

9.4.2 Transverse Instability

The longitudinal modes described in Section 9.4.1 rely heavily on the spatial inhomogeneity of the unperturbed motion (e.g., velocity shear due to space charge in the diocotron instability, and differential rotation in the negative-mass instability as reflected in the effective mass factor $d\omega_0/d\epsilon = (d\omega_0/dr)/(d\epsilon/dr)$). The spatial inhomogeneity of the RF field plays a secondary role. Near synchronism, the growth rates of these modes are proportional to $I^{1/3}$. By contrast, the gain mechanism in both the crossed-field configuration and the peniotron depends crucially on the spatial inhomogeneity of the RF field. This situation arises because the transverse drift of the *favorable* electrons, which are acted on by the RF fields, progressively populate into a region with stronger RF fields, whereas the *unfavorable* electrons drift to a region of weaker RF field. The small-signal gain for the transverse mode is proportional to $I^{1/2}$.

For the present sheet-beam model (Figure 9.9), the transverse modes are characterized by

$$(\omega - l\omega_0) \approx \pm(1 + \gamma_0^2 h^2)^{1/2}\omega_0 \tag{9.33}$$

We can readily show that the frequency $(1 + \gamma_0^2 h^2)^{1/2}\omega_0$ represents the radial oscillation of an electron about its guiding center in a frame co-moving with the wave, under the combined action of E_0 and B_0. For example, in the nonrelativistic limit of the orbitron configuration ($h = 1/\gamma_0^2 \rightarrow 1$), this frequency becomes $\sqrt{2}\omega_0$, which is the frequency of radial oscillation, about the guiding center of a particle under a central force field with strength that behaves as $1/r$ (Figure 9.7). In another limit $h = 0$, $\omega_0 \rightarrow \omega_c$ (9.33) becomes $\omega - \omega_c \approx \pm\omega_c$, which represents gyro-resonance [2, 3, 11, 19, 28–33, 40, 41].

As mentioned earlier, the transverse modes are also governed by the differential equation (9.30). By using the same analytical procedure that leads to (9.31), we obtain the dispersion relation for the transverse modes for a thin electron layer [28]:

$$\begin{aligned}(\omega - l\omega_0)^2 &- (1 + \gamma_0^2 h^2)\omega_0^2 \\ &= \frac{-l\tau}{R(b_+ + b_-)}(l^2\omega_p^2)\left(1 - \frac{\beta_0^2\omega}{l\omega_0}\right)^2\left[\left(\frac{b_+}{l} - q\right)\left(\frac{b_-}{l} + q\right)\right]\end{aligned} \tag{9.34}$$

where

$$q \equiv -\omega_0(1 + h)/[(l - \beta_0^2\omega/\omega_0)(\omega - l\omega_0)] \tag{9.35}$$

Equation (9.34) is the lowest order dispersion relationship in the expansion parameter τ/R. This relationship is valid for general combination of E_0 and B_0. In this case, dc space charge is not critical, so the dispersion relation gives the correct result in the planar geometry limit.

The dispersion relation (9.34) is rather complicated. The wave admittances b_+ and b_- reflect the corrugation of the walls and the RF field inhomogeneities that such corrugation produces [10, 22, 32, 41]. Without going into the details, we make the following comments.

(1) In the absence of wall losses, small-signal growth can occur only when the transverse mode is synchronous with the circuit mode, for which $b_+ + b_- \approx 0$. The growth rate under such a synchronous condition is proportional to $I^{1/2}$. This small-signal growth is typically weaker than that associated with the longitudinal modes.

(2) Synchronism is always required for small-signal growth. Thus, the relative importance among the centrifugal force, Lorentz force, and electrostatic force in supporting the equilibrium seems to be less crucial. Therefore, the transverse mode is likely to be less sensitive to the equilibrium type, but the resistive instabilities are equally serious for both transverse modes and longitudinal modes.

(3) Conventional theories of transverse modes also express the gains in terms of the RF electric fields (and their gradients) at the beam location [34, 41]. Equation (9.33) is a generalization including a radial electric field in a curved geometry.

(4) Numerical simulations indicate that the transverse modes could be very efficient modes of operation, as in peniotrons [22, 29]. The small-signal theory only gives the initial stage of the development. The effects of the spatial inhomogeneities of the RF field are incorporated in b_+ and b_-, which become complicated expressions when ridges are present in the waveguide walls. Much work remains to be done in the study of the transverse modes, especially their possible coupling with the longitudinal modes when the sheath thickness is no longer small.

9.5 REMARKS

A rudimentary theory is given for a class of radiation sources in which the basic electron flow may be regarded as laminar. Emphases have been placed on the dynamic dependence of the equilibrium type and the role of the circuit structure. In general, the spatial inhomogeneities of the RF fields are important when the transverse migration of the electrons is essential, as in crossed-field devices and peniotrons. However, they are less important when the radiation draws upon the charge bunching along the particle orbits, as in the (large-orbit) gyrotron and orbitron. In the latter devices, the spatial inhomogeneities of the *unperturbed* motion (i.e., velocity shear) are far more important than those of the RF fields. Regardless of the dynamic properties, the radiation sources always operate at a frequency close to a natural frequency of the waveguide circuit.

Several interesting properties were predicted from a study of the longitudinal modes on a thin beam. For example, the orbitron configuration offers perhaps the highest small-signal gain, especially at low beam energy. Away from synchronism, both the negative mass and diocotron instabilities may be stabilized by a negative radial electric field of sufficient magnitude. This condition arises as the electron's effective mass changes from negative to positive. (We should stress that instability may still occur when the circuit mode is synchronized with the beam mode, even if the beam has a *positive-mass* behavior.) The most stable configuration against synchronous excitation *and* against resistive wall instability is when the beam is very *rigid*. This occurs when the effective mass is infinite: when the radial electric field is adjusted so that

$$h = -\beta_0^2/2$$

All of the above predictions are yet to be confirmed in a controlled experiment, such as the one proposed in Figure 9.16. This is essentially a two-cavity klystron with a bent drift tube. The orbit is bent by either a magnetic field B_0, an electrostatic field, or both. The polarity and the magnitude of the externally imposed voltage (V) and the external magnetic field B_0 may be adjusted to correspond to various values of h (Figure 9.11). The response of the electron beam may be monitored at the output cavity after an external RF signal is impressed upon the beam at the input cavity. Such an experiment may be done with an electron beam of energy $\lesssim 10$ keV, $B_0 < 100$ G, $E_0 < 5$ keV/cm, and beam current < 0.1 A. It also serves as a proof-of-principle experiment for the "super-bunched" klystron [43], as well as a *controlled* experiment for the orbitron mechanism discussed in this chapter.

In the case of a thin beam, the longitudinal mode is decoupled from the transverse mode, and each mode may be treated separately, which is not so when the beam is thick. In the case of a thick beam, because of the velocity shear, both (9.13a) and (9.13b) may be satisfied for electrons at different radii. Thus, the longitudinal mode of the electrons at one layer may interact with the transverse mode of the electrons at a different layer. In fact, this was noticed in the original stability theory of the Brillouin flow, and Buneman [19], in effect, already interpreted the instability as due to an exchange of the negative energy wave between those electrons satisfying (9.13a) and the positive energy waves of electrons at a different layer where (9.13b) is satisfied! There remains to be extracted a rich amount of interesting physics hidden in (9.30).

There are deficiencies in the laminar flow model. The lack of Larmor motion in such a model was only recently corrected in the treatment of diocotron instability. Kleva, Ott, and Manheimer [35] showed that the diocotron instability cannot be completely removed from an electron sheath, regardless of the relative sizes of the sheath thickness and the Larmor radius. (This prevalence of the diocotron instability has recently been conjectured to occur within the Debye sheaths of a magnetized

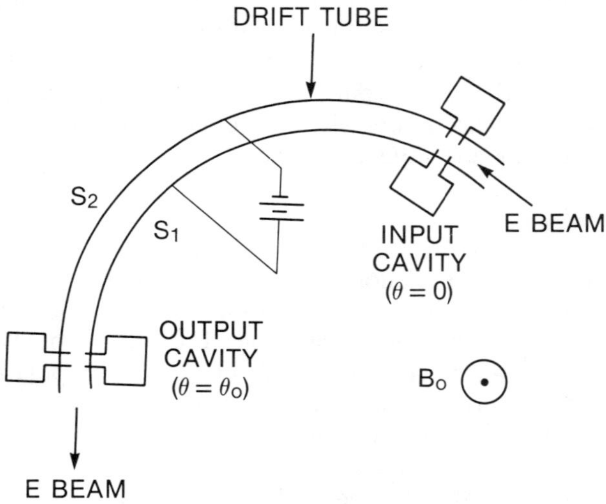

Fig. 9.16 Schematic Drawing of a Proposed Laboratory Experiment to Demonstrate the Effects of Equilibrium Type on the Dynamic Properties of an Electron Beam

Note: Parameters required: beam energy < 10 keV, current < 0.1 A, radial electric field $E_0 < 5$ keV/cm, $B_0 < 100$ G, subtended angle $\theta_0 < 150°$.

plasma, leading to the phenomenon of Bohm diffusion [47].)

Dynamically, perhaps the most obvious deficiency in existing magnetron theories is the failure to take into account the periodic variation in the dc electric field, which results from the wall corrugation. Such a periodic variation should also be introduced in the unperturbed state of the electron flow. If one recalls the importance of the periodic motion in the mean flow of the electrons in free-electron lasers, one cannot fail to worry about the possible omission of significant physics when some built-in "wiggling motions" have been entirely ignored in specifying the unperturbed state of a magnetron in virtually all existing theories.

REFERENCES

1. G.B. Collins, ed., *Microwave Magnetrons* (McGraw-Hill, New York, 1948).
2. E. Okress, *Cross-Field Microwave Devices* ed., (Academic Press,

New York, 1961), Vols. 1 and 2. These two volumes remain unsurpassed in providing broad coverage of cross-field devices. Some other useful texts are listed in Refs. [3, 4].

3. M. Chodorow and C. Susskind, *Fundamentals of Microwave Electronics* (McGraw-Hill, New York, 1964), Chapter 10;
R.G. Huttler, *Beam and Wave Electronics in Microwave Tubes* (D. Van Nostrand, Princeton, NJ, 1960);
J.F. Gittens, *Power Travelling-Wave Tubes* (American Elsevier, New York, 1965);
P.L. Kapitza, *High Power Microwave Electronics* (Pergamon Press, New York, 1964), Vols. 1 and 2.
4. See R.B. Miller, *Introduction to the Physics of Intense Charged Particle Beams* (Plenum, New York, 1982).
5. G. Bekefi and T. Orzechowski, *Phys. Rev. Lett.* **37**, 379 (1976);
A. Palevsky and G. Bekefi, *Phys. Fluids* **19**, 986 (1979). A particle simulation code for relativistic magnetron was developed by A. Palevsky, Ph.D. Dissertation, Department of Physics, Massachusetts Institute of Technology, Cambridge, MA (1980).
6. J. Benford, Chapter 10 in this volume. An extensive amount of Soviet literature is cited there. Chapter 6 of Miller (Ref. [4]) also contains a discussion of relativistic magnetrons.
7. Strictly speaking, therefore, these magnetron theories (including the present one) may be applicable only to smooth bore magnetrons with walls that do not have corrugations. It is not clear whether the effects of wall corrugation on the *unperturbed* motions of the electrons would be dynamically significant.
8. See Chapter 4 by M.J. Baird and Chapter 5 by V.L. Granatstein in this volume.
9. See Chapter 6 by J. Pasour in this volume.
10. Y.Y. Lau and L.R. Barnett, *Int. J. Infrar. and Mill. Waves* **3**, 619 (1982); *Int. J. Electronics,* **53**, 693 (1982).
11. G. Dohler and R. Moats, Int. Electron Device Meetings, *Tech. Digest.,* (1978), p. 400. See also, S. Ono, K. Isutaki, and T. Kageyama, *Int. J. Electr.* **56**, 507 (1984).
12. A. Alexeff and F. Dyer, *Phys. Rev. Lett.* **43**, 351 (1980); I. Alexeff, *IEEE Trans.* **PS-12**, 280 (1984) and *Phys. Fluids* **28**, 1990 (1985). See also, I. Alexeff, Chapter 8 in this volume.
13. R.L. Walker, in Ref. [1], p. 227.
14. E. Ott and R.V. Lovelace, *Appl. Phys. Lett.* **27**, 378 (1975).
15. R.V. Lovelace and T.S.T. Young, *Phys. Fluids* **28**, 2450 (1985).
16. J. Benford, Chapter 10 in this volume.

17. A. Palevsky and A. Drobot, *Proc. 9th Conf. Numerical Simulation of Plasmas*, Northwestern University, Evanston, IL (1980).
18. G.D. Sims, in Ref. [2], p. 183.
19. O. Buneman, in Ref. [2], pp. 209, 380.
20. G.E. Thomas, *J. Appl. Phys.* **53**, 3491 (1982).
21. When the number of vanes is large, it is not easy to tell whether the device operates as a peniotron or as a cyclotron maser. Experimental evidences gathered by various groups (Dohler, Barnett, Namkung) remain inconclusive at this time. See also Ref. [29] below.
22. G. Dohler, D. Gallagher, R. Moats, and F. Scafuri, Int. Electron Device Meeting, *Tech. Digest* (1981), p. 328.
23. W. Namkung, *Phys. Fluids* **27**, (1984). The relativistic version was given earlier by W.W. Destler, R.L. Weiler, and C.D. Striffler, *Appl. Phys. Lett.* **38**, 570 (1981).
24. This theoretical model for orbitrons is certainly the simplest one that comes to mind. Similar to the Brillouin flow in magnetrons, this model should be treated as a working hypothesis. Nevertheless, interesting results emerge from a study of such an idealized model. There are other more complicated theoretical models to describe the operation of orbitrons. For example, Ref. [25] indicates that the TEM mode interacting with electrons with high orbital eccentricity may also lead to orbitron emission—resembling a cylindrical Backhausean oscillator; Ref. [26] invoked a nonlinear wave-wave coupling process.
25. J. Burke, W.M. Manheimer, and E. Ott, *Phys. Rev. Lett.* **56**, 2625 (1986). A nonlinear theory was recently developed by A.K. Ganguly, H.P. Freund, and S. Ahn, Naval Research Laboratory, Memo Report 5979 (1987).
26. R.W. Schmacher and R.J. Harvey *Bull. APS* **29**, 1179 (1984).
27. Y.Y. Lau and D. Chernin, *Phys. Rev. Lett.* **52**, 1425 (1984).
28. D. Chernin and Y.Y. Lau, *Phys. Fluids* **27**, 2319 (1984).
29. G. Dohler, *Int. J. Electr.* **56**, 617, 629 (1984); P.S. Rha, L.R. Barnett, J.M. Baird, and R.W. Grow, Int. Electron Device Meeting, *Tech. Digest* (1985), p. 525; U.A. Shrivastava, R.W. Grow, P.S. Rha, J.M. Baird, and L.R. Barnett, *Int. J. Electr.* **61**, 33 (1986); P. Vitello, *IEEE Trans.* **MTT-32**, 917 (1984); See also, W. Namkung (to be published). In the case of smooth waveguides, emission of the longitudinal modes at cyclotron harmonics has also been investigated by Luhmann's group. See D.B. McDermott *et al.*, *Int. J. Infrar. and Mill. Waves* **4**, 639 (1983). Lau and Barnett [10] gave a detailed description of the experimental observation of harmonic generation.
30. G.C. MacFarlane and H.G. Hay, *Proc. Roy. Soc.* (London), **63B**, 409 (1950); O. Buneman, *J. Electron Control* **3**, 507 (1957); R.L. Kyhle and H.F. Webster, *IRE Trans. Electron Dev.* **ED-3**, 172 (1956);

J.R. Pierce, *ibid.*, 183; R.H. Levy, *Phys. Fluids* **8**, 1288 (1965); O. Buneman, R.H. Levy, and L.M. Linson, *J. Appl. Phys.* **37**, 3203 (1966); V.K. Neil and W. Heckrotte, *J. Appl. Phys.* **36**, 2761 (1965); R.W. Gould, *ibid.* **28**, 599 (1957). See also, J.D. Lawson, *The Physics of Charged Particle Beams* (Clarendon, Oxford, 1977).

31. O. Buneman, in Ref. [2], p. 367.
32. G. Mourier, in Ref. [2], p. 396.
33. R.C. Davidson, *Theory of Non-Neutral Plasmas* (Benjamin, New York, 1974). H.S. Uhm and J.G. Siambis, *Phys. Fluids* **22**, 2377 (1979); see also references therein.
34. See Chapter 6 of Gittens *op. cit.* [3], and Chapter 10 of Chodorow and Susskind, Ref. [3].
35. R.G. Kleva, E. Ott, and W.M. Manheimer, *Phys. Fluids.* **28**, 941 (1985).
36. Y.Y. Lau, *IEEE Trans.* **ED-29**, 320 (1982), and **ED-31**, 329 (1984).
37. See S. Chandrasekhar, *Hydrodynamic and Hydromagnetic Stability* (Oxford University Press, New York), 1960; P.G. Drazin and L.N. Howard, *Adv. Appl. Mech.* **9**, 1 (1966)
38. J.L. Hirshfield and J.M. Wachtel, *Phys. Rev. Lett.* **12**, 533 (1964).
39. The equilibrium quantities h, E_0, and B_0 which appear in (9.15), (9.16), and elsewhere in the text (e.g., in the dispersion relations (9.25) and (9.31)) actually include the dc self-fields. It turns out that, for a thin beam, the dc self-field modifies the dispersion relation (9.25) only by introducing the extra term proportional to $(\tau/R)^2$ in (9.31). See Refs. [27, 28] for more detail.
40. R.J. Briggs and V.K. Neil, *J. Nucl. Energy* **C9**, 209 (1967).
41. G. Dohler and W. Fritz, *Int. J. Electr.* **55**, 505, 523 (1983).
42. C.E. Nielsen, A.M. Sessler, and K.R. Symon, in *Proc. Int. Conf. High-Energy Accelerators and Instrumentation,* Geneva, Switzerland, (CERN, Geneva, 1959), p. 239;
A.A. Kolomenskii and A.N. Lebedev, *ibid.,* p. 115;
See also, R.W. Landau and V.K. Neil, *Phys. Fluids* **9**, 2412 (1966).
43. Y.Y. Lau, *Phys. Rev. Lett.* **53**, 395 (1984).
44. C.K. Birdsall and J.R. Whinnery, *J. Appl. Phys.* **24**, 314 (1953). L.J. Laslett, V.K. Neil, and A.M. Sessler, *Rev. Sc. Inst.* **36**, 436 (1965); see also references therein.
45. A.M. Sessler, private communication (1983).
46. I. Alexeff (private communication, 1984) also alerted us of this possibility.
47. Y.Y. Lau, Naval Research Laboratory, Memo. Rep. No. 5968 (1987).

Chapter 10
RELATIVISTIC MAGNETRONS

James Benford

Physics International Company
San Leandro, California

10.1 INTRODUCTION*

Of the many types of crossed-field devices that have been developed at low output powers, only the magnetron has been extrapolated to the high pulsed power (>100 MW) regime. The magnetron is attractive as a high-power source for the same reasons that it received much attention in its lower power regime some 40 years ago. The magnetron is efficient, compact, robust, and relatively simple. These virtues have led to wide use of the conventional magnetron in such diverse applications as the earliest radar systems and contemporary microwave ovens. Indeed, the power limits to which magnetron technology can be extrapolated are as yet unknown. Because the magnetron has demonstrated the highest pulsed output powers reported to date by any microwave device, at this early stage in its development there is cause for optimism that it may obtain much higher power. This chapter covers the origins of the relativistic magnetron, the principles of its operation, the present state of experiments with respect to theory, and the new effects revealed at high-current levels.

10.2 HISTORY OF THE MAGNETRON

Figure 10.1 shows the stages of magnetron development, the power levels achieved, and some of the principal contributors. The magnetron was invented by Arthur Hull in 1913 to resolve a patent dispute involving

*This chapter is based on work supported by Physics International Company.

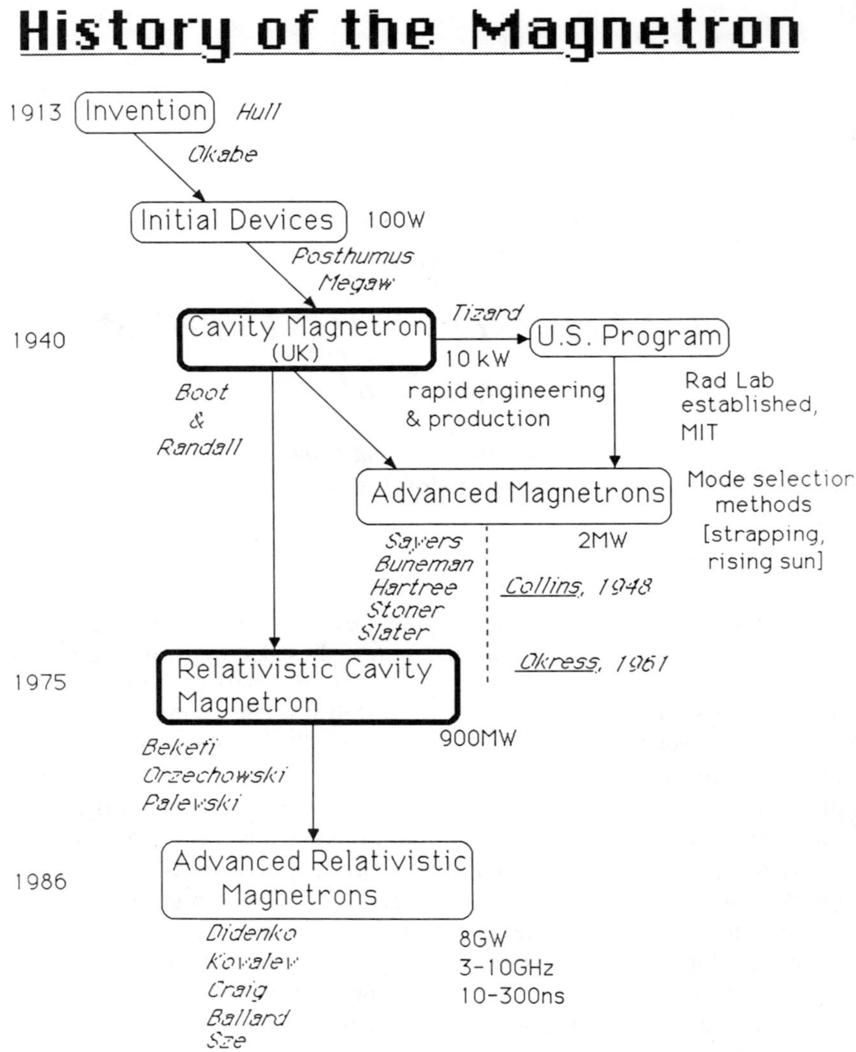

Fig. 10.1 History of Magnetron Development

electron tubes [1]. The initial devices were small and flimsy, and had unexciting output powers and low efficiencies (Figure 10.2). Work by others such as Okabe in the ensuing quarter century brought the output

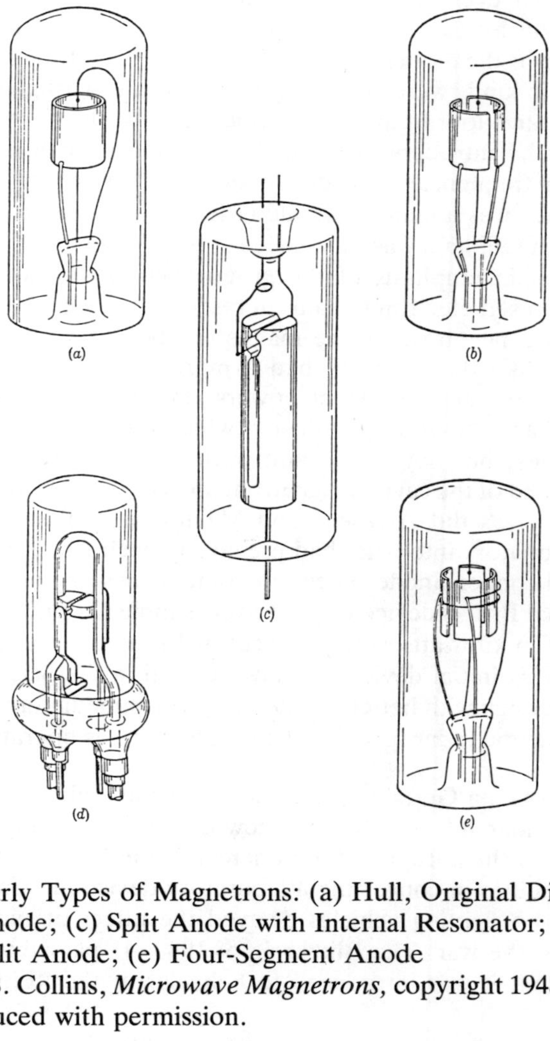

Fig. 10.2 Early Types of Magnetrons: (a) Hull, Original Diode; (b) Split Anode; (c) Split Anode with Internal Resonator; (d) Improved Split Anode; (e) Four-Segment Anode
Source: G. B. Collins, *Microwave Magnetrons,* copyright 1948 by McGraw-Hill, reproduced with permission.

power to the level of ≈100 W. These early devices operated in the UHF region. In the 1930s, Posthumus and Megaw, among others, made advances in the theoretical and hardware aspects of the magnetron, and clarified the occurring types of oscillation [2].

The seminal event in magnetron history was the invention of the cavity magnetron by Boot and Randall in 1940 [3, 4]. In this case, the cliché "necessity is the mother of invention" applies. British radar stations, the first of their kind, were operating in the south of England with low output powers and low frequencies, which gave wide beams that were easily detected by intruders. The need to establish much higher power levels at higher frequencies in order to stem the tide of the German air force (Luftwaffe) was a major national objective. A research team under the leadership of Oliphant was established at the University of Birmingham in 1939. The main emphasis of the group was on the newly invented kylstron, which used electron streams in resonant cavities. Boot and Randall decided to concentrate on the magnetron because it did not need a focus beam, unlike the klystron, had a more favorable geometry, and seemed capable of carrying higher powers. The magnetron at that time was no more than a laboratory curiosity, which was capable of producing centimeter waves, but not at constant frequency nor high power. Posthumous conceived of the cavity magnetron, and Boot and Randall invented it independently, as did Alekseev and Malairov [5]. Boot and Randall introduced resonators into what had previously been a split-anode design and used solid-copper anode blocks to contain these resonators (Figure 10.3). The initial British device was relatively simple to build and it worked immediately. Ten kilowatts were produced at 10 cm, an astonishing power for the time. This initial device can now be seen in the Science Museum in London. A design with better engineering was immediately undertaken, and an experimental radar set using the magnetron was operational in only three months.

In April 1942, a German U-boat was first sunk with the aid of magnetron-driven radar [6]. In 1943, the new device was widely introduced, which resulted in the abrupt and total defeat of the U-boats on the North Atlantic. Difficulties in converting a laboratory device into military hardware installed on fighting vehicles delayed the effect of radar until the middle years of the war. The Blitzkreig of 1940–41 passed without magnetron radar, but it was introduced to help defeat the "Little Blitz" of 1943–44.

Sir Henry Tizard, scientific advisor to the British government, led a mission to the US to demonstrate the magnetron immediately after its invention. The Americans were very receptive and established a program to exploit the magnetron for radar applications, leading to the formation of the MIT Radiation Laboratory as well as other groups at Columbia University and leading industrial laboratories. There ensued rapid engineering and production of a wide variety of magnetrons for radar applications on land, at sea, and in the air. Millions of magnetrons were produced during World War II, and it was the leading and most advanced

Fig. 10.3 Early Copy of the British Cavity Magnetron, Showing an Anode Block of Six Resonant Cavities, with Leads for Current Supply from the Right and for Power Extraction on the Left

microwave device at the time. This effort resulted in 2 MW magnetrons by 1946. The magnetron may well have been the most important scientific contribution to victory in World War II, greater even than the more well known atomic bomb.

After the war, the large body of wartime work was enshrined in the tome, *Microwave Magnetrons*, edited by G. B. Collins and published in 1948 [7]. This was a time of intense theoretical activity on the magnetron, resulting in considerable understanding of its operation, although even now no complete theory exists. Contributors were Buneman, Hartree, Stoner, Slater, and many others. By this time, more advanced magnetrons were being produced, which used mode selection methods such as *strapping* and "rising-sun" geometries. Strapping (connecting alternate resonators to shift the frequency spectrum, preventing mode hopping at high power) was invented by James Sayers. Work on magnetrons progressed at diminished rates. The postwar work was documented in *Cross-Field Devices*, edited by E. Okress and published in 1961 [8]. Magnetron development virtually ceased at this point.

The relativistic cavity magnetron was invented by direct extrapolation of the original cavity magnetron. The work was undertaken by Bekefi, Orzechowski, and Palevsky at MIT in 1975, following an experiment at high voltages on the smooth-bore magnetron, which gave low microwave powers of a few kilowatts. The initial device produced 900 MW output power, and immediately generated interest in the US and the USSR. The basic design was a cavity magnetron, which was uncooled and operated in a single-shot mode, with pulse duration of a few tens of nanoseconds due to limited electrical pulse duration. In the ensuing years, work by Didenko and Kovalev in the USSR and by Craig, Ballard, and Sze in the US has lead to more powerful magnetrons in a variety of designs. Typical power levels are of a few gigawatts. Magnetrons have been operated in the regime of 3–10 GHz, with pulse durations as long as 300 ns.

10.3 PRINCIPLES OF OPERATION

The cavity magnetron geometry is shown in Figure 10.4. The central cathode is surrounded by an anode cavity structure, consisting of identical resonators separated by vanes. The space between the anode and the cathode is referred to as the *interaction space*. A voltage is applied to the cathode, and electrons are emitted from surface plasma formed by the explosion of microprotrusions. An axial magnetic field (i.e., perpendicular to Figure 10.4) inhibits electron propagation across the gap. If the field is sufficiently strong, electrons will be prevented from reaching the anode. The critical field for the *magnetic insulation* is

$$B^* = \frac{M_0 c}{e d_e} \sqrt{\gamma^2 - 1} \tag{10.1}$$

where the effective gap is

$$d_e = \frac{r_a^2 - r_c^2}{2 r_a} \tag{10.2}$$

The second condition for the operation of the magnetron is resonance between waves and the rotating particles. Electrons that satisfy the Hull cut-off condition (10.1), orbit the cathode at a drift velocity given by

$$v_d = \frac{E \times B}{|B|^2} \tag{10.3}$$

The exact nature of the electrons' trajectories is a matter of debate, but detailed analysis seems to show both laminar flow and cycloidal flow. The electron cloud is referred to as the *Brillouin layer.* Depending on the magnetic field, the cloud can extend to near the anode ($B/B^* \approx 1$), or be held close to the cathode ($B/B^* > 1$).

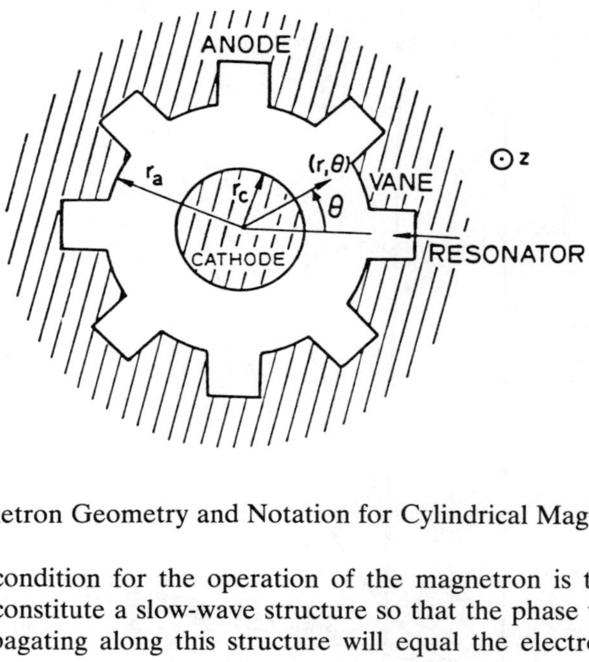

Fig. 10.4 Magnetron Geometry and Notation for Cylindrical Magnetrons

The key condition for the operation of the magnetron is that the anode cavities constitute a slow-wave structure so that the phase velocity of a wave propagating along this structure will equal the electron drift velocity:

$$v_p = \omega/k = v_d \tag{10.4}$$

where v_p is the phase velocity of the cavity wave of frequency ω and wavenumber k. This resonant state is described by the Buneman-Hartree resonance condition:

$$B_{BH} = \frac{2m^2}{e^2}\, eV \frac{r_a^2}{r_a^2 - r_c^2} \frac{n}{\omega_n} \tag{10.5}$$

where ω_n is the frequency of the nth mode. For high voltages, both the magnetic-insulation and Buneman-Hartree conditions are linear with voltage, as shown in Figure 10.5. For voltages greater than cut-off, the magnetron will conduct. For voltages less than the Buneman-Hartree threshold, the magnetron will be nonresonant and it will not emit microwave power. The region of magnetron oscillation is between the two conditions. Efficient operation normally occurs near the Buneman-Hartree threshold. The two conditions are proportional, and their ratio is a function of only the voltage and the phase velocity of the resonant wave. Figure 10.6 shows that the ratio of B_{BH} to B^* varies from one to two, with phase velocities of a fraction of the speed of light.

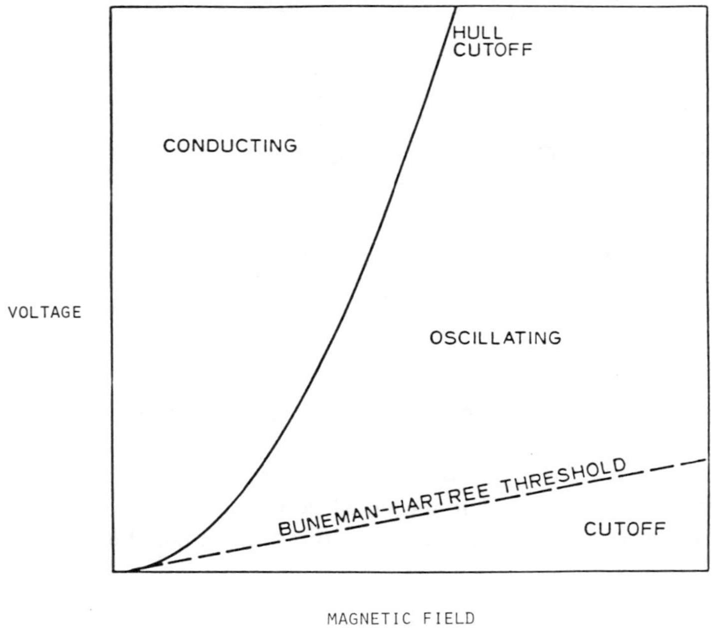

Fig. 10.5 General Features of Magnetron Operating Domain
Note: The magnetron oscillates when the operating point lies between the cut-off and the threshold curves.

A variety of modes occur because the anode slow-wave structure is, in effect, a group of coupled resonators. Therefore, several possible types of oscillations exist, which differ in the way fields are distributed among the cavities. Mode competition is an issue because each mode grows from the ambient noise of the electron cloud in the Brillouin layer. If one mode

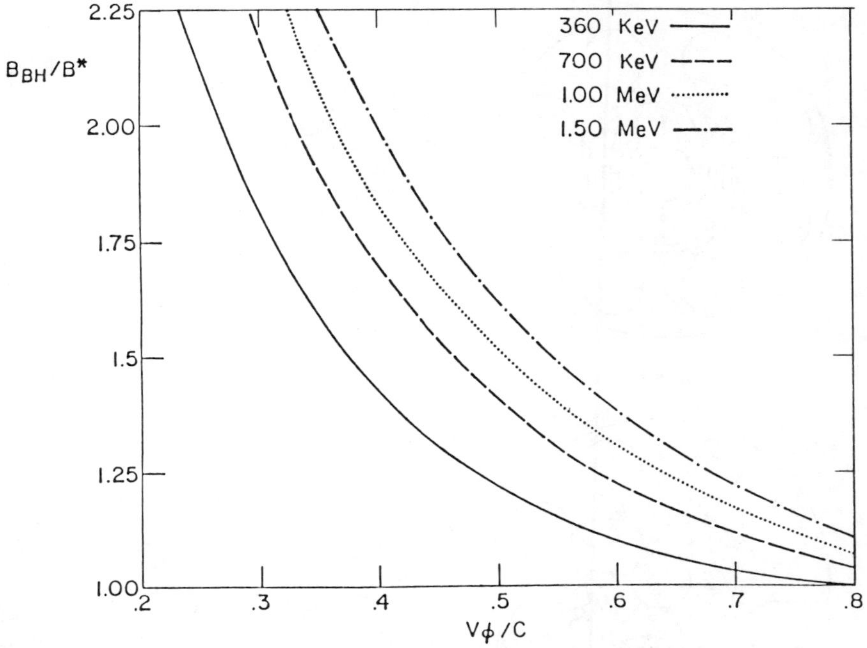

Fig. 10.6 Ratio of Buneman-Hartree to Hull Cut-Off Magnetic Fields

produces a field maximum where another produces a minimum, they will compete with one another and suppress each other's growth. Therefore, the simultaneous presence of many modes leads to low-power operation. The key to high power is to operate in a single mode. Modes are designated by mode number n, which is the number of times the RF field pattern repeats in one rotation around the anode. If each vane in an N-vane system differs in phase from its neighbor by $\Delta\phi$, then

$$N\Delta\phi = 2\pi n \tag{10.6}$$

There are many possible cases, as shown in Figure 10.7. The two most interesting cases are called the π and 2π modes, indicating that the phase difference between alternating resonators is either π or 2π. Conventional magnetrons are usually designed to operate in the π mode, and relativistic magnetrons have been found to operate in both π and 2π modes. Figure 10.8 shows the field geometry for these two cases. Conventional magnetrons are designed to operate in the π mode because it is *nondegenerate*

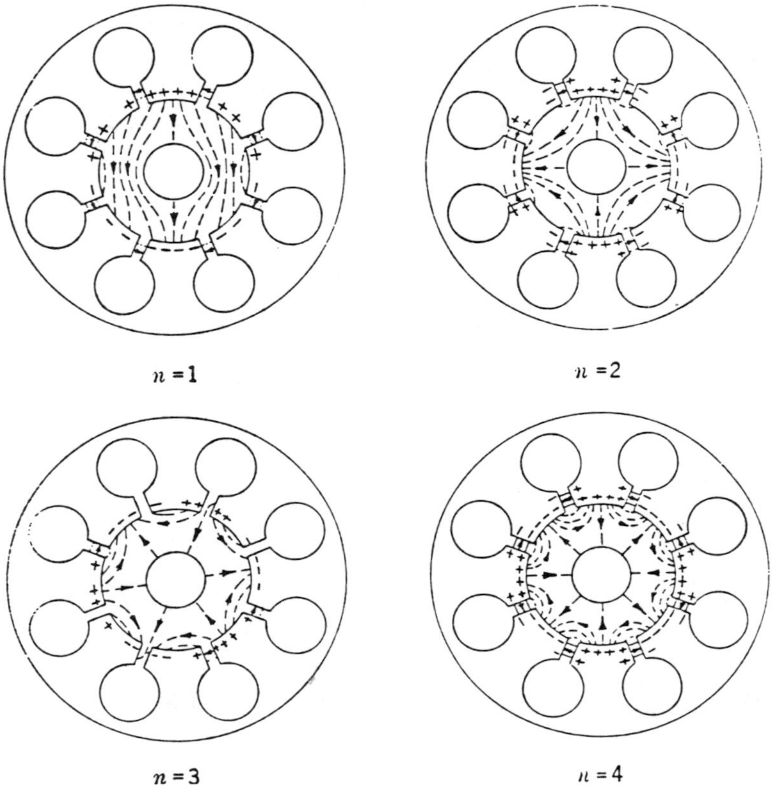

Fig. 10.7 Field and Charge Distributions for the Four Principal Modes of an Eight-Resonator Magnetron [7]

(see below), and because it has the highest frequency. Note that large N leads to the presence of many modes, and consequently makes difficult their separation in frequency and the suppression of mode competition. Figure 10.9 shows an experimental measurement by Ballard of the azimuthal structure of modes for a 12-resonator cylindrical magnetron.

Thus far, we have discussed initial conditions for magnetron operation. The electron achieves an orbit near the cathode, and then moves at a velocity given by (10.3). In this frame of reference, the electron does not experience the imposed dc electric field, but does have a residual drift from the RF field:

$$v'_d = \frac{E_{RF} \times B}{|B|^2} \tag{10.7}$$

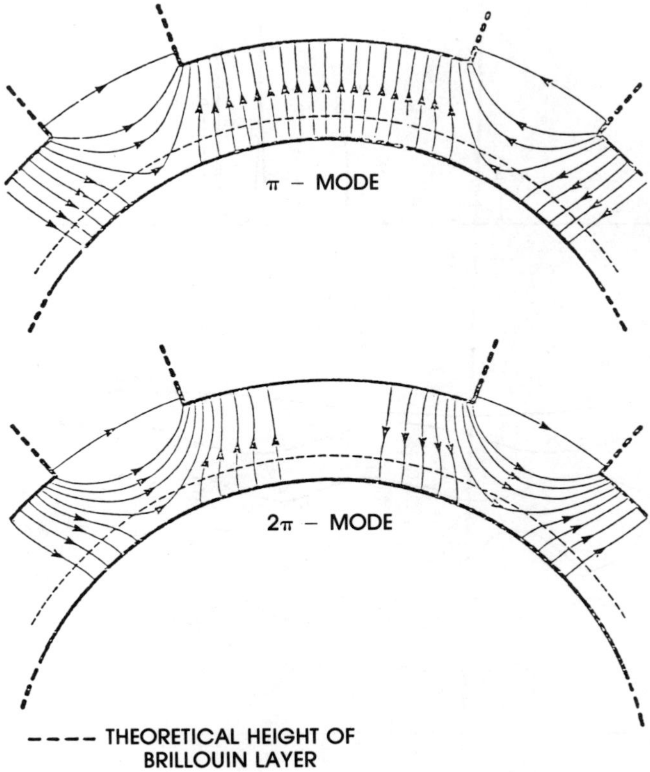

Fig. 10.8 Microwave Electric Field Structure for π and 2π Modes of the A6 Magnetron [9]

The flow will drift so as to form two regions. In one region, particles gain energy from the RF wave through $J_e \cdot E_{RF}$, move toward the cathode and are reabsorbed there. In the other region, the electrons will lose energy to the wave, and move toward the anode. This process forms a pattern of spokes, as shown in Figure 10.10. The RF fields are stronger if near the resonant cavity than they are when closer to the cathode. Therefore, more

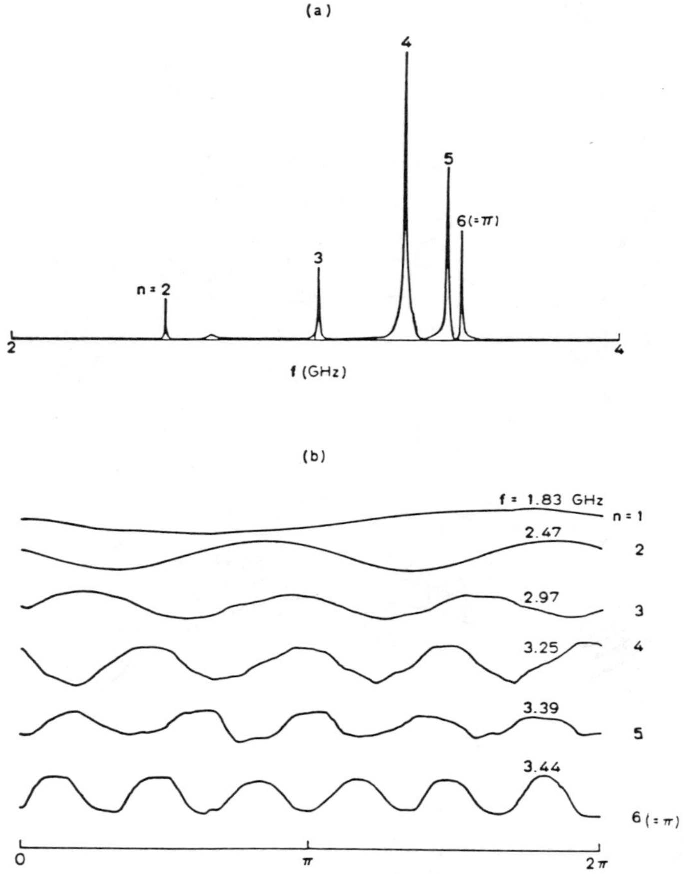

Fig. 10.9 Azimuthal Mode Structure of a 12-Resonator Cylindrical Magnetron, Showing: (a) Resonant Frequencies of the Anode Block; (b) Azimuthal Variation of RF Electric Field of Various Resonances in the Interaction Space

energy is gained by the wave from electrons near the anode than is lost to electrons near the cathode. Consequently, the wave grows. The spoke formation phenomenon has been observed in particle-in-cell (PIC) simulation of the magnetron, as shown in Figure 10.11. The final process is highly nonlinear, and it resists complete theoretical description. Development of the magnetron has thus relied more on empirical design than theoretical results.

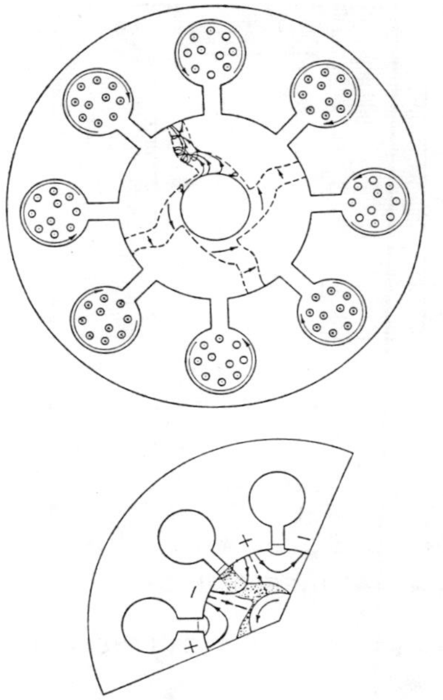

Fig. 10.10 Spoke Formation in the Nonlinear, High-Power, Saturated Regime of the Magnetron

In order to design a magnetron to operate in a specific mode, we make the admittance that looks out of the resonators equal to the admittance of the interaction space, and solve the resulting transcendental equation for the resonances. There are an infinite number of solutions, but only those of the lowest order have high growth rates. Figure 10.12 shows a specific case for the geometry of the magnetron developed at Stanford. The resonator admittance intersects the interaction space admittance for various modes. The lowest order modes are designated as 0_0, 1_0, 2_0, *et cetera*, and the next highest order, 0_1, 1_1, 2_1, *et cetera*. The π mode is very close in frequency to the other modes of the first-order resonances. Figure 10.13 shows the Brillouin diagram of the A6 magnetron of Palevsky [9], which is the most investigated of the relativistic resonators. The lowest order frequencies peak at the π mode, and all other modes are twofold and therefore degenerate. Again, the π mode is not well separated from the other modes. The so-called 0_1 mode in conventional parlance is designated as the 2π mode in the study of relativistic magnetrons. The 2π mode is the lowest frequency of the second-order modes. Note that the

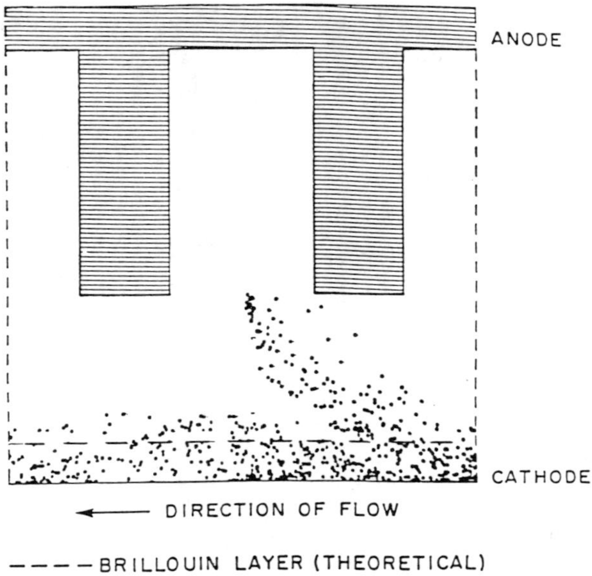

Fig. 10.11 Particle-in-Cell Simulation of Spoke Formation

phase velocity of the mode is given by the slope of a straight line drawn from the origin, given in units of segments per second. By a coincidence of design, the π and 2π modes for the A6 magnetron have the same phase velocity.

The conventional magnetron has been developed in several resonator types. A number of examples are shown in Figure 10.14. During World War II, separation of modes was first produced by use of the strapping technique, in which alternate resonators are attached by conductors. The capacitive coupling thereby generated causes mode separation. A more advanced scheme, called the *rising-sun* geometry, shown in Figure 10.14(e), also separates modes by the use of alternating unstrapped resonators of two different designs. This method is typically used for the high-frequency regime because of the difficulty of strapping at such frequencies. Figure 10.14(f) shows the inverted magnetron, in which the cathode is the outer electrode and the resonator structure is on the interior electrode.

Most relativistic magnetron experiments have been conducted with the basic unstrapped resonator, in which coupling between resonators occurs only through the interaction space, and not as yet with the strapped

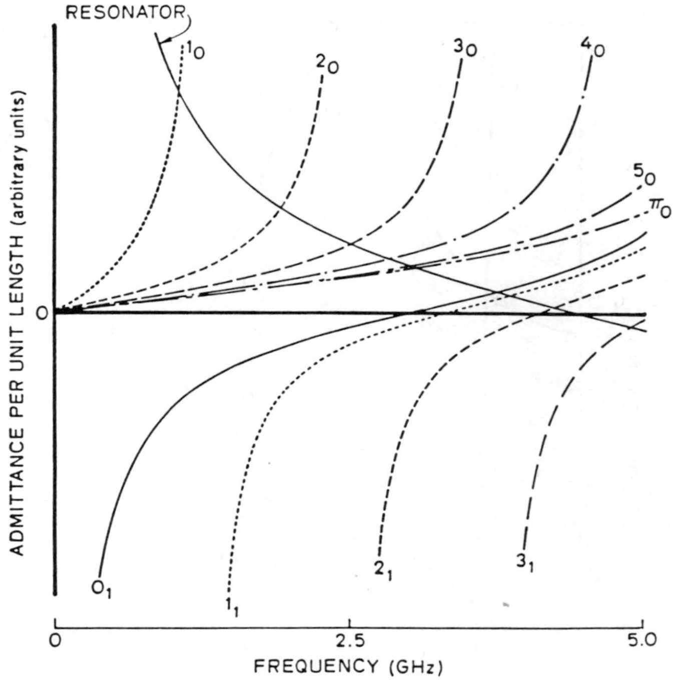

Fig. 10.12 Graphical Solution of the Admittance-Matching Equations for the Geometry of the Stanford Magnetron

Note: Resonance occurs at frequencies where the resonator admittance intersects the admittance of various modes.

resonator. Recently, however, experiments have begun on a rising-sun resonator. Several groups have conducted experiments using inverted magnetron geometries.

We should always remember that the magnetron is a three-dimensional object, and therefore axial modes may exist. The resonators can be open at the ends, or can be shorted by endcaps, which close off the resonators (but not, of course, the interaction region). The frequencies of the axial modes are given by

$$(f_n^l)^2 = f_n^2 + \left(\frac{lc}{4L}\right)^2, \quad l = 0, 1, 2, \ldots \tag{10.8}$$

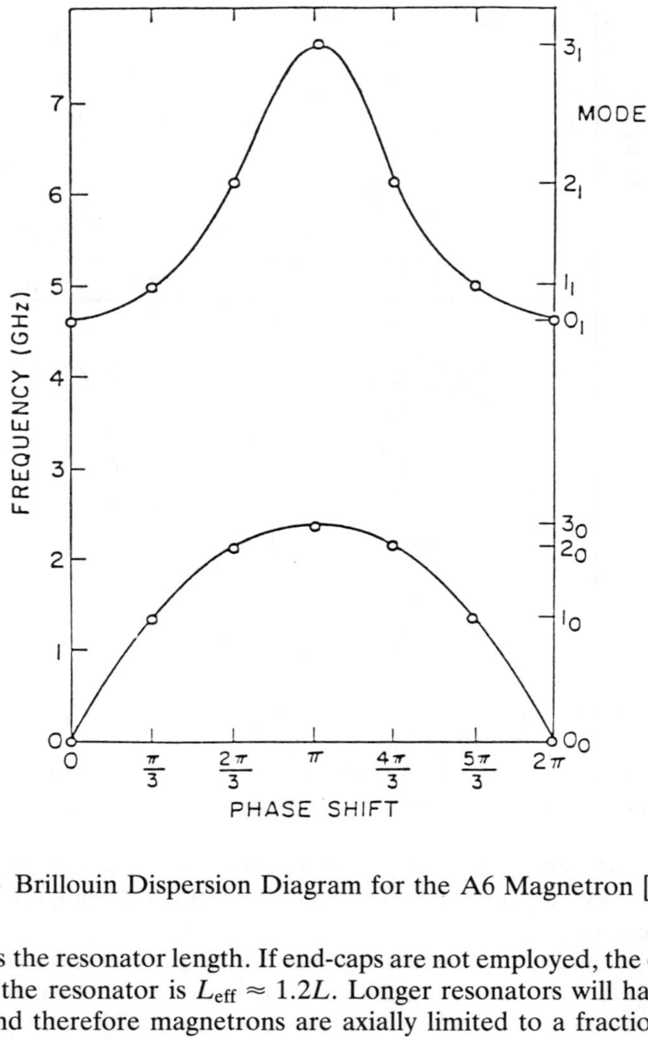

Fig. 10.13 Brillouin Dispersion Diagram for the A6 Magnetron [9]

where L is the resonator length. If end-caps are not employed, the effective length of the resonator is $L_{eff} \approx 1.2L$. Longer resonators will have more modes, and therefore magnetrons are axially limited to a fraction of the wavelength, typically 0.6λ.

10.4 EXPERIMENTAL CONFIGURATIONS

The principal feature of the relativistic magnetron as compared with the conventional magnetron is operation at much higher voltages and currents. The relativistic magnetron is actually a high-current version of

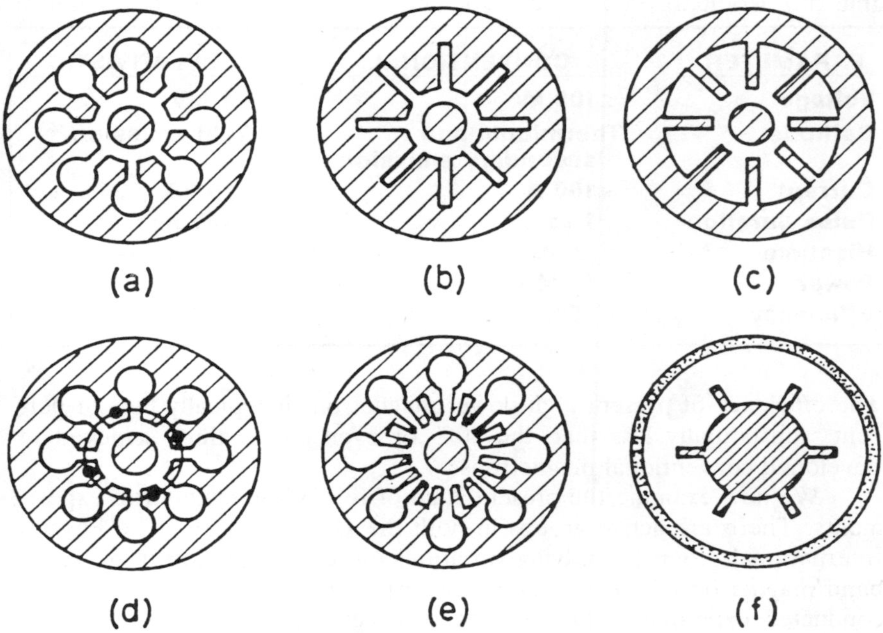

Fig. 10.14 Magnetron Cross Sections: (a) Hole-and-Slot Resonator Based on Hertzian Dipole; (b) Rectangular Resonators Based on Short-Circuited, Quarter-Wave Lines; (c) Sectoral Resonators; (d) Strapped Resonators with Alternate Vanes Connected; (e) Rising-Sun Anode with Alternate Resonators of Different Shapes and Frequencies; (f) Inverted Magnetron with Cathode Exterior to the Anode

Source: G. B. Collins, *Microwave Magnetrons*, copyright 1948, McGraw-Hill, reproduced with permission.

the conventional magnetron, where the relativistic voltages are necessary to drive the high currents. There are no relativistic effects in magnetron operation other than the substitution of the relativistic factor γ in the appropriate equations. Table 10.1 shows the relevant parameters. Relativistic magnetrons operate higher by factors of 10 in voltage, 10^2 in current, and 10^3 in voltage rise time. Output power differs by at least 10^2. The conventional magnetron emits by thermionic emission or secondary emission from the cathode after the Brillouin layer has been established. In the relativistic version, field emission supplies the required high current.

Table 10.1 Comparison of Relativistic and Conventional Pulsed Magnetrons

PARAMETER	CONVENTIONAL	RELATIVISTIC
Voltage	≲100 kV	~1 MV
Cathode	Thermionic and secondary emission	Field emission
Current	~100 A	≳10 kA
Pulse duration	≳1 μs	≲100 ns
Risetime	≲ 200 kV/μs	~100 kV/ns
Power	≲10 MW	≳1 GW
Efficiency	~50%	≲30%

The efficiency of present high-power magnetrons is typically tens of percent, substantially less than the 50% or even 80% values achieved in developed conventional magnetrons.

We now examine the present state of relativistic magnetron experiments. There are active groups in both the US and USSR. The Physics International group is studying variations on, and extraction from, an S-band magnetron of the A6 resonator design [10–12], and this group has conducted experiments on an X-band design [13]. At Sandia National Laboratories, Ballard has experimented with inverted magnetrons extrapolated from the design he developed at Stanford [14, 15], and more recently with an A6 magnetron. The Varian group led by Gary Thomas, with an interest in radar applications, has conducted experiments using a rising-sun resonator design, the first of its kind in the relativistic domain [16]. Clark and coworkers at Sandia are investigating the linear magnetron in the relativistic regime (see Section 10.7).

In the USSR, relativistic magnetron activity occurs at two locations. At the Institute of Applied Physics in Gorki, Kovalev leads a group that has studied a variety of magnetrons, including the first X-band magnetron [17–20]. Other notable experiments are the use of an axial extraction technique and the study of amplification by the relativistic magnetron, although at low power. At the Institute for Nuclear Physics at Tomsk, Didenko and coworkers have studied a variety of resonators, and have paid particular attention to long pulse durations and the phenomenon of gap closure [21–24].

A summary of the operating parameters of relativistic magnetron experiments to date is shown in Table 10.2. Most of the work has been conducted at about 3 GHz, with only two experiments in the upper regime of about 9 GHz. Voltages vary up to 1.7 MV, currents are typically 10 kA, and the corresponding magnetic fields implied by the Buneman-Hartree resonance at these voltages are about one tesla. The output pulse

Table 10.2 Summary of Relativistic Magnetron Operating Parameters

Laboratory	*Parameters* f (GHz)	V (MV)	I (kA)	B (T)	P (GW)	Δt (ns)	E (J)	η (%)	N	$\sigma = r_c/r_a$	r_a (cm)	h (cm)	*Mode*
MIT[a] A6	4.6	0.8	14	1.0	0.5	30	15	4	6	0.75	2.1	7.2	2π
MIT M8	3.7	1.7	4	0.46	0.4	30	12	12	8	1.47	2.9	5.1	π
MIT M10	3.7	1.6	9	0.7	0.5	30	15	3	10	1.22	3.8	4.6	π
LLNL[b]	3.9	0.9	16	1.6	4.5	16	72	30	6	0.75	2.1	7.2	2π
SRINP[c] #1	2.4	1.1	4	1.2	2.0	50	100	45	6	—	—	—	π
SRINP #2	2.4	0.45	6	0.4	0.8	300	240	30	6	—	—	—	π
NRL[d]	3.2	0.6	5	0.2	0.8	30	24	27	54	1.10	19.4	5.1	π
Stanford[e]	3.3	0.42	3	0.15	0.02	40	1	2	12	0.50	5.1	9.5	π
IAP[f] #1	9.1	0.6	7	0.6	0.5	20	10	12	8	—	—	—	π
IAP #2	9.2	0.95	40	1.0	4.0	15	60	11	8	—	—	6.5	π
Sandia[g]	3.1	1.0	10	0.3	0.25	50	12	3	12	1.3	5.22	9.23	π
PI[h]	2.8	0.7	20	1.0	3.0	20	60	20	6	0.6	2.1	7	π
PI[h]	8.3	1.0	30	1.0	0.3	10	3	0.2	6	0.48	1.4	6	π

[a] Massachusetts Institute of Technology
[b] Lawrence Livermore National Laboratory
[c] Scientific-Research Institute of Nuclear Physics, Tomsk, USSR
[d] Naval Research Laboratories
[e] Stanford University
[f] Institute of Applied Physics, Gorki, USSR
[g] Sandia National Laboratory
[h] Physics International Company

tends to be short, a few tens of nanoseconds, and the total power varies from a fraction of a gigawatt to several gigawatts. Total output energy is tens of joules, except for a one-long-pulse case from the Tomsk group. The efficiencies vary greatly, but are typically $\approx$ 10%. The number of resonators is usually a few in order to avoid mode competition, but one experiment (NRL) used a large number of resonators. Typical dimensions for the resonators are 5–10 cm. Most resonators have been designed and found to operate in the π mode with only a few reported cases of 2π mode operation.

The most successful and most studied of the resonator designs is the A6, first reported by Bekefi and Orzechowski [25]. The A6 resonator appears to operate in the π mode, but the early experiments reported the 2π mode. Sze and coworkers [12] have shown that the A6 resonator produces most of its power in the π mode, then lapses into the 2π mode at the end of the pulse, perhaps because gap closure lowers the Buneman-Hartree resonance voltage. (Note that (10.5) can be written $V_{BH} \sim d$, where d is the anode-cathode gap.)

Figure 10.15 shows a photograph of a typical S-band relativistic magnetron experiment. Figure 10.16 shows a cross section of this experiment. The resonator constitutes a small portion of the overall experimental volume, which is dominated by the high-voltage electrical feed and magnets. Most experiments use pulsed solenoids to provide longitudinal magnetic field, but a few employ dc water-cooled magnets. Electrical power is supplied to the cathode, which primarily emits in the region of the resonator, and in this case also has a load diode that allows independent adjustment of the total axial current in the magnetron (see Section 10.6). Rogowski coils are located along the experiment, allowing resolution of any current losses and measurement of the radial current, as distinct from the axial-current portion of the total current. This is useful in measuring efficiency. In this experiment, radiation is radially extracted in waveguides, which is a typical procedure.

Of particular note is the use of a washer to create enhanced radiation in the center of the resonator. This technique was introduced by Craig and coworkers [26], and has been measured by Benford and coworkers [10]. Figure 10.17 shows the operation of an A6 magnetron as a function of the washer diameter. Clearly, the presence of a washer creates a marked improvement of output power, in some cases by as much as an order of magnitude. The washer field emits electrons into the midsection of the resonator, where conversion efficiency may be higher in a radially extracted system with end-caps, which produce an electric field maximum at the midpoint.

Also shown in Figure 10.17 is the influence of end-caps, which result in a distinct improvement in this experiment, although there is some debate

Fig. 10.15 A6 S-Band Experiment at Physics International
Note: The resonator is buried in the central solenoid. Waveguides extract power in radial directions.

in the literature on the use of end-caps. End-caps help to define the mode structure and to prevent axial mode competition. The MIT group [9, 27] and Vintizenko [28] also report improved operation with end-caps. Didenko [22] reports that a single upstream end-cap gives the best results.

Sze and coworkers [12] have studied power extracted in the radial direction as a function of the number of extraction waveguides. The length of the waveguides was longer than the pulse durations so that interference from reflections was not possible. The experiments of Sze *et al.* show that the frequency does not change with the number of open resonators, and operation remains in the π mode for all operating conditions. Figure 10.18 shows that maximum power is obtained for three waveguides, but power per waveguide declines as their numbers increase. Note that the cavity Q is reduced as more waveguides are added because of more openings in the resonator. The optimum number appears to be the three-waveguide geometry, presumably corresponding to the optimum Q value.

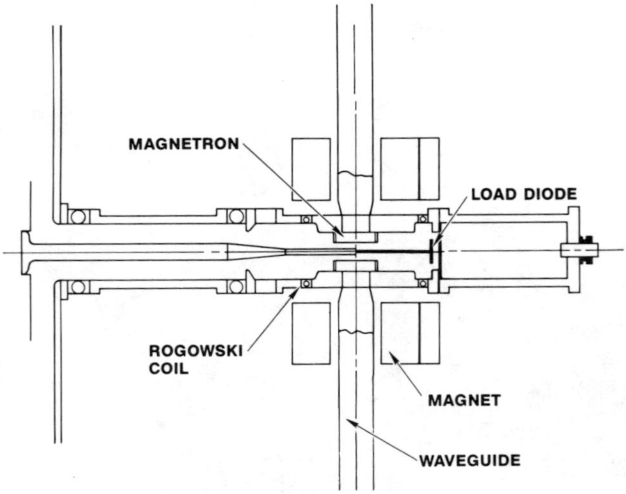

Fig. 10.16 Cross Section of the Experiment in Figure 10.15 [10]

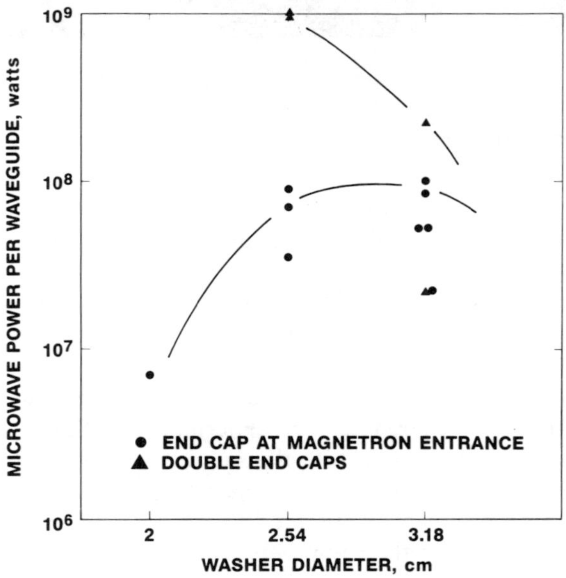

Fig. 10.17 Effect of a Washer Cathode on Magnetron Performance
Note: Double end-caps facilitate high-power operation [10]

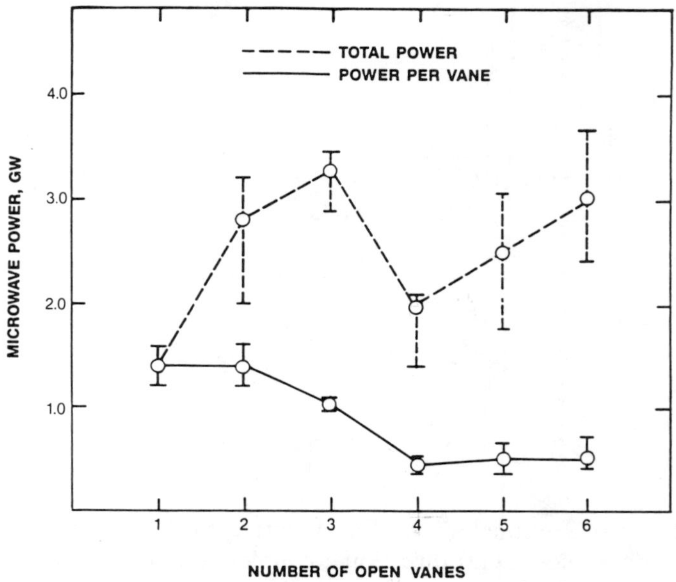

Fig. 10.18 Power Extraction from the Magnetron

While radial extraction is the norm, axial extraction has been used as well. Figure 10.19 shows the inverted magnetron experiment of Ballard. Axial extraction occurs by use of a mode converter.

Another method of axial extraction is shown in Figure 10.20. Here, the magnetron resonator structure is axially tapered to merge into an output waveguide. This method allows a gradual transition from the magnetron geometry to an output impedance matched to free space. In the USSR, this method is referred to as *diffraction coupling*. It has been used in several experiments by the Kovalev group.

Inverted magnetrons have been studied extensively by Ballard (Stanford, SNLA) [14, 15], the MIT group [27], and NRL [29]. In general, a reduction of output power is reported by these groups. The MIT group proposes that their radial extraction method, which requires radiation to propagate through the electron cloud in order to reach the extraction waveguide, is inherently inefficient. Ballard has proposed that the inverted resonator does not contain such high RF fields as the normal resonator [30].

In modern conventional magnetrons, extraction is by a coaxial cavity (Figure 10.21). Slots in the back walls of alternate resonators allow coupling of the π mode RF into the transverse electric (TE) modes of the

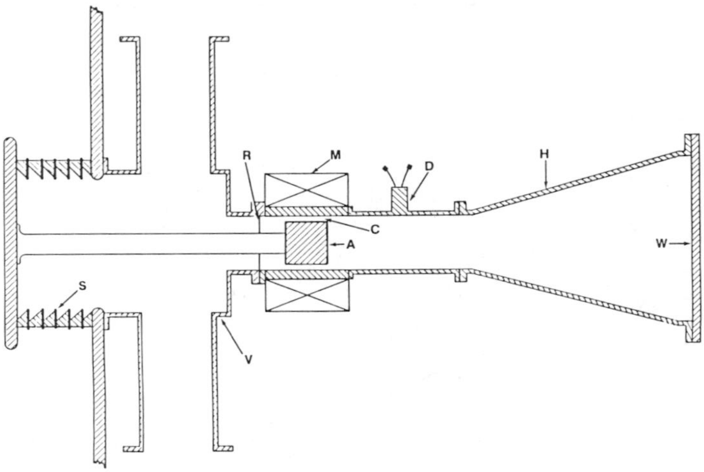

Fig. 10.19 Cross-Sectional View of an Inverted Magnetron [14]
Key: (A) anode block, (D) directional coupler, (H) horn, (M) magnets, (R) Rogowski coil, (S) insulator stack, (V) vacuum chamber, and (W) output window

coaxial cavity. Typically, TE_{011} mode cavity excitation helps to stabilize operation of the magnetron. The resulting resonator structure is larger and less complex than comparable strapped magnetrons, and has largely superceded them in commercial magnetron designs. This coaxial configuration has been used in experiments conducted at NRL with a 54-vane resonator, which is by far the most complex resonator used to date [29]. Radiation, extracted through the apertures at the backs of the resonators, filled a coaxial stabilizing cavity (Figure 10.22). This cavity supported the TE mode, and forced coherence between the 54 resonators on the periphery. Extraction from this geometry is by means of a cylindrical output cavity that feeds a mode converter. The reported output power was 500 MW. The electrical pulse duration was shorter than the filling time of the cavity (Q/ω) so that the central stabilizing cavity did not reach full power.

Once extraction has been achieved, the radiation field can be analyzed by using a waveguide in the far field. A directional coupler with variable attenuation feeds the signal to a crystal, or other diagnostic, located in a screen room. The resulting radiation pattern depends on the method of extraction. The pattern for axial extraction in transverse magnetic (TM) modes is as shown in Figure 10.23. A central null is surrounded by lobes in the forward direction and the pattern is radially polarized.

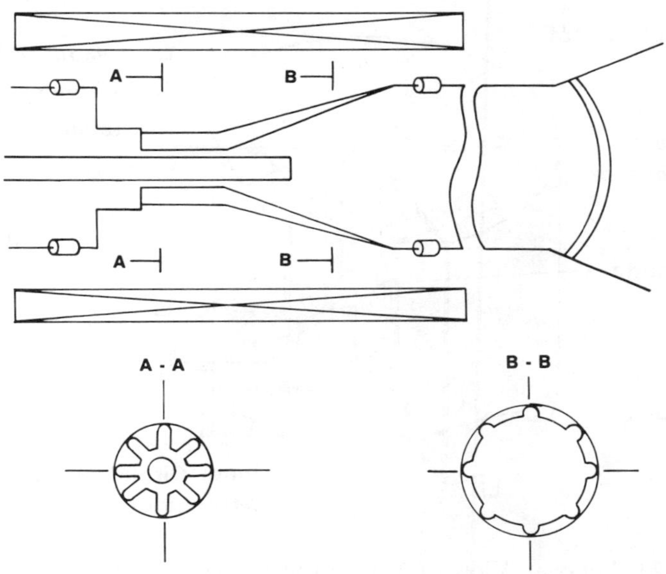

Fig. 10.20 Axial Extraction Method Using Tapered Magnetron Structure

Extraction by radial waveguides produces TE_{10} radiation patterns, as shown in Figure 10.24. Here, the power distribution is smooth and polarization is linear. The axial extraction method of Kovalev (Figure 10.20) produces a TM_{ni} pattern, where n is the mode number and i is the radial wave index. Extraction from a coaxial cavity gives TE_{0n} patterns, which have a central null.

Most magnetron experiments have been conducted at ≈3 GHz. Two experiments at X band have been conducted by the Kovalev group and the Physics International group [18, 13]. Scaling to high frequencies is not favorable for magnetrons. The Buneman-Hartree relation can be written as

$$V_{BH} \sim \frac{B_{BH} r_a^2}{\lambda} \left(1 - \frac{r_c^2}{r_a^2}\right) \tag{10.9}$$

For constant voltage and r_c/r_a:

$$V_{BH} \sim r_a^2 B_{BH} \tag{10.10}$$

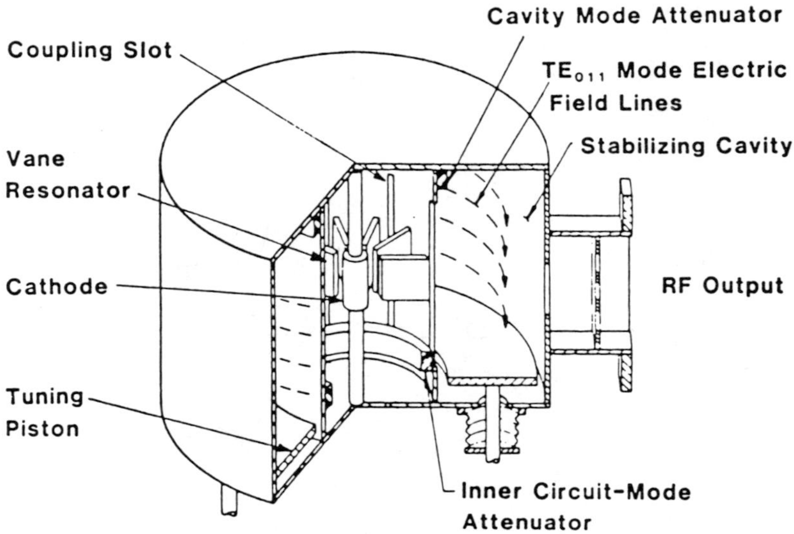

Fig. 10.21 Cutaway View of Magnetron Using Coaxial Cavity for Stabilization

Source: *Microwave Tube Manual* by Varian Associates, US Air Force Publication #TO00-25-251, Oct. 1979.

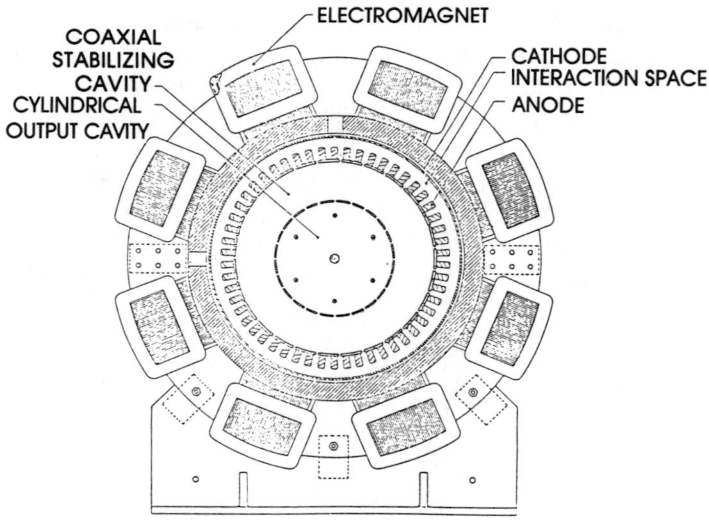

Fig. 10.22 Hybrid Inverted Coaxial Magnetron with 56 Vanes Using a Central Coaxial Stabilizing Cavity and a Cylindrical Output Cavity

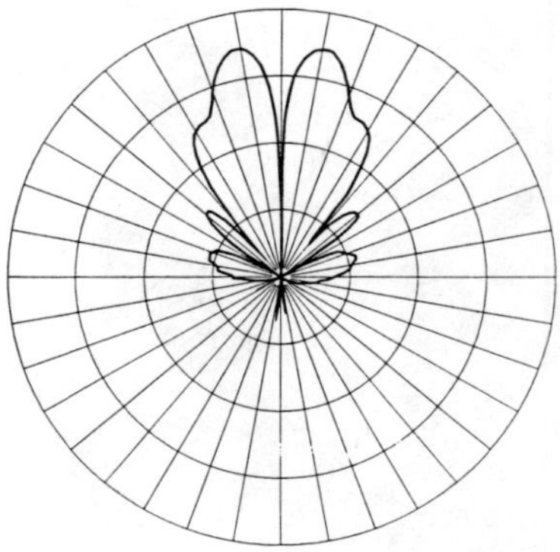

Fig. 10.23 TM-Mode Radiation Pattern Showing a Central Null with Prominent Lobes [14]
Note: Radial circles are 10 dB increments.

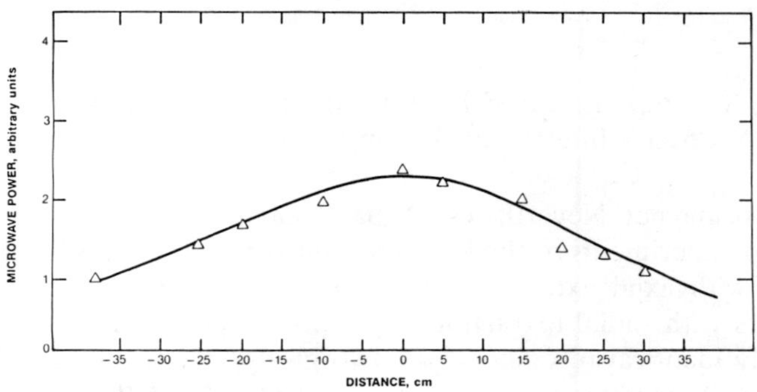

Fig. 10.24 Radiation Pattern from a Radially Extracted Rectangular Waveguide Giving a Smooth Distribution [12]

Because $\lambda \sim r_a$:

$$B_{BH} \sim 1/\lambda \tag{10.11}$$

Therefore, at high frequencies, high magnetic fields are required for insulation and the anode block becomes smaller (Figure 10.25). Both of these factors stress technology because experimental tolerances become more difficult with smaller block size. Heating and anode erosion are more substantial, and a high magnetic field means more energy and complexity

Fig. 10.25 A Comparison of S-Band (A6) and X-Band Resonators (Courtesy of Physics International Company)

in the electromagnet. Nevertheless, X-band magnetrons have been studied, and the first experiments by the Kovalev group produced 4 GW of radiation at 9 GHz with axial extraction [18]. Sze *et al.* have conducted similar experiments with radial extraction and reported powers of 0.5 GW [13]. With nearly identical resonators, the two groups report powers differing by an order of magnitude. The resonators used at X band are quite small, so extrapolation to shorter wavelengths seems doubtful.

The devices that we have described are microwave oscillators, or resonant systems, for which the dominant modes grow from a noise background not influenced by external signals. Amplifiers take an external signal and boost its power to higher levels. Conventional magnetron amplifiers have been developed at low power [31]. These amplifiers achieve power levels of about 1 MW and operating gains of about 10 dB with an amplification bandwidth of 10%. Only one relativistic magnetron amplifier experiment has been reported to date [20]. Zherlitsyn *et al.* injected a 40 kW signal into a relativistic magnetron through the rear of a resonator. Radiation at the injected frequency was collected from the back of an adjacent resonator at a power level of 40 MW, giving a 30 dB gain. The overall efficiency of the amplifier was about 1%.

10.5 EXPERIMENT AND THEORY COMPARISON

The basic features of relativistic magnetrons agree well with calculation, especially as regards voltage, magnetic field, resonant frequency, time scales, and rise times. Figure 10.26 shows a comparison of the operating space of four experiments. All of them operate very near the Buneman-Hartree condition, where we would expect the efficiency to be a maximum (see below). The bars in Figure 10.26 show the range of longitudinal magnetic fields over which the magnetron resonates. The resonator will emit microwaves at fields higher than the Buneman-Hartree condition predicts. Gap closure may be an explanation (see (10.5)). Figure 10.27 shows the electrical characteristics of the A6 magnetron. As the magnetic field increases, the impedance of the diode increases due to inhibition of cross-field diffusion. Figure 10.28 shows the power extracted from the A6 as magnetic field varies. Figure 10.29 shows that the A6 operates close to the Buneman-Hartree condition for a variety of voltages.

The time scale of magnetron operation to date is typically tens of nanoseconds, substantially less than the electrical pulse duration. The shortness of the pulse may be due to several effects. First, there is only a narrow range of voltage for which the magnetron will oscillate. In most experiments, the voltage substantially varies during the pulse. Evidence for this explanation is that, occasionally, two microwave pulses are seen, one occurring during the voltage rise, and the other during the voltage fall. Another explanation is that closure of the anode-cathode gap by cathode gap by cathode plasma can alter the resonance condition, and thus change B_{BH} so that the radiation pulse will be short, even with a flat voltage pulse.

The rise time of the pulse is 10 ns or less. This is consistent with the measured Q of the cavity and the bandwidth of the radiation, which are inversely related to each other. Experimentally, bandwidths are approximately 1% of the resonant frequency, as shown in Figure 10.30. The rise time of the resonator should be

$$\tau_r = Q/\omega \tag{10.12}$$

The Q values are measured to be on the order of 100, which gives rise times of about 5 ns for a 3 GHz resonant frequency.

The efficiency of the conventional magnetron increased steadily during its development, and achieved values as high as 80% or 90% in some designs. The relativistic magnetron, which must be considered as in its early phase of development, has achieved values of a few tens of percent. Early experiments yielded approximately 20% efficiency, and there have

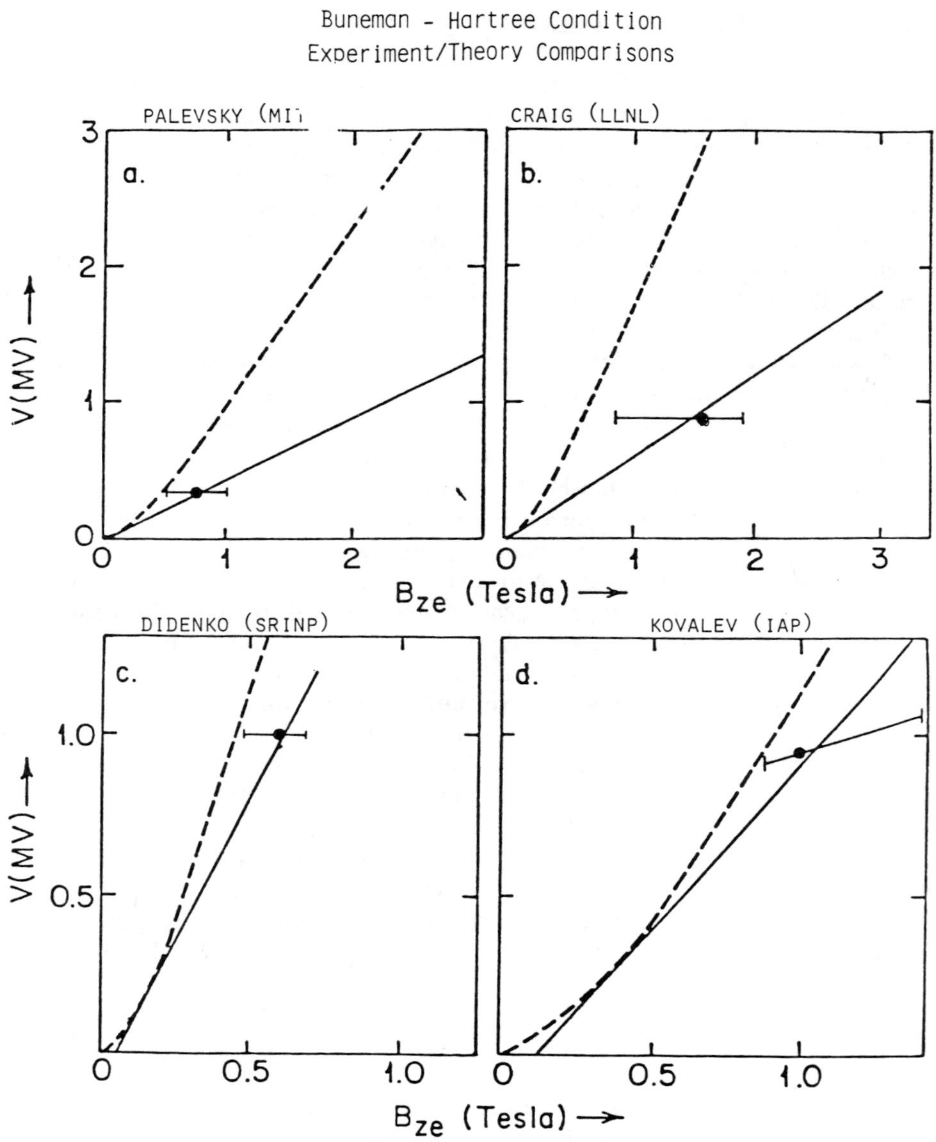

Fig. 10.26 Operating Space of Four Magnetron Experiments [11]
Note: Hartree condition is the solid straight line and Hull magnetic insulation condition is the broken curved line. Dots represent maximum power points, brackets give range of operation.

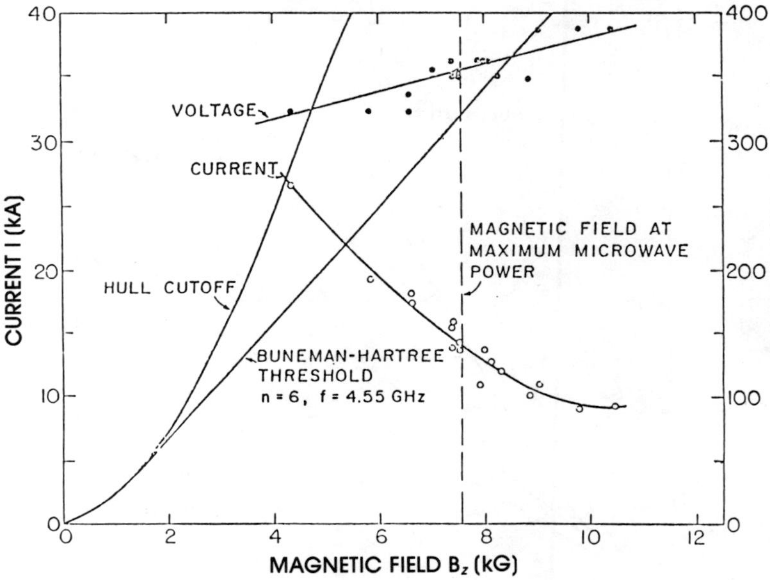

Fig. 10.27 Current and Voltage Characteristics of the A6 Magnetron [9]

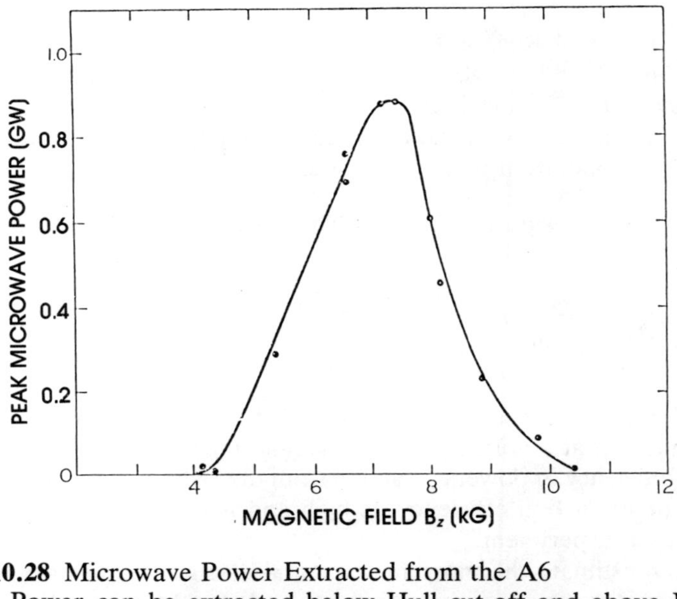

Fig. 10.28 Microwave Power Extracted from the A6
Note: Power can be extracted below Hull cut-off and above Buneman-Hartree magnetic fields.

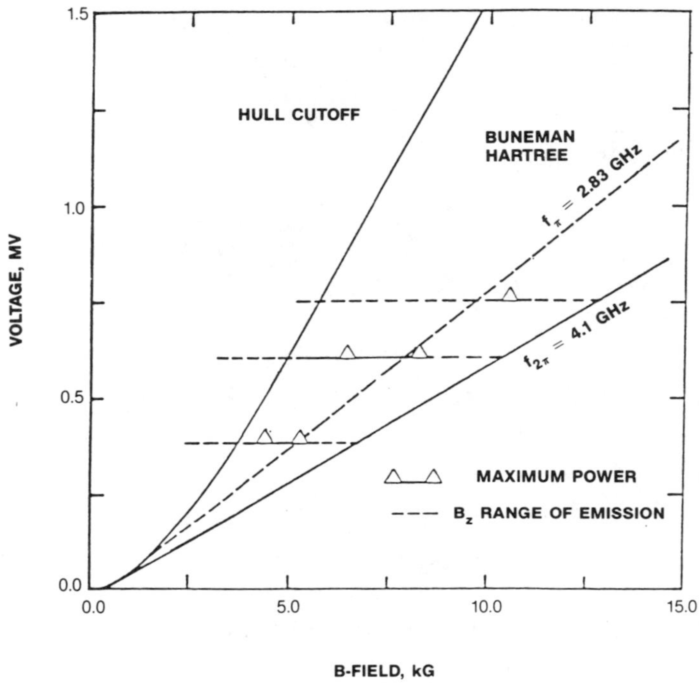

Fig. 10.29 Operating Regime of Washer A6 Magnetron Showing Calculated Hull Cut-Off and Buneman-Hartree Resonance for both π and 2π Modes [12]

Note: This resonator was shown to operate at high power in the π mode with low-power operation in the 2π mode.

been reports of values as high as 30%. The electronic efficiency is defined as

$$\eta_e = \frac{P_{\mathrm{RF}}}{I_r V} \tag{10.13}$$

where P_{RF} is the microwave power and I_r is the current flowing radially in the magnetron at a voltage V. The overall efficiency is the product of η_e and the efficiency of power extraction from the radiator. This RF circuit efficiency, or extraction efficiency, is high and it is a measured quantity specific to each experiment.

We can estimate the more fundamental electronic efficiency. Electrons arriving at the anode convert most of the dc potential into RF energy,

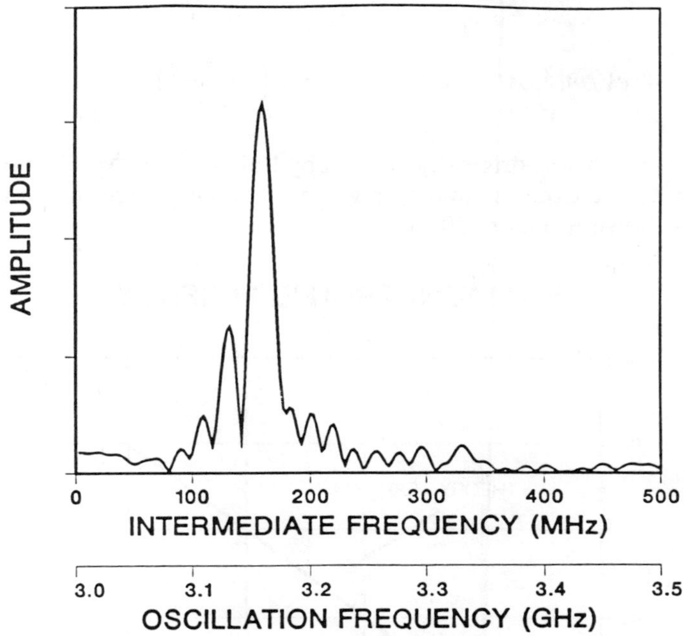

Fig. 10.30 Fourier Transform of Heterodyne Mixer Data from an Inverted Magnetron [14]
Note: Intermediate frequency is the difference between resonator frequency (~3.16 GHz) and local oscillator frequency (3.0 GHz).

with the main loss arising from electron kinetic energy that is deposited in the anode block. Therefore, electronic efficiency is

$$\eta_e = 1 - \frac{(\gamma - 1)mc^2}{eV} \tag{10.14}$$

The kinetic energy of electrons reaching the anode is approximated by assuming that the $E \times B$ drift contributes one component of velocity, and the orbital motion of the electron about the magnetic field contributes an equal velocity. The two velocity components are added by using the relativistic velocity addition law. The worst case is when the two components are parallel, which gives

$$\eta_e = 1 - \frac{2V^*}{A^{*2} - V^{*2}} \tag{10.15}$$

where

$$V^* = eV/mc^2, \; A^* = eA/mc, \; A = B\left(\frac{r_a^2 - r_c^2}{2r_a}\right)$$

This result was originally derived by Ballard [15]. Application of this equation to the experimental data given by Table 10.2 results in the comparisons shown in Figure 10.31.

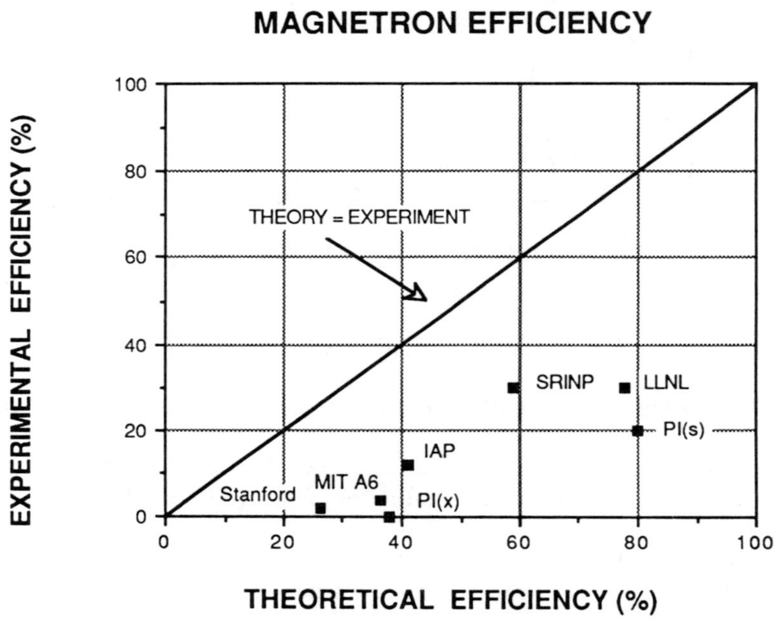

Fig. 10.31 Comparison of Experimentally Observed Power Efficiency and Theoretical Maximum Efficiency (Eq. (10.14))

Experimental efficiencies are substantially less than theoretical efficiencies, typically by a factor of 2. There are several contributors to this lower experimental efficiency. In some experiments, the downstream current loss emitted from the tip of the cathode is not taken into account when calculating the electrical power, thus lowering the experimentally determined efficiency. That is, both the radial and axial currents are used, instead of the radial current alone, as in (10.13). Furthermore, nonlinear effects in the electron plasma, such as plasma wave generation, can dissipate energy, hence subtracting from that available for generating microwaves. Moreover, the resonators used in experiments may not be optimized, and therefore do not realize the theoretical efficiency. Finally, the RF circuit efficiency, while a small factor, is not reported in some experiments, and therefore the true electronic efficiency is not known. The efficiencies shown in Figure 10.31 are derived from the sometimes fragmentary accounts given in the literature.

The total power that can be extracted from the magnetron is

$$P_{\mathrm{max}} = \frac{\omega_n}{Q_L} \int_v \frac{1}{2} \epsilon_0 E_{\mathrm{RF}}^2 \, dv \tag{10.16}$$

where E_{RF} is the peak RF field, v is the volume of the resonator, and Q_L is the loaded Q, or the operating value of Q with a load attached to the resonator. Clearly, low Q_L and high RF electric fields are necessary for high extracted power.

Total power is limited by the effect of the microwave fields on insulation. If the RF magnetic field is strong enough during one part of the RF cycle to reduce the total magnetic field (dc plus RF) to less than B^*, then insulation will fail, electrons will reach the anode without giving up energy to RF fields, and microwave power will fall. The field-limiting condition can be written

$$\Delta B_{\mathrm{RF}} \leqslant \frac{1}{2} (B_z(\mathrm{max}) - B^*) \tag{10.17}$$

where $B_z(\mathrm{max})$ is the experimentally determined magnetic field for maximum power output ($B_z(\mathrm{max}) \approx B_{BH}$). Palevsky [9] and Ballard [30] have shown that the A6 magnetron operating in the 2π mode is limited by this condition to 750 kV/cm, corresponding to 3 GW. Ballard's calculations show higher powers in 2π mode than in π mode due to increased volume in the resonators, with other parameters remaining equal. Young [12] has derived a nonlinear saturation model of the magnetron, which predicts a saturated power level:

$$P_{\mathrm{sat}} = \frac{\omega_n}{Q_L} W_{NL}(1 - X) \tag{10.18}$$

W_{NL} is the saturation level of wave energy, $X = \omega_n/2Q_L\Omega$, where Ω is the magnetron instability growth rate:

$$\Omega = \frac{\omega_c}{16} \tag{10.19}$$

and ω_c is the electron cyclotron frequency. By using (10.18), Young finds agreement with experimental values of extracted power from an A6 resonator with a washer cathode.

10.6 HIGH-CURRENT EFFECTS

Since the relativistic magnetron is, in fact, a high-current extrapolation of the conventional magnetron, we would anticipate unique effects due to the high currents and high voltages necessary to drive such currents. Three such effects have been reported to date. The first is a change in the resonance condition, which causes a drop in magnetron performance at currents higher than about 10 kA. The other effects are gap closure due to plasma emitted from the cathode and erosion of the anode block due to impact by high-voltage electrons.

The Buneman-Hartree condition can be written

$$U_H = \frac{eV}{mc^2} = (1 - \beta_\phi^2)^{1/2} + \beta_\phi b_z - 1 \tag{10.20}$$

where

$$\beta_\phi = \frac{r_a \omega}{c}, \quad b_z = 0.27 B_z \text{ (kG) } r_a \text{ (cm)} \left(1 - \frac{r_c^2}{r_a^2}\right)$$

For high current, the azimuthal magnetic field produced by the axial current flowing in the magnetron produces new drifts. Lovelace and Young [11] have shown that the Buneman-Hartree condition is modified in two different ways, depending on which components of the canonical angular momentum are conserved. If the axial component is conserved, then

$$U_{HA} = [(1 - \beta_\phi^2)(1 + b_\phi^2)]^{1/2} + \beta_\phi b_z - 1 \tag{10.21}$$

where $b_\phi = I/17$ (kA) $2\ln r_a/\mathrm{r}_c$. This condition occurs if the electron achieves azimuthal synchronism with the wave.

If instead the electron achieves both azimuthal and axial synchronism with a wave, which implies that the axial and theta components of angular momentum are conserved, then the condition becomes

$$U_{HB} = (1 - \beta_\phi^2 - \beta_z^2)^{1/2} + \beta_\phi b_z + \beta_z b_\phi - 1 \tag{10.22}$$

where $\beta_z = u_z/c$, and u_z is the axial velocity component.

The Hartree voltages for cases A and B are shown in Figure 10.32. The experimental data indicate that high-power operation occurs for case A, which implies that only the axial component of canonical angular momentum is conserved. Data for several voltages at fixed magnetic field are shown in Figure 10.33. The rapid drop in power with increasing axial current is apparent. A physical picture of this interaction is that the B_θ caused by the axial current interacting with the radial electric field produces a drift in the axial direction, which flushes electrons from the resonator before they can produce high power. This effect limits the amount of current that can be drawn by the magnetron for high-power operation, and thus presents a lower bound to the impedance at which the relativistic magnetron can operate.

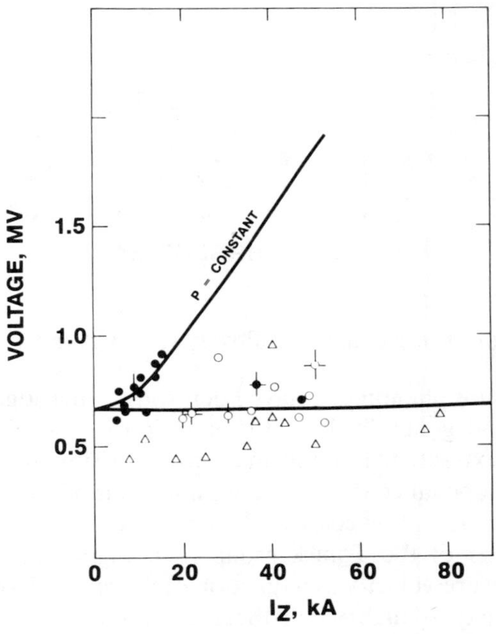

Fig. 10.32 Resonant Voltage *versus* Axial Current (from [11], data from [10])

Field emission by its nature produces a cathode plasma from the explosion of microprojections. This plasma can move across the anode cathode gap, despite the axial magnetic field, because it is dominated by

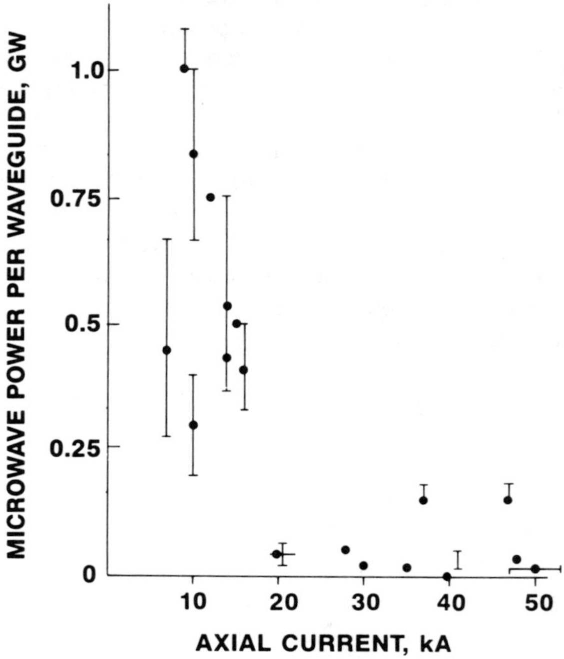

Fig. 10.33 A6 Magnetron Normalized Power *versus* Axial Current [10]

collisions. Gap closure phenomena have been widely investigated in field-emission diodes, and generally limit the duration of the electron beam pulse that can be extracted. For the magnetron, a similar phenomenon appears to occur. Impedance drops as the anode-cathode gap closes, and hence the microwave radiation ceases. The only study of this phenomenon to date is by Gleizer of the Tomsk group [23]. Figure 10.34 shows his measurement of closure velocity as a function of the applied axial magnetic field. Closure velocity is high when there is production of high-power microwave radiation. However, for a coaxial diode without the slow-wave structure on the anode, the closure velocity is lower by a factor of 5. The conclusion is that the high electric and magnetic fields in the resonator cause anomalous cross-field transport, resulting in gap closure.

In Section 9.5, we gave evidence that the microwave pulse duration is short because the voltage in the typical experiment rapidly varies in time so that the resonance condition is met for only a short interval. The phenomenon of gap closure indicates that if the pulse voltage were constant, gap closure would eventually truncate the microwave pulse. Moreover,

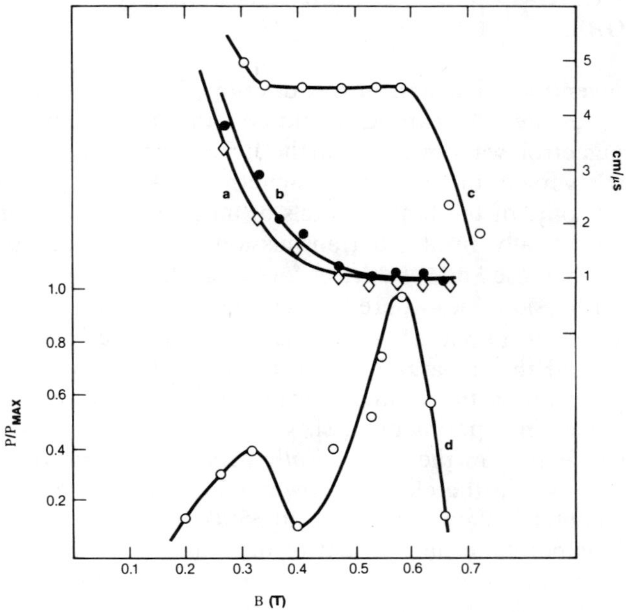

Fig. 10.34 Microwave Power and Gap Closure Velocity as a Function of Axial Magnetic Field: (a), (b) Velocity with Smooth Anodes; (c) Closure Velocity with an Anode Slow-Wave Structure; (d) Normalized Microwave Power with Slow-Wave Structure [23]

because the Buneman-Hartree voltage is proportional to the anode-cathode gap, the resonance condition would be altered as the gap closes so that the pulse would end, even before the gap shortens. Data are limited, and so we need further research.

Experimental magnetron efficiency is a few tens of percent. Part of the energy that is not radiated as microwaves will be deposited by the electrons in the anode. Because the electrons have high energies, they deposit in the interior, and thus cause melting and spallation. Erosion is observed at the entrance of the resonator. The damage is asymmetric because of the $E_r \times B_z$ drift of the spokes. Anode erosion is a major limitation to the relativistic magnetron. Kovalev [17] reports an anode lifetime of less than 500 shots. To date, no advanced techniques, such as the use of Elkonite™ or tungsten in the anode, have been attempted in order to reduce anode erosion, but these materials have shown substantial improvements of high-power gaps in the electrical industry.

10.7 MAGNETICALLY INSULATED TRANSMISSION-LINE OSCILLATORS

The conventional magnetron was developed in both cylindrical and linear geometries (i.e., the cathode could be either a cylinder or a plate). The linear magnetron was developed in the 1950s, primarily as an amplifier. The relativistic version of the linear magnetron is a new type of device, which changes some of the fundamentals of magnetron operation.

The magnetically insulated transmission line has been developed since the 1960s in pulsed power devices for efficiently propagating electron current in transmission lines where the electric field is above the threshold of electron emission (≈300 kV/cm). Such lines operate by generating a magnetic self-field that is large enough to confine the electrons carrying the current. The properties of magnetically insulated lines have been explored extensively in experiments [32].

The new feature in the *magnetically insulated transmission-line oscillator* (MILO) is that the electron flow confines itself with its self-field. The basic geometry is shown in Figure 10.35(a). The slow-wave structure on the anode is perpendicular to that of the standard magnetron, where the vanes extend in the direction transverse to the longitudinal electron flow. Figure 10.35(a) shows electrons emitted from the cathode drifting downstream, early in the development of the RF field. The magnetic field of the current flow bends the electrons away from the anode. The interaction of the electron stream with the slow-wave structure produces growing electromagnetic oscillations. Figure 10.35(b) shows a fully saturated flow with strong spoke formation. The system operates in the π mode with successive cavities 180° out of phase. Radially, each cavity operates at a quarter-wavelength of a TEM wave.

The MILO is a crossed-field amplifier without an applied magnetic field. The elimination of the external field coils simplifies the device. It can be designed for both planar and coaxial geometries. Because the magnetic field that provides insulation is produced by the electron flow itself, there is a continual adjustment of the magnetic field to voltage changes, which means there is no fine tuning in the device. This is an advantage over the relativistic magnetron, which resonates at a voltage that is proportional to the applied magnetic field. One of the objectives of the MILO is to serve as a crossed-field amplifier so that injected microwave signals can be amplified by extracting energy from the electron flow. The amplified wave would be extracted from the slow-wave structure at its end. If amplifier operation can be demonstrated, it would make possible the construction of arrays of high-power microwave sources.

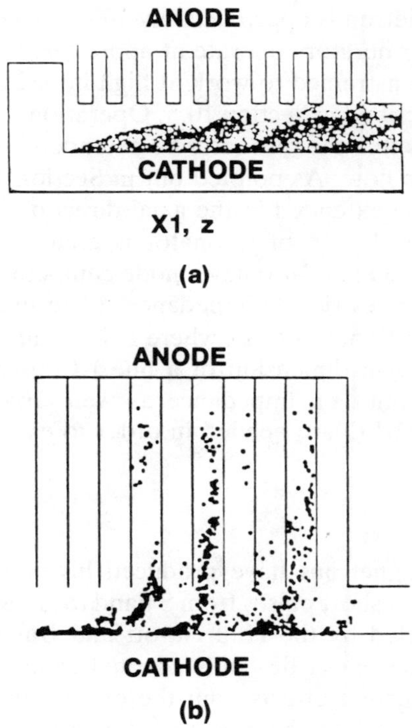

Fig. 10.35 Simulations of MILO Devices: (a) Growing Waves on Electron Flow; (b) Saturated Flow and Spokes

Experimental study of the MILO is in its beginning stages. Experiments by Clark, Bacon, and Marder [33] have produced microwave oscillations at 1.4 and 0.72 GHz, in agreement with the dispersion relation for waves of this geometry. The experimenters report a narrow frequency spectrum, which does not change during the pulse, although the voltage experiences considerable excursion. Simulations of this experiment calculate a 15% efficiency of transfer from pulsed power to RF power. Present observed efficiencies are much lower.

One of the potential virtues of the MILO is that the emitting cathode may be made large, which keeps the power density low and allows operation at low impedance. This would provide a good impedance match to existing low-impedance pulsed-power sources and allow use of their very high electrical powers in microwave production. The impedance of the relativistic magnetron is approximately 100 Ω, whereas contemporary high-pulsed-power generators operate at approximately 1 Ω. The relativistic magnetron is constrained to work at high impedances because of the self-field effects described in Section 10.5. Operation of the MILO at lower impedances requires that the resonator be extended in the direction transverse to the electron flow. As pointed out in Section 10.3, the relativistic magnetron cannot be extended in the axial direction because mode competition occurs if the length of resonator is greater than the oscillation wavelength. The MILO may also have mode competition in the transverse direction, which will restrict its impedance. (The impedance of a transmission line is proportional to D/L, where D is the anode-cathode spacing and L is the transverse dimension of a line.) Consequently, the MILO may have a lower limit on impedance as well. Experiments on mode competition in the MILO are needed in order to explore this question.

10.8 SUMMARY

Relativistic magnetrons have produced high-power microwaves at moderate efficiency in short pulses from S band to X band. Their operation is reasonably predicted by the Hull cut-off and Buneman-Hartree resonance conditions. Resonator design is adapted from that of conventional magnetrons. High-current effects limit the current that can be used in a single resonator as well as the number of pulses such a resonator can sustain. Gap closure appears to limit the microwave pulse duration.

The factor which limits the total output power appears to be growth of RF fields to levels that are sufficient to violate the electric and magnetic field limits described by the Buneman-Hartree and Hull cut-off criteria.

The high-current magnetron is still in its early developmental period, and has already achieved gigawatt power levels. Figure 10.36 shows the growth of microwave power from magnetrons of both the conventional and relativistic varieties. The conventional magnetron power curve is saturated at approximately 10 MW, and has the S-shape typically found in developing technology. For example, Livingston has shown that the energy capability of a given particle-accelerator technology follows the S-shaped curve, and then is succeeded by a more powerful technology, which saturates at a higher power and, in turn, is succeeded by yet another technology. Experiments are underway at Physics International and MIT for phase-locking of high-current magnetrons to an external signal, which would allow their use in high-power arrays as oscillators or amplifiers. Future developments will probably focus on longer pulse durations and repetitive operation. The frequency domain will be extended, probably at the lower end. As for total output power, we cannot say at present what

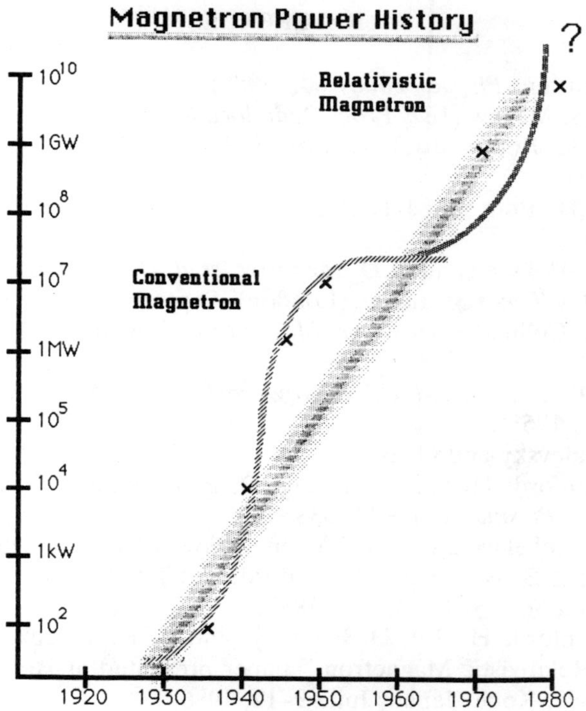

Fig. 10.36 History of Magnetron Peak Power

power levels can be achieved by this technology with more sophisticated resonators or larger resonator volumes. Therefore, the ultimate power limitation of relativistic microwave magnetrons may be much higher, and still awaits discovery.

ACKNOWLEDGEMENTS

I gratefully acknowledge helpful discussions with John Creedon, Henry Sze, Bill Ballard, Maurice Esterson, Richard Smith, and Ted Orzechowski in the preparation of this review. Noreen Curry and Dominic Benford gave invaluable assistance in the preparation of the manuscript. I also wish to thank Physics International Company for support of my work.

REFERENCES

1. J. Brittain, *Physics Today* **60**, (1985).
2. E.C.S. Megaw, *IEE Proc. Radiolocation Conv.*, 977 (1946).
3. H.A.H. Boot and J.T. Randall, *IEE Proc. Radiolocation Conv.*, 928 (1946).
4. H.A.H. Boot and J.T. Randall, *IEEE Trans. Electron Dev.* **ED-23** (7), (1976).
5. N.F. Alekseev and D.D. Malairov, *et al., Proc. IRE* **32**, 136 (1944).
6. *Sunday Times Magazine* (London), Sept. 7, 1975, p. 8.
7. G.B. Collins, *Microwave Magnetrons* (McGraw-Hill, New York, 1948).
8. E. Okress, *Cross-Field Microwave Devices* (Academic Press, New York, 1961).
9. A. Palevsky and G. Bekefi, *Phys. Fluids* **22** (5), 1979.
10. J. Benford, H. Sze, T. Young, D. Bromley, and G. Proulx, *IEEE Trans. Plasma Sc.* **PS-13**, 538 (1985).
11. R.V. Lovelace and T.S.T Young, *Phys. Fluids* **28**, 2450 (1985).
12. H. Sze, B. Harteneck, J. Benford, and T.S.T Young, *IEEE Trans. Plasma Sc.* **PS-15** (3), 327 (1987).
13. J. Benford, H. Sze, D. Bromley, and B. Harteneck, "Progress on the Relativistic Magnetron," paper presented at BEAMS '86 Conference, Kobe, Japan, June 9–12, 1986.
14. W.P. Ballard, L.M. Earley, and C.B. Wharton, Sandia Rep. No. SAND-85-1952, UC-28 (1986).
15. W.P. Ballard, S.A. Self, and F.W. Crawford, *J. Appl. Phys.* **53** (11), (1982).
16. W.O. Doggett, T.A. Tredo, J. Jackson-Ford, G.E. Thomas, D.J. Jenkins, and R. Smith, *IEEE Int. Plasma Science Conf. Record* (IEEE Pub. No. 86-CH-2371-6), Saskatoon, Saskatchewan, 1986.
17. N.F. Kovalev, B.D. Kol'chuqin, V.E. Nechaev, M.M. Ofitserov, E.I. Soluyanov, and M.I. Fuks, *Sov. Tech. Phys. Lett.* **3** (10), (1977).
18. N.F. Kovalev, E.G. Krastelev, M.I. Kuznetsov, A.M. Maine, M.M. Ofitserov, V.A. Papdichev, M.I. Fuks, and L.N. Chekanova, *Sov. Tech. Phys. Lett.* **6** (4), (1980).
19. V.E. Nechaev, M.I. Petelin, and M.I. Fuks, *Sov. Tech. Phys. Lett.* **3** (8), (1977).
20. A.G. Zherlitsyn, G.V. Mel'nikov, A.S. Sulakshin, and Yu.G. Shtein, *Sov. Phys. Tech. Phys.* **49** (11), (1979).
21. A.N. Didenko, A.S. Sulakshin, G.P. Fomenko, Yu.G. Shtein, and Yu.G. Yushkov, *Sov. Tech. Phys. Lett.* **4** (1), (1978).
22. A.N. Didenko, A.S. Sulakshin, G.P. Fomenko, V. Tsvetkov, Yu.G. Shtein, and Yu.G. Yushkov, *Sov. Tech. Phys. Lett.* **4** (7), (1978).

23. I.Z. Gleizer, A.N. Didenko, A.S. Sulakshin, G.P. Fomenko, and V.I. Tsvetkov, *Sov. Tech. Phys. Lett.* **6** (1), (1980).
24. I.I. Vintizenko, A.S. Sulakshin, G.P. Fomeno, L.F. Chernogalova, and Yu.G. Shtein, *Sov. Tech. Phys. Lett.* **9** (4), (1983).
25. G. Bekefi and T. Orzechowski, *Phys. Rev. Letts.* **37** (6), (1976).
26. G. Craig, J. Pettibone, and D. Ensley, in Abstracts for *IEEE Int. Plasma Science Conf. Record* (1979).
27. R.A. Close, A. Palevsky, and G. Bekefi, *J. Appl. Phys.* **54** (7), (1983).
28. I.I. Vintizenko, A.S. Sulakshin, G.P. Fomenko, L.F. Chernogalova, and Yu.G. Shtein, *Sov. Tech. Phys. Lett.* **9** (4), (1983).
29. W.M. Black, W.M. Bollen, R. Tobin, R.K. Parker, L.R. Barnett and G. Franey, *Proc. IEEE Int. Electron Devices Conf.* (1980).
30. W.P. Ballard, private communication.
31. A.S. Gilmour, Jr., *Microwave Tubes,* (Artech House, Dedham, MA, 1986).
32. M.S. DiCapua, *IEEE Trans. Plasma Sc.* **PS-11**, 205 (1983).
33. B.M. Marder, L.D. Bacon, and M.C. Clark, *Bull. APS* **31**, 1616 (1986).

Chapter 11
BASIC THEORY OF CERENKOV RADIATORS

William Case

Department of Physics
Grinnell College
Grinnell, Iowa

11.1 INTRODUCTION

The well known phenomenon of *Cerenkov radiation* was experimentally established by Cerenkov [1] in 1934, although hints of earlier observations exist [2]. The process was correctly explained theoretically by Frank and Tamm [3] three years later. Theoretical work by Heaviside [4] in 1889 also points to the effect. Although this latter work is interesting, it cannot be taken too seriously because it is *nonrelativistic* (prerelativistic) and predicts radiation for charges moving in a vacuum when their speeds are greater than that of light (c).

The Cerenkov radiation indicated above, which we will call *spontaneous Cerenkov radiation,* can be thought of as the electromagnetic shock wave due to a charge moving through a material at a rate greater than the speed of the electromagnetic wave in the medium. This is shown in Figure 11.1. The topic covered in this chapter is *stimulated Cerenkov radiation.* The main features of this phenomenon will be presented by consideration of the very simple system shown in Figure 11.2. A cold homogeneous electron beam passes through a dielectric (dielectric constant ϵ) in the positive z direction. A strong magnetic guide field, also in the z direction, constrains the charges to motions in the z direction. The reader should note that this system is nonphysical in that an electron beam cannot pass

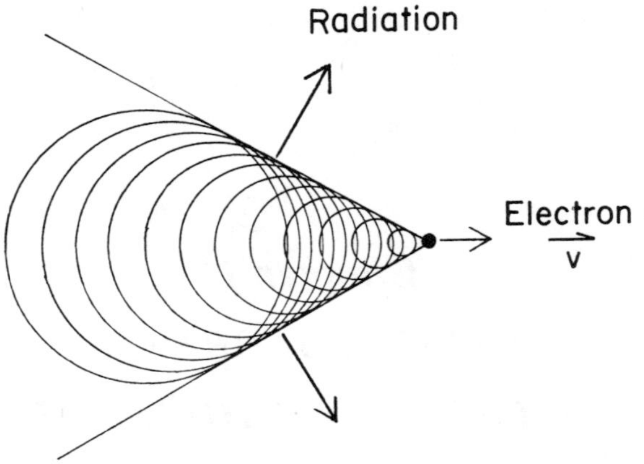

Fig. 11.1 Spontaneous Cerenkov Radiation for an Electron Moving through a Dielectric at a Speed $> c/\sqrt{\epsilon}$

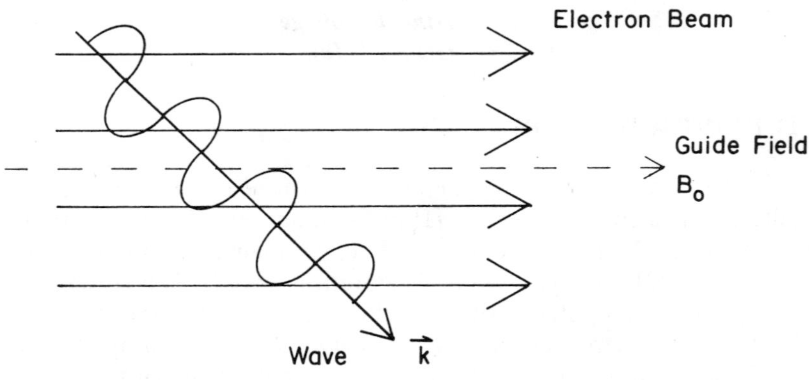

Fig. 11.2 Simple System Used to Study Stimulated Cerenkov Radiation
Note: An electron beam with guide field B_0 passes through a dielectric, which interacts with an electromagnetic wave having a propagation vector of $\vec{k}$.

through a dielectric, except in the extreme case where the dielectric is a gas. In such cases, the beam would have to be highly relativistic ($\gamma_0 > 30$, where $\gamma_0 \equiv (1 - v_0^2/c^2)^{-1/2}$, v_0 = initial beam velocity).

11.2 DERIVATION OF THE DISPERSION RELATION

The dispersion relation for the above system is obtained by the application of Maxwell's, the Lorentz force, and continuity equations. Taking the curl of Faraday's Law and making use of Ampere's law [5] gives

$$\begin{aligned}\nabla \times (\nabla \times \vec{\mathbf{E}}) &= -\frac{1}{c}\frac{\partial}{\partial t}(\nabla \times \vec{\mathbf{B}}) \\ &= \frac{\epsilon}{c^2}\frac{\partial^2 \vec{\mathbf{E}}}{\partial t^2} - \frac{4\pi}{c^2}\frac{\partial \vec{\mathbf{j}}}{\partial t}\end{aligned}$$

Applying the vector identity $\nabla \times (\nabla \times \vec{\mathbf{E}}) = \nabla(\nabla \cdot \vec{\mathbf{E}}) - \nabla^2 \vec{\mathbf{E}}$ followed by the Poisson equation to the left-hand side gives

$$\frac{4\pi}{\epsilon}\nabla\rho - \nabla^2\vec{\mathbf{E}} = -\frac{\epsilon}{c^2}\frac{\partial^2 \vec{\mathbf{E}}}{\partial t^2} - \frac{4\pi}{c^2}\frac{\partial \vec{\mathbf{j}}}{\partial t} \tag{11.1}$$

The space and time-dependent parts of the fields and densities are assumed to be small. These are now expanded, keeping only the *zero*th and first-order terms. The first-order terms (subscript 1) are taken to be of the form $Ae^{i(\vec{\mathbf{k}}\cdot\vec{\mathbf{r}} - \omega t)}$, while the *zero*th-order terms (subscript 0) are constants. In this notation, we have

$$\begin{aligned}\rho &= \rho_0 + \rho_1 \\ \vec{\mathbf{j}} &= \vec{\mathbf{j}}_0 + \vec{\mathbf{j}}_1\end{aligned}$$

where $\vec{\mathbf{j}}_0$ and $\vec{\mathbf{j}}_1$ are entirely in the z direction due to the strong guide field B_0. With these expressions, the z component of (11.1) becomes

$$\frac{4\pi i k_z}{\epsilon}\rho_1 + k^2 E_z = \frac{\epsilon\omega^2}{c^2}E_z + \frac{4\pi i\omega}{c^2}j_1 \tag{11.2}$$

In order to close the equation we must relate ρ_1 and j_1 to E_z.

The z component of the force equation for the electrons is given by

$$\frac{\mathrm{d}(m\gamma v)}{\mathrm{d}t} = -eE_z$$

where m is the electron rest mass and e is the magnitude of the electron charge. As before, the velocity is expanded to first order:

$$v = v_0 + v_1$$

where v_0 is constant. When this is substituted into the force equation, we obtain

$$m\gamma_0^3 \frac{dv}{dt} = m\gamma_0^3 \left(\frac{\partial v_1}{\partial t} + v_0 \frac{\partial v_1}{\partial z}\right) = -eE_z$$

With the $e^{i(\vec{k}\cdot\vec{r} - \omega t)}$ dependence of the first-order quantities, this becomes

$$v_1 = \frac{ie}{m\gamma_0^3(k_z v_0 - \omega)} E_z \tag{11.3}$$

The continuity equation:

$$\frac{\partial \rho}{\partial t} = \nabla \cdot \vec{j}$$

gives the necessary expression for ρ_1. When $\vec{j}$ is expressed in terms of the *zero*th and first-order charge densities and velocities, and if we keep only the first-order terms, we find

$$j_z = (\rho_0 + \rho_1)(v_0 + v_1) = j_0 + j_1$$

$$j_1 = \rho_0 v_1 + v_0 \rho_1$$

When substituted, we obtain

$$\rho_1 = \frac{iek_z\rho_0}{m\gamma_0^3(k_z v_0 - \omega)^2} E_z = \frac{i\omega_p^2 k_z}{4\pi\gamma_0^3(k_z v_0 - \omega)^2} E_z \tag{11.4a}$$

and

$$j_1 = \frac{i\omega_p^2 \omega}{4\pi\gamma_0^3(k_z v_0 - \omega)^2} E_z \tag{11.4b}$$

where we have used the plasma frequency $\omega_p^2 = 4\pi(-e\rho_0)/m$ (note that ρ_0 is negative). We now obtain the dispersion relation by combining (11.2) and (11.4):

$$\left(k^2 - \frac{\epsilon\omega^2}{c^2}\right)\left(k_z - \frac{\omega}{v_0}\right)^2 = \frac{\omega_p^2(k_z^2 - \epsilon\omega^2/c^2)}{\epsilon\gamma_0^3 v_0^2} \tag{11.5}$$

It is instructive to take the limits of the dispersion relation as the beam is turned off (beam density $\rightarrow 0$). These are

$$\omega_p^2 \rightarrow 0 \rightarrow \begin{cases} k^2 = k_z^2 + k_\perp^2 = \epsilon\omega^2/c^2 & \text{(A)} \\ k_z = \omega/v_0 & \text{(B)} \end{cases}$$

where k_z and $k_\perp$ are the longitudinal and transverse components of $\mathbf{k}$. The value of $k_\perp$ will be fixed if the system is enclosed in a cylindrical conducting tube. These limiting dispersion relations are shown in Figure 11.3. Curve A represents the electromagnetic wave and curve B represents the degenerate (since $\omega_p = 0$) space-charge modes of the electron beam. The beam modes can be studied more fully by recalculating with only pure longitudinal modes and $\omega_p \neq 0$, or by considering the $\omega \rightarrow \infty$, $\epsilon\omega^2/c^2 \neq k_z^2 \approx k^2$ limit of (11.5). In the latter case, space-charge modes are decoupled from the electromagnetic waves by examining regions where there is wide separation of the solutions of the dispersion relations. Either approach gives

$$\omega = k_z v_0 \pm \frac{\omega_p}{\gamma_0^{3/2}\epsilon^{1/2}}$$

and is shown in Figure 11.4. The lower branch is called the *slow space-charge mode,* and is so named because its phase velocity, when positive, is always less than that of the beam. The upper branch is called the *fast space-charge mode* for similar reasons.

When we examine the relation between the velocity (v) and the density ($n = \rho/(-e)$) for the slow space-charge mode by using (11.2) and (11.4), we find that the first-order terms are π out of phase, as shown in Figure 11.5. When the wave is absent ($v_1 = n_1 = 0$), the kinetic energy density is given by

$$mc^2(\gamma_0 - 1)n_0$$

When the slow space-charge wave is present, the kinetic energy is reduced because the density is higher where the velocity has been reduced, and the kinetic energy lower where the velocity is greatest. Such waves are called negative energy waves, and they offer the possibility of converting this kinetic energy into electromagnetic energy. The fast space-charge wave is a positive energy wave, and its kinetic energy increases as the wave grows.

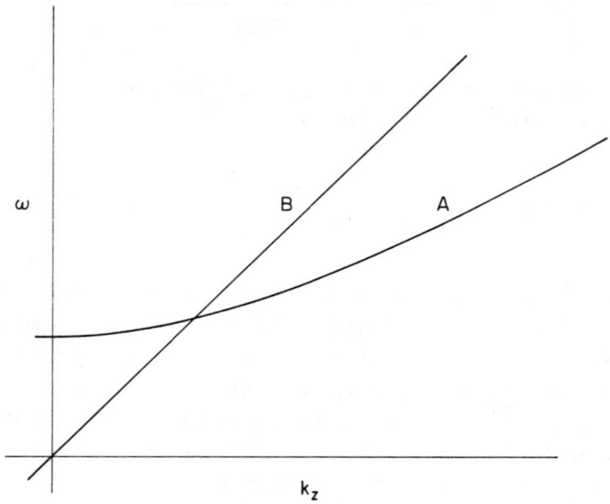

Fig. 11.3 Uncoupled Dispersion Curves
Note: Curve A is the electromagnetic mode and curve B is the uncoupled space-charge mode with $\omega_p \rightarrow 0$.

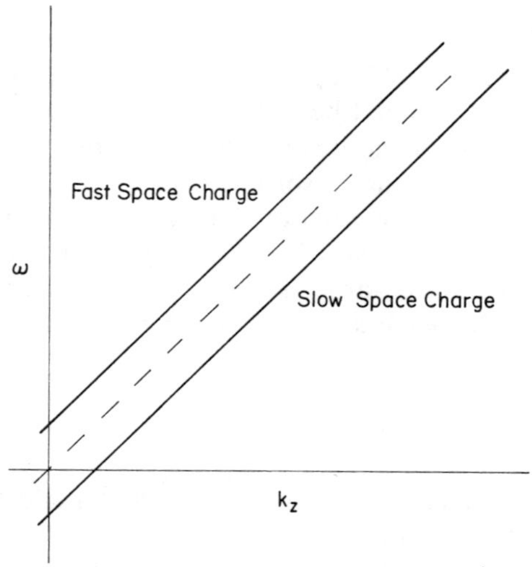

Fig. 11.4 Fast and Slow Space-Charge Modes

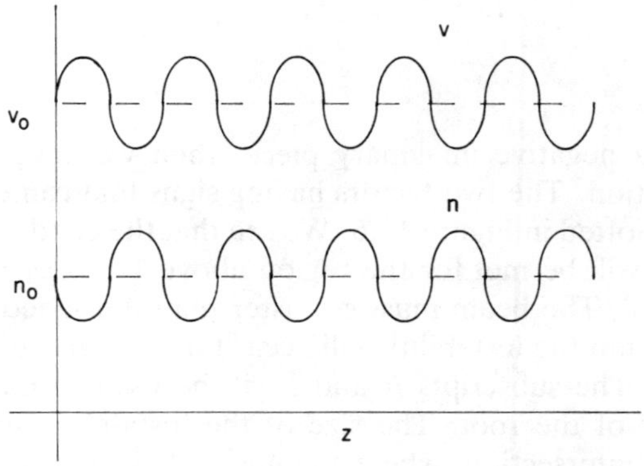

Fig. 11.5 Beam Number Density and Velocity for the Slow Space-Charge Mode

Note: The first-order corrections are π out of phase.

11.3 ANALYSIS OF THE DISPERSION RELATION

Next, we solve the dispersion relation given in (11.5). We can seek solutions for k_z real and ω complex (solutions growing or decaying in time), ω real and k_z complex (growing or decaying in space), or k_z and ω both complex (mixed). The choice depends on the boundary conditions for the solution of the particular problem. We will select the second choice. The solutions could be found numerically with great accuracy by using a computer, but we will try to gain a sense of their behavior by seeking approximate analytical solutions. For low beam density (ω_p small), we expect solutions similar to the uncoupled space-charge and electromagnetic modes given earlier. We seek solutions similar to those of our space-charge mode. Solutions similar to the electromagnetic modes could be obtained by the same approach, but are less interesting. If ω is taken as given and k_z is to be found, then we write [6]:

$$k_z = \omega/v_0 + \Delta$$

where Δ is small. When substituted into the dispersion relation, we obtain

$$\Delta^2 \approx \frac{\omega_p^2\omega^2(1 - \epsilon\beta_0^2)}{\epsilon\gamma_0^3 v_0^4(k^2 - \epsilon\omega^2/c^2)} \tag{11.6}$$

where

$$\beta_0 \equiv v_0/c$$

If Δ has a negative imaginary piece, then we have growth in the positive z direction. The two factors having signs that can effect the character of Δ are plotted in Figure 11.6. We see that the conditions for growth (and damping) will be met for the region above $1 - \epsilon\beta_0^2 = 0$ and below $k^2 - \epsilon\omega^2/c^2 = 0$. The beam must go faster than the speed of light in the medium ($c/\sqrt{\epsilon}$) and the instability will occur for $k^2 - \epsilon\omega^2/c^2 > 0$, as shown in Figure 11.7. The subscripts R and I will be used to indicate real and imaginary parts of the root. The size of the instability increases as one approaches the intersection, where the uncoupled space-charge and electromagnetic modes would have the same phase velocities (synchronism).

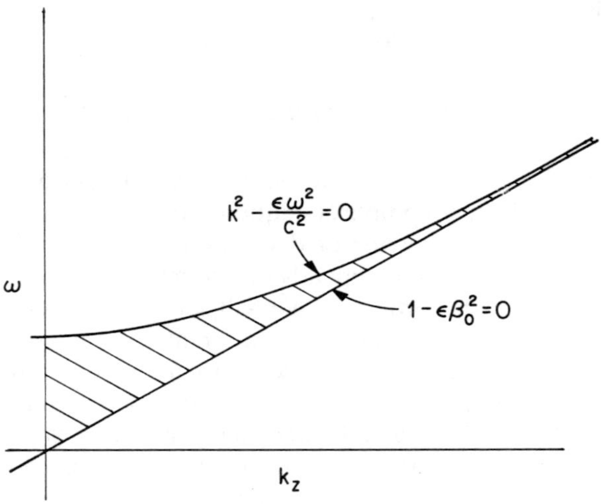

Fig. 11.6 Determination of the Instability Region from the Factors in the Dispersion Relation

Note: Beam lines passing through the shaded region are unstable.

At synchronism:

$$\omega/v_0 = (\epsilon\omega^2/c^2 - k_\perp^2)^{1/2}$$

and our earlier approximation fails. At this point, the approximate solution can be found by writing

$$k_z = \omega/v_0 + \Delta$$

$$k_z = (\epsilon\omega^2/c^2 - k_\perp^2)^{1/2} + \Delta$$

When substituted into the dispersion relation, we obtain

$$\Delta^3 = \frac{\omega_p^2\omega}{c^3}\frac{1 - \epsilon\beta_0^2}{2\epsilon\gamma_0^3\beta_0^3}$$

or

$$\Delta = \left(\frac{\omega_p^2\omega}{c^3}\right)^{1/3}\left(\frac{\epsilon\beta_0^2 - 1}{2\epsilon}\right)^{1/3}\left(\frac{1}{\beta_0\gamma_0}\right)\begin{cases} -1 \\ \frac{1}{2} + \sqrt{\frac{3}{2}}\,i \\ \frac{1}{2} - \sqrt{\frac{3}{2}}\,i \end{cases} \tag{11.7}$$

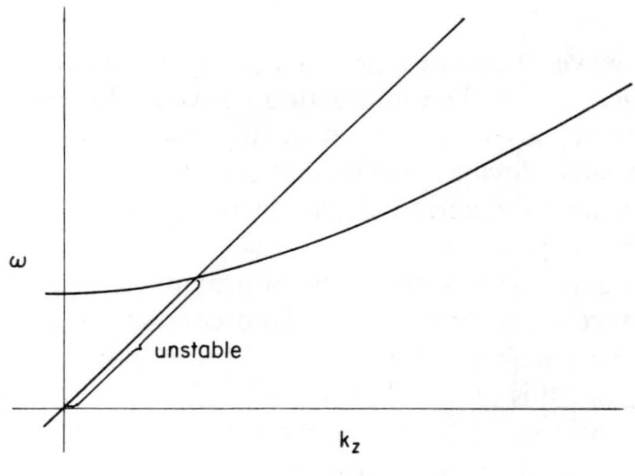

Fig. 11.7 Dispersion Relation of Space-Charge Modes to Indicate Unstable Section

These roots are represented graphically in Figure 11.8. Plots of the solution Δ_R and k_{zR} are shown in Figure 11.9. The growth rate $|\Delta_I|$ at synchronism ($\Delta_{I\max}$) *versus* beam velocity (β_0) is shown in Figure 11.10. The growth rate is seen to decrease at both extremes. As β_0 approaches 1 the growth rate goes to zero due to a relativistic effect. The growth relies on the bunching of the beam from the motion of the electrons in the z direction. As the electrons approach the speed of light, they strongly resist this motion (the relativistic mass increases) and the growth falls. In the other limit, as the beam velocity approaches the speed of light in the dielectric, the

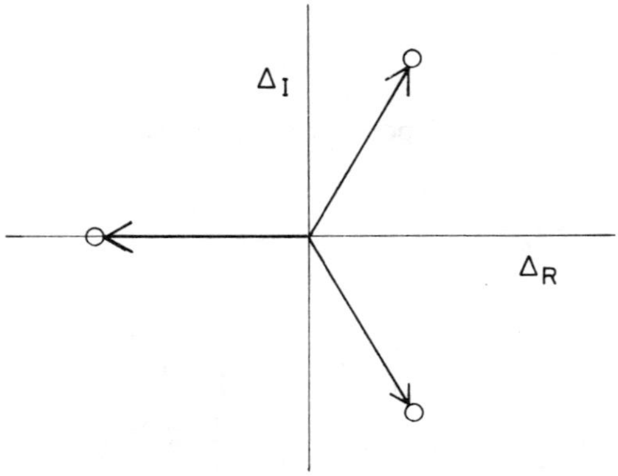

Fig. 11.8 Roots of the Dispersion Relation at Synchronism in the Complex Δ Plane

electromagnetic wave becomes close to a plane wave propagating in the z direction and $E_z/E \rightarrow 0$. The interaction between the electrons and the electromagnetic wave is by E_z, and hence approaches zero in this limit as well. Ideally, in such devices, we should avoid these limits, but, as we shall see, if we want to generate short wavelengths, then we must go to high beam velocities ($\beta_0 \approx 1$).

In order to gain some sense of the approximate size of the gain, we consider a structure consisting of two plane conducting plates separated by 1 cm of dielectric with constant $\epsilon = 4$. This is shown in Figure 11.11. In the lowest mode, this gives $k_\perp = \pi$ cm^{-1}. With a beam of $\beta_0 = 1/\sqrt{2}$ ($\approx$ 200 kV), we have $\gamma_0 = \sqrt{2}$. For a current of 10 A through 1 cm^2, we find $\omega_p \approx 3 \times 10^9$ rad/s. At synchronism $k_{zR} = \pi$ cm^{-1}, $\omega = 6.7 \times 10^{10}$ rad/s ($\approx$ 10 GHz) and the wavelength in free space (λ_{free}) would be 2.8 cm. With these values:

$$\Delta = .127 \times \begin{cases} -1 \\ .5 + .87\, i & (1/\text{cm}) \\ .5 - .87\, i \end{cases}$$

or $|\Delta_I| = .11$ 1/cm, which corresponds to an e-folding length of $\approx$ 9 cm. By allowing the beam to pass through the dielectric, we obtain better coupling than could be achieved in the more physical system described below. Despite this, careful calculations based on these physical systems reveal growth rates of the same order (although smaller) as those we have obtained. Again, the reader should note that the system described above

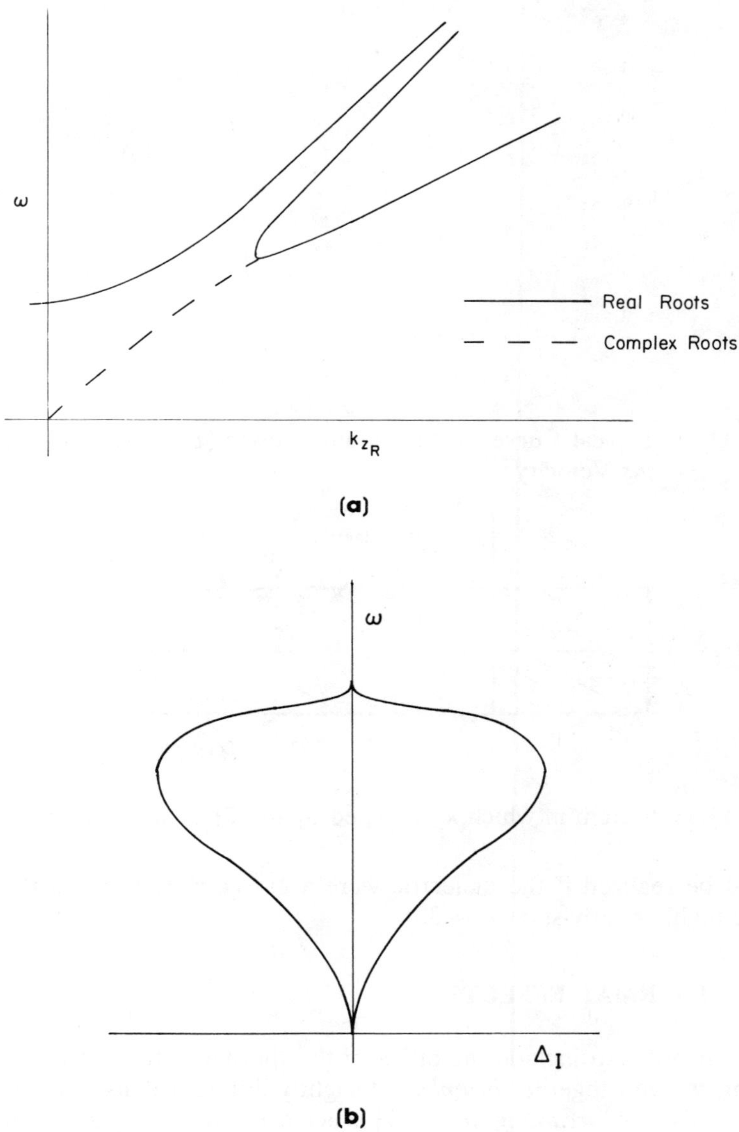

Fig. 11.9 Dispersion Curves for the System
Note: (a) shows ω *versus* the real part of k_z. Regions where the root is purely real are displayed with a solid curve and those where an imaginary part is present are indicated by a dashed curve; (b) shows the corresponding imaginary part.

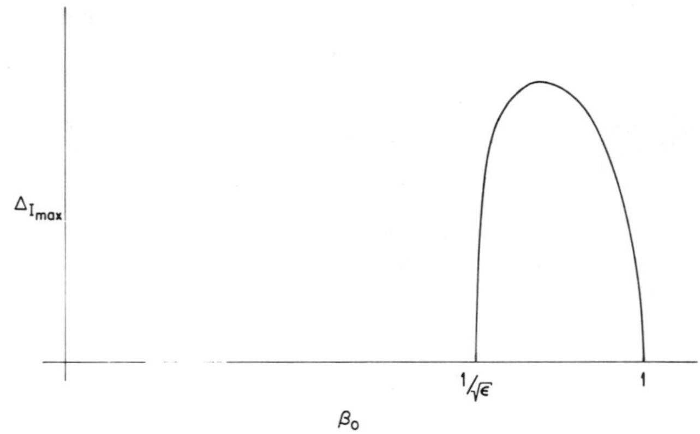

Fig. 11.10 Typical Curve for Maximum Growth (Δ_I at Synchronism) *versus* Velocity

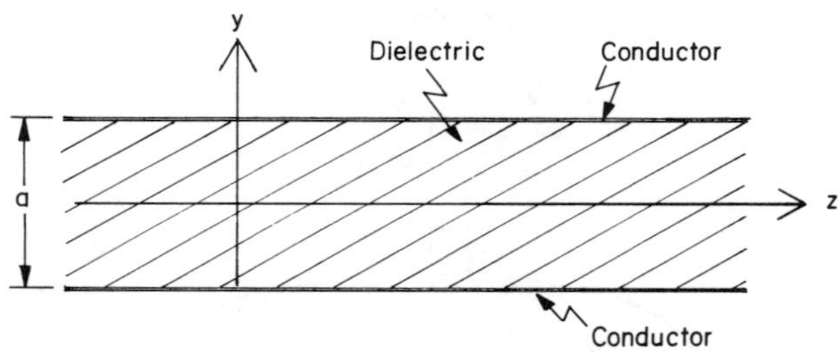

Fig. 11.11 System in which $k_\perp$ is Fixed by the Presence of Conductors

could be realized if the dielectric were a gas ($\epsilon \approx 1.001$) and the beam were highly relativistic ($\gamma_0 \approx 30$).

11.4 THERMAL EFFECTS

In order to include the effect of the thermal spread of the electron beam, we add together *beamlets* of slightly different densities and velocities. This *multistreaming model* [7] is achieved by the replacement:

$$\frac{\omega_p^2}{(k_z v_0 - \omega)^2} \to \omega_p^2 \int \frac{F(v)}{(k_z v - \omega)^2}\, dv$$

where $F(v)$ is the normalized distribution of initial beam velocities:

$$\int F(v)\, dv = 1$$

on the right-hand side of the dispersion relation in (11.5). When $F(v)$ is a Lorentzian [8], $|\Delta_I|$ varies with thermal spread as shown in Figure 11.12. When we superimpose a graph of the beam velocity distribution on the electromagnetic dispersion relation, as shown in Figure 11.13, we see that the *warmer beam* result agrees with the Landau result [9]. We find growth where dF/dv is positive, and damping where it is negative.

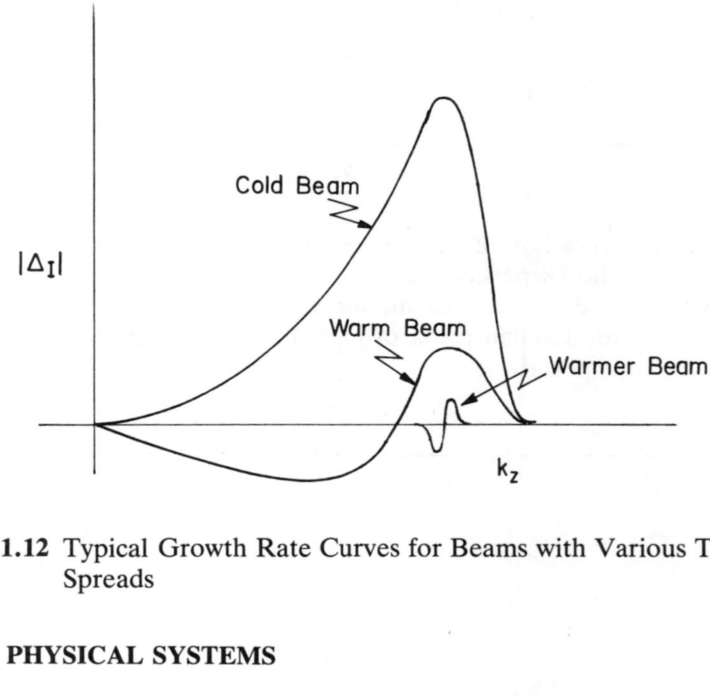

Fig. 11.12 Typical Growth Rate Curves for Beams with Various Thermal Spreads

11.5 PHYSICAL SYSTEMS

11.5.1 Dielectrically Loaded Waveguides and Periodic Structures

In a physical system, the beam cannot pass through the dielectric, except in extreme cases [10]. In order to obtain coupling, we must pass the beam near the dielectric and couple through the exponential tail of the electromagnetic wave. The z component of the electric field for such a system is shown in Figure 11.14. The e-folding length of this tail can be determined by writing the wave equation in vacuum:

$$\omega^2/c^2 = k_z^2 + k_\perp^2$$

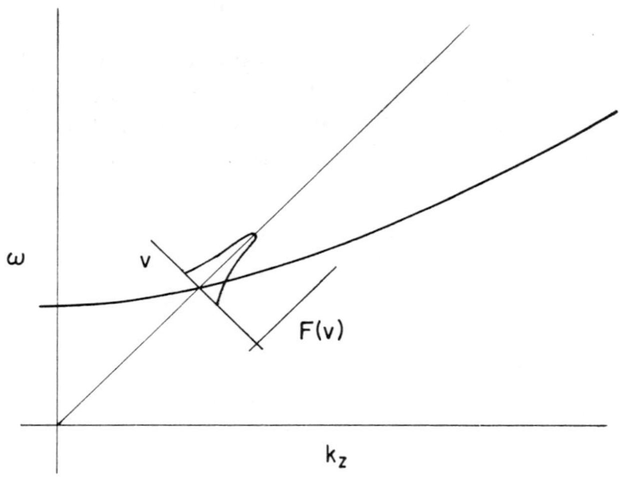

Fig. 11.13 Superposition of the Velocity Distribution Curve of the Beam and the Dispersion Curve

Note: With this we can see that the *warmer beam* curve shown in Figure 11.12 leads to the Landau result of growth where dF/dv is positive and decay where it is negative.

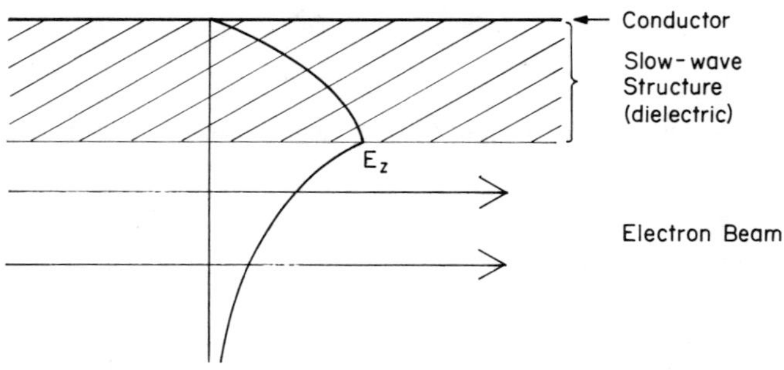

Fig. 11.14 Physical System where the Electron Beam Couples to the Wave through the Exponential Tail of E_z

and adding the synchronism condition:

$$\beta_0 = \omega/ck_z$$

This gives an *e*-folding length of

$$\Gamma = i/k_{\perp} = (\beta_0\gamma_0/2\pi)\lambda_{\text{free}} \tag{11.8}$$

For good coupling, we must have most of the beam within a few *e*-folding lengths of the dielectric. We can improve coupling by increasing γ_0, and if we are interested in very short wavelengths, then high γ_0 is mandatory. If the gap between the dielectric and the edge of the beam is s, we would expect a factor of $e^{-s/3\Gamma}$ in the growth rate at synchronism.

The most experimentally studied system is a circular waveguide with a dielectric liner [11]. In such a system, the field E_z in the hole is given by the modified Bessel function $I_0(pr)$ and its behavior [12] is very similar to the exponential behavior described above, when $pr > 1$. This work is discussed in detail by John Walsh in Chapter 12.

The structure that allows the electromagnetic wave and the electron beam to travel at the same phase velocity (slow-wave structure) need not be a dielectric. Consider a conducting wall that is periodic (with period d) in the z direction. The normal modes of this system will have fields of the form: $F(x, y)g(z)e^{i\omega t}$. The periodic boundary conditions must now be matched, and by Floquet's theorem $g(z)$ must be of the form:

$$\begin{aligned} g(z) &= e^{ik'z} \text{ (periodic with period } d) \\ &= e^{ik'z} \sum_{n=-\infty}^{+\infty} a_n e^{ink''z} \end{aligned} \tag{11.9}$$

where $k'' \equiv 2\pi/d$. If this system is similar to a cylindrical waveguide with only a slight dependence of the cross section on z (e.g., circular waveguide with slightly rippled wall), the solutions and dispersion relation for this system will be similar to that of the cylindrical guide. For the cylindrical guide (no z dependence of the cross section), we have

$$g(z) = e^{ik_z z}$$

with the dispersion relation:

$$\omega^2 = \omega_c^2 + k_z^2 c^2$$

where ω_c is the cut-off frequency. The phase velocity of this structure is always greater than c ($\omega/k_z > c$). The presence of the periodic structure demands the addition of the terms in (11.9). The system can now be thought of as supporting several pieces, or *submodes*, with phase velocities given by

$$v_p = \omega/k_z = \frac{\omega}{k' + nk''} \qquad (n = 0, \pm 1, \pm 2, \ldots)$$

when $a_n \neq 0$. If the ripples are slight, $k'(\omega)$ will be close to the cylindrical guide value $k_z(\omega)$. Many of the various n values give phase velocities that are slower than c which allow synchronism between the wave and the electron beam. The reader should note that these submodes are parts of one mode and other modes will have their own sets of submodes, which lead to a very cluttered dispersion relation [13]. If we achieve synchronism between the electron beam and one of these submodes, the result is much the same as with the dielectric slow-wave structure, often leading to growth. Because only one submode is involved in the extraction of the energy, which must, in turn, drive the entire mode (the boundary condition must be maintained), the growth rate will be small, unless a significant fraction of the wave energy is in the interacting submode. This requires that a_n for the interacting submode be comparable to the larger a_n values in the series of (11.9). A typical dispersion relation for such a system is shown in Figure (11.15). Operation at point 2, where we are close to synchronism for a range of frequencies, would be suitable for an amplifier. Such devices are called *traveling wave tubes* (TWTs). Operation at point 1, where curves cross sharply and they clearly determine the frequency, would be desirable for an oscillator. Because the group velocity ($d\omega/dk_z$) at this point is negative, such devices are called *backward-wave oscillators* (BWOs) [14]. The electromagnetic waves and the electron beam have several synchronisms, and we must carefully select the particular desired mode to excite and to reduce mode competition. This problem can be partially solved by placing the beam so that it coincides with the region of maximum E_z for the desired mode and weak E_z for the undesired modes.

The operation of all of these devices is very similar to the dielectric devices discussed above. All of these devices rely on the coupling between the slow space-charge mode of the electron beam and an electromagnetic mode, which is moving at a velocity less than that of light. The loss of gain at high γ_0, shown in Figure 11.10, is present. The argument leding to (11.8) holds, and the fields are found to fall off exponentially away from the structure.

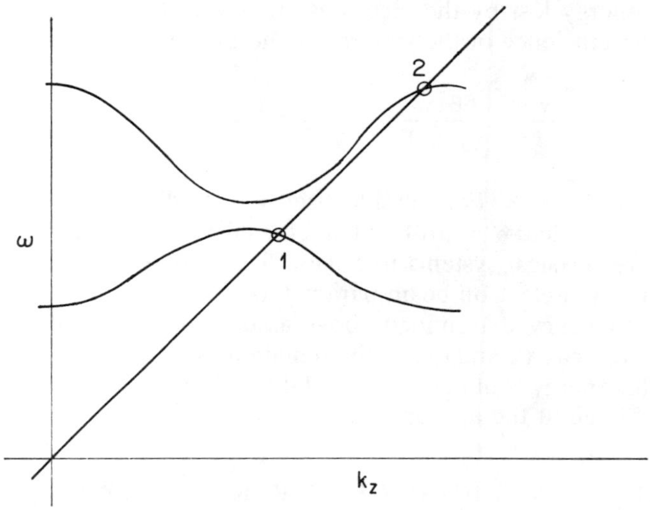

Fig. 11.15 Typical Dispersion Relation for Waveguide with Rippled Wall
Note: Point 1 corresponds to operation as a TWT. Point 2 represents operation as a BWO.

11.6 TRAPPING

The exponential growth determined above cannot indefinitely continue. At some point, the linear approximations made above will break down and growth will cease. One prominent effect leading to loss of growth is *trapping*. The growth is based on the assumption that the velocity is of the form:

$$v = v_0 + v_1$$

where v_0 is greater than the velocity of the electromagnetic wave. As the electromagnetic field grows, the wave will trap the electrons in the electrical potential associated with E_z. The amount of energy removed from an electron can be determined by comparing its original energy $mc^2\gamma_0$ and its energy when trapped $mc^2\gamma_w$ (where $\gamma_w \equiv (1 - v_w^2/c^2)^{1/2}$ and v_w is the velocity of the electromagnetic wave).

The energy lost by the electrons is converted into electromagnetic energy. The efficiency of the system for the process is

$$\text{eff} = \frac{\gamma_0 - \gamma_w}{\gamma_0 - 1} \approx \frac{\delta\beta_0\gamma_0^3}{\gamma_0 - 1} \tag{11.10}$$

where $\beta_0 = \beta_w + \delta$. For our idealized device discussed at the end of Section 11.3, we find $\delta = .014$ and an efficiency of 7%. Again, we should expect more physical systems to have efficiencies of the same order for similar wave and electron beam parameters.

The efficiency determined above assumes dumping of the trapped beam with velocity v_w and loss of the remaining kinetic energy. In practice, much of this energy could be recovered through the use of a suitable beam collector, biased at the appropriate potential.

11.7 COLLECTIVE AND SINGLE-PARTICLE GROWTH

Another way in which our idealized system differs from a real system is indicated in Figure 11.16. In a real system, the electron beam usually enters unbunched, and the interaction occurs in a finite region (length L), after which the electromagnetic wave and electron beam separate. This introduces the boundary conditions $\rho_1 = 0$ and $v_1 = 0$ at the beam entrance ($z = 0$). These boundary conditions are fit by mixing the modes. Because the conditions at $z = 0$ are to be maintained at all times, we must add modes with the same frequencies. Suitable modes are the forward and backward electromagnetic modes and the fast and slow space-charge modes. If we assume, as indicated in Figure 11.16, that no wave enters from the right, we must drop the backward electromagnetic mode. This leaves the three modes shown in Figure 11.17, all with the same ω, but different values of k_z. We consider the system shown in Figure 11.11 with no variation in the x direction. The total E_z, with the common time dependence suppressed, is given by

$$E_z = (E_1 e^{ik_{1z}} + E_2 e^{ik_{2z}} + E_3 e^{ik_{3z}})\cos k_\perp y \tag{11.11}$$

where $k_\perp$ is real, and E_1, E_2, and E_3 must be adjusted to fit the boundary conditions at $z = 0$. When this is done by using (11.3) and (11.4a), we obtain

$$E_1 = \frac{\Delta_3 - \Delta_2}{\Delta_2 - \Delta_1}\frac{\Delta_1^2}{\Delta_3^2}E_3 \tag{11.12a}$$

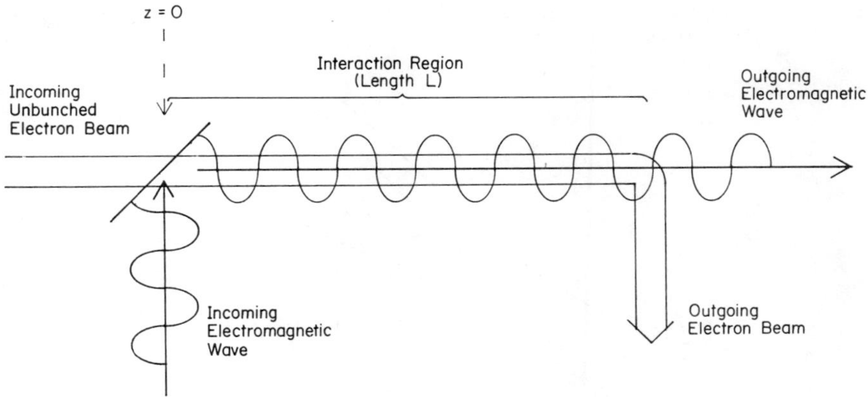

Fig. 11.16 System in which the Electron Beam and Electromagnetic Wave Interaction Occurs in a Finite Length

$$E_2 = \frac{\Delta_3 - \Delta_1}{\Delta_1 - \Delta_2} \frac{\Delta_2^2}{\Delta_3^2} E_3 \tag{11.12b}$$

where $k_j \equiv k_0 + \Delta_j$ and $k_0 \equiv \omega/v_0$.

In the appropriate regime, the slow space-charge mode $E_2 e^{ik_2 z}$ is unstable and will grow exponentially, while the fast space-charge mode $E_1 e^{ik_1 z}$ decays. The electromagnetic mode $E_3 e^{ik_3 z}$ is stable. If the interaction region is long enough and the growth rate large enough, then the growing term will dominate to give

$$E_z \approx E_2 e^{ik_2 z} \cos k_\perp y$$

which is the *collective regime.* If the growth rate is small, the slow space-charge term will not dominate. However, there may still be growth, even when the slow space-charge mode grows very slowly or not at all. Because the three modes have different propagation constants, they will beat against each other. We may have growth or damping, depending on the relative velocities of the three modes and the length of the interaction region. This is the *single-particle regime.*

In order to examine the flow of energy from the electron beam to electromagnetic energy in the single-particle regime, we consider the statement of conservation of energy for complex fields [15] of the form:

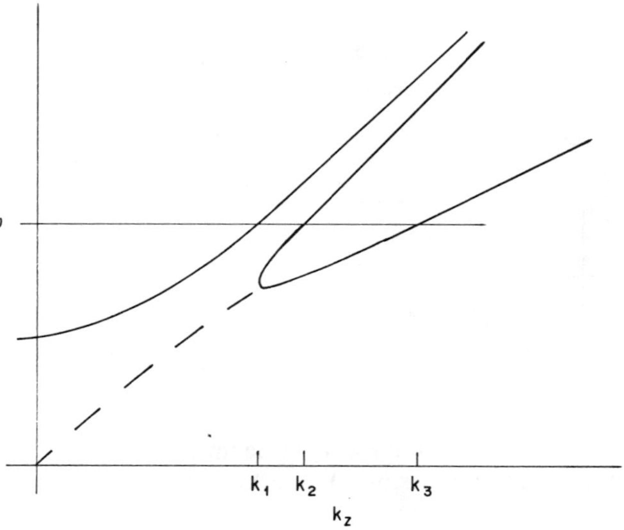

Fig. 11.17 Dispersion Relation to Indicate the Modes to be Mixed to Satisfy the $\rho_1 = v_1 = 0$ Boundary Conditions at $z = 0$

$$F(\vec{\mathbf{r}}, \mathrm{t}) = g(\vec{\mathbf{r}})\mathrm{e}^{i\omega t}$$

thus,

$$(1/2)\int \vec{\mathbf{j}}^* \cdot \vec{\mathbf{E}} dV + 2i\omega \int (W_e - W_m) dV + \oint \vec{\mathbf{S}} \cdot \vec{\mathbf{n}} dA = 0 \quad (11.13)$$

where

$$W_e \equiv (1/16\pi)\vec{\mathbf{E}} \cdot (\epsilon \vec{\mathbf{E}})^*, \quad W_m \equiv (1/16\pi)\vec{\mathbf{B}} \cdot \vec{\mathbf{B}}^*,$$
$$\vec{\mathbf{S}} \equiv (c/8\pi)\vec{\mathbf{E}} \times \vec{\mathbf{B}}^*$$

The real part of (11.13) is a statement of conservation of energy averaged over a period of the electromagnetic wave. Because ω, W_e, and W_m are real, this becomes

$$-\mathrm{Re}\frac{1}{2}\int \vec{\mathbf{j}}^* \cdot \vec{\mathbf{E}} dV = \mathrm{Re}\oint \vec{\mathbf{S}} \cdot \vec{\mathbf{n}} dA \quad (11.14)$$

The term on the left-hand side is the average power extracted from the electron beam, while the right-hand side is the average flow of electromagnetic power out of the volume, which we take as the interaction region.

Because ω is real for our system, the average energy within the volume is constant, and the result is the loss of the second term in (11.13) with the taking of the real part. We identify the right-hand side of (11.14) as the generated power. Because the left-hand side is easier to calculate, we obtain an expression for this power by calculating it. This is fairly straightforward by using (11.11), (11.12), and (11.4b), and the result is

$$-\mathrm{Re}\frac{1}{2}\int \vec{\mathbf{j}}^* \cdot \vec{\mathbf{E}}\,\mathrm{d}V = \frac{\omega_p^2 \omega L^3 a E_3 E_3^*}{16\pi\gamma_0^3 v_0^2} f(\Omega) \tag{11.15}$$

for the power per unit width in the x directions, where

$$f(\Omega) \equiv \frac{1}{\Omega}\left[\frac{\sin^2\left(\frac{\Omega}{2}\right)}{\left(\frac{\Omega}{2}\right)^2} - \frac{\sin\Omega}{\Omega}\right]$$

and

$$\Omega \equiv \Delta_3 L$$

In this derivation, we have assumed $\Delta_2 = -\Delta_1$ and $\Delta_1, \Delta_2 \ll \Delta_3$, which corresponds to the single-particle regime, where the exponential growth rate is small. If the reader wishes to fill in the algebra, he or she is cautioned against making the second approximation at too early a stage in the calculation. The factor $f(\Omega)$ is shown in Figure 11.18. This and similar results have been found by others using a variety of approaches [16].

If we wish to view this as a gain, we must compare it to the power that would flow through the system were the beam absent. The electromagnetic mode:

$$E_3 \cos k_\perp y\, \mathrm{e}^{ik_3 z}$$

carries a power per unit width of

$$\mathrm{Re}\int_{\text{cross section}} \vec{\mathbf{S}} \cdot \vec{\mathbf{n}}\,\mathrm{d}a = \frac{a\epsilon\omega k_3 E_3 E_3^*}{16\pi k_\perp^2} \approx \frac{a\epsilon\omega^2}{16\pi k_\perp^2 v_0} E_3 E_3^* \tag{11.16}$$

Hence, we define the gain (G) as the power generated per power flowing into the device, and obtain

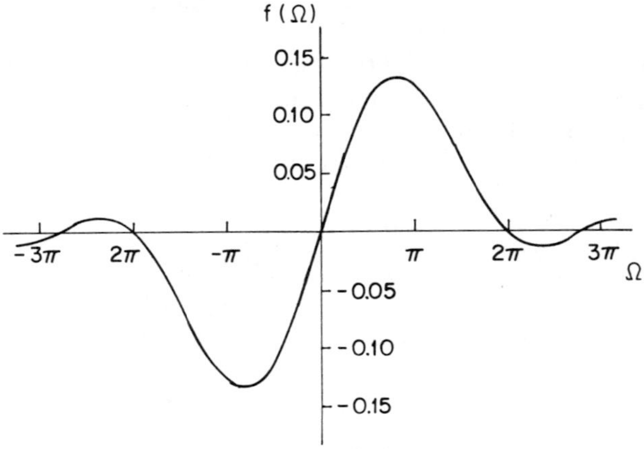

Fig. 11.18 Plot of the Function $f(\Omega)$

$$G = \frac{\omega_p^2 L^3 k_\perp^2}{\gamma_0^3 \epsilon \omega v_0} f(\Omega) \tag{11.17}$$

A parallel calculation can be carried out for the collective regime. We take E in the interaction region to be

$$E_2 \cos(k_\perp y) e^{ik_2 z}$$

and assume the electron beam and electromagnetic mode to be in synchronism. The power generated per unit width is

$$-\mathrm{Re}\frac{1}{2}\int \vec{\mathbf{j}}^* \cdot \vec{\mathbf{E}}\, \mathrm{d}V = \frac{3a\beta_0^2 \epsilon^{3/2} c E_2 E_2^*}{32\pi(\epsilon\beta_0^2 - 1)} (e^{\sqrt{3}\,|\Delta_2| L} - 1) \tag{11.18}$$

From (11.7) and (11.12) we see that $E_2 E_2^* = E_3 E_3^*$ at synchronism. The ratio of the power generated per power in absence of the beam now is

$$G = \frac{3\beta^3 c^2 \epsilon^{1/2} k_\perp^2}{2\omega^2(\epsilon\beta^2 - 1)} (e^{\sqrt{3}\,|\Delta_2| L} - 1)$$

In comparing the two regimes, we note, although the single-particle growth may give rapid initial growth, its ultimate gain in this amplifier configuration is limited to that which can be achieved by the beating of waves having magnitudes which are always equal to or less than that of the input wave. The collective growth is initially slow, but it has the potential to become considerable due to its exponential character. The ultimate gain in the collective regime is also limited due to nonlinear effects, such as trapping.

11.8 ANOTHER APPROACH—MOTIONS PERPENDICULAR TO v_0

From arguments presented earlier, we have seen that for short wavelengths we must consider beams of high energy (high γ). This led to the problem of loss of growth due to relativistic-mass effects, indicated in Figure 11.10. One way to avoid this would be to use motions in which v_1 was partially perpendicular to v_0, and such can be done by reducing B_0. The corresponding calculation is more involved, and the reader is directed to the reference cited in [8] for details. The system now exhibits *Cerenkov cyclotron instability* as well as the ordinary *Cerenkov instability* discussed above. As a special case, we note that for $B_0 = 0$ at synchronism [8, Sec. V]:

$$\Delta_\omega^3 = \frac{\omega_p^2 c^2\left[(1-\beta_0^2)\left(\frac{\epsilon\omega^2}{c^2} - k_z^2\right) + (\epsilon\beta_0^2 - 1)k_\perp^2\right]}{2\epsilon^2\gamma_0\omega}$$

$$\approx \frac{\omega_p^2 c^2 \beta_0^2(\epsilon - 1)k_\perp^2}{2\epsilon^2\gamma_0\omega}$$

The reader may note that this differs somewhat from our earlier Δ whereby k_z is real, $\omega = k_z v_0 + \Delta_\omega$, and the dispersion relation is solved [6] for Δ_ω. As we anticipated, this growth suffers less at high γ_0 than did our earlier system, described by (11.7). The growth is seen to rely on $k_\perp$. Physical arguments for this dependence are also given in the cited reference [8]. The problem arises when we note that the reason for considering high γ was to reduce $k_\perp$ in the beam region. We now see that this may well lead to weak growth for physical systems where the beam travels near the slow-wave structure. Such systems would need to be examined in more detail to see whether they hold more promise than those studied in earlier sections of this chapter.

REFERENCES

1. P.S. Cerenkov, *Dolk. Akad. Nauk. SSSR* **2**, 451 (1934).

2. E. Curie, *Madame Curie* (Heinemann, London, 1941).
3. I.M. Frank and Ig. Tamm, *Dokl. Akad. Nauk. SSSR* **14**, 109 (1937).
4. O. Heaviside, *Phil. Mag.* (1888): Feb., p 130; Mar., p. 202; May, p. 379; Oct. p. 360; Nov., p 434; Dec., p. 488.
5. The author is using Gaussian (cgs) units. The results (11.6), (11.7), (11.10) and (11.17) can be read in MKS units once we make the identification: $\omega_p^2 = -e\rho_0/(m\epsilon_0)$.
6. Had we sought ω for k_z fixed, the results would be similar because the dispersion relation is approximately of the form $F(\omega - k_z v_0)$; setting $\omega = k_z v_0 + \Delta_\omega$ would give solutions: $\Delta_\omega = -v_0\Delta$.
7. T.H. Stix, *The Theory of Plasma Waves* (McGraw-Hill, New York, 1962).
8. Details of this calculation for a more general system are given in W.B. Case, R.D. Kaplan, J.E. Golub, and J.E. Walsh, *J. Appl. Phys.* **55**, 2651 (1984).
9. L.D. Landau, *J. Phys. USSR* **10**, 25 (1946).
10. J.R. Neighbours, F.R. Buskirk, and A. Saglam, *Phys. Rev. A* **29**, 3246 (1984).
11. E.P. Garate and J.E. Walsh, *IEEE Trans. Plasma Sc.* **PS-13**, 524 (1985).
12. M. Abramowitz and I.A. Stegun, *Handbook of Mathematical Functions* (US National Bureau of Standards, Washington, DC, 1964).
13. V.I. Kurilko, V.I. Kucherov, A.O. Ostrovskii, and Yu.V. Tkach, *Sov. Phys. Tech. Phys.* **24**, 1451 (1979).
14. J.A. Swagle, J.W. Poukey, and G.T. Leifeste, *Phys. Fluids* **28**, 2882 (1985); R.A. Kehs, A. Bromborsky, B.G. Ruth, S.E. Graybill, W.W. Destler, Y.C. Carmel, and M.C. Wang, *IEEE Trans. Plasma Sc.* **PS-13**, 559 (1985).
15. J.D. Jackson, *Classical Electrodynamics*, 2nd Ed. (John Wiley and Sons, New York, 1975), p. 242.
16. W.B. Colson, Ph.D. Thesis, Department of Physics, Stanford University, Stanford, CA (1977); A. Yariv and C. Shih, *Opt. Comm.* **24**, 233 (1978); A. Grover and Z. Livni, *Opt. Comm.* **26**, 375 (1978); J.E. Walsh and J.B. Murphy, *IEEE J. Quantum Electr.* **QE-18**, 1259 (1982).

Chapter 12

CERENKOV MASERS: EXPERIMENT

John Walsh

Department of Physics and Astronomy
Dartmouth College
Hanover, New Hampshire

12.1 INTRODUCTION

A review of the experimental work on stimulated Cerenkov sources will be presented in this chapter. The term *Cerenkov radiation* is most often used to describe the wake produced by a charged particle moving with superluminal velocity in a dielectric medium [1, 2]. This emission by a single particle is a spontaneous emission process, but Cerenkov emission can also be stimulated, like all radiation processes. A stream of particles moving in a dielectric material or through a dielectrically loaded resonator will interact with its own spontaneously emitted radiation and produce further stimulated emission [3, 4]. If the strength of the beam or the feedback is sufficient, the beam will bunch on a scale of the wavelength of the emitted radiation and power can be extracted from the system. Sources based on this principle have been operated in the centimeter, millimeter, and far-infrared wavelength ranges.

The earliest reference to electromagnetic wake radiation seems to be the work of Heaviside [5], who was considering the possibility of developing a practical light source using superluminal electrons. A second independent analysis, which, like Heaviside's, assumed that the electron velocity could exceed the velocity of light *in vacuo,* was performed by Sommerfeld [6], and some earlier experimental observations are attributed to M. Curie [7] and to M. Mallet [8]. Cerenkov, in a series of classic experiments, was the first to identify clearly the physical characteristics of the radiation that bears his name, and the subsequent analysis by Frank and Tamm [9] took proper account of the constraints imposed by the special theory of relativity.

The work of Cerenkov and that of Frank and Tamm were approximately coincident with the early stages of microwave tube development. Hence, it was natural to consider this effect as the basis of a practical millimeter-wavelength radiation source, and a large number of theoretical analyses were completed as well as a somewhat smaller number of experiments related to this proposition. A number of further references can be found in [3] and [10]. Among the early experiments cited, the work of Coleman [11] was particularly interesting in that it employed a dielectric coupler driven by a relativistic electron beam. We will show in Section 9.2 that Cerenkov sources work best when beams of moderately high energy are used to drive a resonator. In fact, the occasional mention [12] of dielectric slow-wave structures in the older microwave tube literature is generally accompanied by a statement that such a device is not of interest. However, a more critical evaluation will show that at higher beam energies, materials with low relative dielectric constant can be employed to obtain high coupling impedance. This advantage is particularly apparent at shorter wavelengths.

The remainder of this chapter consists of four sections. In Section 12.2, we present the scaling relations needed to appreciate the experimental advantages and disadvantages of Cerenkov sources. A review of experimental results is given in Section 12.3, and Cerenkov sources are compared with other slow-wave devices in Section 12.4. Conclusions are reached in Section 12.5.

12.5 SCALING RELATIONS FOR CERENKOV MASERS

A typical Cerenkov source consists of five major components [13, 14]: a high-voltage, pulse-forming network; an electron gun; a dielectrically loaded resonator; a beam dump; output coupling elements. These components are shown in schematic form by Figure 12.1. The *pulse-forming network* (PFN), which may be either a lumped or distributed system, is generally terminated in either a high-voltage pulse transformer or a transmission line. The details of such hardware are discussed elsewhere and need not be repeated here. In general, any pulse generator capable of producing voltage levels in the 100–500 kV range will work well for a Cerenkov source in the millimeter-to-submillimeter spectral regime. At shorter wavelengths, somewhat higher voltages are needed in order to maintain adequate coupling. At still shorter wavelengths (far-infrared),

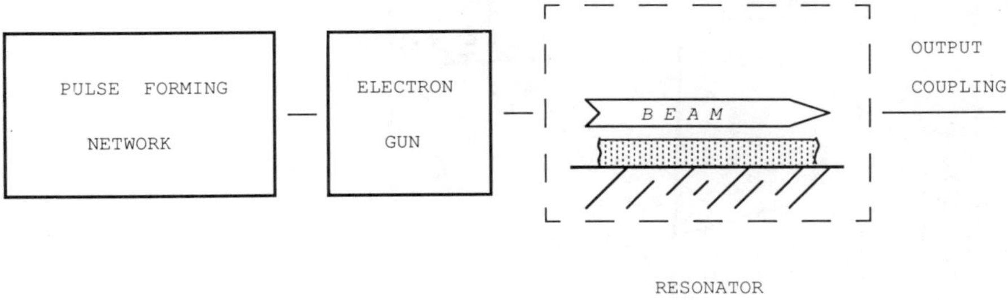

Fig. 12.1 Components of a Cerenkov Source

radio-frequency accelerators can also serve as beam sources, but these systems will not be considered in this chapter [15].

After formation, the mildly to moderately relativistic electron beam drifts along the axis of a dielectrically loaded resonator. In the millimeter-to-submillimeter range, this can be an annular dielectric structure lining the wall of a cylindrical drift tube. A sketch is shown in Figure 12.2. In an azimuthally symmetric structure, the TM modes [16] have an axial component of electric field, which couples to the electron beam. The wave propagates at a velocity less than the speed of light and slightly less than the velocity of the beam. When these conditions are met, the beam bunches in the retarding phase of the field, and gives up energy to the wave. We will consider the further details of this process after we have discussed the dispersion relation of a typical structure.

A sketch of the dispersion relation for a cylindrical drift tube loaded with an annular dielectric is shown in Figure 12.3. This curve can be used to illustrate a number of the basic features of a Cerenkov source. In addition to the dispersion curve itself, which is

$$D(\omega, k, \text{geometry}) = 0 \tag{12.1}$$

Figure 12.1 contains the light line:

$$\omega = ck \tag{12.2}$$

the beam line:

$$\omega = ck\beta \tag{12.3}$$

where $\beta = v/c$ (v is the beam velocity and c is the speed of light), and the light line in the dielectric:

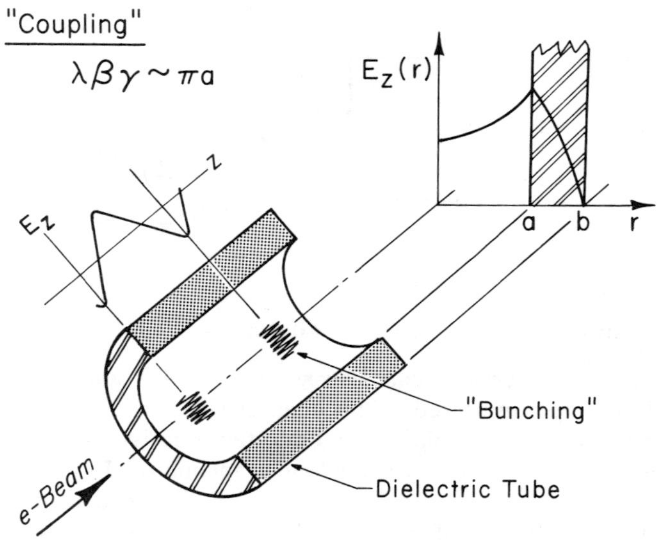

Fig. 12.2 Typical Resonator Cross Section for a Millimeter-Wavelength Cerenkov Oscillator
Note: Also shown are the profile of the axial component of the electric field and the coupling criterion.

$$\omega = ck/\sqrt{\epsilon} \tag{12.4}$$

The cut-off frequency ω_{co} is determined by the transverse dimensions of the structure, the ratio of the cross section that is filled with dielectric, and the relative dielectric constant of the liner material. The numerical value of ω_{co} will, of course, lie between that of an empty and filled (with dielectric constant ϵ) structure. In the region between the cut-off and the light line, the phase velocity of the waves determined by the curve $D = 0$ are fast waves (i.e., they propagate at greater than light velocity). The curve $D = 0$ crosses the light line at a point determined by the geometry and the value of $\sqrt{\epsilon}$. However, unlike the cut-off frequency, the value of which is primarily fixed by the overall transverse dimension, the light-line crossing can be made to occur at frequencies above cut-off, as the filling factor decreases. Hence, operation at high frequency in the fundamental mode of a structure with a convenient transverse dimension is possible.

In the region beyond the light-line crossing, the dispersion curve describes slow waves. They must be evanescent in the channel region. These waves have a minimum on the axis and increase as they approach

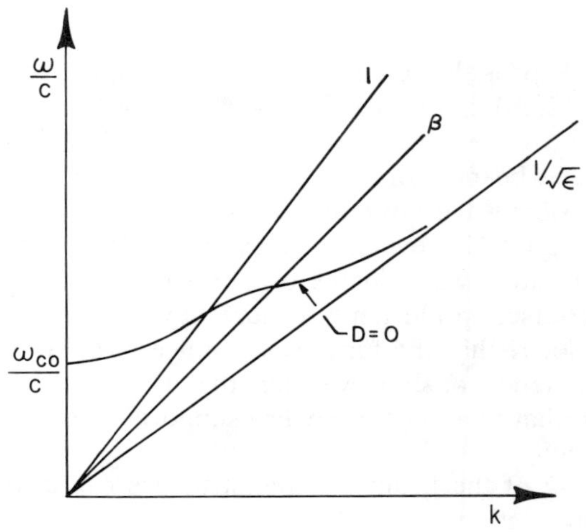

Fig. 12.3 Dispersion Relation for the Uncoupled Beam and Resonator Modes in a Dielectric-Lined Waveguide
Note: The curves labeled 1, β, and $1/\sqrt{\epsilon}$, respectively, are the light line, the beam line, and the light line in the dielectric.

the dielectric interface [16]. In a cylindrical structure, such waves are described by modified Bessel functions and in rectangular geometries, these waves will be hyperbolic or exponential.

The transverse wavenumber:

$$q = \sqrt{k^2 - \frac{\omega^2}{c^2}} \tag{12.5}$$

which determines the scale length of the evanescence, is independent of the choice of geometry (as well as the value of ϵ).

The coupling between an electron beam propagating in the channel region and the modes of the structure will peak near the point where the beam and wave phase velocities are equal. By invoking (12.3) (synchronism), (12.5) can be rewritten in a form that is more convenient for intuitive discussion:

$$q = k/\gamma \tag{12.6a}$$

or

$$q = \omega/c\beta\gamma \tag{12.6b}$$

where $\gamma = 1\sqrt{1 - \beta^2}$ is the relative energy of a beam electron.

Equation (12.6b) can be used to illustrate one of the principal differences between Cerenkov masers and more traditional slow-wave sources. When the beam energy is low, $\beta\gamma \approx \beta$, and the transverse dimension of the slow-wave structure must be small compared with the operating wavelength. However, when $\beta\gamma \gtrsim 1$, this condition is reversed and operation at short wavelengths becomes possible. These kinematic arguments, of course, would apply to any slow-wave structure, but the smooth-bore dielectric has the further advantage of operating on a single space harmonic. Hence, at short wavelengths, its coupling impedance will generally exceed that of a structure which supports many space harmonics at a given frequency.

The coupling of the beam and the structure can also be described with a dispersion relation:

$$D(\omega, k, X_b (\omega, k), \text{geometry}) = 0 \tag{12.7}$$

where

$$X_b = -\frac{\omega_p^2}{\gamma^3} \frac{1}{(\omega - kv)^2} \tag{12.8}$$

is the beam susceptibility (ω_p^2 is the beam plasma frequency). In general, the beam supports two modes, the fast and slow space-charge waves.

The nature of the dispersion relation of the coupled system can be determined from the following argument [17]. At modest beam densities, the dispersion is

$$D = (\omega - \omega_{0k}) \frac{\partial D^0}{\partial \omega} + \frac{\partial D}{\partial X_b} \cdot X_b \tag{12.9a}$$

where

$$D^0 \equiv D|_{X_b=0} \tag{12.9b}$$

and $\omega = \omega_{0k}$ is the solution of $D = 0$ in the absence of the beam (i.e., the unperturbed dispersion relation). Near synchronism (12.3), the roots of (12.9a) are well approximated by

$$\frac{\omega}{kv} - 1 = C\frac{(-1 \pm i\sqrt{3})}{2}, \quad -C \tag{12.10}$$

where

$$C^3 \equiv \frac{\omega_p^2}{\gamma^3(\omega)^3} \frac{\partial D/\partial X|_{x=0}}{\partial D^0/\partial\omega|_{\omega=\omega 0k}} \tag{12.11}$$

The dimensionless parameter C is defined in a manner consistent with traditional traveling wave tube (TWT) theory. The coupled-beam dielectric-resonator system operates in the single-particle or Compton limit. This regime is important for far-infrared and shorter wavelength Cerenkov free-electron lasers, but at millimeter and submillimeter wavelengths, the modes are resolved and given by (12.10). In a system that is long enough for a growing wave to become dominant, the gain per pass will be

$$G = 47\ CN - 9.54 \quad \text{(dB)} \tag{12.12}$$

where $N \equiv kvL/2\pi$ is the number of periods and L is the length of the resonator. The factor of -9.54 dB appears because it is assumed that the power in the initial incident wave is divided equally among the three forward-moving modes [18, 19].

The dispersion diagram for the coupled system is shown on Figure 12.4. The region of instability extends from the origin to the point where the beam and the waves are nearly synchronous. The gain also increases monotonically and reaches a maximum near, but slightly below, the point of synchronism, then rapidly falls to zero. In the wavelength range where the inevitable gap (e.g., δ) between the beam and the dielectric is small ($q\delta \ll 1$), the gain will peak at very high frequency ($qa > 1$, where a is the channel radius). As the operating wavelength is decreased, for example, by decreasing the filling factor, the point of maximum gain will move toward lower frequencies. Axial velocity spread of the beam may also limit the peak gain at short wavelengths. Typical voltage tuning ($D = 0$, $\omega = kv$) and gain curves for a millimeter-wavelength device are shown on Figure 12.5.

The Cerenkov masers discussed in the next section have all been operated as oscillators. At longer wavelengths, this is not the traditional method of using a traveling wave source. However, in the millimeter region, the resonator contains many axial wavelengths and at high gain,

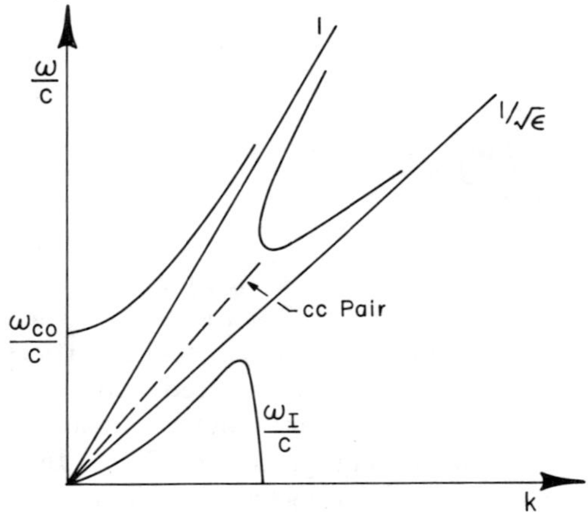

Fig. 12.4 Dispersion Curves for a Coupled-Beam Resonator System
Note: The qualitative features of these curves are independent of the resonator geometry.

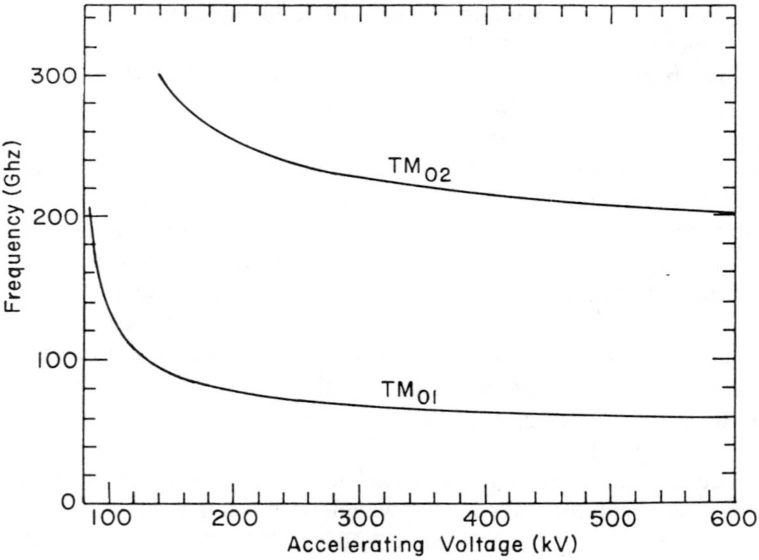

Fig. 12.5 Typical Beam Energy Tuning Curves for a Millimeter-Wavelength Cerenkov Oscillator.
Note: The TM_{01} and TM_{02} curves are both shown.

we can obtain a smooth voltage-tuned output. The output power is given approximately by

$$\frac{I_{(L)}}{I_{sat}} = \frac{[\exp(2\alpha L)] - |\rho|^2/3}{[\exp(2\alpha L)] - 1} \tag{12.13}$$

where

$$2\alpha L = 2\sqrt{3\pi}\, CN \tag{12.14}$$

and $|\rho|^2$ is the reflectivity of the output coupler. Threshold for the oscillator is defined by

$$\exp(2\alpha L) = 1/3\, |\rho|^2 \tag{12.15}$$

and the saturation level is given approximately by

$$I_{sat} \approx 4CP_b \tag{12.16}$$

for modest C ($\lesssim 0.1$). The experimental behavior of the Cerenkov maser oscillators discussed in Section 12.3 is in good agreement with the predictions of these general scaling relations.

12.3 CERENKOV MASER EXPERIMENTS

The parameters for a number of Cerenkov maser experiments [13, 14, 17, 20, 23] are summarized in Table 12.1. These may be conveniently divided into two groups, according to the type of electron beam generator. Entries 1, 5, 6, and 7 employed a *Marx generator and pulse-forming-line* combination, which was terminated in a field-emission diode. Beams of 500–1000 kV with many kA of current are generated by these sources. In a typical Cerenkov experiment, only a portion of the beam current produced by this pulse generator is transported through the resonator. The second type of generator closely resembles the systems used in conventional high-power klystrons [14]: a pulse-forming network is discharged by an *H thyratron* into the primary of a pulse transformer. The high-voltage pulse is delivered to an anode-cathode gap, and thus forms the beam. In these systems, a thermionic cathode is preferred because the voltage pulse may be 1–2 μs in duration. The pulse length of the intense beam source is typically 10–20 ns. Unlike the intense beam source, the tranformer-driven systems can be operated at modest repetition rates. We will discuss representative examples of each type of system.

Table 12.1 Cerenkov Maser Experiments

Beam		*Power*	*Wavelength*	
Voltage (kV)	*Current* (A)	*P* (kW)	λ (mm)	*Ref.*
750–100	500	10^3	4	[13]
100–200	2–20	10	10–3	[14]
100–200	2–20	250–25	10–3, 1	[20]
100–200	2–20	20–10	3–2, 1	[17]
500–100	50–100	10^3–25	2–1	[21]
500–100	10–50	10^2	1–.375	[22, 23]
1000	10^3	—	cm	[24]

The general layout of a transformer-driven system is shown in Figure 12.6. A beam with a current that can be as high as several tens of amperes is formed on an oxide-coated cathode and accelerated across a single gap. The cathode is immersed in a magnetic guide field that confines the beam and helps to transport it through the cylindrical resonator. In an experiment of this type, the resonator is typically 10–20 cm long, and materials such as fused quartz or boron nitride are used for the dielectric liner. Both of these have relative dielectric constant of $\epsilon \approx 4$, and hence a Cerenkov threshold of $\beta \approx 0.5$ (88 kV). Thus, a pulse generator operating in the 100–200 kV range is well matched to these structures.

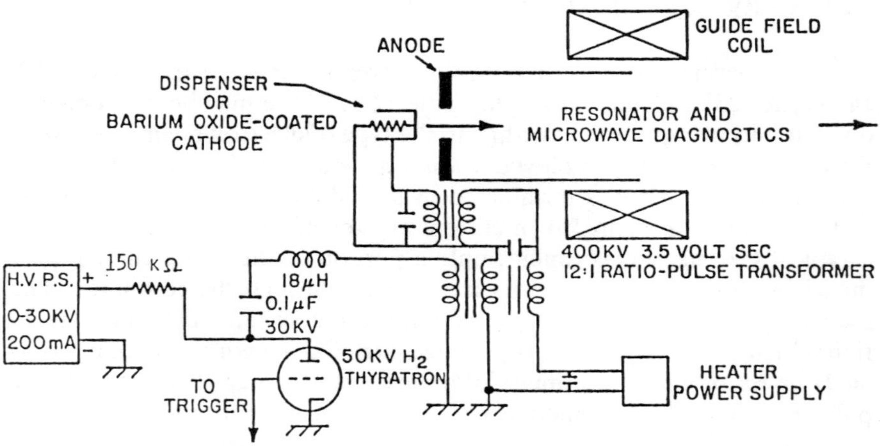

Fig. 12.6 Layout of a Pulse-Transformer-Driven Millimeter-Wavelength Cerenkov Source

A channel diameter of 2–3 mm and a liner thickness of about 1 mm will produce radiation in the 3 mm wavelength range. The experimentally determined tuning curve is shown in Figure 12.7. The open rectangles are measured frequencies. A mechanically swept Fabry-Perot interferometer was used for this purpose because the wide tuning range made the use of a super heterodyne system inconvenient. The dark line is the theoretically expected frequency-voltage-tuning curve for a resonator of the given dimensions. It is a plot of $D^0 = 0$ along the constraint imposed by synchronism. At higher voltages the absolute agreement is very good. As the voltage approaches threshold, however, there appears to be a systematic departure from the theoretically expected result. This may be due to the buildup of some small residual charge on the insulating dielectric. If this buildup occured, it would have the effect of lowering the beam voltage and shifting the experimentally observed points to the right in the figure. At the reptition frequencies employed in these experiments (1–10 Hz), it was not possible to observe directly any charge buildup. On some resonators, a very thin conducting layer added to the surface did not degrade the Cerenkov output, but there were no extensive tests of coated resonators.

The high-voltage pulse and a typical Cerenkov output power curve are shown in Figure 12.8. These pulse are, after attenuation, detected with ordinary millimeter-wavelength diodes. The observed power levels at about 100 GHz are approximately 120 kW from a 10 A, 180 kV beam. The estimated efficiency, 6–7%, is consistent with that expected from a saturated TWT operating with the stated parameters (Section 9.2). Higher efficiencies would be obtained at longer wavelengths, and, if desired, the phase velocity profile of the structure could probably be tapered in such a way as to enhance the saturation level, and hence the efficiency. The unexpended electron beam energy could also be partially recovered if a system of depressed collectors were used in the ground return. However, efficiency enhancement schemes were not investigated during the course of these proof-of-principle experiments.

Operation on higher order (TM_{02}) modes was also observed with the transformer-driven systems. The highest frequencies reached in this way were about 340 GHz. The highest fundamental-mode operation was approximately 140 GHz, but this was a limit set by the mechanical precision necessary to fabricate a boron nitride resonator. On the basis of the obtained results, it is thought that a pulse-transformer-driven system operating in the 200–250 kV range would work well at a wavelength around 1 mm.

The intense-beam sources used in Cerenkov maser experiments operated at higher voltages. Values between 500 kV and 1.1 mV were typical.

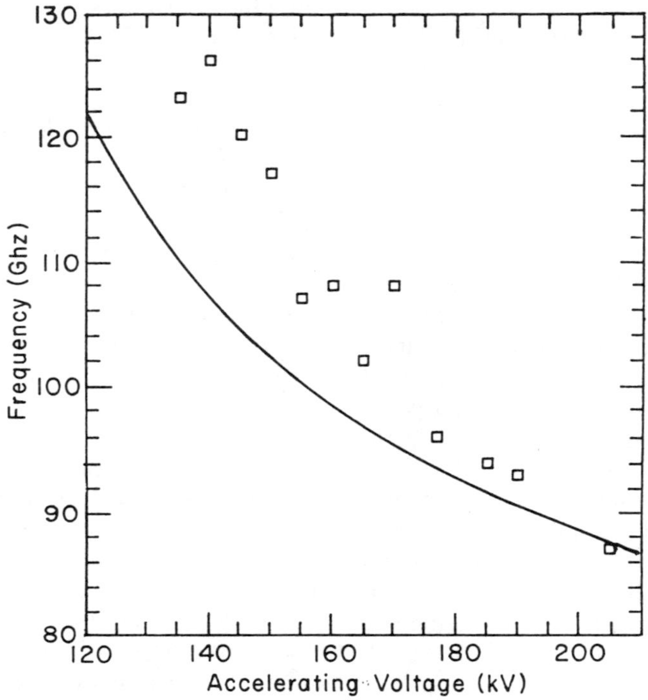

Fig. 12.7 Experimental Tuning Curve for the Source of Figure 12.6. *Note*: The solid line is the theoretical curve for the TM_{01} curve. The inner and outer radii are 1.97 mm and 2.47 mm. The dielectric material is boron nitride ($\epsilon = 4.2$).

A generator of this type would be used to produce either very high levels of Cerenkov radiation in the longer millimeter and centimeter-wavelength ranges or fundamental-mode output at shorter wavelengths. One example of the short-wavelength experiment will be discussed below.

The anode-cathode geometry and the layout of an intense-beam experiment are shown in Figure 12.9. Again, a solenoidal magnetic field was used to transport the beam. The measured frequency-beam-voltage-tuning curve is shown in Figure 12.10. In this case, the frequency is measured both with cut-off filters and with a millimeter-wavelength grating spectrometer. The liner material was PTFE tubing with a 250 μm wall thickness and 35 cm length. Shorter lengths were not tried, but it is suspected that this length was greater than that needed to obtain saturation. The measure of frequencies and the expected tuning are again in very good agreement.

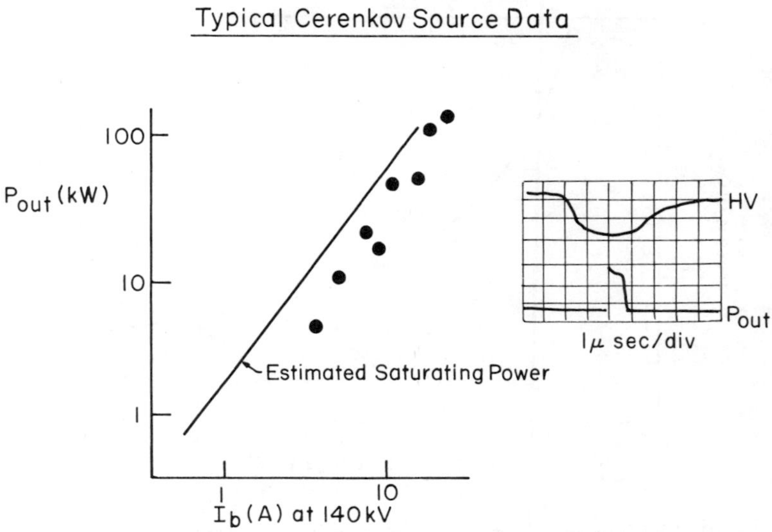

Fig. 12.8 Typical Output Pulse and Power Curve from the Source of Figure 12.6.
Note: Detectors were IN53 diodes in WR-28 mounts.

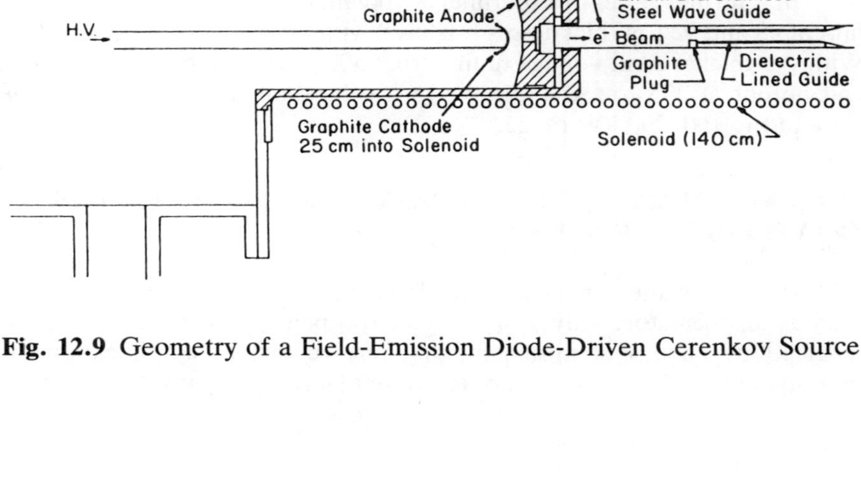

Fig. 12.9 Geometry of a Field-Emission Diode-Driven Cerenkov Source

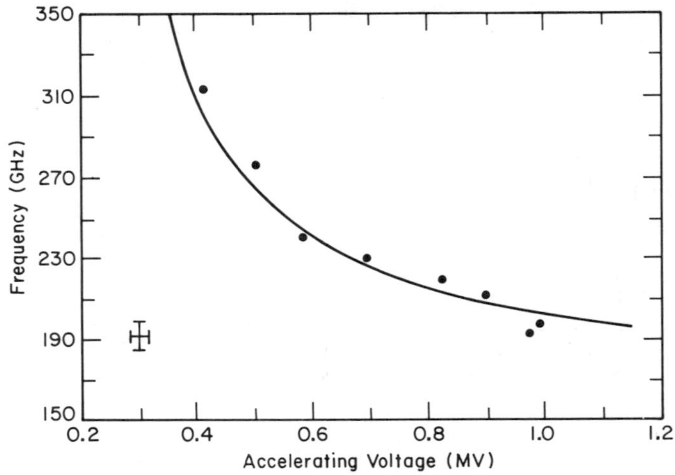

Fig. 12.10 The Tuning Curve for the Device Illustrated in Figure 12.9. *Note*: The solid curve is again the expected TM_{01} tuning. The dielectric material was thin-wall PTFE tubing.

Furthermore, unlike the lower energy results, there is no systematic deviation at lower relative voltages. These experiments are, of course, inherently single-shot and the pulse length was about 10 ns, and very little liner charging is expected during the pulse duration. The estimated efficiency in these experiments was below that observed in the transformer-driven system. Typical levels were several hundred kW (1 mV, 60 A beam). The lower efficiency may be a result of higher beam-voltage ripple and the generally uncontrolled nature of the experiment.

In a further series of experiments, a planar resonator lined with TPX film of 48 and 92 μm thicknesses was used with this same beam generator. With the thicker film, voltage tuning from 500 μm to 1 mm was observed, and output at 375 μm was observed with the thinner film. Power levels were in the 100–200 kW range.

12.4 COMPARISON OF THE CERENKOV SOURCE WITH OTHER TRAVELING WAVE DEVICES

In most of the experiments to date, the Cerenkov source has been used as an oscillator. Waves on the electron beam grow from noise, are reflected by an output mismatch, and receive further amplification on subsequent passes. The process proceeds until it reaches saturation. During the growth period, the system is dominated by the exponentially growing mode. However, the other two forward-moving modes cannot be completely neglected because they contribute to the decoupling and the time required to reach saturation.

The operation of the Cerenkov source is obviously very similar to the conventional traveling wave tube amplifier (TWTA), although that device is not generally used as an oscillator. At longer wavelengths, the TWT oscillator would have a step-tuning characteristic, which would be set by the axial mode spacing. In the Cerenkov oscillator, however, the gain (in the millimeter region) is high, the beam velocity is synchronous with the electromagnetic mode velocity over a wide range, and the number of axial modes is large. These conditions taken together give the device a smooth, voltage-tuned output. Unlike the periodic structures used in the traditional TWT, the smooth-bore Cerenkov resonator does not couple backward modes to the electron beam. Hence, there is not an operating regime of the Cerenkov oscillator that is analogous to the BWO.

The suggestion that an electron beam and an electromagnetic wave traveling in near synchronism could be used as the basis of a practical source was originally made by Kompfner [25, 26, 27]. Initially, TWT development was motivated by the need for microwave amplifiers with bandwidths that were much greater than the ones that could be obtained from multicavity klystrons. The helix, a structure which will support waves traveling at less than the speed of light, also has a region in its dispersion diagram where the velocity of an electron beam and the wave can be matched over a range with a large fractional bandwidth ($\Delta\omega/\omega$). The helical structure is well suited for TWT operation because, at longer wavelengths, the fractional bandwidth can exceed an octave.

Following its introduction, traveling wave sources were the object of intense development programs, and as a result of this work a number of now standard devices became available. At longer wavelengths, the centimeter through the upper millimeter range, the TWT and the BWO are generally regarded as a mature technology, although improvements are still forthcoming. An authoritative guide to these developments is the literature provided by the tube manufacturers, and an excellent general review of millimeter traveling wave sources has been prepared by R. Espinosa [28].

Commercially available millimeter-wavelength TWT are based on either the helix or a series of coupled cavities. The helical structure has the advantage of greater bandwidth, but the coupled-cavity method generally provides a higher level of output power. In either case, the transport of the electron beam is facilitated by either a solenoidal magnet or a *periodic permanent magnet* (PPM) array. In addition to the power the solenoid consumes, it is generally heavier than the PPM system, but the highest power outputs require the solenoid.

A typical pulsed, high-power TWT might produce 150 kW at 10 GHz when driven by a 65 kV electron beam [29]. Furthermore, because the product of the power and the frequency in an optimized design will be roughly constant, power levels of 15–20 kW would be expected at 100 GHz. This level is somewhat optimistic when compared with tube manufacturers' specifications, but it is well below the levels that can be produced with a laboratory [21] Cerenkov source.

The reader must note, however, that the Cerenkov source in the laboratory case was operating at a beam voltage of 140 kV, or twice that used in the coupled-cavity tube (CCTWT) used in the example. If the gun in the CCTWT was space-charge-limited, a doubling of the voltage would result in a five-to six fold improvement beam power. Optimistically assuming that a constant efficiency can be maintained as the operating frequency is increased from 10 to 100 GHz, the output power would increase by the same amount. This would make the output of the hypothesized 100 GHz CCTWT comparable to the Cerenkov source.

In discussing this comparison, the reader should also note that the structure for a 100 GHz CCTWT would be very difficult to produce, whereas the dielectric resonator for a Cerenkov source (in this spectral range) is simple. The dielectric resonator thus enjoys a significant potential advantage for short-wavelength, high-power operation. The principal disadvantage of the Cerenkov source is the performance degradation that might result if the beam were not carefully transported through the dielectric resonator. Charging of the dielectric could be a particularly difficult problem for high-repetition-rate or long-pulse operation. A possible solution for this problem has been investigated in the laboratory. In some experiments, a thin conductive coating added to the dielectric did not seriously affect the Cerenkov oscillator. However, it is not yet known whether the addition of such a coating would be a satisfactory solution for dielectric charging problems at any wavelength and pulse repetition rate.

The Cerenkov oscillator should also be compared with the BWO. We already pointed out that the dielectric resonator couples only forward modes, but the device nevertheless produces a voltage-tuned output. In this respect, the dielectric resonator is very similar to the BWO. If it is based on the coupling of an electron beam to a helix, the BWO is an excellent source of tuned radiation at centimeter and longer millimeter wavelengths. Although the output power is often rather low, because the peak wave amplitude occurs near a minimum of the current modulation (i.e., near the beam input), the BWO is a very useful source for diagnostic purposes or local oscillator applications. At shorter wavelengths, the helical BWO becomes difficult to fabricate, but other slow-wave structures can be put to use. When a grating is used, the device is often called a *carcinotron*, and sources of this type have been operated up into the THz range.

The carcinotron, which is usually operated with modest beam voltages (20–80 kV typically) on a backward mode, is generally a low-power source (<1 W). Grating-based structures [30], however, if operated with higher voltages on a forward mode, can also produce short wavelengths and higher output power levels. A number of variations have been investigated, such as the *millitron* [31] and the extended-interaction oscillator [32], which are design evolutions based on the CCTWT. Others, such as the *oroton* [33–35], are based on a grating-loaded quasioptical resonator. This approach, if used with an electron beam of sufficient energy (see (12.6), Section 12.2) could produce output at very short wavelengths. In an experiment that has since been repeated by a number of researchers, Smith and Purcell [36] obtained spontaneous emission at visible wavelengths from a grating and a 250 kV beam. It appears likely that an oscillator based on this concept could be made to produce useful levels of output power, well down into the submillimeter or even the infrared wavelength regions.

Grating-coupled sources operating on a forward mode are virtually identical to the Cerenkov sources. The dielectric structure will couple the beam to a single space harmonic, and hence will have a slightly higher coupling strength than the grating that supports many space harmonics at the same frequency. However, a metal grating may be more rugged than a dielectric film, and, of course, will not allow any build up of charge. Thus, the use of gratings may be a superior coupling method for CW or high-repetition-rate sources.

12.5 CONCLUSIONS

The data presented and discussed in Section 12.2 support the claim that the Cerenkov oscillator is a good source of coherent radiation at millimeter and submillimeter wavelengths. In this spectral region, the dielectric structure has characteristics that make it superior to the coupled-cavity and the helix, both in tuning ability and power-handling capacity. We also noted that the good performance of the Cerenkov source stems, in part, from the fact that it has been operated at voltages which are two to three times higher than those typically used in commercially available TWTs. In general, at short wavelengths, the performance of any traveling wave source will improve if the operating voltage and current are increased in proportion. However, the performance difference of the Cerenkov and coupled-cavity TWT do not depend entirely on the higher operating voltage. The smooth-bore, single-mode dielectric resonator has a high coupling impedance at short wavelengths and is easy to fabricate.

When CW or high pulse repetition rates are needed, dielectric charging might limit performance of Cerenkov sources and more work is needed on this aspect of the design. In addition to the possibility of using thin conductive coatings to bleed the charge at a reasonable rate, we mentioned the possibility of substituting a grating for the dielectric. Periodic structures, in general, have a long history of application in microwave sources, but gratings have been used primarily in low-power, backward-wave devices. However, when used with higher-energy beams coupled to a forward mode, periodic structures have characteristics that are very similar to the dielectric and may be superior in some applications.

Finally, we should mention that in addition to the oscillator experiments, which involve natural bunching of an electron beam, there has been a recent series of interesting experiments [37–39] wherein the Cerenkov radiation from a prebunched beam has been observed. The immediate goal of this work has not been the production of radiation for other experiments, but rather the examination of the basic process, which might be described as *enhanced spontaneous emission*. This work is similar in spirit to earlier experiments [40–44], but electron beams produced by accelerators have replaced those generated with conventional high-voltage supplies. The emission from the prebunched beam is sensitive to its emittance and energy spread, and therefore such experiments may lead to the development of beam-diagnostic techniques.

ACKNOWLEDGEMENTS

The work described in this chapter has been supported by a number of agencies. The US Army Research Office in Durham, NC, deserves particular mention for supporting both millimeter and far-infrared Cerenkov laser development under contracts DAAG29-83-K-0018 and DAAG29-85-K-0176. In addition, the US Air Force Office of Scientific Research provided support for millimeter experiments under contract 82-0168, as did the US Office of Naval Research for Cerenkov-Raman laser development under contract N00014-79-C-0760. Finally, I should mention that the first experiments on stimulated Cerenkov radiation were jointly supported by Dartmouth College and the US Army Harry Diamond Laboratory under Contract DAAG39-78-C-0032 with a pulse transformer that was provided by the Stanford Linear Accelerator Center (SLAC).

REFERENCES

1. P.A. Cerenkov, *Dok. Akad. Nauk. SSSR* **2**, 451 (1934).
2. J.D. Jackson, in *Classical Electrodynamics*, 2nd Ed. (New York: John Wiley and Sons, 1975) pp. 638–641.
3. J. Walsh in *Novel Sources of Coherent Radiation*, S. Jacobs, M. Sargent, and M. Scully, eds. (Addsion-Wesley, Reading, MA, 1978), pp. 357–381.
4. J. Walsh, in *Free-Electron Generators of Coherent Radiation,* S. Jacobs, M. Sargent, H. Pilloff, M. Scully and R. Spitzer, (Addison-Wesley, Reading, MA, 1980), pp. 255–322.
5. O. Heaviside *Phil. Mag.* (1888): Feb., p. 130; Mar., p. 202; May, p. 379; Oct., p. 360; Nov., p. 434; Dec., p. 488.
6. A. Sommerfeld, *Gotting Nachricht* **99**, 363 (1904).
7. E. Curie, *Madame Curie* (London: Heinemann, 1941).
8. C.R. Mallet, *Acad. Sci.* (Paris) **183**, 274; **186**, 222; **188**, 445.
9. I. Frank, and I. Tamm, *Dok. Akad. Nauk. SSSR* **14**, 109 (1937).
10. J. Van Bladel, *Relativity and Engineering* (Berlin: Springer-Verlag, 1984).
11. P. Coleman, and J. Enderby, *J. Appl. Phys.* **27**, 631 (1956).
12. A.H.W. Beck, *Space Charge Waves*, (New York: Pergamon Press, 1958).
13. J. Walsh, T. Marshall, and S. Schlesinger, *Phys. Fluids* **20**, 709 (1977).
14. K. Felch, *et al., Appl. Phys. Lett.* **38**(8), 601 (1981).
15. J. Walsh, C. Shaughnessy, and B. Johnson *Nucl. Instrum. Meth.* **A250**, 308 (1986).
16. D. Marcuse, *Theory of Dielectric Optical Waveguides* (New York: Academic Press, 1974).
17. E. Garate *et al., J. Appl. Phys.* **58**(2), 627 (1985).
18. J. Pierce, *Traveling Wave Tubes* (New York: D. Van Nostrand, 1950).
19. J. Slater, *Microwave Electronics* (New York: D. Van Nostrand, 1948).
20. S. Von Laven, *et al., Appl. Phys. Lett.* **41**(5), 408 (1982).
21. E. Garate, et al., *Appl. Phys. Lett.* **48**(20), 1326 (1986).
22. E. Garate, et al., paper submitted to *Phys. Rev. Lett.* (1987).
23. E. Garate, C. Shaughnessy, and J. Walsh, *Proc. 8th Int. Free-Electron Laser Conf.*, Glasgow, Scotland, September 1986.
24. R.T. Bush, *Bull APS* **31**, 1482 (1986).
25. R. Kompfner, *Wireless World* **52**, 369 (1946).

26. R. Kompfner, *Proc. IRE* **35**, 124 (1947).
27. R. Kompfner, *Proc. IRE* **35**, 111 (1947).
28. R. Espinosa, *Microwave System Design Handbook* (1983), pp. 126–150.
29. *Microwave System Design Handbook* (1985), pp. 71–82.
30. J. Walsh, T. Buller, B. Johnson, G. Dattoli, and F. Ciocci, *IEEE J. Quantum Electr.* **QE-21**(7), 920 (1985).
31. B. James, *Microwave System News and Communication Technology* (1986) p. 105.
32. Varian Associates Tube Catalogue.
33. R. Leavitt, D. Wortman, and C. Morrison, *Appl. Phys. Lett.* **35**(5), 363 (1979).
34. K. Mizuno, S. Ono, and Y. Shibata, *IEEE Trans. Electron Dev.* **20**, 749 (1973).
35. E.M. Marshall, P. Mukhopadhyay Phillips, and J. Walsh, paper submitted for the IEEE International Conference on Plasma Science, Arlington, VA, June 1987.
36. S. Smith and E. Purcell, *Phys. Rev.* **92**, 1069 (1953).
37. The dielectric structure also has a high coupling impedence at longer wavelengths, but this spectral region has not been explored in the experiments performed to date.
38. J. Neighbours, F. Buskirk, X. Maruyama, *J. Appl. Phys.* (1987).
39. F. Buskirk and J. Neighbours, *Phys. Rev. A* **28**, 1531 (1983).
40. X. Maruyama, *Proc. Conf. on Small Accelerators,* to be published in *Nuclear Instrumentation and Methods* (North Holland, Amsterdam, 1987).
41. G. Mourier, *Travaux du Congrès International Tubes Hyperfréquences*, Paris, 1956, pp. 132–135.
42. J.G. Linhart, *ibid.*
43. H. Lashinsky, *J. Appl. Phys.* **27**, 631 (1956).
44. R. Ulrich, *Z. Phys.* **199**, 171 (1967).
45. M. Danos, *J. Appl. Phys.* **26**, 2 (1953).

Chapter 13
VIRTUAL CATHODE OSCILLATOR (VIRCATOR) THEORY

Donald J. Sullivan

Mission Research Corporation
Albuquerque, New Mexico

John E. Walsh

Department of Physics and Astronomy
Dartmouth College
Hanover, New Hampshire

Evangelos A. Coutsias

Department of Mathematics and Statistics
University of New Mexico
Albuquerque, New Mexico

13.1 INTRODUCTION

Since the discovery of the Child-Langmuir relation [1, 2], it has been known that exceeding the limiting current of a diode leads to the development of a *virtual cathode*. Subsequently, numerous papers were written on experiments and theory relating to space-charge-limited flows. Birdsall and Bridges [3] provide an excellent background and bibliography. More recently, the exact steady-state solutions for electron beams in one-dimensional relativistic diodes [4] and bounded drift spaces [5] were derived in the US and USSR. We can easily see that for sufficiently large currents there are two steady states for an electron beam, only one of which is stable [6]. These two states coalesce at the *space-charge limit* (SCL), and above it they disappear [7]. As current is increased above the SCL, the beam developes a jump instability and relaxes into an oscillating state.

In the early 1960s, computer models were developed, which quantitatively depicted the nonlinear oscillatory nature of the virtual cathode [8–10]. These were one-dimensional, non-relativistic, electrostatic, multiple-sheet models. Birdsall and Bridges [8, 9] qualitatively pointed out many interesting dependencies of the oscillation frequency and the potential minimum position on injected current, thermal spread, and circuit resistance. Dunn and Ho [10] present computer experiments with one and two species.

The phenomenon of virtual cathode formation in intense relativistic electron beams figures prominently in a number of high-interest research areas. Devices used to produce high-current ion beams for inertial confinement fusion—*pinch reflex diode* [11, 12] and *reflex triodes* [13, 14]—depend on the virtual cathode to inhibit electron transport, and use the cathode's potential well to accelerate ions. The recent concept of the *spherical electron-to-ion converter* [15] requires a virtual cathode.

The virtual cathode plays a dominant role in areas other than production of light ion beams for fusion. The virtual cathode is attributed with the main role in collective ion acceleration in neutral gas [16, 17]. Control of virtual cathode motion is the mechanism for acceleration in the *ionization front accelerator* [18, 19] and also in two concepts for *collective-effect accelerators* [20, 21]. However, the virtual cathode's most interesting application is its use in producing high-power centimeter and millimeter wavelength microwaves [22–36]. Experiments using reflex triodes have already produced GW power levels with 12% beam-to-RF efficiency [25]. This application of the virtual cathode will be considered in the present chapter.

The *virtual cathode oscillator* (vircator) in its modern form is an essentially new class of microwave tube. The vircator has the attribute of conceptual simplicity, high-output-power capacity, and wide tuning ability. Experimental work [22–25, 31–36] and the results of particle code simulations [26–31] clearly indicate that the vircator is a source of extraordinarily high power. Some of these experimental results are listed in Table 13.1. Hence, the vircator has some unique applications. The long-term success of the vircator will be measured in part by the position that the device takes in the competitive ranks of standard microwave tubes [37], but in the shorter term the demonstrated success of preliminary experiments assures that vircators will be used for things other than microwave tube research.

The development of sources in the microwave range of the spectrum is a pursuit that almost dates from the origin of modern electromagnetic theory [38], and precursors [39–41] of the modern vircator can be found

Table 13.1 Results of Vircator Experiments

Reference	*System*	*Peak Power*	*Frequency**	*Efficiency*
Mahaffey, *et al.*	Reflex Triode	100 MW	11 GHz (10.0–12.4 GHz)	1.5%
Brandt, *et al.*	Reflex Triode	1 GW	9.8 GHz (7.0–13.0 GHz)	2.0%
Buzzi, *et al.*	Foil Diode	1 GW	10 GHz (9.0–14.0 GHz)	1.25%
Didenko, *et al.*	Reflex Triode	1.4 GW	3.3 GHz (2.1–5.0 GHz)	12%
Scarpetti, *et al.*	Foil Diode	300 MW	6.9 GHz (5.3–12.4 GHz)	1.6%
Clark *et al.*	Foil Diode	1 GW	2–10 GHz Tuned (1.4–18.0 GHz)	2.0%
Didenko, *et al.*	Reflex Triode	600 MW	3.3 GHz	50%

*Detector bandwidth in parentheses.

in the early short-wavelength results of literature in vacuum tube research. During the course of 40 to 50 years of intensive research and development in microwave technology, a very great variety of conceptual approaches have been introduced for the production of electron-beam-generated coherent radiation. A relative few of these have survived, been reduced to practice, and put in a form that can be routinely used in applications. Recent additions to the family are gyrotrons [42] and, possibly, the general class of free-electron lasers [43, 44].

Each successful addition is based in part on a development that relaxes a constraint or expands the general parameter space available for electron beam bunching. In the case of the vircator, the availability of convenient high-power, high-voltage, low-impedance electron beam generators plays a key role. Single-pulse generators of very high peak power have been used in vircator research [22–25, 31–36], and compact, high-power, high-repetition-rate pulsers with suitable beam parameters have been demonstrated for vircator operation [45]. The development of these pulsers is essential to more routine use of the vircator mechanism, and therefore it has a good chance of becoming a standard approach in some applications [46].

The performance of microwave tubes can be summarized in part by a number of general scaling laws. One of these, which is useful in assessing the comparative performance of the vircator, is the relation between output power and operating frequency:

$$P_o = P_r \left(\frac{f_r}{f_o}\right) \alpha$$

where the subscripts r and o stand for reference and output, respectively. The parameter α [47] is a positive number, which for standard microwave tubes, lies in the range of 2–4. If a reference frequency, $f_r = 10$ GHz is adopted, high-power klystron or magnetron technology will typically have P_r in the .1–1 MW range. Experiment [45] has shown that, when driven by a power supply of comparable size, a vircator can easily achieve a power (at 10 GHz) near or above the upper end of this range. Furthermore, the vircator in this range also has a demonstrated wide tuning ability (mechanical tuning of 50% bandwidth has been achieved). Neither the klystron nor the magnetron have this tuning capability. High-power coupled-cavity TWTs have a pulsed peak power limit near the lower end of the stated output power range, and sources of this type will have a more limited bandwidth. Helical TWTs will have a high fractional bandwidth, but the power level available at 10 GHz is typically much smaller (10 kW). However, the TWT is generally used as an amplifier, and vircator-amplifier operation has not yet been demonstrated by experiment.

In a closed-cavity tube (klystron, magnetron, coupled-cavity TWT), the parameter α is generally near the middle to upper end of the range $\alpha = 2$–4, while α is generally smaller in the more open structure (at least one dimension much greater than the operating wavelength). The vircator has a potentially large mode volume [35] (operating wavelength cubed greater than or equal to the interaction volume), and hence it to may decline with frequency at a comparatively slow rate. However, the scaling of the vircator mechanism has not yet been demonstrated over multidecades in frequency. Hence, its competitive position with respect to the gyrotron is not yet fully established, and gyrotron sources, when operated at millimeter wavelengths may exceed by as much as two orders of magnitude the output power available from other standard tubes. As compared with a gyrotron, however, a vircator is based on a simpler technology, and hence may be an attractive option for some millimeter-wavelength applications.

In summary, the results of research to date support the claim that a vircator can perform at or above the level of standard microwave tubes. Anticipated power levels and operating frequencies of the vircator relative to other short-pulse microwave tubes are shown in Figure 13.1.

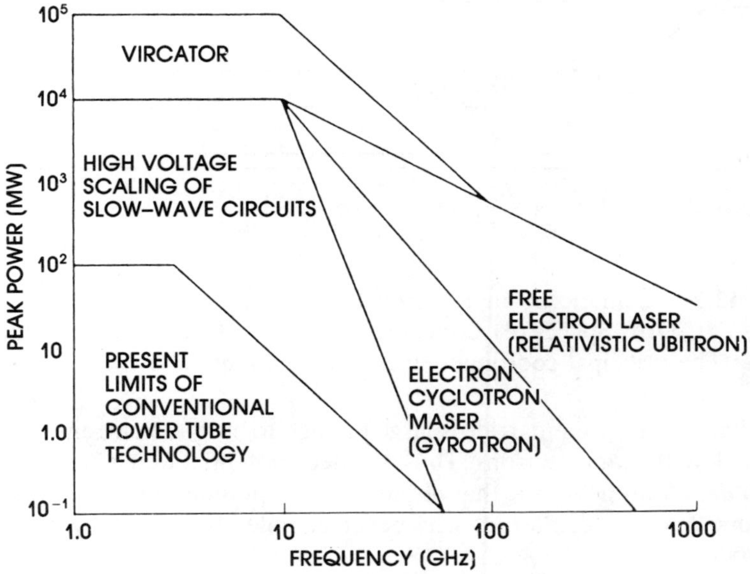

Fig. 13.1 Projected Limits of Ultra-High-Power Microwave Source (<100 ns) Technology (Courtesy of R. Parker).

13.2 THE SPACE-CHARGE-LIMITING CURRENT

13.2.1 Theory

The concept of limiting current can be visualized in the following, oversimplified physical picture. As particles from the electron beam (see Figure 13.2) enter the drift tube, the charge which they carry creates a potential barrier against further transmission of current. If the injected current is greater than the limiting current, the potential energy $e\Phi$ will exceed the kinetic energy of the beam $(\gamma_0 - 1)mc^2$, where

e = electron charge;
Φ = electrostatic potential $\leqslant 0$;
γ = relativistic factor $(1 - \beta^2)^{-1/2}$;
γ_0 = value of γ at injection;
β = v/c;
v = electron velocity;
m = electron rest mass.

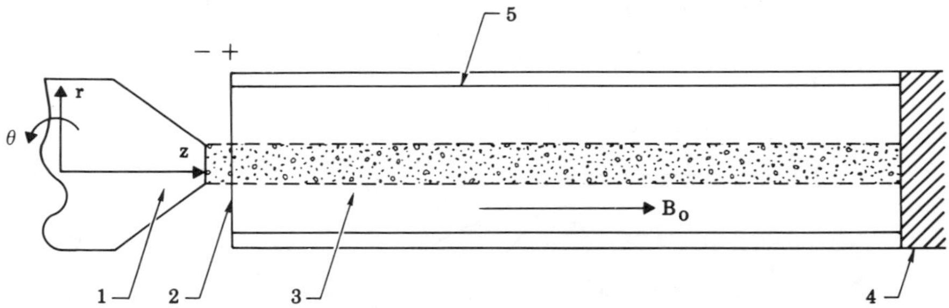

Fig. 13.2 Schematic of an Electron Beam: (1) Cathode; (2) Anode; (3) Drift Space; (4) Collector; (5) Return Current Conductor
Note: The principal coordinate axes z, r, and θ are shown.

At that position, the barrier is large enough to stop the current and cause reflection of beam electrons. Thus, the deep potential well is called a *virtual cathode*. Note, however, that because the amplitude of the well fluctuates in time and space, the beam is never completely cut off by the virtual cathode.

The general space-charge-limiting current problem does not lend itself to analytical solution. Instead, theoretical work on the subject has attempted to solve for limiting currents, which develop in electron beams, in the strictly electrostatic case. The solutions derived are only valid for beams in steady-state equilibrium (i.e., $I < I_{SCL}$). (The subscript SCL denotes the space-charge limit.) Also, the derivations depend on the following assumptions.

(1.) The electron beam is a cold plasma, $T_e = 0$ which implies the electrons are collisionless, and allows us to neglect the $\vec{\nabla} P_e$ and viscosity terms in the momentum transfer equation. Here, T_e is the electron temperature and P_e is the electron pressure.

(2.) The electron beam is axially homogeneous and infinite. In real terms the infinite-beam assumption implies $L \gg R$, where L is the drift tube length and R is the drift tube radius. This allows the beam to reach a steady state far enough downstream from the injection plane to make the equilibrium beam density n_b^0, velocity $\vec{\mathbf{v}}^0$, and electric field $\vec{\mathbf{E}}^0$, independent of z.

(3.) The equilibrium radial density and velocity profiles are azimuthally symmetric about the magnetic axis. In conjunction with assumption 2 this means that n_b^0, $\vec{\mathbf{v}}^0$, and $\vec{\mathbf{E}}^0$ are functions only of the radial distance r.

(4.) The electron particle density and kinetic energy are independent of r at the anode plane. Therefore, the beam is uniform in space and monoenergetic at injection into the drift tube; anode foil scattering is neglected.

(5.) The externally imposed longitudinal magnetic field $\vec{\mathbf{B}}_0$ is effectively infinite. If this assumption is not made, the current-induced-equilibrium self-magnetic fields would play a major role in determining the equilibrium radial profile of the beam. Because $\vec{\mathbf{B}}_0$ is taken as infinite, it guides the electrons and contains them, and hence the self-fields can be treated as perturbations. Another way of expressing this assumption is to state that the Larmor radius of an electron in the $\vec{\mathbf{B}}_0$ field must be much less than the beam radius, $r_L \ll r_0$.

Because assumptions 1 through 5 reduce the problem to the electrostatic case, solving Poisson's equation will give the desired result. By using this approach, several researchers have derived expressions for the space-charge-limited current of electron beams in vacuum and in charge-neutral plasmas. Excellent reviews have been written [48, 49], and the most widely cited results are those of Bogdankevich and Rukhadze [48]. By using the conservation of energy and momentum, these authors were able to write Poisson's equation for a cylindrically symmetric electron beam as

$$\frac{1}{r}\frac{\partial}{\partial r} r \frac{\partial \Phi}{\partial r} = \begin{cases} -\frac{4\pi j}{c}\left[1 - \left(1 - \frac{e\Phi}{mc^2}\right)^{-2}\right]^{-1/2}, & 0 < r \leqslant r_0 \\ 0, & r_0 < r \leqslant R \end{cases} \tag{13.1}$$

with appropriate boundary conditions. Here, j is the current density. The radial cross section of the beam is given in Figure 13.3(a), where R is the drift tube radius and r_0 is the radius of the axially centered beam.

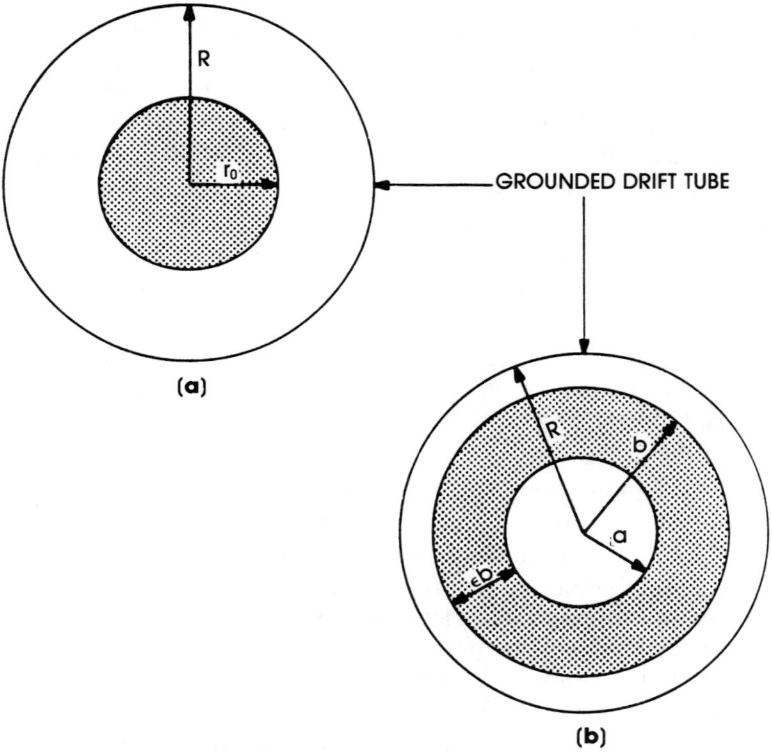

Fig. 13.3 Radial Cross Section of Two Electron Beams
Note: (a) Solid beam, where r_0 is the beam radius and R is the drift tube radius; (b) annular beam, where a is the inner radius, b is the outer radius, R is the drift tube radius, and $\epsilon = b - a/b$.

Equation (13.1) is a nonlinear differential equation. The approach taken by Bogdankevich and Rukhadze [48] is to obtain solutions by making certain analytical approximations. These approximations are related to beam energy at injection (the nonrelativistic limit, $\gamma \approx 1$, or ultrarelativistic limit, $\gamma \gg 1$) and beam geometry (the pencil beam limit, $\ln(R/r_0) \gg 1$, or fat beam limit, $\ln(R/r_0) \ll 1$). The result is a series of first-order analytical approximations, which are restricted in validity to narrow regimes. Indeed, the most often used relation is an interpolation formula, which utilizes the uniform density approximation. The interpolation neglects the radial variations in Φ, γ, and number density n, and this formula consistently underestimates the true value of limiting current.

The work of Genoni and Proctor [50] produces a more accurate, less restricted relation for space-charge-limiting current. In particular, the relation is a second-order analytical expression, which takes into account the radial dependence of γ, Φ, and n. Also, the beam is neither confined to the ultrarelativistic limit, nor is the beam geometry valid in only certain regimes. The general electron beam radial configuration is given in Figure 13.3(b). As we can see from the figure, R is the drift tube radius, a and b are the inner and outer beam radii, respectively, and $\epsilon = b - a/b$.

The derivation also begins with conservation of energy and momentum, so that Poisson's equation for a cylindrically symmetric beam takes the form:

$$\frac{1}{r}\frac{\mathrm{d}}{\mathrm{d}r}\left[r\frac{\mathrm{d}\gamma(r)}{\mathrm{d}r}\right] = \begin{cases} \dfrac{4\nu[(b/R)^2 - (a/R)^2]}{\beta(r)}, & a \leqslant r \leqslant b \\ 0, & \text{elsewhere} \end{cases} \tag{13.2}$$

subject to the boundary conditions:

$$\gamma(R) = \gamma_0 \tag{13.3}$$

where $\gamma(0) \geq 1$ and finite; $\nu = Ie/mc^3$ is dimensionless current, independent of r and z. Relation (13.3) is a nonlinear differential equation, which, as expected, shows the radial dependence of γ. By choosing an appropriate Green's function, we can express (13.2) and (13.3) as an integral equation and obtain a numerical solution. More importantly, by assuming a uniform density radial profile of γ inside the beam as a *zero*th-order guess, we can iterate on the integral equation to obtain a second-order analytical approximation to the current. Thus, not only can we find the limiting current by determining the value of γ that maximizes the current, but also evaluate the current in the drift tube for any value of γ and r. By using this approach, an equivalent integral equation, valid for $a \leqslant r \leqslant b$, is

$$\gamma(r) = \gamma_0 - \frac{4\nu}{b^2 - a^2}\left[\ln\frac{R}{r}\int_a^r \frac{x}{\beta(x)}\,\mathrm{d}x + \int_r^b \frac{x}{\beta(x)}\ln\frac{R}{r}\,\mathrm{d}x\right] \tag{13.4}$$

A first approximation $\gamma(r) \approx \gamma^I(r)$ is obtained by putting $\beta = K$ (constant) in the integrals of (13.4). If K is fixed by requiring $\gamma^I(a) = (1 - K^2)^{-1/2}$, we obtain the following approximation to the limiting current:

$$\nu_L \approx \nu_L^{\mathrm{BR}} \equiv \frac{(\gamma_0^{2/3} - 1)^{3/2}}{1 - f(\epsilon) + 2\ln R/b} \tag{13.5}$$

where

$$f(\epsilon) = \frac{(1 - \epsilon)^2}{1 - \epsilon/2}\left|\frac{\ln(1 - \epsilon)}{\epsilon}\right| \tag{13.6}$$

Equation (13.5), which appears in a paper by Miller and Straw [51], is the logical generalization to annular beams of the widely used Bogdankevich-Rukhadze [48] interpolation formula for solid beams, to which it reduces in the limit $\epsilon \rightarrow 1$. A second approximation, $\gamma^{II}(r)$, is obtained by repeating this procedure using $\beta = [1 - (\gamma^I)^{-2}]^{1/2}$ in the integrals of (13.4). In the special case of a solid beam ($a = 0$), the integrals can be evaluated in closed form, but they must be solved numerically for the general case of an annular beam. Genoni and Proctor [50] have obtained useful analytical approximations to $\gamma^{II}(r)$ for annular beams by performing the second iteration with $\beta = (1 - \Gamma^{-2})^{1/2}$, where $\Gamma(r)$ is a quadratic fit to $\gamma^I(r)$, which yields the exact $\gamma^{II}(r)$ in the solid beam limit. A particular choice of K then yields a definite relation between ν and $\gamma^a \equiv \gamma(a)$. The simplest of these that gives good agreement with numerical results is

$$\nu = \frac{(\gamma_0 - \gamma_a)^2}{(\Gamma_b^2 - K)^{1/2} - (\gamma_a^2 - 1)^{1/2}}\,\frac{1 - g}{(1 - f + 2\ln R/b)^2} \tag{13.7}$$

in which

$$\Gamma_b = \gamma_a + (\gamma_0 - \gamma_a)\frac{1 - g}{1 - f + 2\ln R/b} \tag{13.8}$$

and

$$g(\epsilon) = \frac{8(1 - \epsilon)^2}{\epsilon(4 - 3\epsilon)}\ln\frac{1 - \epsilon/2}{1 - \epsilon} \tag{13.9}$$

In the limit $\epsilon \to 1$, both g and f tend toward zero, and (13.7) reduces to a formula obtained by Thompson and Sloan [52, Eq. (14)] in their analysis of solid beams. The reader should note that different quadratic fits, while producing more complicated analytical expressions, produce more accurate results [50].

13.2.2 Simulation

The space-charge-flow simulations were carried out using a two-dimensional particle-in-cell plasma simulation code, CCUBE, which is fully relativistic and electromagnetic [53, 54]. CCUBE (version 1) self-consistently solves for the time-dependent trajectories of tens of thousands of plasma particles over thousands of plasma periods. The critical portion of the code consists of a "leap-frog" procedure. At each time step, the charge and current associated with the particles are interpolated bilinearly onto a spatial mesh. The quantities then serve as source terms in Maxwell's equations for determining the electromagnetic scaler and vector potentials on the mesh. Finally, the potentials are interpolated back to the particle positions. The basic coordinate system is cylindrical with the assumption of azimuthal symmetry. A uniform axially directed magnetic field can be imposed to ensure beam equilibrium.

All variables are expressed in dimensionless terms. In particular, the depth of the potential well formed by the electronic space charge is given by ϕ, where

$$\phi(r) = \frac{e\Phi(r)}{mc^2} = \gamma(r) - \gamma_0 \leqslant 0 \tag{13.10}$$

The five components in configuration-momentum space are $X1 = z\omega_p^0/c$, $X2 = r\omega_p^0/c$, $V1 = \gamma\beta_z$, $V2 = \gamma\beta_r$, $V3 = \gamma\beta_\theta$, where β_z, β_r, and β_θ are the components of β in the z, r, and θ directions, respectively; $\omega_p^0 = (4\pi n_o e^2/m)^{1/2}$ is the electron plasma frequency arising from the beam density at injection n_0.

In these simulations, the code is made to model the injection of cold, monoenergetic, relativistic electrons with uniform density into a long, evacuated, equipotential, cylindrical drift tube along an effectively infinite magnetic guide field in order to comply with the assumptions of Section 13.2.1. This is accomplished in the following manner.

(1.) A cold beam is obtained by setting the electron thermal velocity components equal to zero.

(2.) Miller and Straw [55] provide a criterion for neglecting the presence of end-plates, and effectively having $L \gg R$:

$$L/R \geqslant 2.58(b/R)^{0.133} \tag{13.11}$$

For the annular beam simulations this yields $L/R \geqslant 2.50$, whereas the worst-case simulation is $L/R = 9.84$. In addition, two runs were made with the downstream axial boundary grounded. The presence of end-plates did not affect the results.

(3.) Azimuthal symmetry is assured by the nature of the code.

(4.) The code injection can be made to neglect anode foil scattering so that the particles are monoenergetic and spatially uniform at injection.

(5.) Thode, *et al.* [56] give a numerical relation to ensure that the imposed magnetic field is effectively infinite. The relation is $\omega_c/\omega_p^0 \geqslant 5$, where ω_c is the beam cyclotron frequency defined as eB_0/mc. This condition is met by running the code's dimensionless, external, magnetic field as

$$B_0 = \omega_c/\omega_p^0 = 8.0 \tag{13.12}$$

The number of cells in the $X1$ and $X2$ direction were 190 and 37, respectively. The $X1$ length was held constant at 50.0 units to give $\Delta X1 = 0.263$ per cell. The $X2$ radius varied from 2.89 to 5.0, depending on the value of injected current. The boundary conditions for the electrons are reflection at the axis and absorption on both the radial and axial surfaces. Potentials are set to zero at the upstream axial and outer radial boundaries in order to represent metal surfaces. At the downstream axial boundary, however, the normal derivatives of the potentials are set to zero in order to simulate an infinitely long cylinder.

In Section 13.2.3, the value of γ_a from the simulation calculation is compared, for a given value of ν, with that given by (13.7), as well as with a numerical solution obtained by iterating on (13.4) with $r = a$ [50].

13.2.3 Comparison of Theory and Simulation

Several CCUBE runs were made by using $\gamma_0 = 5$ and varying values of ν in order to determine the maximum value of $|\phi|$. This consistently occurred at the radial value $r = a$. Therefore, $|\phi|$ maximizes for $\Phi(a) \equiv \Phi_a$. The characteristics of the simulation runs are given in Table 13.2. The last column in the table refers to whether the downstream axial boundary

Table 13.2 Characteristics of Simulation Runs

Run	ν	$\lvert\phi_a\rvert$	*Error* (%)	a/R	b/R	$L(c/\omega_p^0)$	$R(c/\omega_p^0)$	*Downstream Axial Boundary*
1	1.175	3.07	1.5	0	0.5	50.0	4.38	0
2	1.175	3.06	1.2	0	0.5	50.0	4.38	*G*
3	1.293	4.20	2.5	0	0.5	50.0	4.60	0
4	0.982	1.01	0.74	0.4	0.8	50.0	2.89	0
5	1.957	2.05	1.1	0.4	0.8	50.0	4.08	0
6	2.846	3.18	1.2	0.4	0.8	50.0	4.92	0
7	2.846	3.18	1.1	0.4	0.8	50.0	4.92	*G*
8	3.034	4.25	3.1	0.4	0.8	50.0	5.08	0

was grounded or open. Such a condition enabled us to check whether end-plate effects were entering the problem. As we can see in Table 13.2, no end-plate effects were noted. The percent error was derived on the basis of the code's management of conservation of energy, which is denoted by a net energy diagnostic. Because net energy continuously increases due to rounding off and truncation errors, the percent error will increase as a function of computer run time. The percent errors given in Table 13.2 were taken after the run reached a steady state, as denoted by total energy and $|\phi|$ reaching a constant, or following formation of a virtual cathode. A virtual cathode was formed in runs 3 and 8. This can be observed in the table, insofar as $|\phi_a| > (\gamma_0 - 1)$. Thus, the space-charge-limiting current lies between $\nu = 1.175$ and $\nu = 1.293$ for the solid-beam case, and $\nu = 2.846$ and $\nu = 3.034$ for the annular-beam case.

We make a comparison between the numerical, analytical, and simulation models in Table 13.3. All three methods produce consistent results for γ_a from given values of sublimiting injected current. The table lists the values given by the three methods, where

$$\gamma_a = \gamma_0 - |\phi_a| \tag{13.13}$$

is used to define γ_a for the simulations. Note that if a virtual cathode is formed, $|\phi_a| > (\gamma_0 - 1)$, then γ_a as defined by (13.13) does not have a physical meaning. Rather, it is simply a mathematical relation between γ_0 and $|\phi_a|$.

Table 13.3 Comparison of Models

ν	γ_a *Numerical*	γ_a *Analytical*	γ_a *Simulation*	*Beam Type*
1.175	1.893	1.93	1.93	Solid
1.293	Virtual Cathode	Virtual Cathode	Virtual Cathode	Solid
0.982	4.004	4.00	3.99	Annular
1.957	2.976	2.98	2.95	Annular
2.846	1.867	1.92	1.82	Annular
3.034	Virtual Cathode	Virtual Cathode	Virtual Cathode	Annular

A more important comparison can be made between the values for space-charge-limiting current ν_L, as derived from the numerical, analytical, and simulation models, and that produced by the widely used Bogdankevich-Rukhadze interpolation formula (13.5). This is done in Table 13.4. Equation (13.5) predicts values for ν_L which are significantly lower than those of (13.4) and (13.7). However, the analytical results differ from the results of the integral equation by roughly 3%, and ν_L is obtained from (13.4) by numerically determining the largest value of ν for which a solution exists. The limiting current for (13.7) is derived from the value of γ_a that maximizes ν.

Table 13.4 Limiting Current Comparison

Beam Type	ν_L *Numerical*	ν_L *Analytical*	ν_L *Simulation*	ν_L $B - R$
Solid	1.27	1.31	1.175–1.293	1.118
Annular	3.03	3.12	2.846–3.034	2.712

The discrepancy in the limiting value of ν between the Bogdankevich-Rukhadze and Genoni-Proctor formulations is due to the error involved in assuming the uniform density approximation. Figure 13.4 is a plot of the relation between ν and γ_a in the uniform density approximation, and in a second-order analytical expression by using a better quadratic fit than (13.7). The simulation points have been added to the figure. It is clear that the uniform density approximation underestimates the value of ν for all values of γ_a. Finally, Figure 13.5 plots $\gamma(r) - 1$ *versus* r. It shows the dependence of beam kinetic energy on r for the modeled annular beam. The successively more energetic lines of dots correspond to the more recent

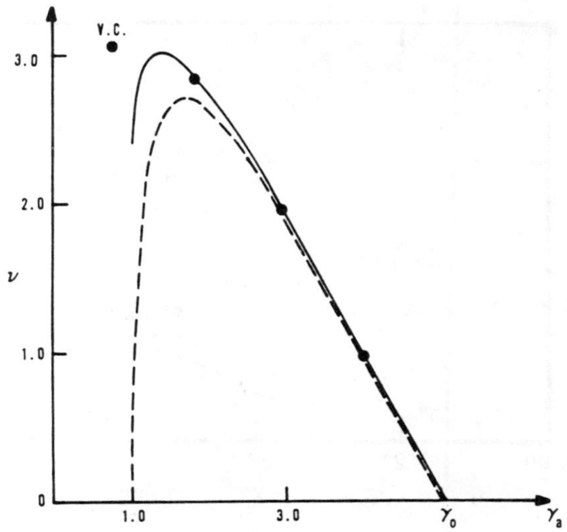

Fig. 13.4 Plot of ν *versus* γ_a for $\gamma_0 = 5$, $a/R = 0.4$, $b/R = 0.8$
Note: The dashed line represents the uniform density approximation. The solid line results from an approximate second-order analytical relation. The points correspond to simulation results. The value γ_a for the point labeled v.c. (virtual cathode) has no physical meaning. It is defined in terms of (13.13).

injections of charge through the anode. As expected, the minimum kinetic energy occurs at the inner beam edge, and the maximum occurs on the outer edge. The uniform density approximation neglects this dependence.

13.3 THE SPACE-CHARGE-LIMIT INSTABILITY

We have visualized the space-charge-limiting current as a point at which the electrostatic potential exceeds the kinetic energy of the beam. However, we may easily note from Figure 13.4 that, at the space-charge limit (SCL), the beam kinetic energy is approximately $\gamma_0^{1/3}mc^2$. Thus, although the onset of virtual cathode formation can be exactly predicted as the point at which the beam exceeds a critical current, we have only recently come to understand the dynamics of its formation and subsequent oscillations.

A small-signal perturbation analysis can be performed, just below the SCL, which seems to predict instability above the limiting current value

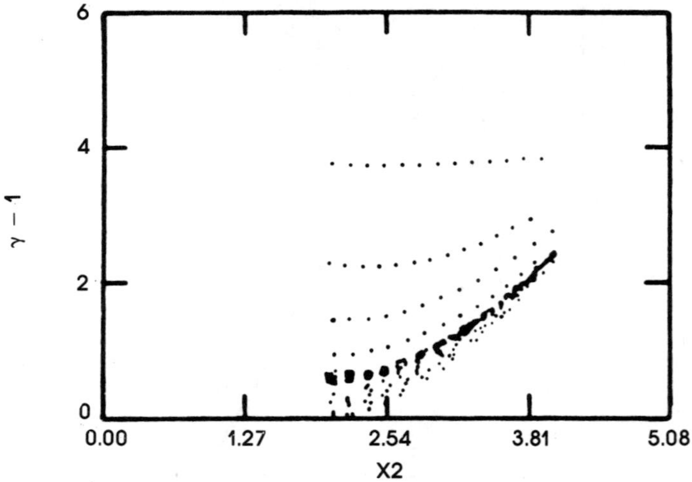

Fig. 13.5 A CCUBE Diagnostic of Particle Kinetic Energy *versus* r for the Annular Beam Showing the Radial Dependence of γ

[6, 9]. However, this analysis is not rigorous, and, in fact, produces misleading results. Instead, we can use multiple-scaling perturbation techniques to study the time-dependent behavior of a beam above the SCL. We derive estimates for the nonlinear growth rate of the ensuing instability, and show that, even below the SCL, the beam is unstable to sufficiently large perturbations. The method can be applied to a wide class of problems, but here we treat the short-circuited, one-dimensional, electrostatic diode (depicted in Figure 13.6) as the simplest model that contains the appropriate physics. We show that, at least in one dimension, an arbitrarily heavy ion background does not alter the qualitative behavior of the beam, and we present numerical results that exhibit virtual cathode oscillations for a neutral beam. Before starting this analysis, it is worthwhile for us to describe the results of numerical simulations in order to present a picture of time-dependent virtual cathode dynamics.

13.3.1 Physical Description of the Virtual Cathode

Simulations were carried out in conjunction with the theory presented in Section 13.3.2 by using a two-dimensional, relativistic, electrostatic, particle-in-cell code CCUBE (see Section 13.2.2). The code can solve self-consistently for the time-dependent trajectories of tens of thousands of plasma particles over thousands of plasma periods. All variables are

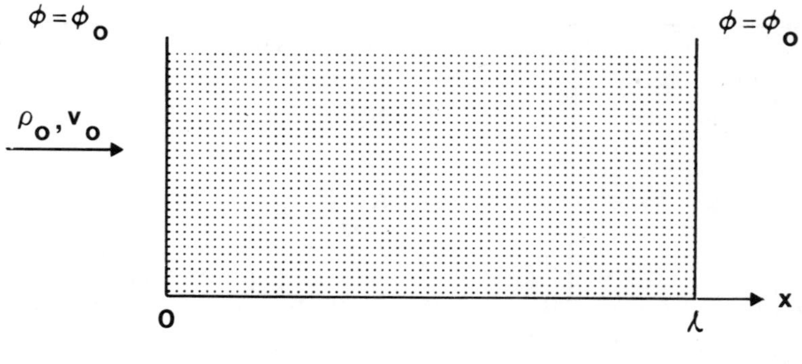

Fig. 13.6 Schematic of the Short-Circuited, One-Dimensional, Electrostatic Diode Modeled

expressed in dimensionless terms. Therefore, length is in units of c/ω_p^0, time is measured in units of $(\omega_p^0)^{-1}$, and particle velocity is given by $v_i = \beta_i\gamma(i = 1, 2, 3)$, where ω_p^0 is the initial electron plasma frequency.

In these simulations, a monoenergetic 51 keV electron beam is injected into a Cartesian geometry. The left and right boundaries are representing grounded a planar short-circuited diode. Periodic boundary conditions in the transverse direction make an effectively one-dimensional configuration space. In general, the simulation had 62 cells in the longitudinal direction, thus modeling a length of 1.0 c/ω_p^0. The time step was 0.0125 $(\omega_p^0)^{-1}$. Twenty particles were injected per cell.

A detailed discussion of the physical dynamics of the virtual cathode based on these numerical results is appropriate here. The usual graph of potential minimum ϕ_m in the diode *versus* electron beam current α is shown in Figure 13.7. (The parameter α will be discussed later.) When α is increased above the space-charge limit, ϕ_m jumps from the stable (normal-C) branch to the oscillatory stable branch. The amplitude and position of ϕ_m, while on the oscillatory branch, describe a limit cycle, as expected for a relaxation oscillation, which is what this represents. Typical limit cycles are depicted in Figure 13.8. As α is increased further the oscillation frequency ϕ_m and virtual cathode position within the diode asymptotically approach limiting values. If α is decreased, the oscillation amplitude $\Delta\phi_m$ decreases and the position of ϕ_m moves toward the diode center. The electron flow reverts to the equilibrium steady state when the perturbation, which is due to the rate of change of diode current below the space-charge limit, is sufficiently large. This normally occurs before reaching the bifurcation point. The entire process forms a hysteresis loop, which is depicted in Figure 13.7.

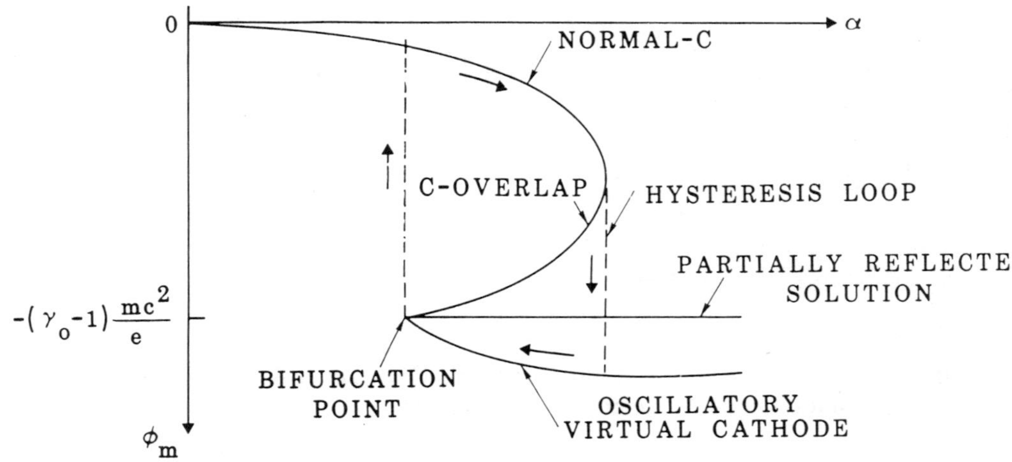

Fig. 13.7 Electrostatic Potential Minimum as a Function of Current α
Note: The plot depicts the various possible solutions such as normal-C flow (stable), C-overlap (unstable), the partially reflected solution (unstable), and the oscillatory virtual cathode (stable). The motion around the hysteresis loop is denoted by arrows.

The virtual cathode originates at the bifurcation point. This is the intersection of the oscillatory state with the C-overlap [7] and partially reflected solution branches [5]. The bifurcation point cannot be reached in the short-circuited diode. Of the three branches emanating from it, two (the steady-state ones) are physically unstable, while the oscillatory branch is numerically unstable at this point. This results because the limit cycle is infinitesimally small at the bifurcation point so that simulation codes lose resolution before they reach it. Loss of resolution creates a small-amplitude, high-frequency oscillation, observed in this study and previously [9], and this result is numerical, not physical.

The problem of loss of resolution can be overcome if we eliminate the hysteresis loop. Then, the C-overlap branch disappears, and we reach the bifurcation point along the stable normal-C branch. This can be accomplished in several ways. The most appropriate for this study is to have a retarding potential difference across the diode that is equal to the injected electron kinetic energy. Then, the C-overlap solution vanishes, and the bifurcation takes place at the space-charge limit α_{SCL}. The oscillation can be described as a small perturbation on the beam, rather than the radical change that results in the short-circuited diode when we exceed α_{SCL}. This analysis indicates that the onset of virtual cathode formation occurs when the electron velocity in the steady state vanishes at some point inside the diode. For the short-circuited diode, this occurs at the diode center; for the biased diode, it occurs at $\leqslant l$. Because the oscillation is a small perturbation on the steady-state fields in the biased diode, the reader can readily see that the virtual cathode oscillation period at onset is the electron transit time from the injection plane to the position where velocity vanishes. This concept will be revisited in Section 13.4.

Finally, consider the particle dynamics during the oscillation period for counterclockwise limit cycles ($\alpha \leqslant \alpha_{SCL}$) as in Figure 13.8(c). At the point where the virtual cathode position is a minimum and the potential well is starting to move to the right, the amplitude is too small to stop the electrons. When virtual cathode motion is to the left, the cathode opposes the electron beam and causes particle bunching. Because the well is deeper, the stream velocity will vanish at some location, and then become negative. Here, the second derivative of the velocity (d^2u/dx^2) is also negative. In this process, the stream is continuously deformed to create a double-valued negative velocity protrusion. The entire system is three-valued (Figure 13.9), as in a collisionless shock wave [57]. Here, the region of triple flow is not limited, it is in usual collisionless shocks, by the presence of a transverse magnetic field [58], but rather the flow is limited by the presence of the walls. Indeed, the reflected part detaches from the main beam, exits periodically through the anode, and is thus responsible for the onset of oscillatory behavior in beam characteristics. As the potential minimum reaches the end of its left motion, the two "lips" of the back-reflected stream close. At this point, no more electrons are reflected and the well moves to the right, and hence repeats the cycle.

For larger values of α, the limit cycle is distorted into a figure eight, where one lobe has clockwise and the other with counterclockwise motion (Figure 13.8(b)). The transition continues until the motion is completely clockwise (Figure 13.8(a)), which indicates a change in the particle-bunching process, and this transition is related to the fraction of current that is reflected, as opposed to transmitted, from the injected electron beam. In Figure 13.8(a), most of the beam is reflected, whereas most electrons are transmitted in Figure 13.8(c).

13.3.2 Scaling Relations

Qualitative scaling relationships for the virtual cathode dynamics in one-dimensional flows can be derived from the Vlasov equation:

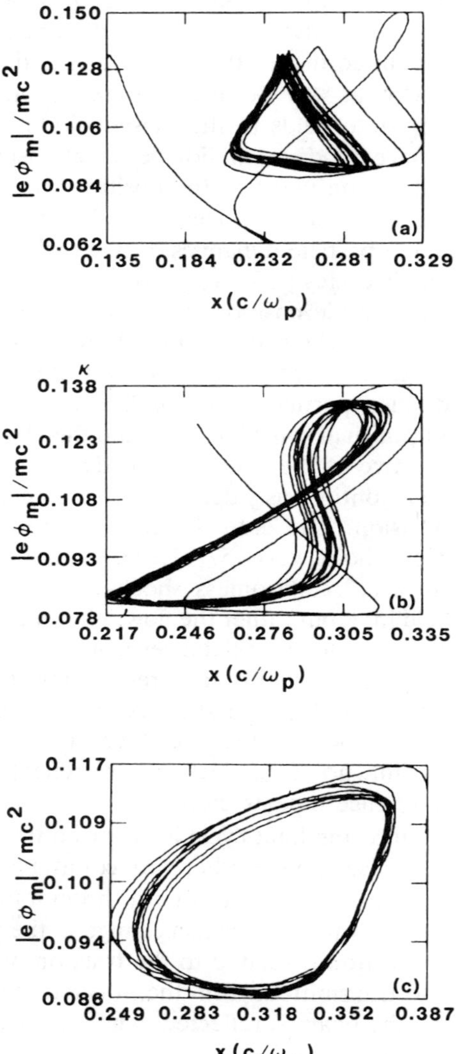

Fig. 13.8 Typical Virtual Cathode Limit Cycles in the Classical Short-Circuited, One-Dimensional Diode with an Injection Energy of 51 keV: (a) $\alpha = 2.5$; (b) $\alpha = 2.0$; (c) $\alpha = 1.4$

Note: The motion in (a) is clockwise and that in (c) is counterclockwise; $l = 1.0\ c/\omega_p^0$.

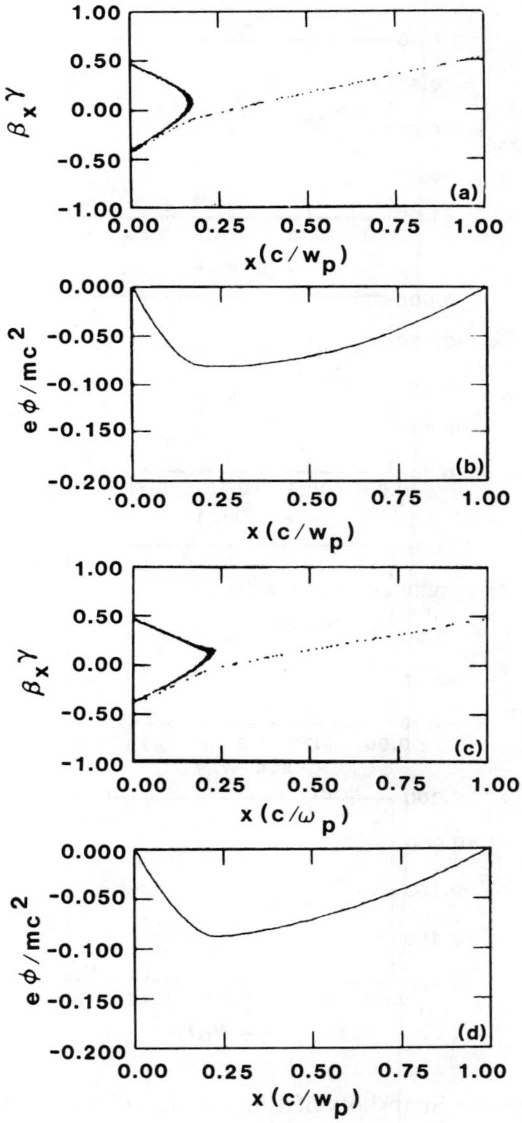

Fig. 13.9 Successive Snapshots of Electron Beam Momentum Space and Corresponding Potential Shape in the Diode for $\alpha = 2.0$, $l = 1.0\ c/\omega_p^0$

Note: The time between frames is $0.5(\omega_p^0)^{-1}$ and the initial beam kinetic energy is 51 keV.

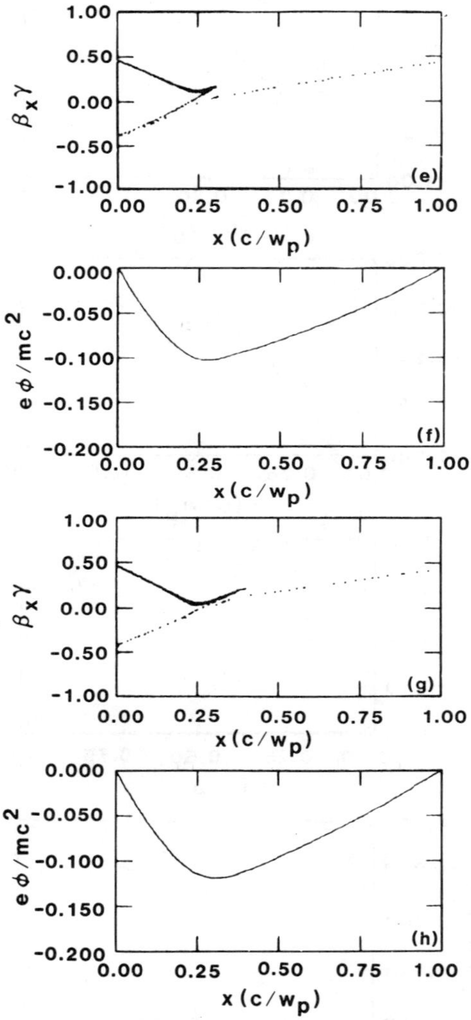

Fig. 13.9 Successive Snapshots of Electron Beam Momentum Space and Corresponding Potential Shape in the Diode for $\alpha = 2.0$, $l = 1.0\ c/\omega_p^0$

$$(\partial_t + v\partial_x + E\partial_p)f(x, p, t) = 0 \tag{13.14}$$

where x is position, t is time, p is momentum, v is velocity, and E is the electric field. The subscripts denote derivation with respect to each variable, and f is the particle distribution function. Although the same results will follow from a fluid formulation, this approach is used to alleviate any ambiguities about their applicability to time-dependent three-velocity flows. Relativistic beams can be treated as well, if the relation between p and v is utilized, thus,

$$p = \frac{v}{(1 - v^2/c^2)^{1/2}} = \gamma v \tag{13.15}$$

Finally, we include Poisson's equation:

$$E_x = \rho \tag{13.16}$$

In order to simplify the nondimensionality, consider only cold flows with injection velocity v_0, injection momentum $p_0 = \gamma_0 v_0$, and charge density ρ_0. We will separately treat the semi-infinite and finite cases. By setting $\gamma_0 = 1$, the classical formulas can be deduced without any further analysis.

In semi-infinite geometry, the boundary conditions are given by

$$\rho(0, t) = \rho_0, f(0, p, t) = p_0 \delta(p - p_0) \tag{13.17}$$

Introduce dimensionless variables $\rho = |\rho_0| n$ and $p = p_0 P$ so that we have fixed boundary conditions. Next, define unknown scales for the remaining variables as $x = \xi X$, $t = \tau T$ and $E = \epsilon e$. By substituting these relations in (13.14) and (13.16), and using (13.15), we obtain

$$\frac{1}{\tau}\,\partial_T + \frac{p_0}{\xi}\,\frac{P}{1 + \left(\frac{p_0}{c}\right)^2 P^2}\,\partial_X + \frac{\epsilon e}{p_0}\,\partial_P \cdot F(X, P, T) = 0 \tag{13.14a}$$

and

$$\frac{\epsilon}{\xi}\, e_X = |\rho_0| n \tag{13.16a}$$

along with the transformed boundary conditions $n(0, t) = -1$, $P(0, t) = 1$ and $F(0, P, T) = -\delta(p - 1)$, where $F(X, P, T)$ is the dimensionless distribution function. If we demand $\epsilon = \xi|\rho_0|$, and we choose $\tau = \sqrt{\gamma_0/|\rho_0|}$, $\xi = v_0\sqrt{\gamma_0/|\rho_0|}$, then (13.14a) and (13.16a) become

$$\left[\partial_T + \frac{P}{\sqrt{1 + \left(\frac{p_0}{c}\right)^2 P^2}}\partial_X + e\partial_P\right] F(X, P, T) = 0 \tag{13.18}$$

$$e_X = n \tag{13.19}$$

which, together with

$$n(0, t) = -1, F(0, P, T) = -\delta(P - 1) \tag{13.20}$$

is the full dimensionless description of the problem. The only parameter still present is p_0/c. Thus, all the characteristics of semi-infinite flow are functions of only this parameter. In the classical limit, $p_0/c \rightarrow 0$, all classical flows when scaled are identical.

In finite geometry, the analysis is the same, except that if X is scaled by ξ, as noted above, the second boundary condition is given at $x = l$ by

$$X = l/\xi = \frac{l}{v_0}\left(\frac{|\rho_0|}{\gamma_0}\right)^{1/2} \tag{13.21}$$

This introduces a second parameter into the problem so that in finite geometry all characteristic quantities will be functions of

$$p_0 = \gamma_0 v_0 \tag{13.22}$$

and

$$\alpha = \frac{l\omega_p}{\beta_0\sqrt{\gamma_0}c} \tag{13.23}$$

where ω_p is the beam plasma frequency at injection, and $\beta_0 = v_0/c$. In the classical limit, $\gamma_0 \rightarrow 1$ and $\beta_0 \rightarrow 0$, the only governing parameter is $\alpha = l\omega_p/v_0$. In the classical case, $\alpha_{SCL} = 4/3$.

13.3.3 Nonlinear Theory

A quantitative analysis of the nonlinear, one-dimensional motion of electrons in the diode is presented in [59], and is given by the equations of continuity and momentum conservation for the electrons plus Poisson's equation. These equations are expressed here as

$$\rho_{\tilde{t}} + (\rho v)_{\tilde{x}} = 0 \tag{13.24a}$$

$$v_{\tilde{t}} + v v_{\tilde{x}} = -\frac{e}{m}\xi \tag{13.24b}$$

$$\epsilon_0\, \xi_{\tilde{x}} = \rho + \rho_h \tag{13.24c}$$

where $0 \leqslant \tilde{x} \leqslant l$. The subscripts denote differentiation with respect to that variable. Electron and heavy ion charge density, electron velocity, electric field, time, and position are indicated by ρ, ρ_h, v, ξ, $\tilde{t}$, and $\tilde{x}$, respectively. The appropriate boundary conditions are $v(0, \tilde{t}) = v_0$, $\rho(0, \tilde{t}) = \rho_0$ and $\int \xi d\tilde{x} = 0$. An ion component is placed in Poisson's equation in order to discuss two-species space-charge flow. Conservation equations of mass and momentum for the heavy ions are not included because it is assumed that their velocity does not change appreciably during their passage through the diode.

In order to simplify the mathematics, we introduce dimensionless equations for our model. These equations are

$$n_t + (nu)_x = 0 \tag{13.25a}$$

$$u_t + u u_x = -\alpha E \tag{13.25b}$$

$$E_x = \alpha(n + I) \tag{13.25c}$$

where $n = \rho/|\rho_0|$, $u = v/v_0$, $E = (e\epsilon_0/m|\rho_0|)^{1/2}\, \xi/v_0$, $t = v_0\tilde{t}/l$, $\tilde{x} = x/l$, $\alpha = (e|\rho|/\epsilon_0 m)^{1/2}\, l/v_0$ is a dimensionless parameter related to current, and I is the ratio of heavy ion to electron charge densities, $I = \rho_h/|\rho_0|$. Alternatively, α may be written as $l\omega_p/v_0$, where ω_p is the beam plasma frequency. The boundary conditions for electrons become $u(0, t) = 1$, $n(0, t) = -1$, and $\int_0^1 E\, dx = 0$. For an ion beam $n(0, t) = 1$, otherwise the following derivation is the same.

Unneutralized Beams

For an unneutralized beam, by setting $I = 0$ and solving by the method of characteristics, we find

$$n^{-1} = -\frac{\alpha^2}{2}(t - s)^2 + \alpha E_0(s)(t - s) - 1 \tag{13.26}$$

where s is the entry time for the particle occupying position x at time t, and $E_0(s)$ is the electric field at $x = 0$. The particle trajectories are found by utilizing (13.25a), from which it follows that

$$\left(\frac{\partial x}{\partial s}\right)_t = n^{-1} \tag{13.27}$$

which thus yields

$$x = \frac{\alpha^2}{6}(t - s)^3 + \alpha \int_t^s E_0(s)(t - s)\,ds + (t - s) \tag{13.28}$$

Integration of the trajectory equation is hard for general time-dependent situations, because imposing the proper boundary conditions leads to a nonlinear integral equation for $E_0(t)$. However, exact solutions can be obtained for several special cases. The problem of injection into an empty diode can be integrated until the formation of a singularity in n, which indicates the crossing of trajectories. In this case, the stream velocity becomes three-valued, and we must use a Vlasov equation description [60], rather than system (13.25), which is derived assuming a single stream of monoenergetic particles. As described in Section 13.3.1, this multistreaming is characteristic of the oscillatory state created when α exceeds its SCL value.

By using these equations, we can derive a similar representation for Figure 13.7 in terms of E_0 and α. For steady states, $E_0(t) = E_0$ (constant), we find

$$u = -n^{-1} = \frac{\alpha^2}{2}(t - s)^2 - \alpha E_0(t - s) + 1 \tag{13.29a}$$

$$x = \frac{\alpha^2}{6}(t - s)^3 - \frac{\alpha}{2} E_0(t - s)^2 + (t - s) \tag{13.29b}$$

If we impose the conditions $x = 1$, $u = 1$ at $t - s = t_0$, the particle transit time, then we note that t_0 must satisfy

$$\frac{\alpha^2}{12} t_0^3 - t_0 + 1 = 0 \tag{13.30}$$

This equation has two positive solutions for $0 \leqslant \alpha \leqslant 4/3$, which coalesce at $\alpha = 4/3$. The largest one, for $0 \leqslant \alpha < 2\sqrt{2/3}$, does not correspond to a real flow. In Figure 13.10, we show E_0 ($= \alpha t_0/2$) *versus* α. This representation will be used in the discussion of nonlinear stability.

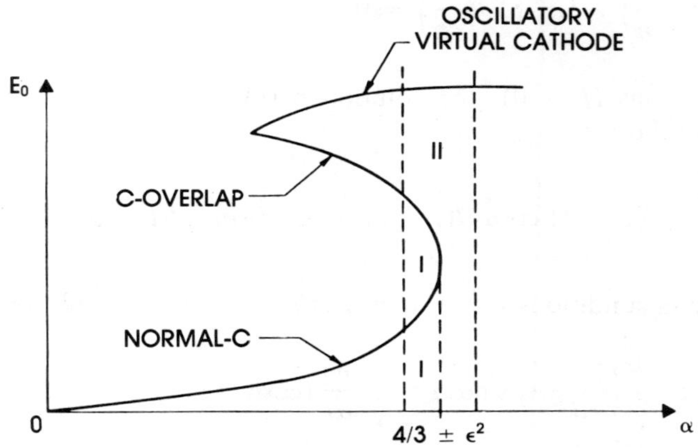

Fig. 13.10 Electric Field at the Injection Plane *versus* Current α for $I = 0$

Note: The plot depicts the normal C-flow (stable) and C-overlap (unstable) solutions. The oscillating virtual cathode (stable) solution is also shown. Regions I and II define the domains of attraction of the normal-C and virtual cathode solutions near the SCL, $\alpha = 4/3$.

Neutralized Beams

The Pierce instability occurs when electron and ion space-charge flow is considered in finite geometries, where there is no potential difference across the boundaries [61]. The ions can be either stationary or in motion with respect to electrons. Charge neutrality is maintained at the injection plane. This instability may have ramifications for charged-particle inertial confinement fusion because of its effect on neutralized beam propagation in the reactor [62]. In this section, we show that the Pierce instability is a special case of electron and ion space-charge flow. In general, two-species flow has steady and oscillatory states, which are analogous to one-species space-charge flow.

The steady-state behavior for the case of arbitrary I can be found in a manner similar to $I = 0$. By rewriting system (13.25) in characteristic coordinates, we arrive at

$$\frac{d^2}{dt^2}\left(1 + \frac{I}{n}\right) + \alpha^2 I \left(1 + \frac{I}{n}\right) = 0 \tag{13.31}$$

For positive ions ($I > 0$), the solution of (13.31), after satisfying the boundary conditions, is

$$1 + \frac{I}{n} = (1 + I)\cos\alpha\sqrt{I}(t - s) + E_0\sqrt{I}\sin\alpha\sqrt{I}(t - s) \tag{13.32}$$

Imposing conditions $x = 1$, $u = 1$ at $t - s = t_0$, we find the system:

$$1 = \frac{1}{I}t_0 - \left(\frac{1 - I}{\alpha I^{3/2}}\right)\sin\alpha\sqrt{I}\,t_0 + \frac{E_0}{\alpha I}(\cos\alpha\sqrt{I}\,t_0 - 1) \tag{13.33a}$$

$$1 = \frac{1}{I} - \frac{1 - I}{I}\cos\alpha\sqrt{I}\,t_0 - \frac{E_0}{I}\sin\alpha\sqrt{I}\,t_0 \tag{13.33b}$$

For $I = 1$, which implies charge neutralization, these equations reduce to the relations given by Godfrey [63] for the Pierce instability. However, the curves shown in Figure 13.11 are obtained by varying I. These are cuts at constant I through a three-dimensional contiguous surface. The space is defined by the axes $\tilde{E} = E_0 I^{1/2}$, $A = \alpha I^{3/2}$, and I. The surface is 2π periodic in A with the vertical plane at $A = 2\pi$ being common for all values of I. For given A, a linearized analysis establishes that the equilibria denoted by the curves are stable (unstable) for the lowest (highest) value of $\tilde{E}$. At $I = 1$, exchange of stability takes place at odd multiples of π. For $I < 1$, exchange of stability occurs at the points where $dE_0/d\alpha \to \infty$.

It is evident from Figure 13.11 that, for $I < 1$, there are no stable equilibrium solutions in the neighborhood of $A = \pi$. Therefore, we expect a virtual cathode to form when $I < 1$ and A adiabatically increases to π. We have found by using numerical simulation that in this case the beam settles to an oscillatory state similar to the virtual cathode for unneutralized beams [60]. By slowly increasing I beyond the neutral beam value of 1 in our simulation, we have established that this oscillation persists. Indeed, finding this oscillatory state for $I > 1$ by other means would have been difficult because the simulation tended to follow the stable steady branch that is present for all values of current.

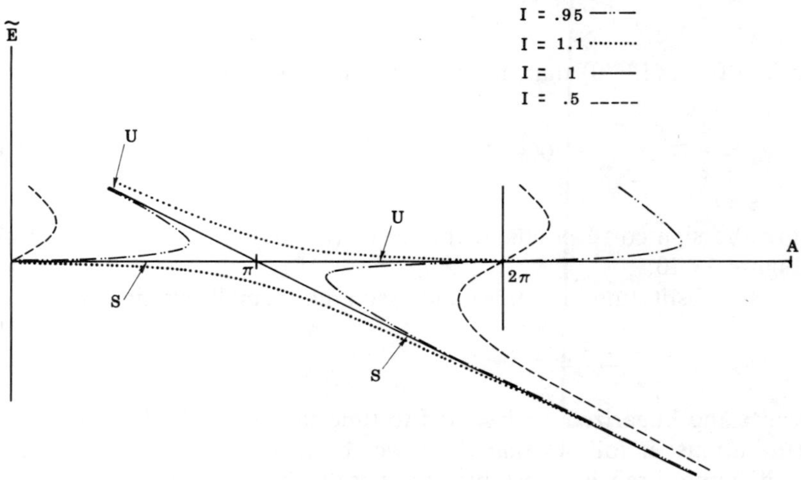

Fig. 13.11 Curves of Scaled Electric Field at the Injection Plane **Ẽ** *versus* Scaled Current *A* for Various Values of Charge Neutralization *I*

Note: The curves represent slices through a three-dimensional surface. *S* and *U* respectively indicate stable and unstable branches for the $I = 1$ slice.

Nonlinear Stability Analysis

For $I < 1$, it is of interest to establish the properties of the beam instability at the SCL, generalized for $I \neq 0$ to mean the point where $d\alpha/dE_0 = 0$. We shall carry out the analysis for $I = 0$, but our method can be applied to any similar "jump" phenomenon.

A linear stability analysis [6] about the steady state described by (13.30) results in the dispersion relation:

$$(2 + \beta)\, e^{-\beta} = 2 - \beta + \frac{\beta^3}{\alpha^2 t_0^3} \tag{13.34}$$

where $\beta = i\omega t_0$. We have written the expression derived by Lomax [6] in terms of our dimensionless variables. For α near the SCL value, we let

$$\alpha = \frac{4}{3} - \epsilon^2, \quad \epsilon \ll 1. \tag{13.35}$$

and find from (13.30) that, near this value, t_0 is approximately

$$t_0 \approx \frac{3}{2} \mp \epsilon \frac{3}{2\sqrt{2}} + 0(\epsilon^2) \tag{13.36}$$

where the sign corresponds to the lower (minus) or upper (plus) branch in Figure 13.10.

By substituting in (13.34) and assuming β small, we find that

$$i\omega = \beta \approx \mp 2\sqrt{2}\,\epsilon + 0(\epsilon^2) \tag{13.37}$$

Because the linearized analysis led to time factors of the form $e^{i\omega t}$ in the perturbations, it follows that the lower branch in Figure 13.10 is stable and the upper branch is unstable, while at the SCL ($\epsilon = 0$) we have neutral stability.

Above the value $\alpha = 4/3$, linearized theory is not applicable. By utilizing multiple-scaling perturbation theory [64], we can carry out a nonlinear stability analysis near $\alpha = 4/3$. In system (13.25), we set $\alpha = 4/3 \pm \epsilon^2$. In this neighborhood, perturbations evolve on a "slow" time scale, depicted by $\tau = \epsilon t$.

If we eliminate the electric field by combining (13.25b) and (13.25c) and utilizing τ, then system (13.25) becomes

$$\epsilon n_\tau + (nu)_x = 0 \tag{13.38a}$$

$$(\epsilon u_\tau + uu_x)_x = -\left(\frac{4}{3} \pm \epsilon^2\right)^2 n \tag{13.38b}$$

where the conditions $u(0, \tau) = 1$, $n(0, \tau) = -1$, and $\int_0^1 E\, dx = 0$ are rewritten as

$$\epsilon \int_0^1 u_\tau\, dx + \frac{1}{2}[u^2(1, \tau) - u^2(0, \tau)] = 0 \tag{13.39}$$

By substituting the asymptotic expansions:

$$u \sim \sum_{i=0}^{j} \epsilon^i u_i(x, \tau) + 0(\epsilon^{j+1}), \quad n \sim \sum_{i=0}^{j} \epsilon^i n_i(x, \tau) + 0(\epsilon^{j+1}) \tag{13.40}$$

for u and n into (13.38), and equating coefficients of various powers of ϵ, a hierachy of equations results for the u_i and n_i.

Solution of the 0(1) system gives

$$\left(u_0 - \frac{1}{2}\right)(u_0 + 1)^2 = 2(2x - 1)^2 \tag{13.41a}$$

$$n_0 = -1/u_0 \tag{13.41b}$$

To solve the 0(ϵ) system, we introduce a new variable q by

$$x = \frac{16}{9}\left(\frac{q^3}{6} - \frac{3}{8}q^2\right) + q \tag{13.42}$$

so that

$$u_0 = \frac{16}{9}\left(\frac{q^2}{2} - \frac{3}{4}q\right) + 1 \tag{13.43}$$

We then find

$$n_1 = \frac{1}{u_0^2}u_1 \tag{13.44a}$$

$$u_1 = C\frac{q(q - 3/2)}{u_0} \tag{13.44b}$$

where C is a constant of integration, which is, in general, a function of the slow time τ. To find $C = C(\tau)$, which determines the slow evolution of the perturbation $u_1(x, \tau)$, we need to go to the next order, 0(ϵ^2). By substituting in the expressions for u_0, u_1, n_0, n_1 and eliminating n_2, we find that $u_2(x, \tau)$ satisfies

$$(u_0u_2)_{xx} + \frac{16}{9}\frac{1}{u_0^2}u_2 = \frac{16}{9}\frac{C_\tau}{u_0}\int_0^q \frac{q(q - 3/2)}{u_0^2}\,dq + C^2\frac{q^2(q - 3/2)^2}{u_0^3} - C_\tau\left(\frac{q(q - 3/2)}{u_0}\right)_x \pm \frac{8}{3}\frac{1}{u_0} \tag{13.45a}$$

$$u_2(0, \tau) = 0, \quad u_2(1, \tau) = +\frac{9}{16}C_\tau \tag{13.45b}$$

The solution to this inhomogeneous, two-point boundary value problem exists, provided that a certain orthogonality condition is satisfied (Fredholm alternative theorem) [65] between the right-hand side and the solution of the adjoint problem that takes account of the boundary conditions. This leads to the desired equation that determines $C(\tau)$:

$$aC_\tau + bC^2 \pm c = 0 \tag{13.46}$$

where a, b, and c are found to be

$$a = -\int_0^{3/2} \frac{q(q - 3/2)}{u_0^2} \left[\frac{16}{27} q^2(q - 9/4) + \frac{3}{2}\right] dq = 1.6850 \tag{13.47a}$$

$$b = \frac{8}{3} \int_0^{3/2} \frac{q^3(q - 3/2)^3}{u_0^4} dq = -3.7968 \tag{13.47b}$$

$$c = \frac{8}{3} \int_0^{3/2} q(q - 3/2)\, dq = -1.5 \tag{13.47c}$$

In (13.46) the plus or minus signs indicate that we are above or below the SCL, respectively.

Above the SCL, we find

$$C(\tau) = -\frac{c}{b} \tan\left[\sqrt{cb}\left(\frac{\tau + \tau_0}{a}\right)\right] \tag{13.48}$$

and below the SCL, we have

$$C(\tau) = \frac{c}{b} \tanh\left[\sqrt{cb}\left(\frac{\tau + \tau_0}{a}\right)\right], \quad \text{if } |C(0)| < \sqrt{c/b} \tag{13.49}$$

$$C(\tau) = \frac{c}{b} \coth\left[\sqrt{cb}\left(\frac{\tau + \tau_0}{a}\right)\right], \quad \text{if } |C(0)| > \sqrt{c/b} \tag{13.50}$$

where τ_0 is a constant of integration. In general, small initial perturbations will lead to the solution:

$$u(x, t) \sim u_0 + \sum_{i=1}^{j} C_i(\tau)\, e^{\omega_i t} u_i(x) + 0(\epsilon^2) \tag{13.51}$$

where ω_i are the various distinct solutions of the dispersion relation (13.14) at $\alpha = 4/3$ [64, 66]. It is straightforward to show that all modes are such that $\mathrm{Re}(\omega_i) < 0$, except one for which $\omega = 0$. Thus, all other modes will decay in the fast time scale, and only the neutral mode ($\omega = 0$) will persist. Our solution after a short time will look like so

$$u \sim u_0 + \epsilon C(\tau) \frac{q(q - 3/2)}{u_0} + 0(\epsilon^2) \tag{13.52}$$

From the given initial conditions, it is easy to determine the initial condition for the neutral mode. Below the SCL, if the initial conditions are such that $C(0) > -\sqrt{c/b}$, the solution will evolve to the stable lower branch in Figure 13.10 (region I), while if $C(0) < -\sqrt{c/b}$, $C \rightarrow \infty$ in finite time (Figure 13.10, region II). A blow-up in finite time also occurs above the SCL for any $C(0)$. This does not mean that the actual solution blows up, just that it evolves to a final state far away from the two steady branches shown in Figure 13.10, and thus is not accessible by using perturbation theory.

As we can see in (13.48), the blow-up above the SCL is described by a tangent function, and therefore we must appropriately interpret the growth rate that we find for this case. Note that the linearized dispersion relation seems to suggest an imaginary exponential growth rate above the space-charge limit [3, 6]. In view of our results, we see that this is actually misleading; moreover, we find that, even below the SCL, the stable steady branch can be destabilized by sufficiently large perturbations.

Our results agree with the linear theory, provided that we consider the limit where the latter becomes applicable. Thus, we must compare the linear theory with (13.49) as $\tau \rightarrow +\infty$ (near the stable branch) and (13.50) as $\tau \rightarrow -\infty$ (near the unstable branch). To demonstrate this, we set $\tau = \epsilon t$ in (13.49), and consider the limit $t \rightarrow +\infty$. Then,

$$C(t) = \frac{c}{b} \tanh\left[\frac{\sqrt{cb}}{a}(\epsilon t + \tau_0)\right]$$

$$= \frac{c}{b}\left\{\frac{1 - \exp - \left[\dfrac{2\sqrt{cb}}{a}(\epsilon t + \tau_0)\right]}{1 + \exp - \left[\dfrac{2\sqrt{cb}}{a}(\epsilon t + \tau_0)\right]}\right\}$$

$$\approx \frac{c}{b}\left\{1 - 2\exp - \left[\frac{2\sqrt{cb}}{a}(\epsilon t + \tau_0)\right] + \ldots\right\} \tag{13.53}$$

We see that the decay rate of the perturbations as $t \to +\infty$ is equal to

$$\frac{2\sqrt{cb}}{a}\epsilon \approx 2.83\epsilon \tag{13.54}$$

which is the same as that found by the linearized analysis [6]. This quantity is important because it also determines an initial *growth rate* for the jump instability above the SCL described by (13.48). If we substitute the original dimensional time variable into our expressions and write the deviation of α from its value at the SCL as

$$\epsilon = (\alpha - \alpha_{SCL})^{1/2} = \left(\alpha - \frac{4}{3}\right)^{1/2} \tag{13.55}$$

we find that the growth rate is given by

$$D = \left(\alpha - \frac{4}{3}\right)^{1/2}\frac{\sqrt{cb}}{a}\frac{v_0}{l} \tag{13.56}$$

or

$$D = \left(\frac{l\omega_p}{v_0} - \frac{4}{3}\right)^{1/2}\sqrt{2}\,\frac{v_0}{l} \tag{13.57}$$

Of course, for the expansions in (13.51) to be valid, we must have $C(\tau) \ll 1/\epsilon$. However, when C is not too large, (13.48) gives a reliable estimate for the growth rate of the instability.

13.4 A LINEAR THEORY OF VIRCATORS

In the previous sections, the virtual cathode and space-charge-limit instability have been modeled as one-dimensional electrostatic phenomena. The linear theory presented in this section addresses electromagnetic effects, which give rise to the generation of microwaves. Microwave tubes can be broadly divided into devices where an electron beam interacts with the electromagnetic fields in a "gap," and devices where the beam interacts with a traveling wave. In some devices, the division can become blurred, but the vircator is rather clearly a member of the first group.

The basic vircator configuration is shown in Figure 13.12(a). Two resonator sections, l_1 and l_2, are separated by a screen mesh, which also serves as an anode (A) for an electron beam formed at the cathode (K). A schematic view of the (zero-order) kinetic energy (ϵ_e) of a typical electron is also shown in Figure 13.12(b). The electron is accelerated across the first A-K gap, and passes through the mesh anode into the second resonator. If the beam current density is above the space-charge limit, a virtual cathode will be formed in the second resonator and the beam will reflex.

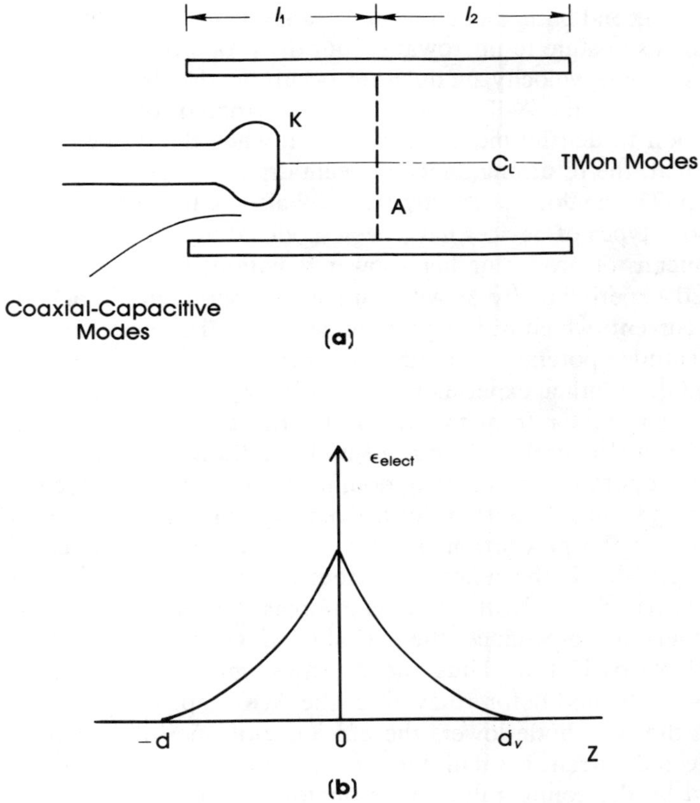

Fig. 13.12 (a) Basic Vircator Configuration and (b) Electron Beam Energy Profile

13.4.1 Vircator Operating Modes

It is possible to identify two general operating modes and a variety of subclasses. In the first, which may be called a *simple vircator*, the A-K gap provides the electron beam, but plays no other role in generating radiation. The screen mesh is assumed as opaque to microwaves, and the properties of the system depend on the resonator l_2 and the electron-beam–virtual-cathode configuration. Oscillating fields in resonator l_2 will first modulate in velocity and then bunch the electron beam. Microwave energy can be extracted, given the proper phase relation between the bunched electrons and the fields. This single (A-K_v) gap system might be expected to resemble a monotron in its operation, and the analogy will be further examined below.

The second basic mode resembles a klystron. In this case, the screen still remains opaque to microwaves, but the properties of resonator l_1 are such that some velocity modulation occurs in the first (A-K) gap. The beam reflexes in the A-K_v region with an associated conversion of velocity modulation to density modulation. Again, when the proper relations are maintained, the returning bunched beam drives the oscillation on the real A-K gap. The analogy to the klystron will also be further explored below.

Both types of devices have been modeled numerically, and they are viable means of producing high-power radiation. Both modes have been observed experimentally as well. In some experiments, blocking the reflexing current, which occurs predominantly on the beam axis, where the virtual cathode potential is a minimum, drastically reduces the microwave signal [33]. In other experiments, the microwave generation is observed to be much greater from the virtual (A-K_v) diode than the real (A-K) diode [36]. In general, a mixture of both modes is present.

The operating mode that dominates depends on which feedback mechanism is most important for a given experimental configuration. This is related to the proportion of beam current, which is reflected by the virtual cathode. If the reflected current is large, the presence of the reflected space charge in the A-K gap forces the diode to operate with a time-dependent impedance, the period of which is determined by the virtual cathode oscillation. Thus, the electrons emitted from the cathode are strongly modulated before they enter the A-K_v gap. Preventing this feedback in the real diode lowers the efficiency of microwave generation. If the reflected current is small, the feedback to the virtual cathode must be supplied by the configuration of resonator l_2. The resonator need not be a true cavity. Impedance mismatches in the waveguide interact with free space, or an antenna may supply sufficient feedback. The larger is the impedance mismatch, then the stronger is the feedback to the virtual cathode, which results in greater beam modulation.

The amount of reflected current is determined by the ratio of the injected beam current to the space-charge-limiting current (I/I_{SCL}). The higher is the ratio, then the more important the reflexing mechanism becomes to the generation of microwaves. Indeed, the nature of the virtual cathode oscillation changes as the I/I_{SCL} increases. We can see this in the virtual cathode limit cycles, depicted by Figure 13.13. Near the space-charge limit, a well defined virtual cathode exists, which repeats its periodic cycle. Well above the space-charge limit, the cycle reverses direction. Rather than the original potential minimum repeating its motion by moving back toward the injection plane, a second virtual cathode is formed near the anode. Because the virtual cathode loses its "memory" in this case, a beam premodulated which enters the A-K_v gap will be more efficient in generating microwaves, if other feedback mechanisms are not present.

In addition to the options described here, many other possible variations exist. The screen mesh will generally not be completely opaque to microwave radiation, and hence resonators l_1 and l_2 will couple. Among the many possibilities thus created, one highly nonlinear variation deserves specific mention. If fields in resonator l_2 create a certain amount of first velocity modulation, then density modulation, then the drive coupled into the real A-K gap may further modulate the incoming beam. Once again, assuming a favorable accumulation of phase relations, the end state would be a highly modulated beam that is driving the fields of resonator l_2. In this limit, a considerable fraction of the incoming beam energy may be converted to radiation. In some respects, this option resembles a two-cavity klystron.

13.4.2 Small-Signal Theory

In order to assess quantitatively the concepts introduced in the preceding subsection, we must further consider the interaction of an electron beam with the fields in a gap. This will be done in two steps: first, we will evaluate the energy exchanged in the small-signal limit; subsequently, we will deduce an expression for the beam-related gap admittance.

If an electron beam enters a gap across which there is a spatially uniform oscillating voltage, the beam will first velocity modulate, then, if the gap is long enough, density modulate (Figure 13.14). The energy exchange may be computed from

$$\frac{d\epsilon}{dt} = \frac{1}{2} \operatorname{Re} \int_{\text{beam}} \mathbf{j}^* \cdot \mathbf{E} \, dV \tag{13.58}$$

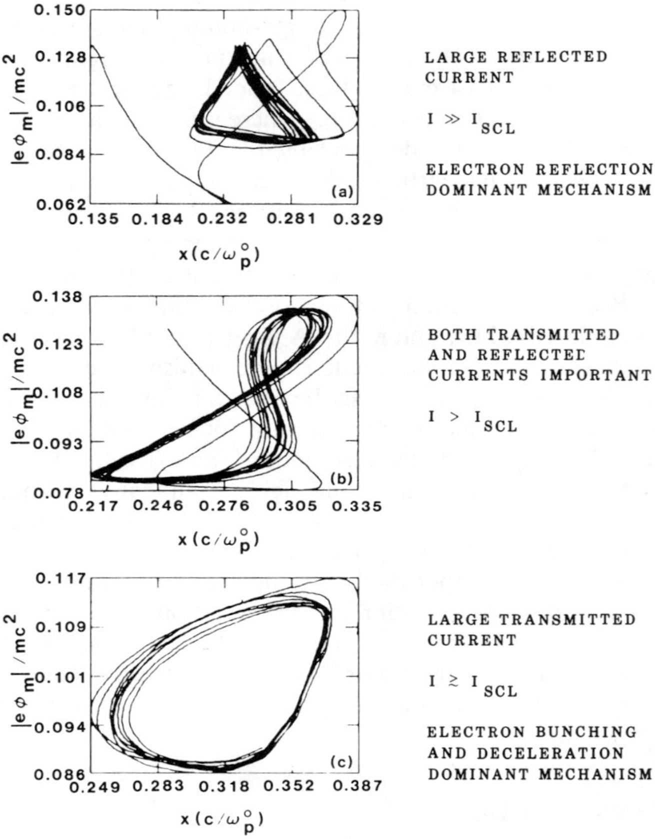

Fig. 13.13 Ratio of Beam Current to Space-Charge-Limiting Current Determines Whether Electron Reflexing or Bunching is the Dominant Microwave Generation Mechanism

where $\mathbf{j}^*$ is the complex conjugate of the current modulation produced by the gap field $\mathbf{E}$, and the integration is over the volume of the beam in the gap. The rate of energy exchange may be conveniently scaled in units of the energy stored in the resonator ϵ_s and the angular frequency ω of the oscillating field. We define

$$\frac{1}{Q} \equiv -\frac{1}{\omega\epsilon_s}\frac{d\epsilon}{dt} \tag{13.59}$$

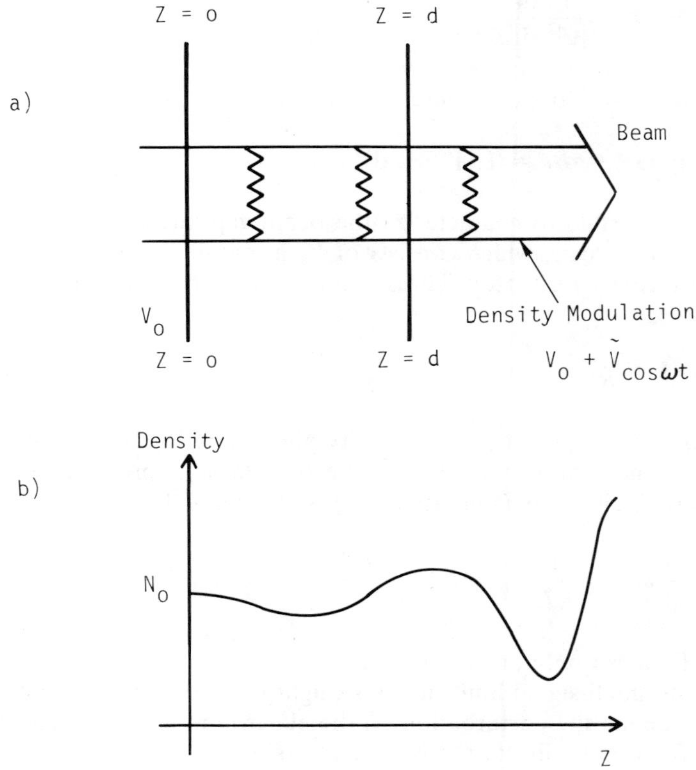

Fig. 13.14 (a) Electron Beam-Gap Configuration and (b) Beam Density Profile as a Function of Distance

The threshold for oscillation:

$$\frac{1}{Q_b} < 0 \text{ and } |Q_b| < Q_L \tag{13.60}$$

where Q_b and Q_L are the beam and loaded-cavity quality factors. The oscillation threshold may be computed with aid of the solution of the equations governing the electron dynamics.

Also useful is the definition of a beam admittance Y_b in terms of the energy exchange rate and the oscillating voltage $\tilde{V}$ on the gap:

$$\text{Re } Y_b = -\frac{1}{|\tilde{V}|^2}\frac{d\epsilon}{dt} \tag{13.61}$$

The complete admittance is determined from

$$Y_b = G_b + iB_b = \int_V \mathbf{j}^* \cdot \mathbf{E}\, dV \tag{13.62}$$

Thus, it is possible to characterize the beam-gap interaction in terms of an equivalent circuit, which consists of Y_b in parallel with the equivalent circuit of a cavity resonator. Threshold is now alternatively defined as

$$\text{Re } Y_b > \frac{1}{R_L} \tag{13.63}$$

where R_L is the equivalent total (cavity plus useful load) resistance. Furthermore, when the oscillation may be treated as approximately steady state, there is also a dynamic frequency shift, defined by

$$B_b = \omega C - \frac{1}{\omega L} \tag{13.64}$$

which will vanish only on resonance.

In the small-signal limit, it is a straightforward matter to evaluate $1/Q_b$. If we define the perturbation on the electron momentum distribution function δf, we obtain the following expression:

$$j = -n_0 e \int dp\, \delta f \mathbf{v} \tag{13.65}$$

(n_0 = average density of the incoming beam). The Vlasov equation may be used to obtain an expression for δf. Then, with the aid of a series of steps, which need not be evaluated in detail here, we obtain the expression:

$$\frac{1}{Q_b} = \frac{\nu}{2(\beta\gamma)^3}\frac{d^3|E|^2}{\epsilon_g(1 + \epsilon_{ext}/\epsilon_g)} \text{Re } \Gamma'(\theta) \tag{13.66}$$

The parameters not yet identified are

d = gap length;
ϵ_g = energy stored in the gap;
E = oscillating electric field in the gap;
ϵ_{ext} = energy stored in the resonator volume, which is external to the gap.

Also,

$$\text{Re } \Gamma'(\theta) = \frac{2(1 - \cos\theta) - \theta \sin\theta}{\theta^3} \tag{13.67a}$$

where

$$\theta = \omega\tau_0 = \frac{\omega d}{v_0} \tag{13.67b}$$

The two expressions (13.67) describe the dependence of the gain (and the form of Y_b) in terms of the undisturbed transit angle θ (τ_0 is the time that an electron, which is undisturbed by the oscillating field, takes to cross the gap). Note that θ has the same form as α discussed in Section 13.3. Equation (13.67) is sketched in Figure 13.15.

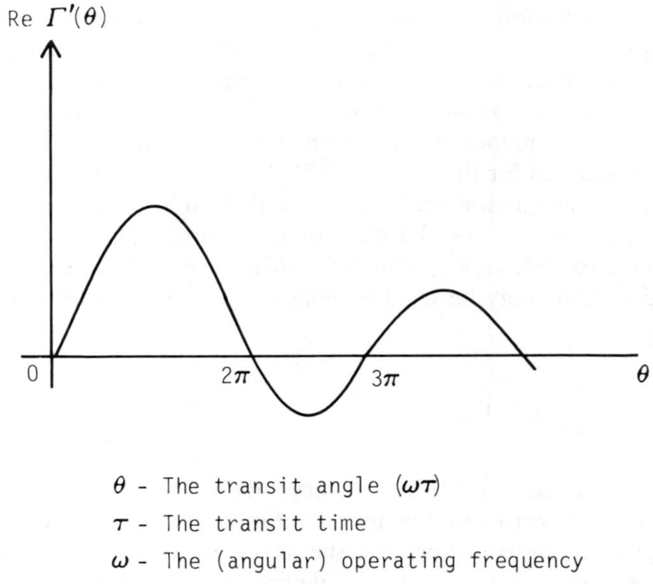

Fig. 13.15 The Function Re $\Gamma'(\theta)$

The Vircator-Monotron

When the transit angle is increased, the function $\Gamma'(\theta)$ has regions where it is negative, and hence instability becomes possible if $1/|Q_b| > 1/Q_L$. The peak gain (max $1/|Q_b|$) occurs near

$$\omega\tau \approx 2\pi(n + 1/4) \tag{13.68}$$

Hence, if τ is known, we can compute the frequency of the radiation produced in this single-gap system.

If the beam is not subjected to an accelerating field in the gap (a drifting beam), then (13.68) predicts

$$f = \frac{(n + 1/4)v_0}{d} = 30(n + 1/4)\,\frac{\beta_0}{d} \quad \text{(GHz)} \tag{13.69}$$

An essentially similar result is obtained when a constant accelerating field is present across the gap. The factor of 30, however, becomes 15 in this limit because the average velocity determines τ. This class of instability, usually termed the *monotron oscillator*, will also show a $V^{1/2}/d$ dependence of the operating frequency, as has been seen experimentally [22] and in PIC code simulations for the vircator [67]. If a system with a virtual A-K_v gap were undergoing a monotron type of oscillation, the transit time would be determined by the zero-order electron dynamics, and it would be generally expected to scale as $j_0^{1/2}$ (j_0 the zero-order current density). Equations (13.61) and (13.66) may be used to obtain τ in the nonrelativistic limit, $\gamma \rightarrow 1$; thus,

$$\text{Re } Y_b = \frac{I}{2V_0}\,\theta\Gamma'(\theta) \tag{13.70}$$

The result summarized in (13.70) is very simple. The gap admittance is proportional to the zero-order beam admittance. In a conventional microwave tube, this quantity is low, and the implied high threshold generally relegates the monotron to the class of devices that are theoretically interesting, but unimportant in practice. The low admittance also generally means that the saturation amplitude will be low, and this is a further objection to practical monotrons if they are based on conventional microwave tube technology. When low-impedance generators are employed, however, not all of these conclusions apply.

The Vircator-Klystron

In the introduction, we described a vircator-klystron mode, and for better understanding of this possibility, a few remarks about the conventional reflex klystron will be of use. The latter device has a single cavity, which first provides velocity modulation, and the condition $\omega\tau_0 \ll 1$ usually applies in this gap. The beam leaves the modulating gap, then moves into a retarding field region and begins to bunch. The various potentials are arranged so that the bunch arrives back at the gap during the maximum retarding phase of the oscillating voltage, and energy is extracted from the beam.

The small-signal admittance of this system may be described by

$$Y_b(\theta) \approx \frac{I}{V_0}\,\theta e^{i(\pi/2-\theta)} \tag{13.71}$$

where θ is again a transit angle. Note that $Y_b(\theta)$ in (13.71) is approximately θ^3 larger than the monotron admittance. Furthermore, θ itself can now be large because it is associated with a decelerating region and not a gap transit angle.

The angle θ where maximum gain occurs is given by

$$\theta = 2\pi(n - 1/4) \tag{13.72}$$

and this gain will increase with n (lower threshold). In the conventional klystron, the magnitude of the transit angle is controlled by the size of the retarding voltage, while in a vircator-klystron, this potential will be equal to the cathode voltage. Hence, a vircator-klystron might be expected to operate in an $n = 1$ mode. Due to the high threshold current required, conventional reflex klystrons do not generally make use of the lowest order mode.

For operation in the $n = 1$ mode, however, the operating frequency will be

$$f = \frac{3}{4}\frac{1}{\tau} \approx \frac{1}{2}\sqrt{\frac{eV_0}{md^2}} \tag{13.73}$$

which is nonrelativistic, where the last relation follows from the Child-Langmuir law. Again, the frequency depends on the real gap, but the frequency will be lower [15 $(n - 1/4)$ β_0/d (GHz)].

13.4.3 Evaluation of the Vircator Frequency with Maximum Gain

Based on this linear theory and the virtual cathode scaling laws derived earlier, we may now evaluate the vircator frequency at which we can obtain maximum gain. The scaling laws (Section 13.3.2) indicate that the vircator's fundamental oscillation frequency is related to the relativistic beam plasma frequency, given by

$$\omega_p^{\text{rel}} = \left(\frac{4\pi n_0 e^2}{\gamma m}\right)^{1/2} \tag{13.74}$$

For a nonrelativistic beam in a semi-infinite, one-dimensional drift space, the vircator frequency ω_{osc} is directly proportional to the beam plasma frequency, given by (13.74) with $\gamma = 1$. This is a good approximation for most nonresonant vircator experiments being conducted to date because the beam is of low energy, and the drift tube length is significantly greater than its radius. Therefore, $\omega_{\text{osc}} \propto n_0^{1/2}$. The current density extracted from a planar diode is defined by the nonrelativistic Child-Langmuir relation, which is

$$j_{\text{CL}} = \frac{4}{9}\left(\frac{2e}{m}\right)^{1/2} \frac{V_0^{3/2}}{4\pi d^2} = n_0 e v_0 \tag{13.75}$$

where V_0 is the diode voltage, d is the anode-cathode (A-K) gap spacing, and v_0 is the electron velocity at injection. An expression for velocity is obtained from conservation of energy. It is

$$v = \left(\frac{2e}{m}\right)^{1/2} V^{1/2} \tag{13.76}$$

Note that we have consistently ignored relativistic effects in ω_{osc}, j_{CL}, and v_0 in this derivation. If we combine (13.75) and (13.76), and solve for n_0, then we obtain

$$n_0 = \frac{1}{9\pi e}\frac{V_0}{d^2} \tag{13.77}$$

or

$$\omega_{\text{osc}} \propto \omega_p = \frac{2}{3}\left(\frac{e}{m}\frac{V_0}{d^2}\right)^{1/2} \tag{13.78}$$

Thus, $\omega_{osc} \propto (V/d^2)^{1/2}$. This is precisely the dependence seen in planar-diode vircator experiments to date [22, 25].

We can now include the linear theory. Note that (13.69) and (13.73) assume one-way, unperturbed (i.e., small-signal) transit times across the gap, as given by d/v_0, where v_0 is the initial velocity of the electron, which are equal to $(2eV_0/m)^{1/2}$. These equations predict frequencies that are high for the vircator in comparison with simulation results [67]. This is because the vircator oscillation frequency depends on the two-way transit time across the A-K_v gap. That is, the periodicity of the electrons is from the anode to virtual cathode and back again. When combined with the strong electric field, which the electrons observe over one cycle, first retarding, then accelerating, this implies that $\tau = 4\tau_0$. By using (13.72) with $\theta = \omega\tau = 4\omega\tau_0$, $\tau_0 = d/v_0$, $v_0 = (2eV_0/m)^{1/2}$, and setting $n = 1$, we obtain

$$\omega_{osc} = \frac{\pi}{2} \cdot \frac{3}{4} \cdot \left(\frac{2eV_0}{md^2}\right)^{1/2} \tag{13.79}$$

A comparison of (13.78) with (13.79) yields

$$\omega_{osc} = \frac{\pi}{2} \cdot \frac{3}{4} \cdot \frac{3\sqrt{2}}{2}\,\omega_p = \frac{5}{2}\,\omega_p \tag{13.80}$$

which is precisely the result seen experimentally in our work [45] as well as from nonlinear one-dimensional electrostatic [60] and two-dimensional electromagnetic [26, 67] simulations at both nonrelativistic and relativistic voltages. This appears to indicate that the simple vircator indeed works like a reflex klystron with $n = 1$ and $\tau_0/\tau \sim 0.25 \ll 1$, rather than as a monotron.

It is important for the reader to note that, while maximum gain occurs at 5/2 ω_p^{rel}, the bandwidth of the vircator over which gain occurs is large. Thus, waves at frequencies other than 5/2 ω_p^{rel} may grow if the vircator configuration allows feedback at those frequencies. Such is the case when the virtual cathode is surrounded by a cavity. The cavity supplies strong feedback at its modes, which results in the vircator frequency shifting away from maximum gain, or microwave growth at multiple frequencies, unless there is proper tuning of the cavity.

13.4.4 Signal Efficiency

If space-charge forces due to bunching are neglected (a questionable assumption if the bunching efficiency is high), it is a straightforward matter to determine also the dependence of Y on the oscillating voltage. In general terms, as V increases, the magnitude of Y drops, and this roll-off determines the *steady-state operating point*:

$$\text{Re } Y \approx \frac{J_1(X)}{X} \tag{13.81a}$$

where J_1 is a Bessel function of first order, and X is the bunching parameter:

$$X \approx \frac{e\tilde{V}}{V_0} \tag{13.81b}$$

Following the klystron analogy, $\theta = 3\pi/4$ for a vircator-klystron, and in this limit the peak efficiency could reach

$$\eta \approx \frac{1.25}{\theta}$$

or about 53%. We do not expect that this peak efficiency could be immediately achieved in practice, but it indicates the potential advantage to be gained from a detailed understanding of the relevant impedances and how they relate to beam-field coupling.

13.5 MICROWAVE GENERATION FROM THE NONRESONANT VIRCATOR

Before we consider microwave generation from the nonresonant vircator, it is useful to summarize the results from the earlier sections. The virtual cathode oscillator occurs when the beam current injected into a waveguide or cavity resonator exceeds the space-charge-limiting current. The limiting current is defined by the beam energy and the waveguide or resonator geometry. We can visualize simplistically the space-charge-limiting current as that at which the electrostatic potential exceeds the kinetic energy of the beam. Above the space-charge limit, no stable steady state for the electron beam exists. Instead, the beam experiences a jump instability, whereby it transits from a stable steady state to an oscillatory, but nonetheless stable, equilibrium. This oscillatory state is the virtual cathode. The space-charge-limiting current for a solid beam in vacuum, in cylindrical geometry, is given approximately by the following interpolation formula [48] (for more accurate formulas, see Section 13.2.1):

$$I_{\text{SCL}} = \frac{mc^3}{e} \frac{(\gamma^{2/3} - 1)^{3/2}}{1 + 2 \ln R/r_0} \quad (mc^3/e = 17 \text{ kA}) \tag{13.82}$$

where γ is the beam relativistic factor at the anode ($\gamma = 1 +$ voltage (kV)/511), R is the radius of the waveguide or cavity, r_0 is the electron beam radius, c is the speed of light, and e and m are the electronic charge and rest mass, respectively.

The virtual cathode oscillates stably at a set frequency in both time and space. This fluctuating potential barrier acts as a gate to reflect some electrons and transmit others. The motion of the gate causes charge to bunch. The charge bunch and virtual cathode are spatially separated as shown in Figure 13.16. By analogy, this configuration represents an *LC* oscillator. The virtual cathode acts as a capacitor to store the beam kinetic energy. During that portion of the cycle in which the potential is greater than the injected beam energy, charge is constrained to remain near the anode. This starves the virtual cathode so that its amplitude decreases below $(\gamma_0 - 1)\, mc^2/e$. Some of the charge bunch is transmitted once this occurs. The electron motion represents a large, time-varying current through an inductor. The presence of charge away from the anode re-establishes the virtual cathode, and the cycle repeats. The oscillating current generates microwaves. The wave frequency is that for oscillation of the virtual cathode. Maximum growth occurs at $5/2\, \omega_p^{\text{rel}}$. This result can be rewritten for planar diodes by using the Child-Langmuir law, as

$$f(\text{GHz}) = 10.2 \sqrt{j(\text{kA/cm}^2)/\beta\gamma} \tag{13.83}$$

A large number of the vircator experiments to date have been conducted in a geometry where the electron beam is injected into a long waveguide [22, 24, 25, 31, 33, 35, 36, 68]. The microwaves are diagnosed either in the waveguide terminated by a calorimeter [31, 68], or in open air after extraction [22, 24, 25, 33, 35, 36]. In either case, there is usually insufficient feedback to shift the vircator frequency from that predicted for maximum gain. These experiments are generally characterized by broad-bandwidth microwave generation, which usually excites the waveguide mode closest to cut-off. These characteristics will now be examined.

13.5.1 Vircator Waveguide Mode

The vircator's choice of waveguide mode is evident in simulations where there is no axial magnetic field. If a cold beam is injected and azimuthal symmetry is assumed, the only nonzero fields are E_z, E_r, and

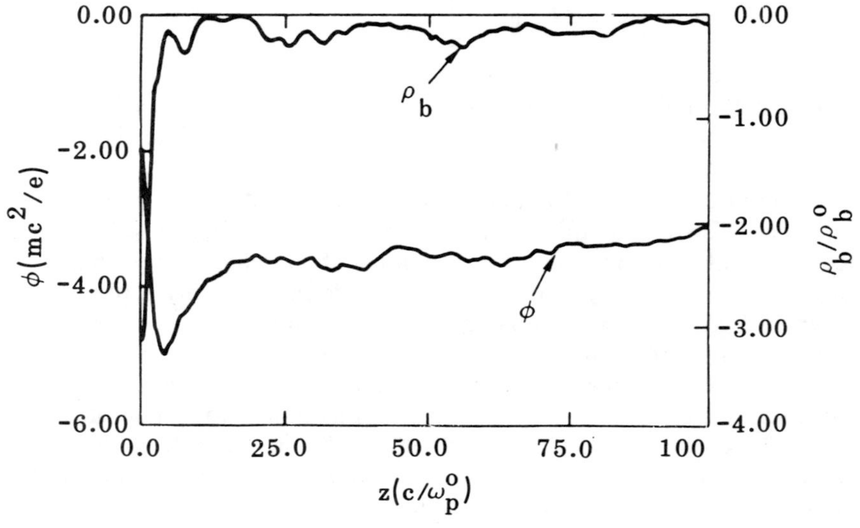

Fig. 13.16 CCUBE Diagnostic of Charge Density ρ_b and Potential ϕ as a Function of a Longitudinal Distance z for an Electron Beam in a Two-Dimensional Drift Space with an Imposed (Large) Axial Magnetic Field

Note: ρ_b is normalized to its value at injection. The value of $(\gamma_0 - 1)$ is 4. The minimum charge density and potential are separated in z.

B_θ in cylindrical geometry. These three fields define a transverse magnetic or TM wave. The direction of wave propagation is, of course, determined by the Poynting flux $(\vec{\mathbf{E}} \times \vec{\mathbf{B}})$. In this case, there will be both axial and radial components of the wave vector $\vec{\mathbf{k}}$. This explains why both polarizations are observed in axially extracted vircator experiments [33, 35]. The wave vector in this case is given by

$$\vec{\mathbf{k}} = \frac{|E_r|}{|E|}\,\hat{\mathbf{k}}_z + \frac{|E_z|}{|E|}\,\hat{\mathbf{k}}_r \tag{13.84}$$

Because the latter term dominates, the Poynting vector is predominantly directed radially outward. The phase velocity of the wave can be expressed as

$$\frac{v_p}{c}\sqrt{1 + \tan^2\Psi} = \sqrt{1 + \left(\frac{k_r}{k_z}\right)^2} \tag{13.85}$$

where Ψ is the angle between the $\vec{\mathbf{k}}$ vector and the z-axis. Because E_z is usually substantially greater than E_r, the vircator preferentially excites the mode closest to cut-off, where the phase velocity can be much greater than c. This does not preclude excitation of lower order TM_{0n} modes. This is particularly true, if the beam cross section overlays a particular node of the J_0 Bessel function upon injection into the waveguide. Note that, whereas in axially extracted vircators a symmetrically configured diode excites a TM_{0n} mode, in radially extracted vircators the output mode will be TE_{1n}.

13.5.2 Vircator Bandwidth

In the absence of a resonant structure, the vircator bandwidth tends to be large due to three factors. These are *nonuniform voltage, A-K gap closure,* and *beam temperature*. This can be understood by examination of (13.78). The dependence of vircator frequency on voltage explains the importance of a flat-top voltage pulse in producing a narrow bandwidth. The closure of the A-K gap, which is due to plasma motion inherent in any foil system, will make the frequency increase during the pulse length as d decreases. This is an effect witnessed in vircator experiments with no resonant cavity [33, 35, 36, 45]. This also makes the spectrum appear broader. An expression for these observations, which is obtained by differentiating (13.78), is

$$\frac{\Delta\omega_{\text{osc}}}{\omega_{\text{osc}}} = \frac{\Delta V}{2V} + \frac{\Delta d}{d} \tag{13.86}$$

Finally, consider the effect of beam temperature on the microwave generation process where no resonant cavity is used, and therefore feedback is minimal. Beam temperature affects both bandwidth and efficiency. In the case of a foil diode, the electrons suffer energy losses and are scattered by the foil. This is in addition to the inherent beam temperature induced by the emission process, which is equivalent to having electrons of widely varying voltage interacting with the virtual cathode. Thus, an extremely broad RF spectrum is produced, which greatly reduces the detectable signal in any particular frequency band.

13.5.3 Nonresonant Vircator Design

A major impediment to constructing an efficient vircator is the effect of heating on the efficiency of microwave generation. As noted some time ago by Birdsall and Bridges [3], beam temperature significantly damps out the amplitude of the potential oscillation. This can be understood in the following way. For a monoenergetic beam, all of the charged particles bunch at the same location. Mathematically, this represents a singularity, where the charge density goes to infinity. In reality, the charge bunch is not infinitely dense, but it does become several times greater than the initial beam-injection density. The severity of the charge bunching leads to efficient microwave generation. If, conversely, the beam has a spread in axial momentum, then the electrons will stop at different locations in the potential well, as shown in Figure 13.17 [69]. This tends to limit the charge bunching and the amplitude of the oscillating electric and magnetic fields. The effect of beam temperature in reducing the RF efficiency of a broad-bandwidth vircator has been witnessed in electromagnetic simulations [70]. A beam spread of less than 3% in energy reduces the microwave generation efficiency by an order of magnitude. Under these conditions, the vircator is nothing more than a Barkhausen oscillator [39]. Indeed, the low efficiency and broad bandwidth observed in nonresonant vircator experiments may be attributed in part to the effects of electron scattering from the foil and resultant increase in beam temperature.

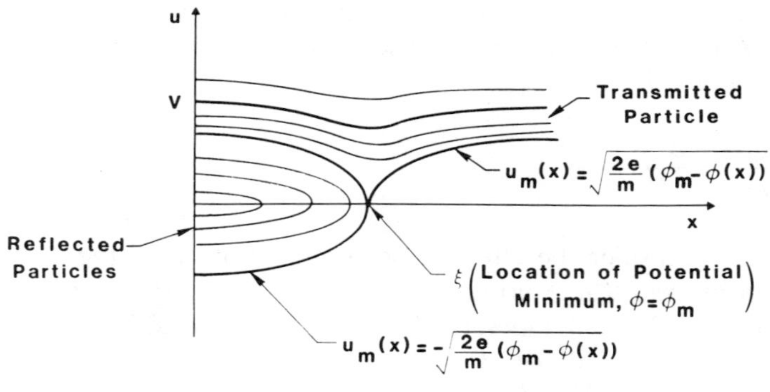

Fig. 13.17 Electron Flow Diagram for a Warm Beam

On the basis of our current understanding of vircator theory, the following characteristics should be designed into any experiment. The voltage pulse should be constant. For centimeter wavelengths, a foil diode or reflex triode can be used, employing a high-transparency mesh for the anode. To mimimize beam scattering, the mesh should be as thin as possible. In a foil diode or reflex triode, tuning is accomplished by changing the A-K gap spacing. Strong axial magnetic fields should be avoided in order to prevent electron reflexing over multiple passes, which can lead to beam heating via the two-stream instability.

For an injected current $I > I_{SCL}$, the oscillation frequency is given by (13.83). Because the oscillating beam is equivalent to a deformable dipole, the preferred waveguide mode for an axisymmetric beam in a straight-walled cylindrical guide is TM_{0n}, where $n \sim D/\lambda_0$, D is the waveguide diameter, and λ_0 is the free-space wavelength. This will place n at or near the mode closest to cut-off. Once n is known, we can determine the phase velocity, group velocity, wavelength, and impedance of the wave in the guide. Similar considerations hold for a rectangular waveguide. Note that both I_{SCL} and the cut-off wavelength λ_c depend on the guide's dimensions and geometry. Both must be considered when choosing an experimental configuration.

13.5.4 Foilless Diode Vircators

Foilless diodes represent a means of achieving millimeter wavelength microwaves from the vircator mechanism. The following points must be considered in the design of a coherent high-frequency vircator. First, generation of high microwave frequencies requires large beam densities. Extremely high beam densities ($n_0 > 10^{14}$ cm^{-3}) have been obtained from foilless diodes. Unlike foil diodes, the beam plasma frequency scales linearly with the electron-cyclotron frequency due to the magnetic field [71]. This is significant because it implies that a single vircator can be tuned over an order of magnitude in frequency (e.g., 10–100 GHz), simply by changing the axial magnetic field strength, without changing the physical structure of the device.

Second, as noted earlier, using a very large value of I/I_{SCL} does not substantially increase frequency or RF efficiency. A foilless diode in a strong axial magnetic field produces a very thin annular beam. Because I_{SCL} for an annular beam is larger than for a solid beam of the same area, the value of I/I_{SCL} will be smaller for the same beam current.

In order to ensure narrow-bandwidth and high-efficiency microwave generation in this configuration, minimal reflexing of electrons in the region between the real and virtual cathodes must occur. The two-stream instability greatly heats the beam electrons, which are tied to the B_z field lines. The axial magnetic field must be shaped to divert the electrons, or *flux excluders* must be employed to confine the magnetic field to the diode region. This arrangement allows the radial space-charge electric field to perform the role of expelling electrons to the waveguide wall. In addition, if the beam is annular, a collimator may be used to help prevent reflexing of electrons back to the cathode.

The requirement that the electron beam must be cold is met by the use of a foilless diode. Experimental [72] and theoretical [71] results indicate that foilless diodes create low-emittance beams. Laminar flow, where the electron's Larmor orbit is smaller than the beam thickness, is obtained when [72]:

$$\omega_c > (\gamma_0 - 1)^{1/2} \frac{c}{\sqrt{a\delta/2}} \tag{13.87}$$

where a is the orbital radius, δ is the radial spacing between the cathode and drift tube wall (which acts as the anode), and ω_c is the electron-cyclotron frequency given by eB_z/mc. Low beam scatter is also ensured because of the lack of a foil.

Finally, the diode voltage and injected current must be constant. More appropriately stated, the *impedance* must be constant. Flat-top voltage pulse can be attained in a variety of ways for several diode configurations. However, at high voltages, the foilless diode purely operates as a resistive load [71], and therefore $\omega_{osc} \propto \sqrt{I/V} = 1/\sqrt{Z}$ is constant. Also, the absence of diode closure in some foilless diode experiments makes a long-pulse device possible.

It is evident from our discussion that the foilless diode in a strong axial magnetic field represents the optimal configuration for a high-frequency vircator. This configuration optimizes microwave power and efficiency, while generating coherent high-frequency radiation. A design that incorporates all of these features was proposed several years ago [67, 73, 74], and is shown in Figure 13.18. The concept has recently been resurrected and is being pursued, as discussed by Kwan [75].

13.5.5 Electromagnetic Particle Simulations

The ultimate tool for designing vircator experiments is an electromagnetic particle-in-cell or PIC code. The 2-1/2 dimensional code CCUBE has been used extensively for this purpose. In conjunction with the detailed theory presented in this chapter, a small number of simulation runs will optimize the microwave generation from a given experiment. Figures 13.19 to 13.23 show typical output from a CCUBE simulation used to design the original MGX vircator experiment at Lawrence Livermore National Laboratories [68]. Such computer runs help define frequency, bandwidth, waveguide mode, and expected microwave power. Detailed comparisons of vircator experiments and simulations are presented in Chapter 14 by L. Thode.

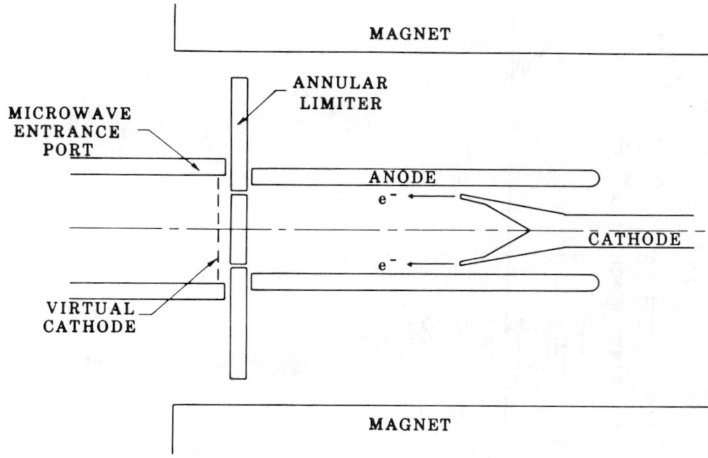

Fig. 13.18 Foilless Diode Vircator Design

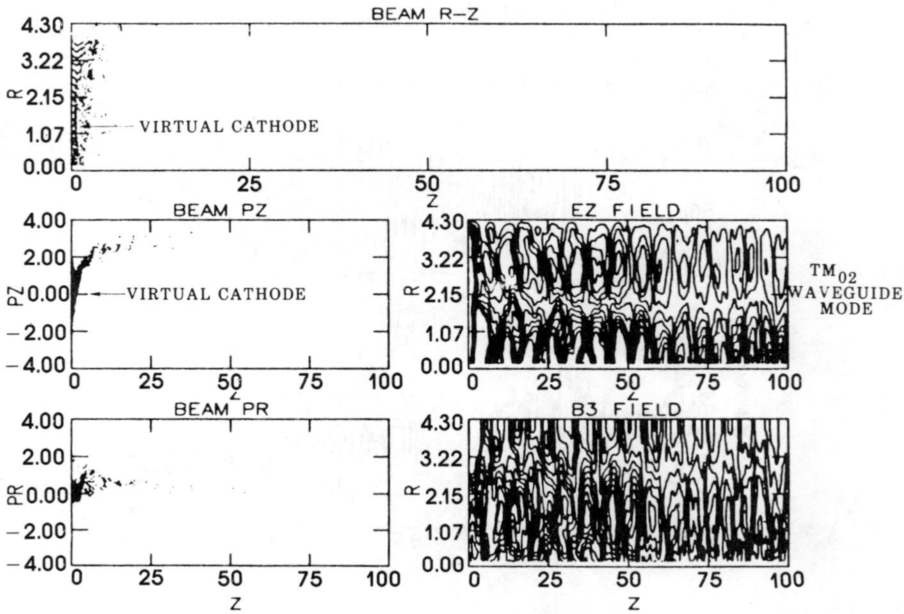

Fig. 13.19 Two-Dimensional EM Simulation of Multigigawatt Virtual Cathode Oscillator (Vircator) Tube (700 kV, 70 kA)

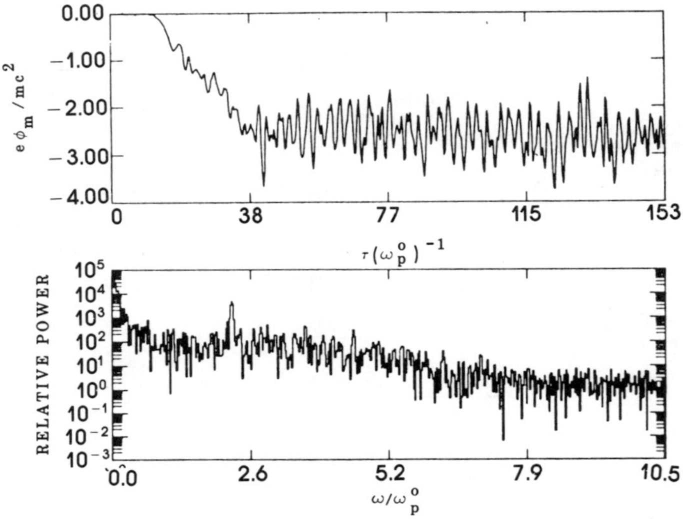

Fig. 13.20 Potential Variation and Power Spectrum in Virtual Cathode Region (700 kV, 70 kA)

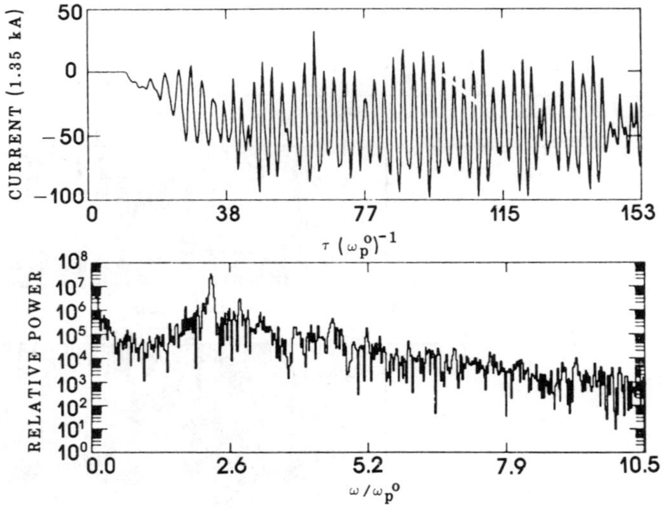

Fig. 13.21 Current Variation and Power Spectrum in Virtual Cathode Region (700 kV, 70 kA)

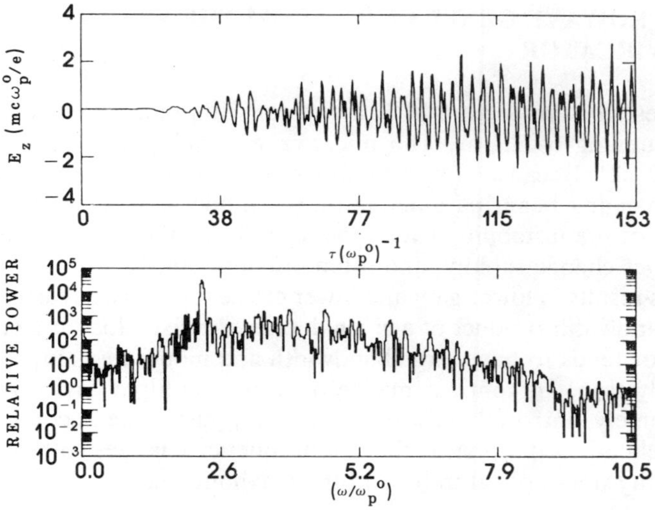

Fig. 13.22 Axial Electric Field in Waveguide (TM_{02} Mode) and Power Spectrum (700 kV, 70 kA)

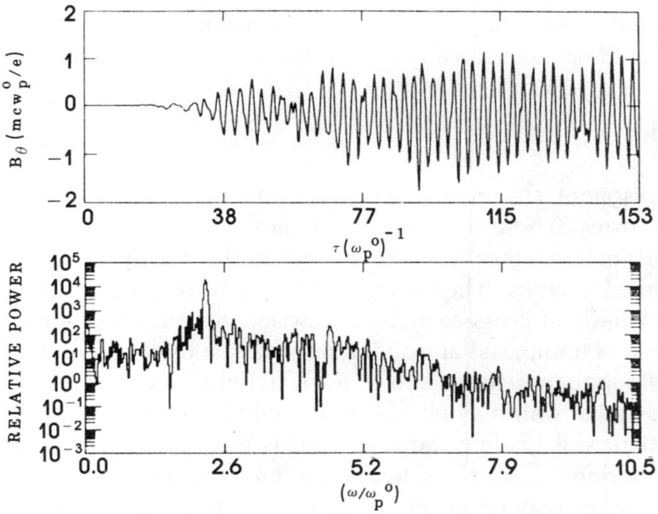

Fig. 13.23 Transverse Magnetic Field in Waveguide (TM_{02} Mode) and Power Spectrum (700 kA, 70 kA)

13.6 MICROWAVE GENERATION FROM THE RESONANT CAVITY VIRCATOR

As described in Section 13.5, a vircator without a resonant structure is a free-running oscillator. The microwave signal grows from noise at roughly 5/2 ω_p^{rel}. Because of beam temperature, nonuniform beam current density, changing beam parameters during diode closure, and the large bandwidth of the instability where gain can occur, the vircator frequency can *and does* change rapidly, and often discontinuously, during the beam pulse. This results in lower gain and lower efficiency for two reasons. First, the gain-bandwidth product of a given device with fixed load is a constant. The vircator tends to have large bandwidth and moderate gain, similar to a reflex klystron. In contrast magnetrons and multiple cavity klystrons have extremely narrow bandwidths and high gain. The second reason, which is related to the first, is that the frequency changes so rapidly that the instability does not saturate, and the maximum gain is never obtained as a result.

In order for the vircator to reach its potential as a high-power source, a resonant structure must be used in order to trade off bandwidth for gain. Ideally, the resonant structure would allow only one cavity mode to grow over the bandwidth in which the vircator would normally have gain. The result is a single operating frequency in a given configuration. The resonant structure would be mechanically adjustable to the frequency of the source.

13.6.1 Theory

The resonant structure used in conjunction with the vircator may take many forms. These may include cylindrical cavities, iris loaded cavities, re-entrant coaxial cavities, re-entrant radial cavities, and radially or axially coupled cavities. These cavities may be of rectangular, circular, or generally cylindrical cross section. Likewise, the output or load may be variable. For example, radial wall loading may be used by employing loop or monopole antennas connected to a coaxial line, waveguide, or another cavity. The radial wall may be loaded directly by a waveguide or another cavity. Radial wall loading, however, leads to asymmetries in the mode pattern, especially if there is heavy output loading. Alternatively, the microwave power may be axially extracted by stub coupling, which maintains symmetry and has the added advantage of forming a natural coaxial transmission line for propagation of the microwaves in a transverse electromagnetic or TEM mode. For cylindrical geometry, this implies that only radial electric E_r and azimuthal magnetic B_θ fields would be present on the output line.

Of the various designs for cavities with output coupling considered to date, the simplest appears to be the most useful. This is a pill-box cavity with a cylindrical cross section. The unloaded resonant vircator is depicted in Figure 13.24. For simplicity, let us assume that the cross section is circular, rather than arbitrary. In the figure, *d* is the anode-cathode (A-K) gap spacing, *r* is the cathode radius, *b* is the diode wall radius, *h* is the height of the resonator, and *r* is the radius of the resonator. The microwaves may be radially extracted through a waveguide, or axially extracted by coupling to a coaxial stub. In order to allow for coupling between the A-K gap region and the cavity, the anode may not be a continuous screen mesh or foil. For instance, it may have a hole in the center, or an open annular region at the outer edge where it connects to the radial wall.

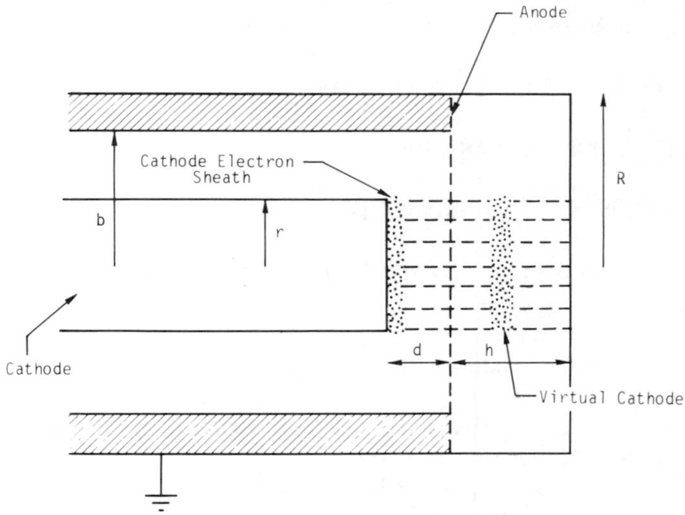

Fig. 13.24 Simple Resonant Cavity Structure for Vircator

The design of the resonant cavity vircator would proceed in the following way. Although it is not necessary, and may not be allowed by physical constraints on its size and shape or the voltage breakdown thresholds of materials, it is nonetheless desirable that the impedance of the diode match as closely as possible the impedance of the pulse-forming line or network in order to maximize power flow. If we assume that the diode may be approximated as planar ($r \gg d$) and that the voltage is not highly relativistic, then the Child-Langmuir relation may be used to define *r* and *d*:

$$Z = \frac{V}{I} = \frac{V}{j_{CL} \cdot \text{area}}$$

$$= \frac{V}{2.34 \times 10^{-6} \dfrac{V^{3/2}}{d^2} \pi r^2} \tag{13.88a}$$

or

$$Z(\Omega) = 136V^{-1/2}(d/r)^2 \tag{13.88b}$$

where V is in megavolts, and d and r are in the same units. Other formulations should be used for highly relativistic beams [4].

The values of r and d are determined by the choice of frequency. The frequency of oscillation can be written as

$$f_{osc}\ (\text{GHz}) = 10.2\ \sqrt{\text{j}(\text{kA/cm}^2)/\beta\gamma} \tag{13.89}$$

by using (13.74) and (13.80), the relation:

$$n_b = 2.08 \times 10^{11}\ \text{j}(\text{kA/cm}^2)/\beta \tag{13.90}$$

and the values of e and m. By substituting the Child-Langmuir relation in the form:

$$j_{CL}\ (\text{kA/cm}^2) = 2.34V^{3/2}\ (\text{MV})/d^2\ (\text{cm}) \tag{13.91}$$

for the current density, we have uniquely defined the value of d, and therefore r, for the desired frequency.

The cavity geometry is set by two factors. The first is to have the resonant frequency of the cavity equal to the frequency of maximum gain for the space-charge instability. This is given by (13.89). The second factor is to have only one mode with a resonant frequency in the bandwidth where the space-charge instability has significant gain. The resonant cavity frequency for vacuum cylindrical cavity of circular cross section is

$$f_{mnp}\ (\text{GHz}) = 4.775 \sqrt{\frac{X_{mn}^2}{r^2} + \frac{p^2\pi^2}{h^2}} \tag{13.92}$$

where r and h are in centimeters; m, n, and p are integers, which can take on values of $m = 0, 1, 2, 3, \ldots$, $n = 1, 2, 3, \ldots$, and $p = 0, 1, 2, 3,$

. . . ; X_{mn} is the nth root of $J_m(X) = 0$ for TM and $J'_m(X) = 0$ for TE cavity modes. Of course, J_m is the Bessel function of order m, and J'_m is its derivative.

One way to ensure that only one cavity mode grows is to choose cavity dimensions whereby only the lowest order mode has a frequency in the gain bandwidth. Vircators in the cylindrical geometry discussed here can generate TM modes. Because of the strong charge bunching in the virtual diode section, the $p = 0$ mode (constant E_z field) is not possible. Finally, given azimuthal symmetry, we expect $m = 0$. Therefore, TM_{011} is the lowest order cavity mode of interest. Another method would be to attempt to place the higher order modes far enough apart so as to have large growth in only one mode.

Finally, there is the load on the resonator system. The load should be set to allow as much microwave power output from the cavity as possible, consistent with the provision of sufficient feedback to have high gain for the instability and the prevention of RF vacuum breakdown. The entire question of optimizing the cavity load is a major one for the system efficiency, and needs to be researched in detail.

To summarize, a resonant cavity vircator would work as follows. An electron beam is accelerated in the real diode, and injected into the resonant cavity with a certain energy and current density. This determines the frequency of oscillation. Because the beam is first accelerated, it can be seen that the vircator is more analogous to a klystron than other space-charge microwave generators, such as microwave triodes or lighthouse tubes [76]. As the electrons form a virtual cathode and begin to oscillate, they receive a very strong feedback from the cavity at one particular frequency. This forces the oscillator to be locked in frequency. The feedback overcomes the natural tendency of the vircator toward frequency shifting and mode hopping due to the large bandwidth of the space-charge instability. With a narrow effective bandwidth and the mode competition eliminated, the gain of the device is much larger, and it is allowed to saturate at a single frequency. The proximity of the output load to the virtual cathode allows for strong coupling.

13.6.2 Reflex Triode Resonant Cavity Experiment

A resonant vircator experiment was conducted on the Pulserad PR112A. The experimental configuration is schematically represented in Figure 13.25. The anode and cathode were arranged in a reflex triode geometry (cathode at ground, anode at positive voltage). The experiment

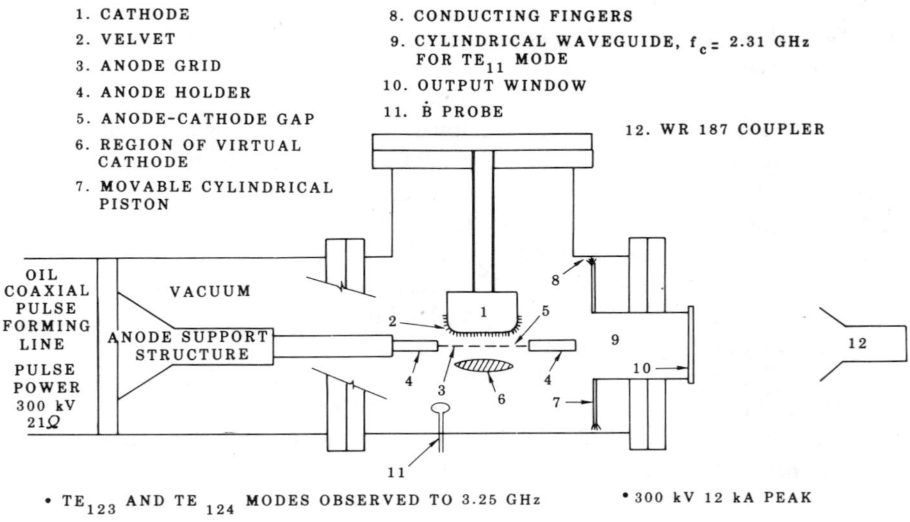

Fig. 13.25 Schematic of Reflex Triode Geometry for Resonant Vircator Experiment on the PR112A

is similar to the Soviet one, which was accomplished in rectangular waveguide [25]. The major difference is the addition of a moveable piston to tune the cavity.

At the proper spacing, the TE_{124} mode was observed with a resonant frequency of 3.25 GHz. By moving the piston toward the pulser by a half-wavelength, the TE_{123} mode was witnessed with the same frequency. The usual frequency-chirping characteristic of a free-running vircator returned when the beam voltage was changed so that the vircator was no longer resonant with the cavity. This behavior is shown in Figure 13.26, recorded by a *polar frequency discriminator* (PFD). When the two traces (horizontal and vertical signal components) maintain the same ratio of amplitudes, the frequency is a constant as in Figure 13.26(b, c). In contrast, the frequency in Figure 13.26(d) is changing from 3.30 to 3.67 GHz. This implies a bandwidth of greater than 10% over the microwave pulse.

The hollow waveguide in the center of the piston was chosen to couple out a TE_{11} waveguide mode from the TE_{123} cavity mode. The radiation pattern for a TE_{11} mode is a solid, horizontally polarized, almost uniformly illuminated circle, which is ideal for testing. The TE_{11} output mode was verified by the light pattern excited in fluorescent bulbs and the plane of polarization of the *E*-field. The total output power was not measured

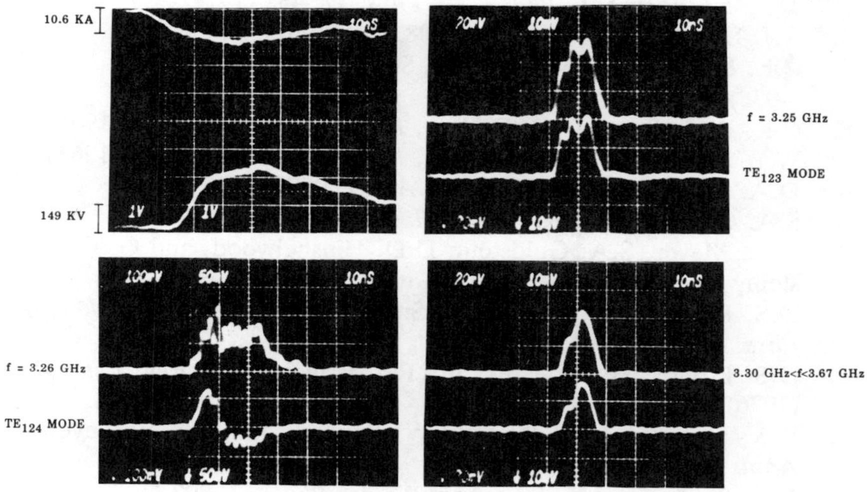

Fig. 13.26 (a) Current (top) and Voltage (bottom) on the Triode; PFD Data Showing Resonant Oscillation (b) at 3.26 GHz and (c) 3.25 GHz; (d) Frequency Sweeping from 3.30 GHz to 3.67 GHz

because the lucite forming the vacuum-air interface over the three-inch diameter waveguide breaks down, and air breakdown has been witnessed as well. However, the microwave detector measures 110 kW/cm^2 at 25 cm distance and the fluorescent bulbs light up uniformly over a diameter of two feet, thus indicating high power. Also, a B-dot probe on the cavity wall to measure B_z indicates a power density in excess of 10 MW/cm^2. This is especially significant because the total input beam power is only 3.6 GW (300 kV, 12 kA). The efficiency could conceivably approach 10% as in the Soviet experiment [25].

REFERENCES

1. G.D. Child, *Phys. Rev. Ser. I* **32**, 492 (1911).
2. I. Langmuir, *Phys. Rev.* **21**, 419 (1923).
3. C.K. Birdsall and W.B. Bridges, *Electron Dynamics of Diode Regions* (Academic Press, New York 1966), Chapter 3.
4. H.R. Jory and A.W. Trivelpiece, *J. Appl. Phys.* **40**, 3924 (1969).
5. V.S. Voronin, Yu. T. Zozulya and A.N. Lebedev, *Zh. Tekh. Fiz.*

42, 546 (1972); *Sov. Phys.-Tek. Phys.* **17**, 432 (1972).
6. R.J. Lomax, *Proc. IEE Pt. C* **108**, 119 (1961).
7. C.E. Fay, A.L. Samuel, and W. Shockley, *Bell Syst. Tech. J.* **17**, 49 (1938).
8. C.K. Birdsall and W.B. Bridges, *J. Appl. Phys.* **32**, 2611 (1961).
9. W.B. Bridges and C.K. Birdsall, *J. Appl. Phys.* **34**, 2946 (1963).
10. D.A. Dunn and I.T. Ho, *A.I.A.A.J.* **1**, 2770 (1963).
11. S.A. Goldstein and R. Lee, *Bull. APS* **23**, 763 (1978).
12. R.A. Meyer, S.A. Goldstein, D.D. Hinshelwood, and G. Cooperstein, *Bull. APS* **24**, 977 (1979).
13. D.S. Prono, J.M. Creedon, I. Smith, and N. Bergstrom, *J. Appl. Phys.* **46**, 3310 (1975).
14. D.S. Prono, J.W. Shearer, and R.J. Briggs, *Phys. Rev. Lett.* **37**, 2 (1976).
15. P.A. Miller, J.A. Halbleib, J.W. Poukey, and J.T. Verdeyen, *J. Appl. Phys.* **52**, 593 (1981).
16. J.W. Poukey and N. Rostoker, *Plasma Phys.* **13**, 897 (1971).
17. C.L. Olson, "Collective Ion Acceleration with Linear Electron Beams," in *Collective Ion Acceleration-Springer Tracts in Modern Physics*, Vol. 84, (Springer-Verlag, 1979).
18. C. L. Olson, *IEEE Trans. Nucl. Sc.* **NS-26**, 4231 (1979).
19. C.L. Olson, J.R. Woodworth, C.A. Frost, and R.A. Gerber, *IEEE Trans. Nucl. Sc.* **NS-28**, 3349 (1981).
20. R.B. Miller, in *Proc. 2nd Int. Top. Conf. on High Power Electron and Ion Beam Research and Technology* (Cornell University, Ithaca, NY, 1977), p. 613.
21. R.B. Miller, in *Collective Methods of Acceleration*, N. Rostoker and M. Reiser, eds. (Harwood Academic Publishers, Geneva, 1979), p. 675.
22. R.A. Mahaffey, P. Sprangle, J. Golden, and C.A. Kapetanakos, *Phys. Rev. Lett.* **39**, 843 (1977).
23. H.E. Brandt, A. Bromborsky, H.B. Bruns, and R.A. Kehs, in *Proc. 2nd Int. Top. Conf. on High Power Electron and Ion Beam Research and Technology* (Cornell University, Ithaca, NY, 1977), p. 649.
24. J.M. Buzzi, H.J. Doucet, B. Etlicher, P. Haldenwang, A. Huetz, H. Lamain, C. Rouille, J. Cable, J. Delvaux, J.C. Jouys, and C. Peugnet, *ibid.*, p. 663.
25. A.N. Didenko, G.P. Fomenko, I.Z. Gleizer, Ya.E. Krasik, G.V. Melnikov, S.F. Perelygin, Yu.G. Shtein, A.S. Sulakshin, V.I. Tsvetkov, and A.G. Zerlitsin, *ibid.*, p. 683.
26. D.J. Sullivan, in *Proc. 3rd Int. Top. Conf. on High Power Electron and Ion Beam Research and Technology* (Institute of Nuclear Physics, Novosibirsk, USSR, 1979), p. 769.

27. D.J. Sullivan, *Bull. APS* **25**, 948 (1980).
28. D.J. Sullivan, D.E. Voss, W.M. Bollen, R.H. Jackson, and E.A. Coutsias, AMRC-R-451 (1983), unpublished.
29. D.E. Voss, D.J. Sullivan, G.F. Kiuttu, R.J. Adler, and J. Walsh, *Bull. Am. Phys. Soc.* **29**, 1285 (1984).
30. T.J.T. Kwan and L.E. Thode, *Phys. Fluids* **27**, 1570 (1984).
31. S. Burkhardt, R. Scarpetti, and R.L. Lundberg, *J. Appl. Phys.* **58**, 28 (1985).
32. A. Bromborsky, H. Brandt, and R.A. Kehs, *Bull. APS* **26**, 165 (1981); private communication.
33. M.C. Clark, private communication.
34. A.N. Didenko, A.G. Zherlitsyn, A.S. Sulakshin, G.P. Fomenko, V.I. Tsvetkov, and Yu. G. Shteyn, *Pisma v. Zhur. Tek. Fiz.* **9**, 1510 (1983).
35. H.A. Davis, R.R. Bartsch, L.E. Thode, E.G. Sherwood, and R.A. Stringfield, *Phys. Rev. Lett.* **55**, 2293 (1985).
36. J. Benford, H. Sze, W. Woo, and B. Harteneck, *Phys. Rev. Lett.* **56**, 334 (1986).
37. R. Espinosa, *1984 Microwave System Design Handbook* (1983), p. 126.
38. H. Hertz, *Wied. Ann.* **19**, 87 (1893).
39. H. Barkhausen and K. Kurz, *Phys. Zeit.* **21**, 1 (1920).
40. W.E. Benham, *Phil. Mag.* **5**, 641 (1928).
41. F.B. Llewellyn and A.E. Bauen, *Bell Syst. Tech. J.* **18**, 280 (1939).
42. V.A. Flyagin, A.V. Gaponov, M.I. Petelin, and V.K. Yulpatov, *IEEE Trans.* **MTT-25**, 514 (1977).
43. S. Jacobs, *et al., Physics of Quantum Electronics*, Vol. 5, (Addison-Wesley, Reading, MA, 1978).
44. S. Jacobs, *et al., Physics of Quantum Electronics*, Vol. 7, (Addison-Wesley, Reading, MA, 1980).
45. D.J. Sullivan, G.F. Kiuttu, R.J. Adler, D.E. Voss, and J. Walsh, *Phase I Report: Resonant Vircator Study*, Mission Research Corporation, AMRC-R-654, (1984), unpublished.
46. In addition to the obvious communication and research applications, there is a very large range of industrial processing applications (See Ref. [37]).
47. The general value of α follows from the ratio of losses to the useful output.
48. L.A. Bogdankevich and A.A. Rukhadze, *Sov. Phys. Upsekhi* **14**, 163 (1971).
49. B.N. Brejzman and D.C. Ryutov, *Nucl. Fusion* **14**, 875 (1974).
50. T.C. Genoni and W.A. Proctor, *J. Plasma Phys.* **23**, 129 (1980).
51. R.B. Miller and D.C. Straw, *J. Appl. Phys.* **47**, 1897 (1976).

52. J.R. Thompson and M.L. Sloan, *Proc. 2nd Int. Conf. on High Power Electron and Ion Beam Research and Technology* (Laboratory for Plasma Studies, Cornell University, Ithaca, NY, October 1977), Vol. II, p. 734, Eq. (14).
53. B.B. Godfrey, *J. Comp. Phys.* **15**, 504 (1974).
54. B.B. Godfrey, *J. Comp. Phys.* **19**, 58 (1975).
55. R.B. Miller and D.C. Straw, *J. Appl. Phys.* **48**, 1061 (1977).
56. L.E. Thode, B.B. Godfrey, and W.R. Shanahan, *Phys. Fluids* **22**, 747 (1979).
57. R.Z. Sagdeev, in *Reviews of Plasma Physics*, Vol. 4, M.A. Leontovich, ed. (Consultants Bureau, 1966), p. 23.
58. S.S. Moiseev and R.Z. Sagdeev, *Plasma Phys.* **5**, 43 (1963).
59. E.A. Coutsias and D.J. Sullivan, *Phys. Rev. A* **27**, 1535 (1983).
60. D.J. Sullivan and E.A. Coutsias, in *High Power Beams '81*, H.J. Doucet and J.M. Buzzi, eds. (Ecole Polytechnique, Paris, 1981), p. 371.
61. J.R. Pierce, *J. Appl. Phys.* **15**, 721 (1944).
62. J.W. Poukey, J.P. Quintenz, and C.L. Olson, *Appl. Phys. Lett.* **38**, 20 (1981).
63. B.B. Godfrey, Mission Research Corporation, AMRC-R-282 (1981), unpublished.
64. J. Kevorkian and J.D. Cole, *Perturbation Methods in Applied Mathematics* (Springer-Verlag, Berlin, 1981).
65. G. Birkhoff and G. Rota, *Ordinary Differential Equations* (Ginn, New York, 1962).
66. A.V. Paschenko and B.N. Rutkevich, *Fiz. Plazmy* **3**, 774 (1977); *Sov. J. Plasma Phys.* **3**, 437 (1977).
67. D.J. Sullivan, *IEEE Trans. Nucl. Sc.* **NS-30**, 3426 (1983).
68. R. Scarpetti, R.L. Lundberg, and S.C. Burkhardt, *Bull. APS* **29**, 1180 (1984).
69. E.A. Coutsias, *J. Plasma Phys.* **31**, 313 (1984).
70. M.A. Mostrom, T.J.T. Kwan, and C.M. Snell, *Bull. APS* **27**, 1075 (1982).
71. M.E. Jones and L.E. Thode, *J. Appl. Phys.* **51**, 5212 (1980).
72. R.B. Miller, K.R. Prestwich, J.W. Poukey, and S.L. Shope, *J. Appl. Phys.* **51**, 3506 (1980).
73. D.J. Sullivan, "A High Power Microwave Generator," US Air Force Invention No. 13,692, US Patent Pending, 1979.
74. D.J. Sullivan, "High Power Microwave Generator Using Relativistic Electron Beam in Waveguide Drift Tube," US Patent No. 4,345,220, August 17, 1982.
75. T.J.T. Kwan, *Phys. Rev. Lett.* **57**, 1895 (1986).

76. D.R. Hamilton, J.K. Knipp, and J.B. Horner Kuiper, *Klystrons and Microwave Triodes*, (Dover, New York, 1966).

Chapter 14
VIRTUAL-CATHODE MICROWAVE DEVICE RESEARCH: EXPERIMENT AND SIMULATION

Lester E. Thode

Los Alamos National Laboratory
Los Alamos, New Mexico

14.1 INTRODUCTION

This chapter is a review of virtual cathode microwave device experiments. Also included are some results of numerical experiments using particle-in-cell plasma simulation techniques. High-power virtual cathode microwave generation has been investigated experimentally since the 1970s, but the bulk of the research has been performed in the 1980s. To date, nearly 30 experiments [1–27] have been conducted and several others are planned.

Initially, the author intended to provide a summary of each experiment. However, that approach made it difficult to understand the status of the overall research effort. Consequently, we decided to summarize the three types of virtual cathode microwave devices based on their operational characteristics. This approach provides a better overview of the subject, although the results of individual experiments are somewhat obscured. However, three somewhat unique experiments are individually described [22–27].

When an electron beam is injected into an evacuated volume, space charge reduces the initial electron beam energy. If the injected beam current exceeds the so-called *space-charge-limiting current*, the kinetic energy of some of the electrons is reduced to zero. These decelerated electrons are then reaccelerated away from the space-charge cloud, back toward the evacuated volume boundary. A space-charge cloud, or *potential*, sufficient to reflect incoming electrons is referred to as a *virtual cathode*. In general, it is difficult to determine when a virtual cathode will form: the space-charge-limiting current depends on the initial beam energy and current-density profile as well as the details of the evacuated volume boundary.

If part of the evacuated volume boundary forms the anode of an electron-beam diode, the resulting configuration is a virtual cathode microwave device. Typically, the evacuated volume is a cavity or waveguide with an output window. Microwave power can be extracted either parallel or transverse to the cathode axis. In a virtual cathode device there are two sources of microwave radiation: (1) oscillations in the position of the virtual cathode [13, 28–41, 45] and (2) trapped reflexing electrons in the potential well formed between the real and virtual cathodes [1, 13, 14, 35, 37, 39, 40, 42–45].

Determining the dominant radiation mechanism has been a major effort in virtual cathode device research. Although the virtual cathode microwave device is simple in geometry, it is difficult to understand theoretically because the feedback mechanisms are nonlinear and time dependent. Understanding the virtual cathode microwave device experimentally is equally difficult. The microwave output is often somewhat chaotic because the nonlinear feedback is associated with the self-electric and self-magnetic fields of the electron beam, which are difficult to control.

Typically, the frequency of the virtual cathode device varies because of diode voltage variation, ion formation, and diode closure. However, the frequency variation is not a basic property of the virtual cathode microwave device itself. Rather, the frequency variation appears to be associated with the pulsed-power generator used to energize the diode. Indeed, it has been demonstrated theoretically that a simple virtual cathode device is significantly more coherent and efficient when the diode voltage is constant in time [45]. Moreover, even with a variation in diode voltage, experiments have achieved frequency control by forming the virtual cathode in a cavity with a resonance near the natural oscillation frequency of the uncontrolled configuration [24].

From the theoretical and experimental investigations, it appears that a simple, uncontrolled, virtual cathode microwave device may not be an ideal microwave tube. However, with diode control, coupled with a resonant output cavity, development of a very high-power microwave device based on either the oscillating or reflexing mechanism may still be possible.

Three types of virtual cathode microwave devices have been investigated: *reflex triode, vircator,* and *reditron*. The *reflex diode* [21] and *turbutron* [42–43] are alternative names used to describe the vircator configuration. In a reflex triode, the anode is pulsed positive and the cathode is grounded. In contrast, the cathode is pulsed negative and the anode is grounded in the vircator. When a magnetic field is applied, both the reflex triode and vircator exhibit significantly reduced power output. In many respects, the reditron is similar to the vircator. However, the reditron uses a thick anode with a narrow slit in conjunction with an applied magnetic field to prevent reflected electrons from re-entering the diode region. If the thick anode is replaced by a thin anode, little microwave radiation is observed until the applied magnetic field is removed. At this point, the reditron has been converted to a vircator. This transformation indicates that the source of microwave radiation is different in the vircator and the reditron.

Table 14.1 provides a basic summary of the operational characteristics of the reflex triode, vircator, and reditron. The information is not representative of a single device, but rather is a composite of all the experiments. The demonstrated range in power, frequency, and pulse width are presented in the first three rows. The fourth row indicates the beam-to-microwave power conversion efficiency. The beam-to-microwave energy conversion efficiency is generally lower because the microwave pulse width is typically shorter than the electron-beam pulse width. The fifth row states whether chirping, or frequency variation, is a characteristic of most experiments. If a repetitively pulsed experiment has been performed for a particular type of device, it is indicated in the sixth row. The seventh row indicates whether the optimal operation is achieved with an applied magnetic field.

Table 14.1 Virtual Cathode Microwave Device Characteristics

Characteristic	*Reflex Triode*	*Vircator*	*Reditron*
Power (GW)	0.1–1.4	0.1–40.0	1.0–2.0
Frequency (GHz)	2–13	0.5–17	4–6
Pulse Length (ns)	10–1300	3–120	10–45
Efficiency (%)	1–46	1–3	2–3
Chirping	Yes	Yes	Unclear
Repetitively Pulsed	No	Yes	No
Magnetic Field	Unclear	No	Yes

Although the vircator has by far the highest demonstrated power level, both the reflex triode and reditron have generated microwave power in excess of 1 GW. The domain of frequency operation is relatively large. As we might expect, the larger are the device dimensions, the lower is the frequency. For example, the 500 MHz vircator experiment had a 61 cm diameter cathode with a 76 cm diameter output waveguide [15]. At the high-frequency end, the 17 GHz vircator experiment used a 2 cm diameter cathode with a 6 cm diameter output waveguide [14]. Low-frequency operation is limited by the field-emission threshold. With a high-voltage generator, frequency operation down to about 100 MHz appears possible, provided that velvet is used on the cathode surface. High-frequency operation is limited by ion formation and diode closure, effects that are aggravated by a small anode-cathode gap spacing and its associated large beam current density. At present, it is not clear where the high-frequency operational limit is located.

Typically, high-power output is associated with a short pulse width, and the microwave pulse width is often shorter than that of the electron beam. However, one reflex triode experiment used a special multiple-pointed carbon cathode to extend the pulse width to 1300 ns [7]. With the exception of the reflex triode, the efficiency of the virtual cathode device is quite low. Increasingly, there is some skepticism concerning the very high efficiency claimed for the reflex triode. The inability to increase the vircator and reditron efficiency above a few percent is a major unresolved problem. However, in all other respects, virtual cathode microwave device development has shown steady progress.

In Section 14.2, we summarize the three basic virtual cathode device configurations. Also, we present a brief discussion of potential-well formation, radiation mechanisms, and extraction techniques. In many experiments, significant mode conversion may have occurred, due to waveguide-to-antenna mismatches. Such mode conversion compounds the already difficult problem of understanding the virtual cathode device characteristics.

The characteristics of seven reflex triode experiments [1–7] are reviewed in Section 14.3. Relative to the vircator and reditron, the reflex triode results are somewhat sketchy. It is clear that the power, frequency, and pulse width are all sensitive to the cathode shape. Also, the frequency output depends strongly on the anode-cathode gap spacing. An applied magnetic field appears to decrease microwave power, but the highest efficiency design may utilize an applied magnetic field. The importance of a resonant cavity for frequency control and efficiency is unclear. Finally, there are a number of questions concerning the power measurements and stated high efficiency.

Axial vircator experiments [8–21] are discussed in Section 14.4. In this case, *axial* refers to configurations where the microwave power is extracted through an output waveguide that has its axis aligned with the cathode axis. With the axial vircator, microwaves are typically radiated from an antenna. In general, the axial vircator has many characteristics in common with the reflex triode. For example, the power, frequency, and pulse width are all sensitive to the cathode shape, and the frequency depends strongly on the anode-cathode gap spacing. Moreover, the microwave power, frequency, and pulse width depend on the anode thickness. It is clear that an applied magnetic field is detrimental to the performance of the axial vircator. To date, the efficiency is consistently only a few percent, but the axial vircator has been scaled to very large power and high energy output [21].

The experimental results of a repetitively pulsed axial vircator [22] are discussed in Section 14.5. Experiments were conducted at 1–10 Hz. Approximately 100 pulses were generated at 1 Hz, with the length of the pulse train limited by anode screen melting. In all cases, microwave power ceased after the sixth pulse, but then reappeared after the eleventh pulse. The characteristics of the individual pulses were similar to those of single-pulse experiments.

In Section 14.6, we present the results of the first *transverse vircator* experiment [23, 24]. In this particular experiment, the electron beam entered through the side of a rectangular output waveguide. Consistent with axial vircator operation, the transverse vircator exhibits a significant reduction in microwave power when a magnetic field is applied. In fact, many of the characteristics of the transverse vircator are similar to the axial vircator. Enclosure of the virtual cathode in a small cavity provided frequency control with a slight increase in efficiency. Another interesting aspect of the experiment was the investigation of microwave power as a function of the position of the virtual cathode within the output rectangular waveguide.

The output characteristics of related transverse vircator experiments [21, 25] are also summarized in Section 14.6. In these experiments, the transverse vircator is similar to an axial vicator, but the output waveguide is truncated to form a cavity. Microwave power is then extracted from the cavity in the transverse direction, perpendicular to the cathode axis, through rectangular waveguides.

Finally, we discuss the reditron experiment [26–27] in Section 14.7. Although the reditron concept was suggested some time ago [32–34], a high-power, proof-of-principle experiment has only recently been performed. The initial configuration generated about 1 GW with a 45 ns pulse width at 6 GHz. More importantly, the operational characteristics of the reditron were found to be consistent with theoretical expectations: (1) power ouput is sensitive to the applied magnetic field strength, (2) there is narrower bandwidth than in the vircator, and (3) output power is concentrated in a single mode.

14.2 CONFIGURATIONS, MECHANISMS, AND EXTRACTION TECHNIQUES

14.2.1 Basic Configurations

A conceptual sketch of the standard reflex triode configuration appears in Figure 14.1. In a reflex triode, the anode, indicated by the dashed line, is pulsed positive by a TEM wave traveling from the insulator of a pulsed power generator. The anode is supported by the center conductor of a vacuum transmission line running from the insulator into the reflex triode, while the outer conductor of the transmission line is attached to the side of the cavity. Both thin foils and highly transparent screens have been used for the anode. In Figure 14.1, the position of the virtual cathode is indicated by the cross-hatched region. The outer conductor of the transmission line, the cathode, and the enclosing cavity are grounded, and microwaves are extracted through the window located in the side of the cavity. Trapped reflexing electrons in the potential well, formed between the real and virtual cathodes, are thought to be the source of microwave radiation.

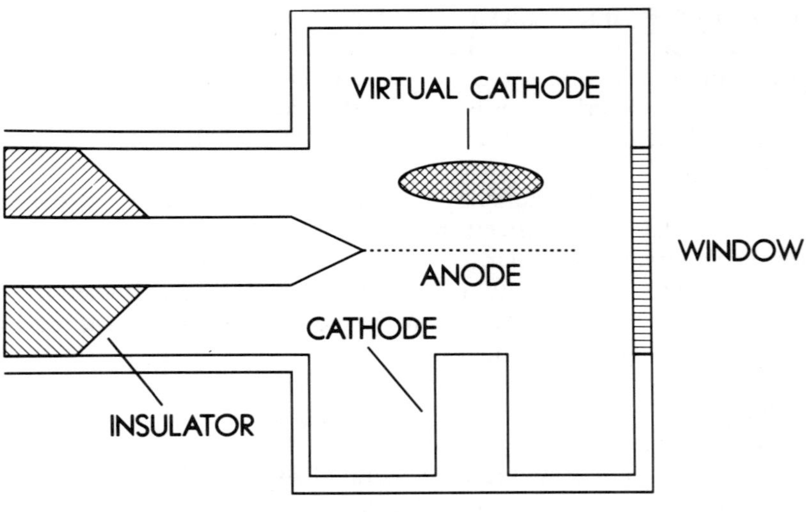

Fig. 14.1 Conceptual Sketch of the Reflex Triode
Note: The virtual cathode is indicated by the cross-hatched region.

A number of different cathode shapes and configurations have been investigated. One reflex triode experiment [7] investigated whether using two closely spaced cathodes would extend the microwave pulse width, as opposed to using the standard single cathode [1, 5, 6] shown in Figure 14.1. Although the standard configuration positions the cathode at the end of the cavity, a configuration with the cathode located on the side of the cavity has also been investigated [2, 3]. There is some indication from the data that this alternative configuration did not perform as well as the standard configuration, but such data are insufficient to support this conclusion.

Transverse extraction occurs with the window located perpendicular to the cathode axis [1–7], as shown in Figure 14.1. Conversely, axial extraction has the window located along the cathode axis. This approach has also been investigated [2, 3]. Of the two approaches, the transverse extraction design appears superior with respect to radiated energy. In either case, the output window is located close to the virtual cathode.

The azimuthally symmetric axial vircator configuration is shown in Figure 14.2. A negative TEM wave travels from the insulator of the pulsed power generator along a coaxial transmission line to energize the axial vircator. The truncated center conductor of the transmission line becomes the cathode. As with the reflex triode, a number of different cathode shapes have been investigated. Both the anode and output waveguide are grounded. The location of the anode is represented by the dashed line. Once again, the position of the virtual cathode is indicated by the cross-hatched region. In contrast to the reflex triode, the output window of the vircator is located far from the virtual cathode. Most vircator experiments have an antenna attached to the output waveguide.

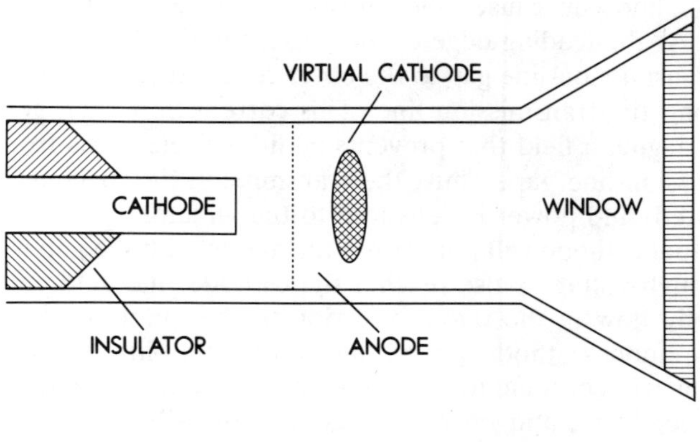

Fig. 14.2 Conceptual Sketch of the Axial Vircator
Note: The virtual cathode is indicated by the cross-hatched region and forms to the right of the anode in the output waveguide.

The center conductor of the vacuum transmission line connecting the insulator of the pulsed power generator to the vircator supports the cathode. In a number of experiments, the actual cathode diameter is larger than the diameter of this center conductor. In the axial vircator, the grounded outer conductor of the vacuum transmission line is connected to the output waveguide at the anode. Thus, the output waveguide diameter may differ from the transmission-line diameter, but in many experiments both the transmission-line and output waveguide diameters are identical, as depicted in Figure 14.2.

In the axial vircator [8–22] microwaves are removed from the virtual cathode region through an output waveguide oriented along the cathode axis. In contrast, the transverse vircator [23, 24] has a rectangular output waveguide located perpendicular to the cathode axis. This configuration is similar to an early reflex triode experiment [4], where the virtual cathode was formed within a rectangular waveguide. Another version of the transverse vircator is similar to the axial vircator with the coaxial output waveguide truncated to form a cavity [21, 25]. In this configuration, rectangular waveguides are attached to the side of the cavity to extract the microwave energy.

To avoid output-window breakdown in axial vircator experiments, a transverse magnetic field is often placed across the output waveguide to remove any residual electrons that have passed through the virtual cathode region. Although most of these electrons are lost to the output waveguide wall, some residual electrons will impact the output window if the distance between the virtual cathode and the output window is too short.

In the vircator, the high-voltage pulse propagating down the vacuum transmission line can cause field emission of electrons from the center conductor. At the leading edge of the pulse, these field-emitted electrons cross the transmission-line gap to the outer conductor, and cause a current to flow along the transmission line. This current flow then generates an azimuthal magnetic field that prevents additional electrons from crossing the transmission-line gap. Thus, the transmission line becomes magnetically insulated, and power is delivered to the vircator.

A virtual cathode will not form if the anode-cathode gap is too large. When the high-voltage pulse reaches the cathode, the field-emitted electrons initially flow in the radial direction to the outer conductor. If the ratio of the anode-cathode gap to the transmission-line gap is too large, the electrons will continue to flow to the outer conductor wall, an electron beam will not form, and microwave generation will not occur. Roughly speaking, this inability to form a virtual cathode occurs when the vircator impedance exceeds the transmission-line impedance. As the anode-cathode gap is reduced, the vircator impedance is reduced. Thus, for a sufficiently small anode-cathode gap, the magnetic field can force the electron flow through the anode foil, a virtual cathode can form, and microwave generation begins.

In many experiments, the virtual cathode does form until late into the high-voltage pulse. Futhermore, under certain conditions, the increase in diode impedance caused by the formation of the virtual cathode can cause the magnetically insulated electron flow to become unstable, resulting in a loss of power to the vircator. Under these conditions, the microwave radiation bursts as the electron flow oscillates between the transmission-line wall and the anode, as indicated in Figure 14.3. Many vircator experiments exhibit a bursting microwave output, and this periodic power loss is observed in particle-in-cell simulations [45]. However, the interplay between the vircator and magnetically insulated transmission line has not been investigated by experiment.

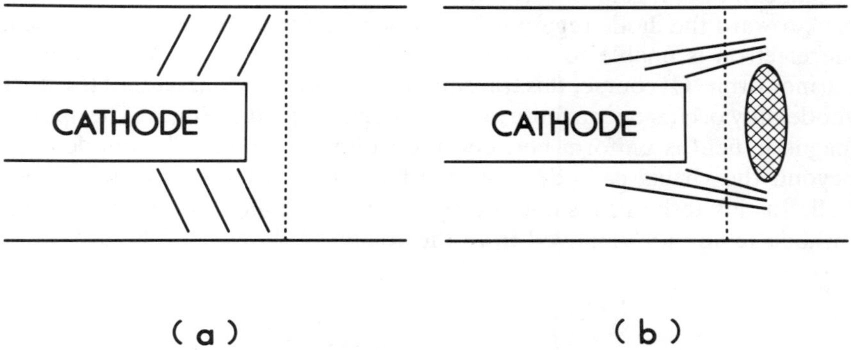

Fig. 14.3 Periodic Motion of the Electron Flow Can Occur under Certain Conditions: (a) Electron Flow to the Outer-Conductor Wall and (b) Electron Flow to the Anode, Leading to Virtual Cathode Formation

Note: The electron flow in state (b) is disrupted and returns to state (a) when the vircator impedance exceeds the vacuum transmission-line impedance.

In an effort to control the frequency and to enhance the beam-to-microwave conversion efficiency, a number of axial vircator experiments [10–14] have investigated the effect of placing a coaxial cavity between the anode and output waveguide. Although the coaxial cavity does not exhibit significantly increased efficiency, there is some indication that the cavity does provide frequency control. The effect of a cavity was also investigated in a transverse vircator experiment [24]. In this case, a small coaxial cavity with an output aperture on its side was placed inside a larger output rectangular waveguide. Frequency control was demonstrated with the addition of the cavity, and a slight enhancement in efficiency was observed.

The most recent virtual cathode device is the reditron [26–27] shown in Figure 14.4. Except for two features, the reditron is identical to the axial vircator. The first difference lies in the design of the anode, which is an electron range in thickness but contains a narrow annular slit. The slit diameter is slightly smaller than the cathode diameter. Only electrons passing through the slit can provide communication between the diode and virtual cathode. The second difference is the use of an applied axial magnetic field to guide the electron beam from the diode, through the anode, and into the virtual cathode region. When an electron encounters the virtual cathode, it gains transverse energy. Therefore, an electron reflected back toward the diode region will be absorbed by the thick anode because the electron is unable to pass through the slit, due to an increase in its Larmor orbit. Of course, this transmission-absorption character of the thick anode only occurs over a finite range of magnetic field strength. The applied magnetic field is uniform between the cathode and virtual cathode, but, beyond the virtual cathode, the field flares out to the output waveguide wall. In this fashion, residual beam electrons passing through the virtual cathode region are removed from the system.

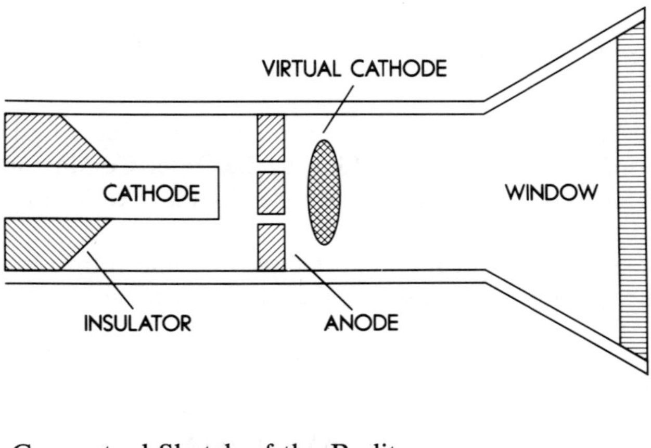

Fig. 14.4 Conceptual Sketch of the Reditron
Note: The reditron uses a thick anode with a narrow annular slit in conjunction with an applied magnetic field to prevent reflected electrons from re-entering the diode.

14.2.2 Mechanisms

In the reflex triode, an electron emitted from the cathode accelerates toward the positive anode. As the electron enters the anode, the kinetic energy of the electron equals the potential difference between the anode and cathode. In passing through the anode, however, the electron loses a small amount of energy. For example, if the anode is a thin foil, the energy loss results from ionization of foil atoms. If the anode is a semitransparent screen, the electron will see fringing electric fields and lose energy through radiation. As an electron leaves the anode, space-charge deceleration further reduces its kinetic energy. The collective process leads to the formation of a virtual cathode. Both anode and space-charge energy loss ensure that electrons cannot escape from the potential well formed between the real and virtual cathodes. This potential well is a two-dimensional region, if we ignore the connection point of the anode to the supporting center conductor of the vacuum transmission line. The confinement is electrostatic in both the axial and radial directions. As electrons bunch in the potential well, they continue to loose energy with each pass through the anode. Eventually, electrons are absorbed in the anode, leading to a net current flow.

The cathode is negative and the anode is grounded in the vircator, and the formation of the potential well between the real and virtual cathodes is somewhat different than in the reflex triode. For the vircator, a space-charge-limiting current flows through the virtual cathode region. Because no electrostatic restoring force exists radially, electrons that are trapped in the axial direction in the potential well between the real and virtual cathodes would quickly escape in the radial direction. With the combination of the azimuthal magnetic field and radial electron motion, however, the resulting electromagnetic force $v_r B_\theta$ provides the necessary radial confinement. Therefore, the electrons are again trapped in a two-dimensional region between the real and virtual cathodes.

Although the formation of the potential well is different for the reflex triode and vircator, the importance of this difference is unclear. For the reflex triode and vircator, the axial dimension of the potential well is about twice the anode-cathode gap spacing. The radial dimension of the potential well is equal to the anode diameter for the reflex triode and to the electron beam diameter in the vircator. In both cases, there is an axial and radial motion in the potential well, so both devices share many characteristics in common.

Published results state that the reflex triode beam-to-microwave power conversion efficiency exceeds that of the vircator by an order of magnitude. The reflex triode would be expected to be more efficient than the vircator, insofar as all the beam electrons are trapped in the potential well, but the difference should be less than a factor of two. A speculation attributes the efficiency difference between the two devices to the ability of microwaves to escape from the virtual cathode region without being reabsorbed by residual electrons. In fact, particle-in-cell simulations of the axial vircator indicate that significant reabsorption occurs as microwaves propagate through the residual electron cloud [45]. The ability of the reflex triode to extract power with the output window relatively close to the virtual cathode might also be related to the absence of residual electrons.

Given the formation of a potential well, a one-dimensional analysis [1] has shown that the trapped reflexing electrons can bunch if $d\omega_0/d\gamma < 0$, where ω_0 is the electron oscillation frequency in the potential well and $(\gamma - 1)\, mc^2$ is the electron kinetic energy. Therefore, electron bunching will occur if the beam is realtivistic or if the potential well is nonparabolic, and both conditions are satisfied in reflex triode and vircator. For this mechanism the radiation frequency is associated with the electron transit time in the two-dimensional potential well [13, 14, 35, 37, 39, 40, 42–45].

If the incoming electron beam exhibits a small energy divergence [36], then the position of the virtual cathode oscillates and generates microwaves [13, 28–41, 45]. For this mechanism, the microwave frequency depends on the beam plasma frequency, $f_b = (4\pi q^2 n_b/m\gamma)^{1/2}/2\pi$, and the ratio of the diode current to the space-charge-limiting current. Here, q is the electron charge, n_b is the average electron density of the beam, m is the electron rest mass, and γ is the Lorentz factor. Theoretically, the maximum frequency of an oscillating virtual cathode is $f_{\max} = (2\pi)^{1/2} f_b$, a frequency obtained when the diode current is much larger than the space-charge-limiting current. A nearly one-dimensional, oscillating, virtual cathode is thought to be the radiation mechanism in the reditron [26].

Although one-dimensional analysis provides understanding of the reflexing and oscillating mechanisms, the reflex triode and vircator experimental results indicate that the radiation mechanism is two-dimensional in nature. In addition, both radiation mechanisms are nonlinear because the feedback is linked to the self-electric and self-magnetic fields of the electron beam, and the two mechanisms appear to be competitive. As a result, particle-in-cell plasma simulation has been used extensively to investigate the axial vircator and reditron because both devices are azimuthally symmetric. Generally, the simulations agree with the experimental results. Conversely, the reflex triode and transverse vircator have not been investigated with particle-in-cell simulation because of their three-dimensional spatial geometry.

14.2.3 Extraction Techniques

A number of techniques have been used to absorb or radiate microwaves in virtual cathode device experiments. First, an output window is located in the side of a cavity in close proximity to the virtual cathode. This is the typical method used in reflex triode experiments [1–7]. Closely related to this approach is simply to locate a window across the output waveguide of an axial vircator [15]. Both of these methods cause significant mode conversion and reflection, thus to obscure somewhat the characteristics of the microwave source itself. Second, a "get-lost pipe," or long waveguide, is used to avoid reflections over the time scale of interest [12, 13, 17]. A variant of the get-lost approach is to place a totally absorbing calorimeter at the end of the output waveguide [20–22]. In reality, of course, some reflection will occur in the case of the absorbing calorimeter.

With most of the vircator experiments and the reditron experiment, an antenna is attached to the output waveguide: a "leaky pipe" [10], rectangular horn [23, 24], or conical horn [10, 11, 14, 19, 27] antenna. With this method, radiation patterns can be measured to obtain energy and mode pattern. With a sufficient number of measurements, the radiated energy can be estimated by using this technique, provided that the large-angle radiation is sufficiently low for the integral to converge. The ability to deduce the radiated mode by using this technique is less clear, however, because the mode dependence of the virtual cathode device is often modified by waveguide-to-antenna mismatch.

Figure 14.5 shows the transition from a cylindrical output waveguide to a conical horn antenna. In the output waveguide, the wave solution is cylindrical, while the wave solution is spherical in the conical antenna. To avoid mode conversion, the transition from the cylindrical region to the spherical region must be carried out slowly. Typically, unless the antenna is very long, conical antennas with a constant taper cause mode conversion. It is possible to design an antenna with a variable taper to reduce significantly mode conversion [46], but none of the virtual cathode device experiments have used such an antenna.

Given the dimensions of a coaxial antenna and incident mode, the magnitude of mode conversion can be calculated, but such an approach is beyond the scope of this chapter. However, it is possible to determine whether significant mode conversion has occurred from a simple estimate. Specifically, to minimize mode conversion, the mismatch Δ, indicated in Figure 14.5, should be small relative to a quarter-wavelength $\lambda/4$. Given the cylindrical waveguide radius R, and the taper angle of the horn θ, the mismatch is determined by

$$\Delta = R\,\{[1 + \tan^2(\theta)]^{1/2} - 1\}/\tan(\theta) \tag{14.1}$$

In Table 14.2, the simple mode conversion estimate is summarized for experiments employing a cut-off waveguide or horn antenna, with the mismatch calculated by using (14.1). In all of the experiments, mode conversion of 30–40% might be present, since $4\,\Delta/\lambda \sim 0\,(1)$. It may be inferred from these estimates that the experimentally measured mode patterns may be unreliable.

Table 14.2 Estimates of Mode Conversion

Experiment	$4\,\Delta/\lambda$
Buzzi *et al.* [8, 9]	0.5
Clark [10]	1.5
Sze *et al.* [11]	0.6
Davis *et al.* [14]	1.6
Saunders *et al.* [15]	2.5
Davis *et al.* [19]	0.5
Benford *et al.* [23]	0.5
Davis *et al.* [27]	0.4

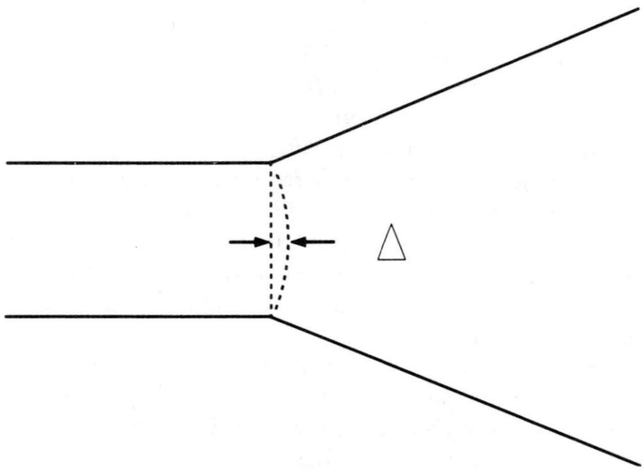

Fig. 14.5 The Transition between a Cylindrical Waveguide and a Conical Antenna Can Cause Mode Conversion

14.3 REFLEX TRIODE EXPERIMENTS

A chronological summary of the reflex triode experiments appears in Table 14.3. The first column is a reference for each experiment. The acronym preceding the reference is used in the text to refer to that particular experiment. Note that most of the reflex triode research has been performed by a single group, and both RT4 and RT5 have reported two separate experiments. In some cases, the entries in Table 14.3 are estimates.

Table 14.3 Summary of Reflex Triode Experiments

Experiment	V(keV)	I(kA)	t_b(ns)	f(GHz)	P(MW)	t_μ(ns)	ϵ(%)
RT1:Mahaffey *et al.* [1]	335	22	50	8–12	100	50	1.5
RT2:Brandt *et al.* [2–3]	1000	20	25	9–13	150	10	0.8
RT3:Didenko *et al.* [4–5]	450	25	80	2–5	1400	60	12
RT4A:Didenko *et al.* [6]	83	2.3	700	2–3	87	600	46
RT4B:Didenko *et al.* [6]	250	6	400	2–3	550	300	37
RT5A:Zherlitsyn *et al.* [7]	93	3.5	1700	2–3	120	1300	37
RT5B:Zherlitsyn *et al.* [7]	270	14	110	2–3	800	50	21

In the second, third, and fourth columns, we present the diode voltage V, electron beam current I, and electron-beam pulse width t_b. The microwave frequency f, microwave power P, microwave pulse width t_μ, and beam-to-microwave power conversion efficiency ϵ, are given in the fifth through eighth columns, respectively. With the exception of RT1, the microwave pulse width is shorter than the beam pulse width. Thus, the beam-to-microwave energy conversion efficiency is lower than stated in the eighth column. In experiment RT1, 10 MW of Ku-band radiation was also observed, and some evidence of irregular, or "spiky," output was reported. Similarly, in RT2, significant microwave power was observed in Ka band, and short microwave bursts of up to 3 GW were observed. In contrast, the data reported in RT3, RT4, and RT5 show no indication of irregular output, and no frequency content above S band was reported. If the output is not irregular, some form of frequency control is implied, possibly provided by the cavity, which is an important observation.

14.3.1 Applied Magnetic Field Experiments

The effect of an applied magnetic field on reflex triode operation is a reflection of the radiation mechanism. For example, if a predominantly one-dimensional axial current oscillation is the source of microwave radiation, an applied magnetic field would not be expected to cause significant reduction of power. However, an applied magnetic field could modify the frequency through modification of the beam plasma frequency or potential well. In contrast, if the source of radiation exhibits a two-dimensional character, with radial current flow along the anode being important, an applied magnetic field should significantly reduce the radiation output.

From the available experimental data, the effect of a weak applied magnetic field on the reflex triode is unclear. However, there is some indication that microwave power drops significantly when a strong magnetic field is applied. In this context, it has been suggested that an applied magnetic field would lead to the development of a two-stream instability between the incoming beam and the reflexing beam component [32]. This collective interaction destroys beam quality, a disruption which causes a loss in microwave power.

In RT1, an applied magnetic field caused a significant reduction in X-band power, with the output power exhibiting a resonance behavior with magnetic field strength. It was also reported that the emitted frequency increased linearly with the applied magnetic strength.

In contrast, RT2 appeared to generate the highest power at 40–100 *G*, but deteriorated significantly as the applied field strength was increased above that level. In this experiment the output power was not very reproducible. Thus, as a point of caution, it is not clear whether sufficient data were obtained to infer that a weak applied magnetic field provides improved operation.

The effect of an applied magnetic field was not reported in RT3. In RT4A, a magnetic field was stated to be present in order to focus the electron beam. The importance of this magnetic field for microwave power output was not discussed, if indeed the field had any effect at all. A sketch of this configuration, with a stated efficiency of 46%, is shown in Figure 14.6. Finally, for RT5 the effect of an applied magnetic field was not discussed, although one of the experiments, RT5A, was similar to the configuration shown in Figure 14.6.

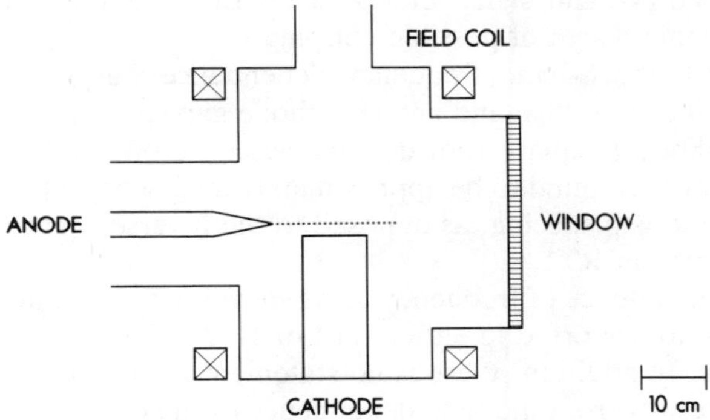

Fig. 14.6 Cross Section of Most Efficient Reflex Triode
Note: The view is along the plane passing through the cathode and coaxial vacuum transmission-line axes.

14.3.2 Frequency Dependence

The frequency is strongly dependent on anode-cathode gap spacing and diode voltage. This dependence can be understood in terms of the transit time in the potential well formed between the real and virtual cathodes. A higher voltage and a smaller anode-cathode gap spacing leads to increased frequency operation. The frequency is also influenced by the shape of the cathode and anode.

Diode Voltage and Anode-Cathode Gap

The X-band frequency dependence on both the diode voltage and the anode-cathode gap spacing was investigated in RT1. The frequency scaled as the square root of the time-averaged diode voltage, and roughly inversely proportional to the square root of the anode-cathode gap spacing. In RT1, there was a significant change in voltage during the microwave pulse, and frequency chirping is to be expected. In fact, because of the length of the microwave pulse relative to the dispersed line length, the large bandwidth observed is most likely due to chirping. Radiation was also observed in Ka band, but its dependence on diode voltage and anode-cathode gap spacing was not reported.

Both X-band and Ku-band radiation were observed in RT2. The dependence of the frequency on various parameters was not investigated.

The dispersed X-band signal indicated multiple frequency components, which was again a sign of possible chirping.

In RT3, the S-band frequency dependence was investigated as a function of diode voltage and anode-cathode gap spacing. As in RT1, the frequency shows a square root dependence on diode voltage. However, the frequency was found to be approximately inversely proportional to the anode-cathode gap spacing, as opposed to the inverse square root dependence observed in RT1.

The dependence of frequency on diode voltage or anode-cathode gap spacing was not reported in either RT4 or RT5. These were both S-band experiments. In addition, there is no statement about chirping, an occurrence that results from the time dependence of the diode voltage or diode closure.

Cathode Shape and Anode Material

With the exception of RT3, all of the reflex triode experiments indicated that the shape of the cathode played a role in determining the frequency. Three different cathode shapes were investigated in RT1: solid cathode, thick annular cathode, and thin annular cathode. For a given diode voltage, the solid cathode generated the highest frequency, while the thin annular cathode gave the lowest frequency. In both RT3 and RT4, it was stated that the anode shape was also important, but no details were reported. A mesh screen was used as an anode in RT1, RT3, RT4, and RT5. The anode screen material and transparency varied. An aluminized Mylar foil anode was used in RT1 and RT2. There are no reported data on the effect of anode screen transparency or anode foil thickness on microwave output power.

To minimize diode closure, a *many-pointed, explosive-emission cathode* (MEC) was utilized in RT5A: 2 mm diameter graphite rods were arranged in a honeycomb distribution with 2 cm separation. A 1300 ns microwave pulse width was achieved with this cathode. Note that, except for the MEC cathode, the reflex triode configuration in RT5A was reported to be the same as in RT4A, which is depicted in Figure 14.6. To reduce the beam current density on the anode, two separate solid cathodes were used in RT5B. In this configuration, the anode grid was common to both cathodes. The dual-cathode configuration was also reported to be successful in increasing the microwave pulse width.

14.3.3 Power Dependence

Measurements

There are a number of questions concerning the actual power achieved in all of the reflex triode experiments. In RT1, the output power was obtained by assuming the radiation to be isotropic. In particular, the microwave power was determined by assuming $P = 4\pi R^2 I_\mu$, where I_μ is a local measurement of the microwave intensity and R is the distance from the virtual cathode to the point where the measurement was made. As in RT1, the stated power in RT2 was not extracted, but rather calculated on the basis of local measurements by assuming a $\cos^2\theta$ dependence for the microwave power distribution, an assumption which yields $P = 2\pi R^2 I_\mu$.

It is not clear how the total power was determined in RT3, RT4, or RT5. If the procedure used in RT1 and RT2 were being applied, the high beam-to-microwave conversion efficiency associated with the reflex triode experiments would be questionable. For example, defining A_w as the window area, then $4\pi R^2/A_w \approx 8$–15, given the window dimensions of RT3, RT4, and RT5. It follows that the efficiency based on extracted power through the window is reduced to $(A_w/4\pi R^2)\epsilon \approx 3$–5%, where ϵ as summarized in Table 14.3 is assumed to be an isotropic efficiency. If we estimate the beam-to-microwave power conversion efficiency in this manner, then the efficiency of the reflex triode is about the same as that of the vircator and reditron. In other words, all virtual cathode microwave devices exhibit an efficiency of a few percent.

Angular Distribution

The power measurements in RT1 indicated an approximately isotropic distribution. However, the measurements in RT2 were more consistent with a dipole radiation field. Thus, the measured power was 10–20 dB down in the direction along the cathode axis, relative to the power observed transverse to the axial reflexing current. In addition, the field polarization was determined to be parallel to the reflexing current. A dipole power distribution was also reported in RT3, and again the field polarization was reported parallel to the motion of the electron beam. No power distribution or field polarization measurements were reported for either RT4 or RT5.

Voltage Dependence

In both RT1 and RT2, the diode voltage changed significantly during the microwave pulse. In contrast, RT3, RT4, and RT5 all exhibited relatively flat diode voltage during the microwave pulse. It was clearly stated in RT3 that constant diode voltage was important for achieving high beam-to-microwave conversion efficiency.

14.4 AXIAL VIRCATOR EXPERIMENTS

A chronological summary of the axial vircator experiments is given in Table 14.4. In the first column is an acronym preceding the reference for each experiment. This acronym is used in the text to refer to that experiment. Diode voltage V, electron beam current I, and electron beam pulse width t_b, are given in the second, third, and fourth columns, respectively. The microwave characteristics are summarized in the fifth through eighth columns: microwave frequency f, microwave power P, microwave pulse width t_μ, and beam-to-microwave power conversion efficiency ϵ.

Comparing the fourth and seventh columns, the microwave pulse width is typically shorter than the beam pulse width. As a result, the beam-to-microwave energy conversion efficiency is lower than stated in the eighth column. In many experiments, one contribution to the short microwave pulse is the late formation of the virtual cathode. Often, however, a sudden loss in microwave power is observed subsequent to virtual cathode formation. At present, the reason for this premature power loss is not understood. It has been speculated that ion formation, electron beam pinching, or diode closure may be responsible for the loss in microwave power, but no definitive theoretical or experimental investigation has been conducted to understand this phenomenon.

Although all axial vircator experiments are similar in their basic geometry, Table 14.4 shows that there is substantial variation in the electron beam parameters and microwave frequencies. Roughly speaking, the higher is the frequency, the smaller are the vircator dimensions. For example, in VC7, the 500–800 MHz experiment, a 61 cm diameter cathode and 76 cm diameter output waveguide were used. At the high-frequency end, in VC5, the 15–19 GHz experiment, a 2 cm diameter cathode and a 6 cm diameter output waveguide were used. Thus, although the experimental configurations are similar in basic geometry, the parameter regime encompassed by the experiments is quite broad.

Table 14.4 Summary of Axial Vircator Experiments

Experiment	V(keV)	I(kA)	t_b(ns)	f(GHz)	P(MW)	t_μ(ns)	ϵ(%)
VC1:Buzzi *et al.* [8, 9]	350	220	150	9–14	1000	50	1.3
VC2:Clark [10]	900	40	35	6–8	1000	30	2.7
VC3:Sze *et al.* [11]	1000	60	65	8–12	1000	50	1.7
VC4:Scarpetti *et al.* [12, 13]	500	40	35	7–9	300	50	1.5
VC5:Davis *et al.* [14]	1800	90	80	15–19	500	12	1.2
VC6:Clark *et al.* [16, 18]	400	40	80	2–3	170	60	1.1
VC7:Saunders *et al.* [15]	500	50	150	0.5–0.8	600	120	2.4
VC8:Burkhardt [17]	1750	111	60	5–7	4000	40	2.1
VC9:Clark *et al.* [20]	1500	300	40	2–4	4000	5	0.9
VC10:Davis *et al.* [19]	1800	90	80	4–6	3500	20	2.1
VC11: Bromborsky *et al.* [21]	8000	250	120	0.7–1.0	40000	50	1.8

14.4.1 Particle-in-Cell Simulation Overview

A number of theoretical calculations of axial vircator experiments will be presented in the section to follow. Because some of the conclusions about the axial vircator are based on these theoretical results, the present section is intended to provide the reader with an overview of the details included in the calculations. Unlike the reflex triode, which exhibits a three-dimensional geometry, the axial vircator can be studied with two-dimensional, azimuthally symmetric, particle-in-cell codes. For our present discussion, the particular code used is CCUBE [47]. Particle-in-cell codes solve for the self-consistent motion of charged particles in conjunction with Maxwell's equations, including complicated boundary conditions. This first-principles approach is ideally suited for problems such as the axial vircator that are nonlinear and time dependent. By using simulation, the plan is to discover fundamental patterns from which to construct a physical model of the microwave generation mechanism. Although significant progress has been made in understanding the axial vircator, a physical model has not been formulated, which would be capable of predicting more efficient configurations. As a result, particle-in-cell simulation remains the best theoretical tool with which to investigate the axial vircator.

To obtain agreement between simulation and experiment, it is necessary to include the interplay among the magnetically insulated transmission line, diode, and virtual cathode. Therefore, all simulations include the coaxial vacuum transmission line, diode, and output waveguide that are consistent with a particular configuration. As might be expected, the drawback associated with the simulation approach is the requirement for a significant amount of computational power. Moreover, the vircator problem is difficult to address with simulation because of space-zoning and time-zoning constraints. Specifically, to obtain accurate predictions for the diode impedance, microwave frequency, and mode distribution requires an anode-cathode gap resolution of 10 to 20 cells, and finer spatial resolution is required for accurate power measurements.

The microwave output associated with the axial vircator is found to be very complex. On the whole, the frequencies are time dependent and the power is often distributed among a number of modes, thus causing spiky or modulated output. Consequently, standard simulation diagnostics are wholly inadequate for axial vircator calculations. To gain useful information, it is necessary to use extensive postprocessing capability, coupled with a number of specialized diagnostics. As a starting point, both the electric and magnetic fields are expanded in terms of Bessel functions associated with TM_{0n} waveguide eigenmodes. Depending on the spatial zoning, the first 10 to 15 mode amplitudes, or coefficients, are calculated. Given these time-dependent coefficients, the microwave frequency, wavelength, and power associated with a given mode can be determined. Filtering is used to remove contributions below cut-off for each mode.

The instantaneous microwave power is often chaotic. As such, a clear mode pattern begins to emerge only after time-averaging. Average microwave power is calculated by three separate techniques. First, the radial electric field E_r and the azimuthal magnetic field B_θ associated with a particular mode are used to calculate individual mode power. At the same time, the mode power can be determined by using the axial electric field E_z, once the frequency is known. With either approach, the total power is then obtained by summing up the separate mode contributions. Alternatively, the total poynting flux can be filtered and integrated across an output waveguide cross section. In test runs with a single-frequency, constant-amplitude wave, the separate techniques yield results within about

5% of one another. However, with the chaotic character exhibited by the axial vircator, agreement between the techniques is far worse, 30–40%, provided that the space-charge contribution is properly removed. Basically, as is true experimentally, it is very difficult to obtain an accurate power measure for the axial vircator.

Another set of diagnostics is used to obtain the time-dependent current-density components at the anode. Both ac and dc components are obtained with filtering. The axial and radial beam current densities are also expanded into a Bessel series. The time-dependent diode voltage, beam current, and diode impedance are monitored, in addition to the current-density diagnostics, along with the current flow down the output waveguide. Of course, the standard phase-space and field information are also available.

Given the beam current density and diode voltage, the time-dependent beam plasma frequency can be calculated and compared to the microwave frequency. This information is used to gain insight into the source of the microwave radiation. For example, recall that the microwave frequency is related to the beam plasma frequency if the microwave generation results from oscillations in the position of the virtual cathode [30, 36].

To determine the energy exchange between the electron beam and microwaves, radial mode integrals of j_zE_z and j_rE_r are monitored in the diode region, the virtual cathode region, and throughout the output waveguide. These diagnostics indicate that energy is removed from the electron beam between the cathode and the virtual cathode, and microwave energy is then reabsorbed by electrons as the microwaves propagate through the output waveguide. These diagnostics are difficult to interpret, but the axial current-density component appears to be the source of microwave generation.

Although it is not possible to show each experimental configuration, an example of the details included in a simulation is indicated in Figure 14.7. Recall that all axial vircator experiments exhibit the same basic geometry with the anode foil placed across the end of an evacuated coaxial waveguide. Microwaves are extracted along the waveguide in the direction opposite to the diode. In some cases, the output waveguide diameter differs from the outer conductor diameter associated with the coaxial transmission line that feeds the diode. Still, due to ease in fabrication, many of the experiments have a common coaxial transmission line and output waveguide diameter. Finally, in a number of experiments, the cathode diameter is larger than the inner conductor diameter of the coaxial transmission line.

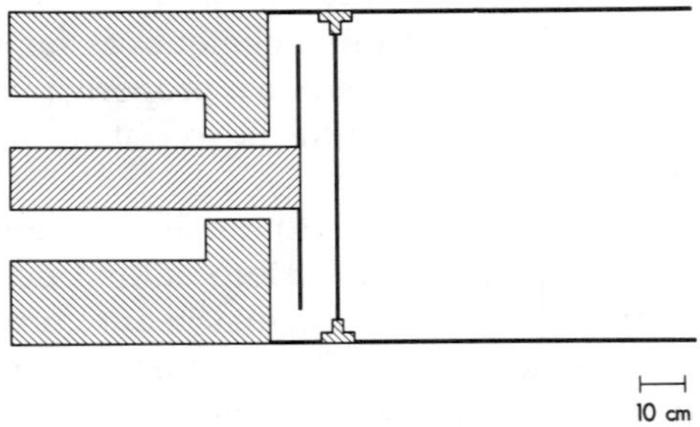

Fig. 14.7 Sketch of the Simulation Configuration Corresponding to the 500–800 MHz Experiment VC6

Refer to Figure 14.7, where a TEM wave is launched from the left-hand boundary onto the coaxial transmission line. Impedance mismatch between the transmission line and the diode is consistently calculated. For the most part, a voltage pulse with a fast rise time to a constant value was used to simulate the axial vircator because a constant diode voltage allows for a more accurate measurement of the microwave frequency and power. However, some long-rise-time voltage pulses were used to investigate the effect of voltage inhomogeneity and frequency chirping. It was found that a time-dependent diode voltage causes chirping, which leads to a large bandwidth, and lowers the beam-to-microwave power conversion generation efficiency [45].

The simulation provides other details in addition to obtaining basic information about an experiment, such as diode impedance or microwave frequency. This additional detail gives us confidence that the simulation results are realistic. For instance, note the complicated coaxial transmission line in Figure 14.7. It was thought that the transmission line was the cause of a 20 ns modulation observed on the output microwave power through a resonance that resulted from the transmission-line impedance mismatches. However, the simulations indicated that the radiation was being generated in both TM_{01} and TM_{02} modes with a difference frequency of about 100 MHz. Thus, the beating of these two modes appears responsible for the observed modulation, with the center frequency determined by the anode-cathode gap spacing and diode voltage.

Although the transmission-line discontinuities were not the cause of the observed modulation in VC7, the discontinuities caused a disruption in magnetic insulation and loss of beam energy, in the region where the smaller coaxial transmission line made its transition to the output waveguide diameter. This energy loss occurred with a field-emission threshold of 300 keV/cm on the inner conductor of the transmission line. At the same time, it was necessary to lower the field-emission threshold to 20 keV/cm on the cathode face before significant microwave radiation was obtained. Similarly, velvet was used to cover the cathode surface in the experiment. Finally, the simulations indicated that the anode foil holder acted as a beam aperture in the experiment.

14.4.2 Simulation Results

Three of the experiments summarized in Table 14.4 have not been simulated: Buzzi *et al.* [8, 9], Burkhardt [17], and Clark [20].

Frequency

For two slightly different anode-cathode gaps, axial current-density profiles are shown as a function of radial position relative to the waveguide radius, R_w, in Figure 14.8. The profiles are measured at the anode and time-averaged over a 1 ns interval. In this case, the simulation data corresponds to the VC5 experiment, with a 2 cm cathode diameter and a cathode shape similar to that depicted in Figure 14.16(b). The cathode edge is located at $r = 0.33R_w$. Note that a negative current density corresponds to electrons moving from the anode toward the virtual cathode. For this experiment, the peak current density is located at a radius somewhat smaller than that of the cathode. In general, however, the radius of the electron beam as it passes through the anode depends on the diode geometry and anode-cathode gap spacing.

The dc current-density profiles presented in Figure 14.8 are typical of all simulated experiments, with an annular electron beam entering the virtual-cathode region and return current flowing back into the diode region near the axis. The corresponding ac current-density profiles peak on axis and drop to zero at the cathode edge. Recently, these profiles have been experimentally measured in VC6 by using a small, directional, Faraday cup to determine the ac and dc axial current-density components [16].

There is a negligible amount of current returning to the diode in the region of the incoming electron beam. For instance, with the 0.400 cm anode-cathode gap, the incoming beam is located between $r_- = 0.22R_w$ and $r_+ = 0.60R_w$. With this information, an estimate of the average beam electron density can be calculated as

$$\langle n_b \rangle = \left[\int_{r_-}^{r_+} J_z(r) r \mathrm{d}r\right] \Big/ \left[q\langle v_z \rangle \int_{r_-}^{r_+} r \mathrm{d}r\right] \tag{14.2}$$

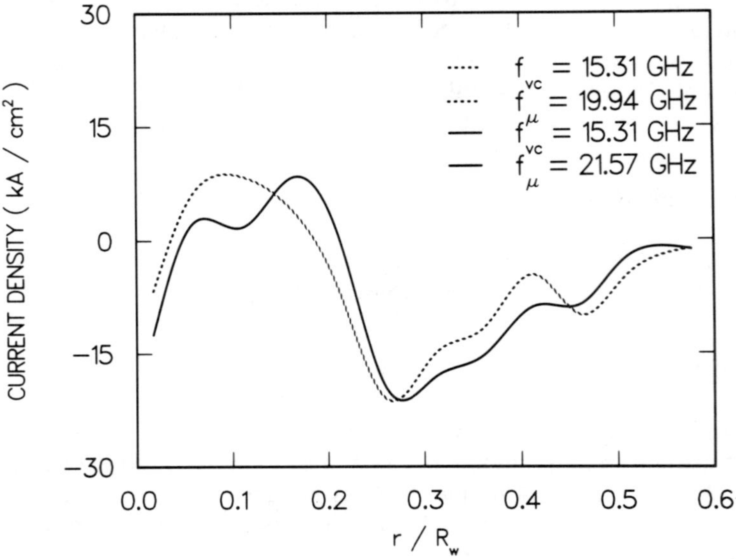

Fig. 14.8 Axial Current Density as a Function of Radial Position for Two Slightly Different Anode-Cathode Gaps: 0.400 cm (solid line) and 0.475 cm (dashed line)

Note: The average beam plasma frequency is the same for both profiles.

where $J_z(r)$ is the axial current density, q is the magnitude of the electron charge, and $\langle v_z \rangle$ is the time-averaged axial velocity. The time-averaged diode voltage $\langle V \rangle$, which takes into account the transit time across the anode-cathode gap, is used to determine the time-averaged axial velocity: $\langle V \rangle = (\langle \gamma \rangle - 1)\, mc^2$ and $\langle \gamma \rangle = (1 - 1/\langle v_z \rangle^2)^{-1/2}$. Given the beam density and kinetic energy, the average beam plasma frequency can be determined from

$$\langle \omega_b \rangle = (4\pi q^2 \langle n_b \rangle / \langle \gamma \rangle m)^{1/2} \tag{14.3}$$

By using (14.2) and (14.3), both profiles shown in Figure 14.8 yield a maximum oscillating virtual cathode frequency of 15.3 GHz, although the current-density profiles and diode voltages are slightly different. In contrast, the microwave frequencies differ by 8%, with $f_\mu = 21.6$ GHz and $f_\mu = 19.9$ GHz for anode-cathode gaps of 0.400 cm and 0.475 cm, respectively. Note that both microwave frequencies are significantly above the maximum frequency attributed to the oscillating virtual cathode,

$$f_{vc} = \langle \omega_b \rangle \, / \, (2\,\pi)^{1/2} \tag{14.4}$$

Because it has been shown theoretically that the reflexing frequency exceeds the oscillating frequency in an unmagnetized axial vircator [13, 39], trapped reflexing electrons are the dominant source of microwave radiation in VC5.

On the basis of simulation results, trapped reflexing electrons are the dominant source of microwave generation in all axial vircator experiments. A summary of the simulation frequency data is presented in Figure 14.9, where f_{μ} is compared with f_{vc}. In some cases, more than one simulation exsits for a given experimental configuration, and the data do not include simulations for VC1, VC8, or VC9. For each simulation, the maximum oscillating frequency was calculated by using (14.2) through (14.4).

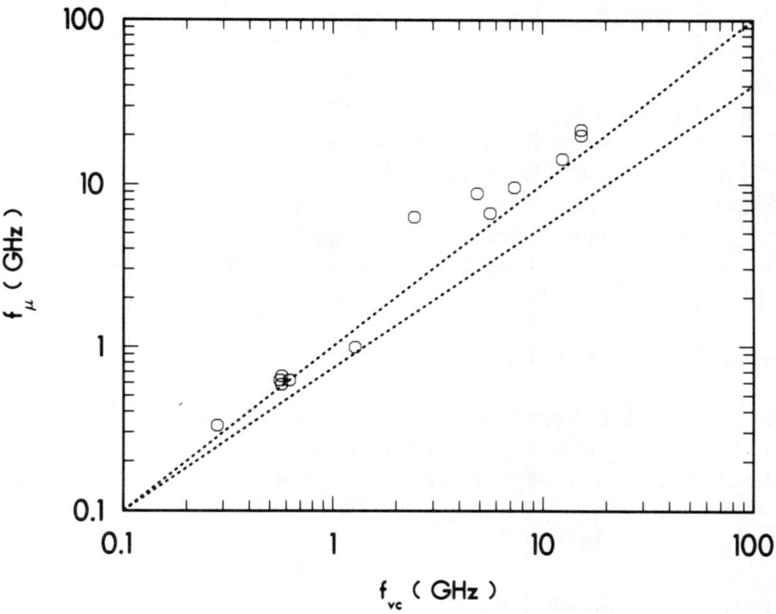

Fig. 14.9 Summary of the Microwave Frequencies Obtained from Simulations of the Experiments Listed in Table 14.4, Excluding VC1, VC8, and VC9

Two experiments exhibit a microwave frequency at or below f_{vc}. In the first experiment, VC7, the foil support acted as a beam aperture with the resulting beam located very close to the output waveguide wall (see Figure 14.7). As a result, the beam current did not greatly exceed the space-charge-limiting current. In the second experiment, VC6, only the cathode face was allowed to have field emission of electrons and, again, the beam current did not greatly exceed the space-charge-limiting current. Based on previous scaling studies [36], the microwave frequency that could be attributed to the oscillating virtual cathode should be closer to f_b, as opposed to $(2\pi)^{1/2} f_b$ for both experiments.

Simulations of VC6 showed an increase in microwave frequency, with $f_\mu / f_{vc} > 1$, when the cathode edge is allowed to have field emission of electrons. In VC6, sandpaper was placed on the cathode edge to reduce the field-emission threshold, and both the frequency and power significantly increased with cathode edge emission. More recently, velvet was used on the cathode shank and edge to increase the microwave power in VC11.

Wavelength

Simulations indicate that power is distributed into a number of TM_{0n} modes, but there generally is a dominant mode that might contain as little as 50% or as much as 90% of the total power, depending on the geometry. In some experiments, a number of modes have the same frequency, while the mode frequencies differ in other experiments. Given the dominant mode and its corresponding frequency ω_μ, the wavenumber k_z can be determined from the relation:

$$(k_z c)^2 = \omega_\mu^2 - (x_{0n} c / R_w)^2 \tag{14.5}$$

In (14.5), x_{0n} are the zeros associated with the Bessel function J of zero order. By defining D as the cathode diameter and using (14.5), the wavenumber is found to be approximately bounded by

$$\pi / D < k_z < 2\pi / D \tag{14.6}$$

independently of the configuration.

This characteristic of the axial vircator is indicated in Figure 14.10, where the wavelengths associated with modes that contain at least 20% of the power are plotted as functions of their corresponding frequencies. The basic configuration of the simulation corresponds to VC6. This particular data set was chosen because it includes variations in diode voltage, anode-cathode gap, coaxial transmission-line impedance, and output waveguide radius over a substantial range. Despite the changes in diode voltage and geometry, the wavelength remains approximately in the range defined by (14.6), while the dominant mode ranges from TM_{01} to TM_{06}. In Figure 14.10, the wavelength range corresponding to (14.6) is bounded by the dashed lines.

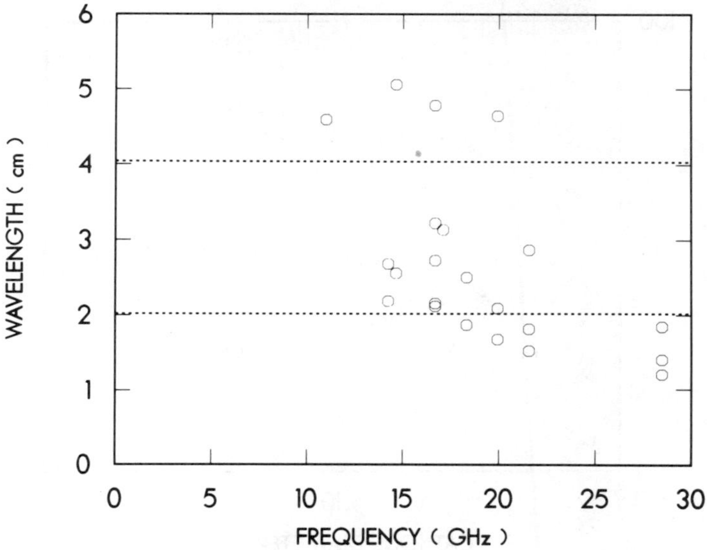

Fig. 14.10 Summary of the Wavelengths Obtained from Simulations for Modes with at least 20% of the Power
Note: The dashed lines correspond to the cathode diameter and twice the cathode diameter.

Following the same procedure, the wavelength corresponding to the dominant mode was determined for the experiments listed in Table 14.4, again excluding experiments VC1, VC8, and VC9. The results are summarized in Figure 14.11, but this time the wavelength is plotted as a function of the cathode diameter. Again, the wavelength is approximately bounded, as defined by (14.6), with the boundary indicated by the dashed lines.

Radiation Mechanism

A conceptual sketch of the microwave source region is depicted in Figure 14.12. The electron beam is born on the cathode edge, and an annular electron flow enters the virtual cathode region, where it separates into reflexing and net current components. The net current component generates the azimuthal self-magnetic field that provides radial confinement of the reflexing electrons. Axial confinement is provided by the electrostatic well that forms between the real and virtual cathodes. The

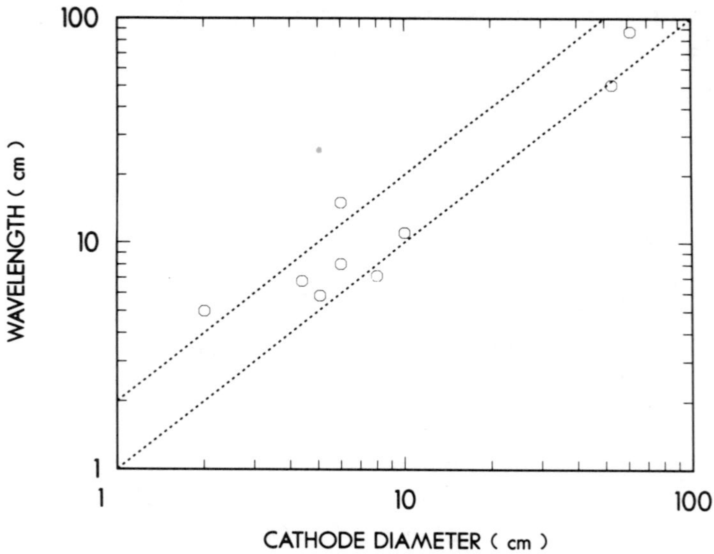

Fig. 14.11 Summary of Wavelengths Obtained from Simulations of the Experiments Listed in Table 14.4, Excluding VC1, VC8, and VC9

Note: The dashed lines correspond to the cathode diameter and twice the cathode diameter.

axial reflexing current, interior to the incoming beam radius, is the source of radiation, as depicted by the arrows in Figure 14.12. The source region, which contains both an axial and radial current distribution, generates microwaves with a frequency determined by the reflexing electrons, and at a wavenumber characteristic of the radial extent of the source. Because the source radiates into a waveguide, the mode that is most consistent with the source frequency and wavelength becomes the dominant mode.

This interpretation of the axial vircator operation implies that the shape of the cathode and anode are of fundamental importance. Moreover, the vircator does not operate close to waveguide cut-off. With the source localized about the cathode axis, it should be clear why an applied magnetic field or a range-thick disk located on the anode along the cathode axis would suppress microwave radiation. Furthermore, if the output waveguide radius, microwave frequency, or cathode diameter increase, then the dominant mode must increase to remain consistent with the waveguide constraint (14.5), since $k_z = 2\pi/D$. Indeed, these output characteristics of the axial vircator are observed experimentally.

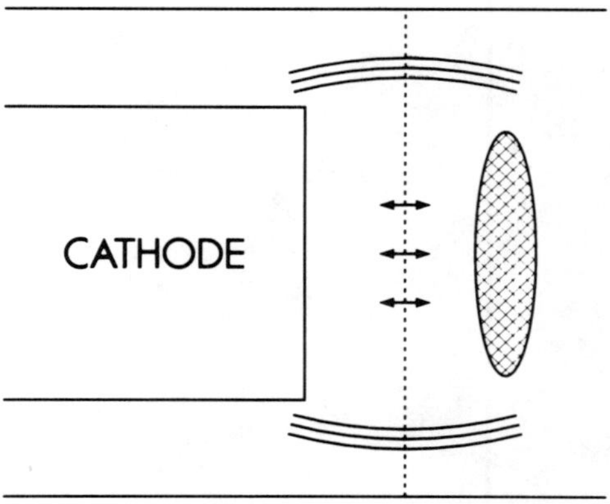

Fig. 14.12 Conceptual Sketch of the Microwave Source Region in the Axial Vircator

Note: The electron beam enters the virtual-cathode region and separates into reflexing and net current components. Microwave radiation is generated by the reflexing current component concentrated about the cathode axis, which is depicted by the arrows.

14.4.3 Applied Magnetic Field Experiments

The effect of an applied magnetic field on microwave generation was investigated in VC1, VC4, VC5, and VC10. Microwave generation is suppressed by an order of magnitude when the strength of the applied magnetic field is about equal to the self-magnetic field of the electron beam. As a specific example, Figure 14.13 shows the microwave power as function of applied magnetic field strength for VC4. Given the accuracy of the data, the slight increase in power observed around 0.3 kG should be interpreted with caution. In any case, maximum power occurred with no applied magnetic field. This result is an indication that two-dimensional motion in the potential well is important for microwave generation in the axial vircator.

In VC1, microwave radiation was observed, even with a 10 kG applied field. At first this result appears to be in variance with other axial vircator experiments. However, because of large diode current in VC1, the self-magnetic field associated with the electron beam was 22 kG, compared with only 2 kG self-magnetic field in VC4. Given the trend shown in Figure 14.13, reduction of about a factor of four in microwave power would be expected in VC1, not a complete suppression of the microwaves. In fact, it was only stated that microwave radiation was still observed with a 10 kG applied magnetic field in VC1.

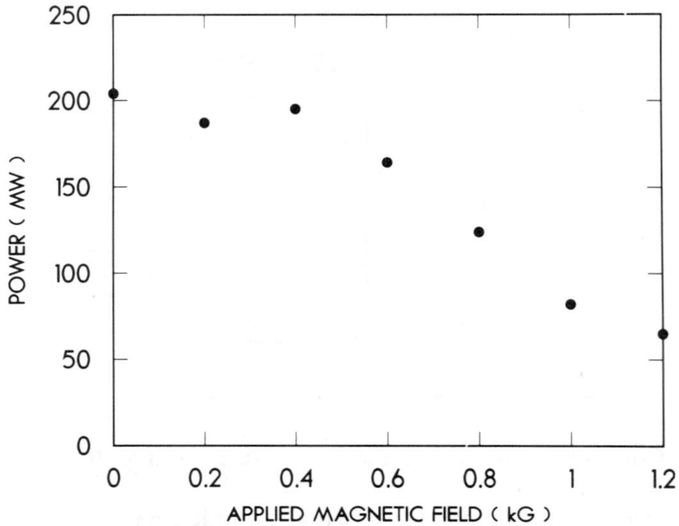

Fig. 14.13 Microwave Generation is Suppressed when an Axial Magnetic field is Applied

Note: The data are from experiment VC4.

14.4.4 Range-Thick Anode Disk Experiments

In VC2, VC5, and VC10 electron-range-thick disks, aligned with the cathode axis, were attached to the virtual cathode side of the anode in order to block reflexing electrons, as depicted in the inset of Figure 14.14. In all three experiments, microwave suppression is observed with increasing disk radius. Typically, an order of magnitude reduction in power occurred when the disk radius became about equal to one-half the cathode radius. Simulations of VC10 and VC11 showed a loss of microwave power in the presence of a thick anode disk [45]. Data from experiment VC2 are shown in Figure 14.14, with relative microwave power plotted as a function of the ratio of disk radius to cathode radius. In the absence of the thick anode disk, experiment VC4 generated about 1 GW by using a 0.00254 cm thick aluminum-foil anode.

All three experiments found that the diode behavior was more stable with an anode disk present because reflected electrons were prevented from re-entering the diode. In addition, reduced diode closure was observed in experiment VC2. These results are an indication that reflexing electrons play an important role in the axial vircator.

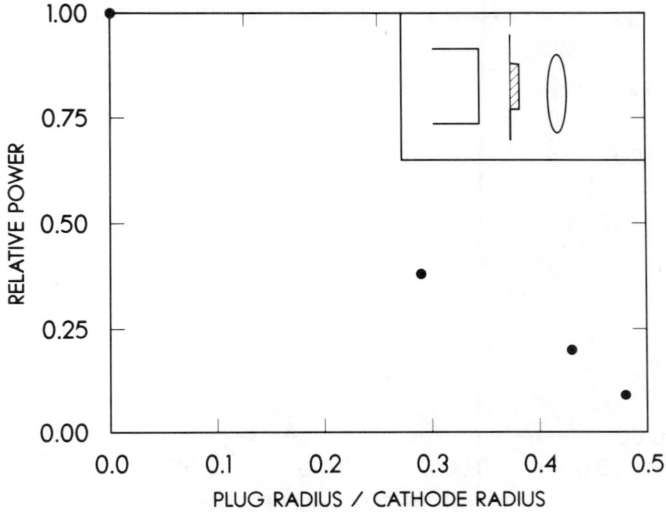

Fig. 14.14 Microwave Generation is Suppressed by a Range-Thick Disk Placed on the Anode

Note: The data are from experiment VC2.

14.4.5 Anode Thickness Experiments

The dependence of microwave power on anode foil thickness was investigated in VC2, VC4, VC5, VC8, VC10, and VC11. Unfortunately, the data are sketchy from most of the experiments. At first glance, the effect of anode thickness appears to vary significantly between experiments. For instance, in VC4 a 0.00254 cm thick aluminum-foil anode reduced the power by a factor of eight, relative to that obtained with a fine steel-mesh screen anode, but it was necessary to use a 0.0106 cm thick titanium-foil anode in VC10 to achieve a similar reduction in microwave power. This result is shown in Figure 14.15, with relative microwave power given as a function of anode foil thickness.

If we take account of differences in the diode voltage among the experiments, then a significant reduction in microwave power occurs when single-pass anode foil scattering reaches 300–400 mrad. Thus, the experimental results are consistent, and they indicate that the radiation mechanism is sensitive to angular scatter of the electron beam. Although the data are limited, the frequency appears to be weakly dependent on anode

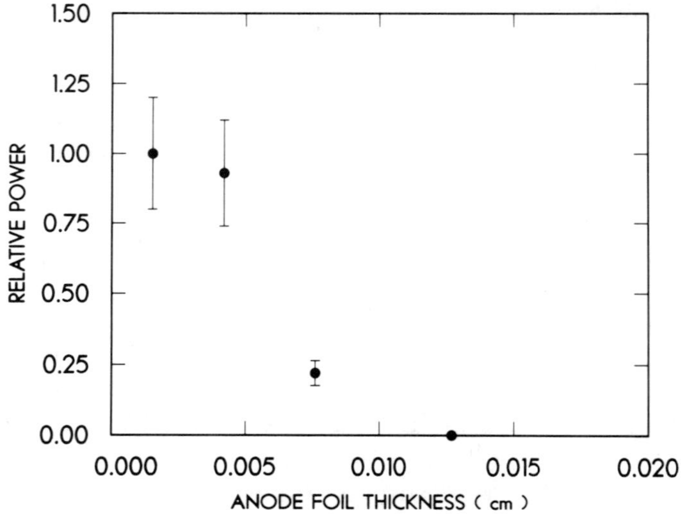

Fig. 14.15 Microwave Generation is Suppressed by a Sufficiently Thick Anode Foil
Note: The data are from experiment VC10.

foil thickness. Simulations indicate that the frequency increases slightly with increasing anode foil thickness. In this context, varying the anode foil thickness in simulations is an alternative method used to isolate the dominant radiation mechanism [13, 45].

14.4.6 Frequency Dependence

Every experiment has observed that the frequency depends on the anode-cathode gap spacing: the frequency decreases with increasing anode-cathode gap spacing. At least over a limited range, the frequency appears to be roughly inversely proportional to the anode-cathode gap spacing. Within experimental error, particle-in-cell simulations agree with this observed frequency dependence on anode-cathode gap spacing, but the predicted frequency dependence is not simple.

Variations of cathode shape were investigated in VC2, VC4, VC5, VC8, and VC10. As an example, we consider the different cathode shapes investigated in VC10, as depicted in Figure 14.16 (a,b). For these two cathode shapes, the frequency as a function of anode-cathode gap spacing is shown in Figures 14.17 and 14.18, respectively. In both cases, the dashed

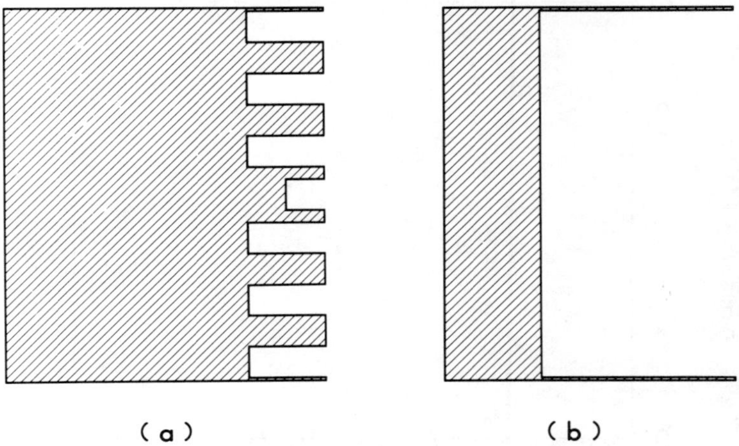

Fig. 14.16 Two Cathode Shapes were Investigated in VC10
Note: The complex cathode (a) is more efficient than the annular cathode (b) at large anode-cathode gap spacing.

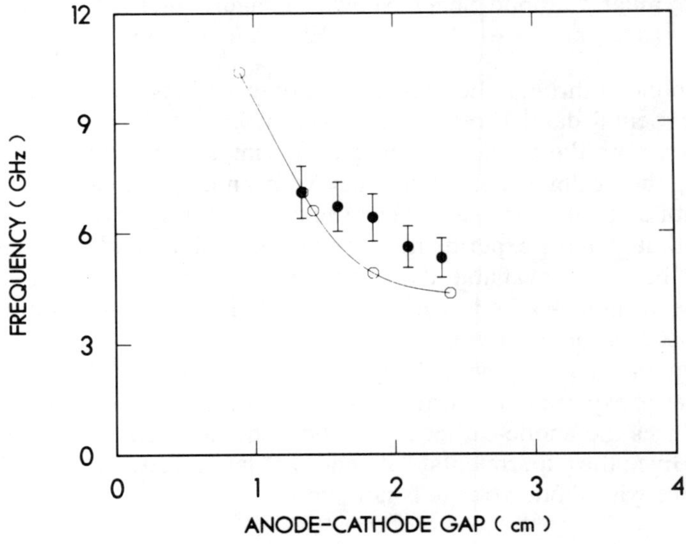

Fig. 14.17 Frequency as a Function of Anode-Cathode Gap Spacing for the Complex Cathode Shown in Figure 14.16 (a): Simulation (open circles/solid line) and Experiment (solid circles)

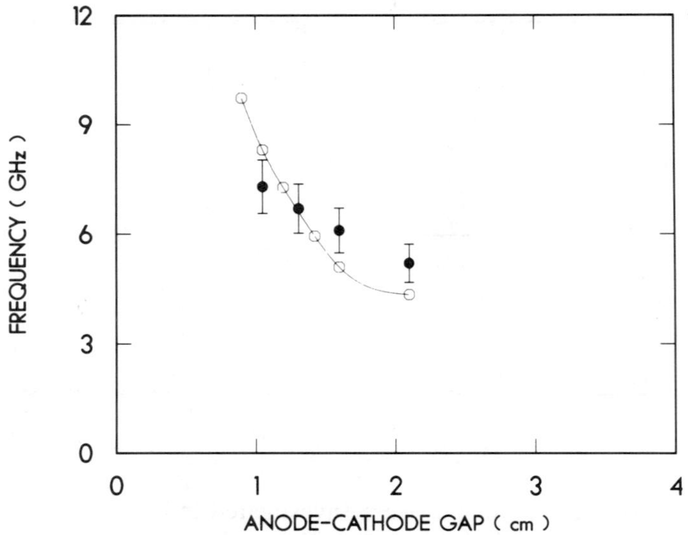

Fig. 14.18 Frequency as a Function of Anode-Cathode Gap for the Annular Cathode Shape Shown in Figure 14.16 (b): Simulation (open circles/solid line) and Experiment (solid circles)

line is a spline fit through the simulation results, and the solid dots indicate the experimental data. From these results, it is clear that the frequency depends on both the anode-cathode gap spacing and cathode shape. Furthermore, the scaling of the frequency with anode-cathode gap spacing differs with cathode shape, as we can see by comparing Figures 14.17 and 14.18. Although most experiments have not been simulated as extensively as VC10, based on available data and simulation results, the frequency sensitivity to anode-cathode gap spacing and cathode shape is a general characteristic of the axial vircator.

Frequency chirping with changing diode voltage is observed in most axial vircator experiments. In the absence of diode closure, which effectively reduces the anode-cathode gap, the frequency increases with diode voltage. Given the experimental data and simulation results, the frequency dependence with diode voltage is not simple.

14.4.7 Power Dependence: Anode-Cathode Gap and Cathode Shape

Microwave power as a function of anode-cathode gap spacing has been investigated in most axial vircator experiments: the microwave power depends on the anode-cathode gap spacing. On the basis of simulation results, the highest beam-to-microwave power conversion efficiency occurs when the electron beam is formed on the cathode shank, or edge, and propagates across the anode-cathode gap without excessive expansion or pinching. As we previously mentioned, the effect of cathode shank (or edge) emission on microwave frequency and power was investigated in VC6 and VC11 [16, 21].

If the anode-cathode gap spacing is too large, the beam expands toward the outer conductor wall of the coaxial transmission line. Conversely, if the anode-cathode gap is too small, the beam pinches toward the axis. Microwave suppression due to beam pinching was shown quite clearly in experiment VC9. In this low-impedance experiment, microwave generation began when the electron beam became magnetically insulated, and the experiment ended as the electron beam became strongly pinched.

The dependence of microwave power on the cathode shape has been most extensively investigated in VC4 and VC10. As an example, we again consider the results using the two cathode shapes shown in Figure 14.16 (a,b). Here, the complex cathode consists of a nested set of quarter-wavelength transmission lines designed to resonate with the trapped reflexing electron flow along the anode foil. This cathode design was motivated by simulation results which suggested that the reflexing electron current interior to the incoming electron beam was the source of microwave radiation. A cathode shape similar to the complex cathode was also found to provide the highest beam-to-microwave power conversion efficiency in VC4. At present, detailed investigation of the cathode shape appears to hold promise for increased efficiency.

Simulation results for the microwave power as a function of anode-cathode gap spacing are shown in Figure 14.19. The open circles are complex cathode results, while the open squares correspond to the annular cathode results. The impedances of the two cathodes are nearly identical, but the complex cathode is significantly more efficient at large anode-cathode gaps. Yet, both cathodes exhibit similar efficiencies at smaller anode-cathode gaps. This behavior appears to be the result of electron shielding, due to field emission from the nested quarter-wavelength transmission lines when the anode-cathode gap is small. Thus, the two cathodes appear similar to the reflexing electron flow for small anode-cathode gaps. In contrast, for large anode-cathode gaps, significantly less emission is observed from the cathode face, correspondingly less shielding occurs, and the interior shape of the cathode becomes important. In terms of scaling, the simulation results are consistent with the experimental results shown in Figure 14.20 for the complex cathode.

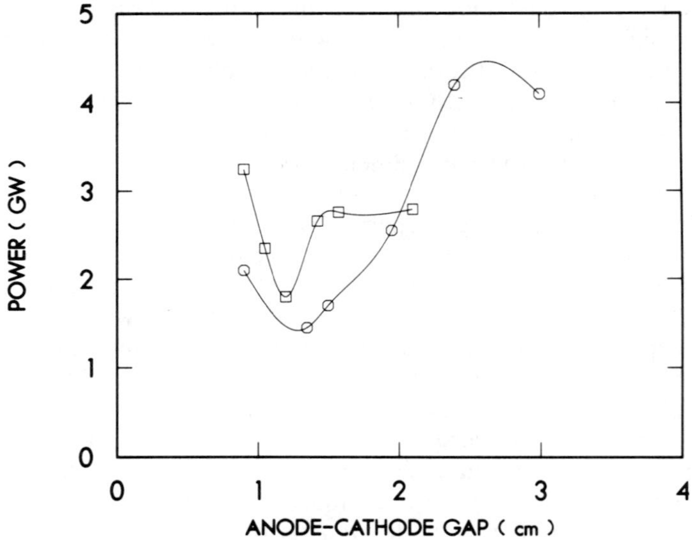

Fig. 14.19 The Microwave Power Depends on the Anode-Cathode Gap Spacing and Cathode Shape

Note: The circles are simulation results for the complex cathode, Figure 14.16 (a), and the squares are simulation results for the annular cathode, Figure 14.16 (b). A spline fit through the data are indicated by the solid lines.

14.4.8 Output Waveguide and Cathode Diameter Experiments

The most extensive investigation of the effect of the output waveguide diameter was carried out in experiment VC5. Neither the microwave power nor frequency were found to be sensitive to the output waveguide diameter. As the output waveguide diameter increased, so too did the dominant mode number. The sensitivity of the mode number to output waveguide radius has also been observed in other experiments. Note that these results are consistent with (14.5), provided $k_z = 2\pi/D$.

The effect of cathode diameter on microwave power and frequency was investigated in VC2, VC6, and VC8, but the data are not extensive. It appears that the frequency is rather insensitive to the cathode diameter. However, it is unclear whether the power is sensitive to the cathode diameter in an uncontrolled configuration.

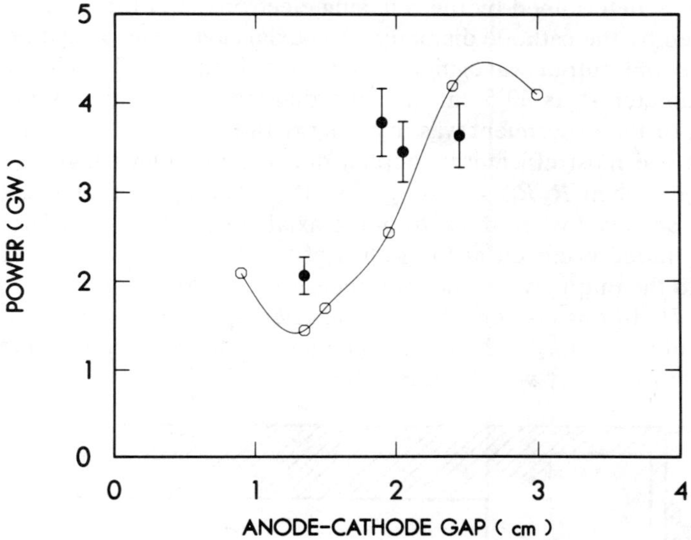

Fig. 14.20 Comparison of Simulation Results (open circles/solid line) and Experimental Data (solid circles) for the Complex Cathode, Figure 14.16 (a)

14.4.9 Coaxial Cavity Experiments

To control frequency chirping, a number of experiments were conducted with a coaxial cavity located between the anode foil and the output waveguide: VC1, VC2, VC3, VC4, and VC5. More importantly, it was hoped that the coaxial cavity would increase the beam-to-microwave power conversion efficiency by suppressing unwanted field components, while, at the same time, providing some feedback at the natural oscillation frequency of the virtual cathode configuration. In fact, the data indicate that some frequency control was achieved, but no significant increase in the beam-to-microwave power conversion efficiency was observed. Despite these initial and, in the author's opinion, promising observations, this line of investigation was not pursued past the "look-and-see" stage of experimentation. As of this writing, the coaxial cavity, in conjunction with a shaped cathode, should be reinvestigated.

Recently, the VC6 experimental configuration was modified to investigate the wavelength dependence of the reflexing electron source: the frequency is determined by the reflexing electrons, and the wavelength is determined by the cathode diameter. A coaxial cavity was placed between the anode and output waveguide as shown in Figure 14.21. The coaxial-cavity diameter R_c is 30.5 cm, and the diagram is to scale. A basic assumption of the experiment was that energy radiated into the TM_{02} mode would be the most efficiently coupled into the output waveguide in the TM_{01} mode when $R_c/R_w = x_{02}/x_{01}$. In other words, maximum coupling would be achieved when the zero of the axial electric field associated with the TM_{02} mode occurred at the radius of the output waveguide. Energy coupled to the output waveguide was measured as a function of the cathode diameter D. In this fashion, the fraction of energy radiated into the TM_{02} mode would vary if $k_z = 2\pi/D$. Independent of the cathode diameter, the microwave frequency was 2.5 GHz.

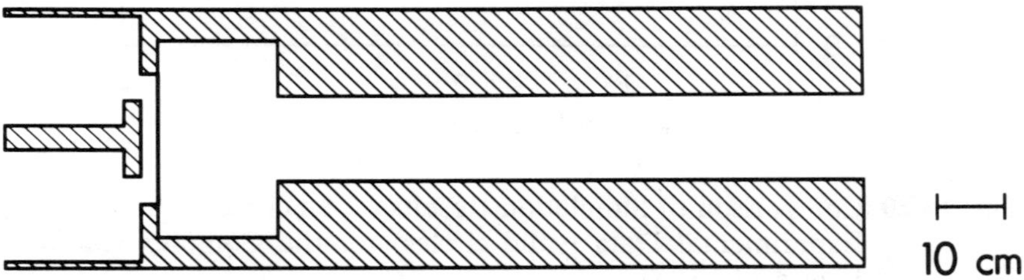

Fig. 14.21 Sketch of the Coaxial-Cavity Configuration Used in Experiment VC6

With this configuration, (14.5) predicts that maximum power should be coupled to the output waveguide for a 16.6-cm cathode diameter when $k_z = 2\pi/D$. Note that, in this geometry, the coaxial cavity wall sets the boundary in the virtual-cathode region, and R_c must be used in (14.5), rather than the output waveguide radius R_w. The ratio of the coupled microwave energy to the peak electron beam power is shown as a function of cathode diameter in Figure 14.22. A spline fit to the data is given by the solid line, while the position of the predicted maximum is indicated by the dotted line. The predicted maximum is about 20% above the observed maximum. Although the agreement is not as close as might be desired, the results are an indication that the axial vircator is behaving as the simulations suggest.

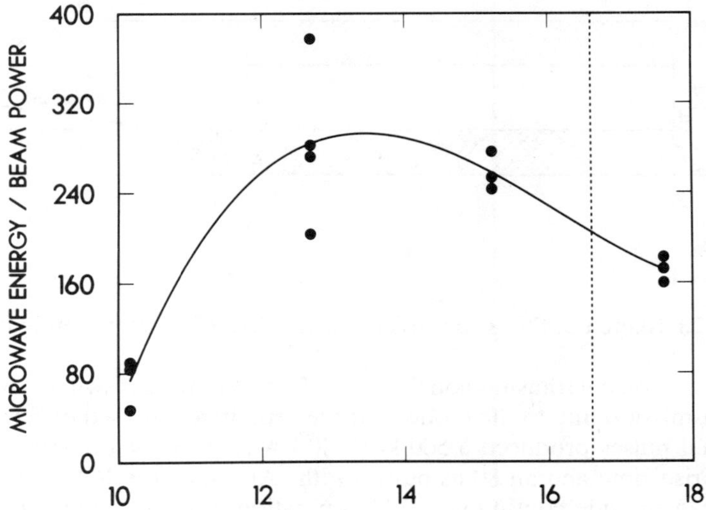

Fig. 14.22 Microwave Energy Divided by Peak Electron Beam Power as a Function of Cathode Diameter for the Coaxial-Cavity Experiment Shown in Figure 14.21

Note: The predicted maximum, indicated by the dashed line, occurs 20% above the observed maximum.

14.5 AXIAL VIRCATOR EXPERIMENT—MULTIPLE PULSE

14.5.1 Experimental Configuration

One multiple-pulse vircator experiment has been performed [22]. The repetitively pulsed configuration is shown to scale in Figure 14.23. For the most part, the vircator is identical to the configuration used to measure anode foil current-density profiles [16]. However, for the repetitively pulsed experiment, the anode was an 80% transparent copper (Cu) screen, and a calorimeter was positioned at the end of the output waveguide to absorb the microwave energy.

The *RAVEN* accelerator uses a primary capacitor bank coupled with a 1:40 transformer in a dual-resoinance configuration to charge a 5.3 Ω,

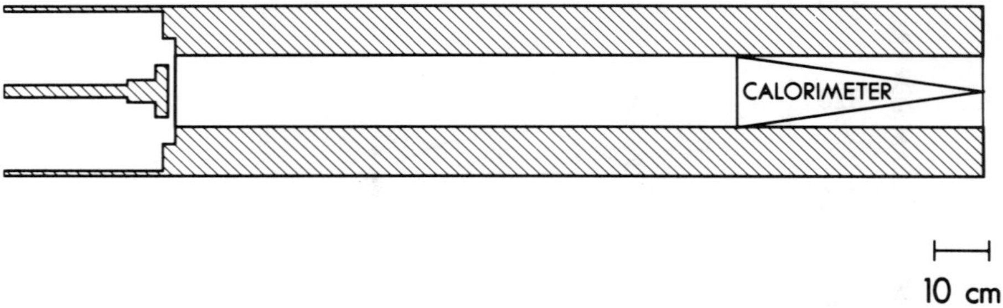

Fig. 14.23 Sketch of the Repetitively Pulsed Axial Vircator Configuration

50 ns water coaxial transmission line. A self-breakdown gas switch connects the transmission line to the diode. For the repetitively pulsed experiment, individual pulses produced a 500 keV, 40 kA electron beam with a 30 ns current rise time and an 80 ns pulse width. Although the RAVEN accelerator can provide continuous 10 Hz operation, the total number of microwave pulses was limited by anode screen melting.

14.5.2 Frequency Dependence

In this particular experiment, both the current and the frequency exhibited step-like changes during the voltage pulse. Between the step-like current behavior, continuous frequency chirping was observed, a typical characteristic of most single-pulse vircator experiments. The frequency dependence with anode-cathode gap spacing was not investigated.

The experiment was simulated by assuming an infinitely long output waveguide as an approximation to the calorimeter. At peak voltage, the simulations predicted a frequency of 2.5–3.5 GHz, the variation depending on the anode-cathode gap spacing and the field-emission threshold. Consistent with the simulations, the experimental data showed a frequency variation from 2.2 to 3.6 GHz. Anode current-density profile measurements demonstrated that the incoming beam was annular as it passed through the anode foil, and the current exhibited a reflexing component which was localized about the cathode axis [16]. This measured current-density profile was also predicted by the simulations. Finally, the simulations indicated that the radiation occurred at the beam plasma frequency.

14.5.3 Power and Energy Dependence

Individual pulses exhibited fast rise time and spiky power output, which are characteristics of single-pulse experiments. Simulations of the experiment indicated that the power was predominantly in the TM_{01} mode. For 1.2 Hz operation, the microwave energy per pulse is shown in Figure 14.24. Note the loss in microwave energy beginning at the sixth pulse and ending at the eleventh pulse. This characteristic was observed independently of repetition rate. Futhermore, the diode exhibited no indication of diode shorting during the absence of microwave energy. After recovering, the microwave energy increased to a level somewhat higher than that achieved initially.

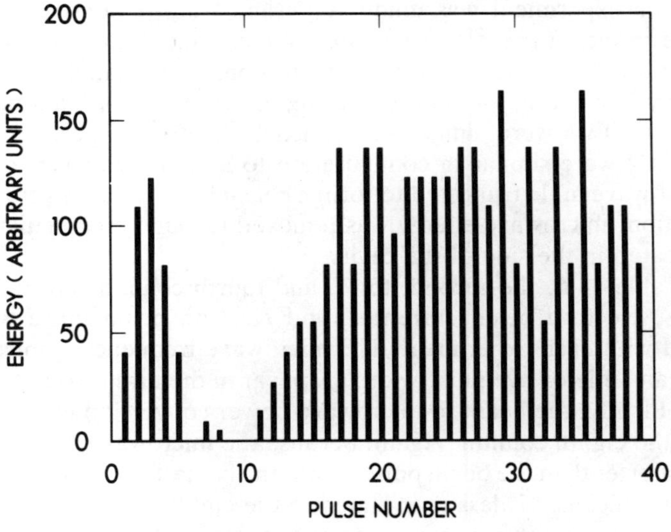

Fig. 14.24 Microwave Power as a Function of Pulse Number at a Repetition Rate of 1.2 Hz
Note: Observe the loss of energy between the sixth and eleventh pulses.

It is suggested [22] that the microwave energy loss is caused by anode foil outgassing, with the subsequent ionization of the gas leading to microwave attenuation. After a number of pulses, the foil outgassing ceases, and the microwave energy is no longer attenuated. Approximately 100 microwave pulses were obtained at 1.2 Hz operation. At 10 Hz operation, the number of pulses was reduced to about 20, a limitation caused by anode screen melting.

14.6 TRANSVERSE VIRCATOR EXPERIMENTS

A chronological summary of the transverse vircator experiments is given in Table 14.5. Again, the acronym appearing before the reference in the first column is used in the text to refer to that particular experiment. Because the experiment was uniquely designed and extensively investigated, the results of the transverse vircator experiment TV1 are discussed in this section. The alternate diode configurations investigated in TV1 were particularly interesting, but also confusing. Overall, the operational characteristics of TV1 were similar to the axial vircator. Experiments TV2 through TV4 were similar in configuration to an axial vircator, but with the output waveguide truncated to form a coaxial cavity. In this transverse configuration, microwave energy was removed through rectangular waveguides located in the side of the cavity.

In Table 14.5, the second, third, and fourth columns are the diode voltage V, electron beam current I, and electron beam pulse width t_b associated with each experiment. The microwave frequency f, microwave power P, and microwave pulse width t_μ appear in the fifth through seventh columns. Finally, the beam-to-microwave power conversion efficiency ϵ is stated in the eighth column. Again, because the microwave pulse width is typically shorter than the beam pulse width, the beam-to-microwave energy conversion efficiency is less than stated in the eighth column. In general, the efficiency of the transverse vircator is less than the axial vircator. At the same time, the advantage of the transverse vircator is the TE radiation pattern generated.

14.6.1 Experimental Configuration

Similar to the axial vircator, the virtual cathode formed inside the output waveguide in TV1. In contrast, the electron beam entered through the side of the rectangular output waveguide, rather than through the end of a coaxial output waveguide. Consequently, the microwave power is in

Table 14.5 Summary of Transverse Vircator Experiments

Experiment	V(keV)	I(kA)	t_b(ns)	f(GHz)	P(MW)	t_μ(ns)	ϵ(%)
TV1: Benford *et al.* [23, 24]	1000	60	65	8–12	350	50	0.50
TV2: Price *et al.* [25]	550	150	100	0.4–2.0	200	40	0.25
TV3: Price *et al.* [25]	1000	80	100	2–5	200	40	0.25
TV4: Bromborsky *et al.* [21]	8000	250	120	0.8–1.0	15000	50	0.75

a TE mode, instead of the TM mode that is characteristic of the axial vircator. Nevertheless, with the same 1 MV, 5 Ω generator and diode configuration [11], the 350 MW output with transverse extraction is lower than the 1 GW output obtained with axial extraction. In addition to transverse extraction, there are other unique features associated with this experiment. For example, a moveable waveguide stub was used to modify the shorted end of the rectangular output waveguide, while the opposite end of the waveguide was connected to a rectangular horn antenna. The moveable stub was intended to investigate any resonance that might occur due to reflections of energy back into the virtual cathode region. In fact, it was found that neither the microwave frequency nor power depended on the location of the stub, provided that its position did not interfere with the formation of the virtual cathode. To investigate the effect of the axial-boundary condition on virtual cathode formation, a moveable wall was located along the cathode axis in the drift chamber. Shorting of the virtual cathode was observed as the moveable wall approached to the anode. The moveable waveguide stub and wall are depicted in Figure 14.25.

14.6.2 Diode-Waveguide Configurations

To investigate the interplay between the diode and the output waveguide, the four diode configurations shown in Figure 14.25 were investigated. In the recessed-diode configuration (Figure 14.25 (a)), the anode is located in the output waveguide wall on the same side as the generator and transmission line. Consequently, only radiation generated in the region between the anode and virtual cathode can be extracted. For the standoff-diode configuration (Figure 14.25 (b)), the anode, which is supported by thin return-current rods, is located in the center of the output waveguide.

The supporting rods were arranged to minimize the perturbation of radiation fields between the cathode and anode. With this configuration, radiation generated between the cathode and virtual cathode can be extracted. In the shielded-standoff-diode configuration, the anode foil is again located at the center of the output waveguide, as in the standoff diode. At the same time, the supporting rods are replaced by a solid return-current conductor that prevents radiation generated between the cathode and anode from being extracted, as indicated in Figure 14.25 (c). In short, the shielded-standoff diode is similar to the recessed diode with the anode translated to the center of the output rectangular waveguide. Finally, in the inverse-recessed-diode configuration (Figure 14.25 (d)), the anode is located in the output waveguide wall opposite the generator and transmission line.

In all frequency bands, the highest power was generated with the standoff-diode configuration. For example, with a 1.05 cm anode-cathode gap, a 2.54 cm cathode radius, and the moveable wall located 10 cm from the cathode, the standoff diode power was 1.7 times that of the recessed diode, which, in turn, was twice that of the shielded-standoff diode. Little radiation was generated with the inverse-recessed-diode configuration. Given that radiation is being generated on both sides of the anode, the standoff diode would provide the most symmetric generation configuration, assuming that the virtual cathode formed at distance from the anode equal to the anode-cathode gap. This symmetry argument is consistent with trapped reflexing electrons between the real and virtual cathodes as the source of radiation.

Recall that trapped reflexing electrons are thought to be the source of microwave generation in the axial vircator. Furthermore, the transverse vircator exhibits similar operational characteristics to the axial vircator, and thus supports the reflexing electron mechanism as the source of microwave radiation. However, investigation of the inverse-recessed-diode configuration concluded that the region between the anode and virtual cathode must be the source of radiation. Given this result, it was concluded that the oscillating virtual cathode was the source of radiation. In summary, the conclusions that can be drawn from the four diode configuration experiments are inconsistent.

14.6.3 Frequency Dependence

Consistent with axial vircator results, the frequency chirps throughout the voltage pulse, with the mean frequency increasing in time. The observed frequency dependence is approximately inversely proportional to

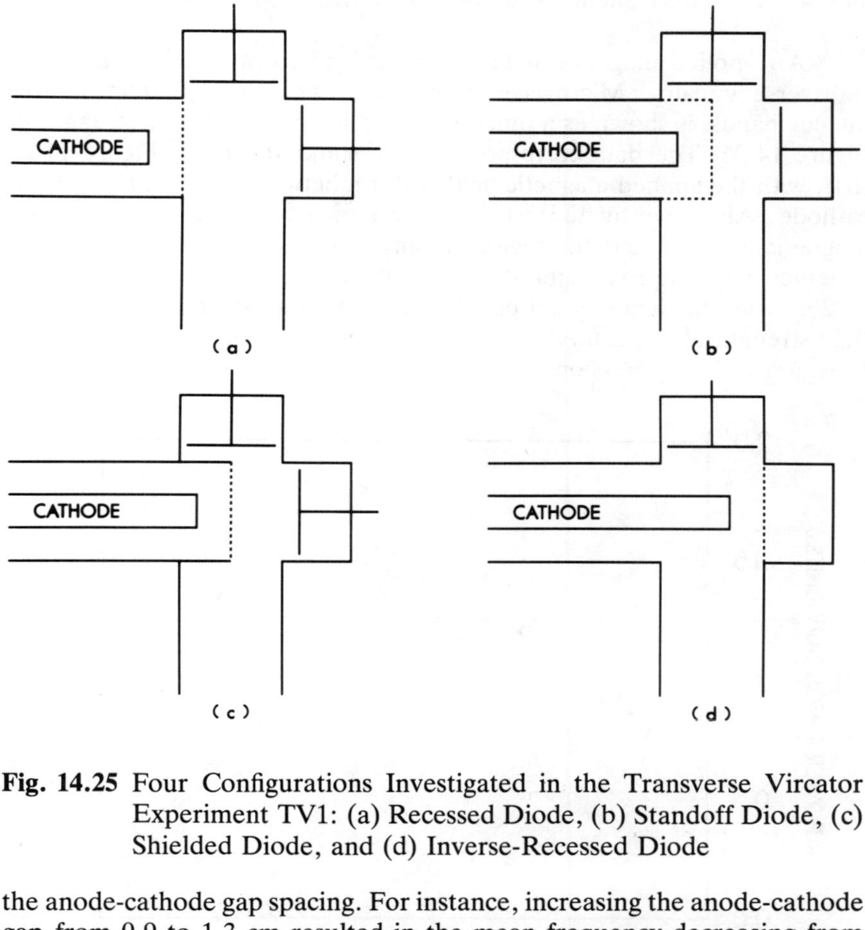

Fig. 14.25 Four Configurations Investigated in the Transverse Vircator Experiment TV1: (a) Recessed Diode, (b) Standoff Diode, (c) Shielded Diode, and (d) Inverse-Recessed Diode

the anode-cathode gap spacing. For instance, increasing the anode-cathode gap from 0.9 to 1.3 cm resulted in the mean frequency decreasing from 10.5 to 8.4 GHz. The frequency dependence with voltage is approximately linear. Little sensitivity was observed in frequency to the position of the moveable waveguide stub. There was study of the effect of anode foil thickness on microwave frequency and power. In fact, the maximum anode foil thickness considered was 0.00254 cm (aluminum). Given this anode foil thickness, a 1 MeV electron beam would scatter about 130 mrad in passing through the foil. However, based on axial vircator results, this magnitude of scattering would not be expected to modify significantly the microwave frequency or power, and little effect was observed.

14.6.4 Applied Magnetic Field and Anode Disk Dependence

An applied magnetic field suppressed microwave radiation in the transverse vircator. Microwave suppression, which occurred in all frequency bands, is shown as a function of applied magnetic field strength in Figure 14.26. The data correspond to the standoff diode, (Figure 14.25 (b)), with the applied magnetic field uniform between the real and virtual cathode. Also investigated was the effect of a converging or diverging magnetic field in the diode region. In any case, microwave suppression is observed with approximately the same dependence as shown in Figure 14.26. From the data it is not possible to determine whether the magnetic field strength at the cathode, anode, or virtual cathode is responsible for the microwave suppression.

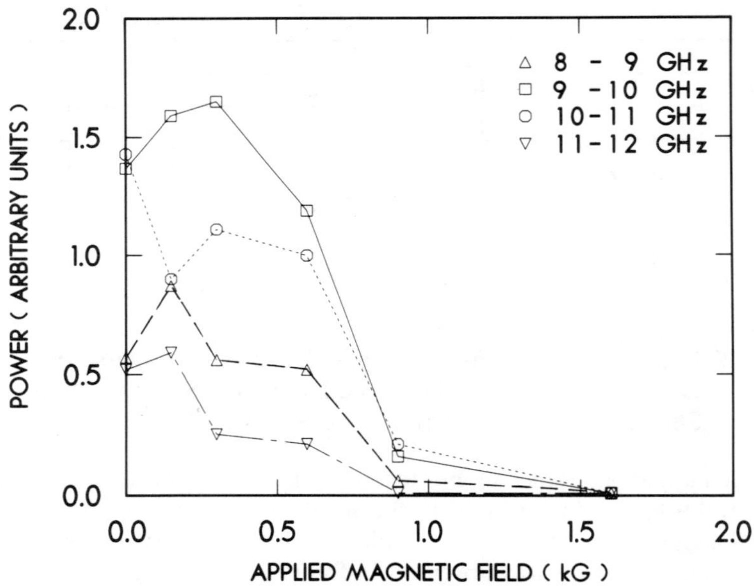

Fig. 14.26 Microwave Power is Suppressed in all Frequency Bands when an Axial Magnetic Field is Applied

Note: The lines are only intended as a visual aid.

It was suggested that microwave suppression by an applied magnetic field results from the prevention of beam pinching in the diode. However, the basic diode dimensions of TV1 are similar to VC3, an experiment that has been simulated. The simulation results show no evidence of beam pinching. Moreover, the experimental and theoretical evidence associated with the axial vircator indicates that microwave generation is suppressed when the beam pinches. Thus, when a magnetic field is applied to the transverse vircator, microwave suppression most likely follows from the destruction of the reflexing electron component localized about the cathode axis. In addition, it has been suggested that the two-stream instability between the incoming and reflected electron streams in one-dimensional limit associated with a strong magnetic field can cause loss of beam coherence [32].

The effect on the microwave power of range-thick disks of varying diameter was investigated. As in axial vircator experiments, the disks were placed on the virtual cathode side of the anode and aligned with the cathode axis. The presence of a thick anode disk suppressed microwave radiation in the transverse vircator. The higher frequency components were suppressed with a small diameter disk, while the lower frequency components were suppressed with a disk diameter approaching one-half of the cathode diameter. The sensitivity of output power to the absorption of reflected electrons by a thick anode disk is another indication that trapped reflexing electrons are the source microwave radiation in the transverse vircator.

14.6.5 Cavity Investigation

Frequency control was demonstrated by enclosing the virtual cathode inside a right-hand circular cavity located inside the output rectangular waveguide. To extract microwaves, a rectangular aperture was located in the side of the cavity. With the standoff diode (Figure 14.25 (b)), the cavity axis was aligned with the cathode axis, and the anode foil was used as one end of the cavity. Cavity resonances were modified by changing the cavity length. When a completely closed cavity with no extraction aperture was used, little radiation was observed.

When radiation is extracted from the cavity, distinct resonance behavior is observed as the cavity is tuned to match the natural frequency associated with the uncontrolled vircator. Furthermore, frequencies that are not resonant with a cavity mode appear to be suppressed. With proper tuning, the observed bandwidth was reduced by a factor of five, down to about 2%. There is some indication that the overall efficiency may have increased with the cavity present, but the increase is less than a factor of two.

14.7 REDITRON EXPERIMENT

The reditron is the most recent virtual cathode device. Although the basic concept was suggested some time earlier [32–34], a high-power, proof-of-principle experiment was carried out only recently. As discussed earlier, there are two competitive mechanisms responsible for microwave generation in the reflex triode and vircator: (1) an oscillating virtual cathode and (2) trapped reflexing electrons in the potential well formed between the real and virtual cathodes. In the reditron, electrons reflected back toward the diode region are removed by a range-thick anode with a narrow slit. Thus, the source of microwave radiation is an oscillating virtual cathode.

The basic operation of the reditron appears consistent with theoretical expectations [26]. As predicted, the initial experiment found the output microwave power to be sensitive to the applied magnetic field strength. At 4.0 GHz, an output power of 1.4 GW was observed in a TM_{02} mode with a bandwidth of about 2%. In addition, several hundred megawatts were observed in higher harmonics. The frequency dependence on the applied magnetic field appeared to be weak in the experiment. In the 4–6 GHz range, a microwave pulse of 10–45 ns was obtained, a width considerably shorter than the 80 ns electron beam pulse.

14.7.1 Experimental Configuration

Figure 14.27 shows the azimuthally symmetric reditron configuration that was investigated [27]. With the exception of the annular cathode wall thickness and the narrow anode slit, the configuration is shown to scale. The cathode diameter is 6 cm, and the annular cathode wall thickness is 0.05 cm. To take into account beam pinching, which results from shorting of the radial electric field at the anode, the 0.50 cm anode slit is actually at a slight angle with respect to the cathode axis. In other words, the diameter of the slit is slightly larger on the diode side of the anode than on the output waveguide side of the anode. At the present time, the importance of this feature is unclear. Three spokes were used to support the center of the 2.54 cm thick, high-density graphite anode. These spokes intercepted about 6% of the beam current. No investigation was made of alternative spoke configurations.

Starting from the graphite anode, a 125 cm long, 18 cm diameter output waveguide was connected to a 125 cm long conical antenna. Microwaves were removed through a 60 cm diameter vacuum-air interface. The applied magnetic field is uniform in the diode, the anode, and 16 cm

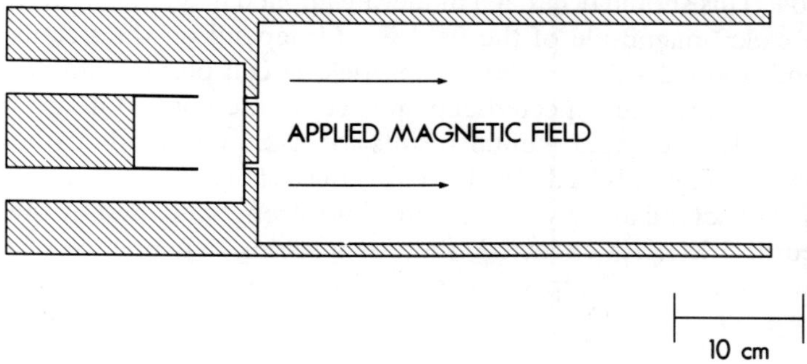

Fig. 14.27 Sketch of the Reditron Configuration

past the anode in the output waveguide; beyond that point, the magnetic field flares to the output waveguide wall. The reditron was energized with a 1.9 MV, 80 ns voltage pulse. The diode current was approximately 40 kA with a 2.6 cm anode-cathode gap.

14.7.2 Power Dependence

Unlike the reflex triode and vircator, the reditron does not generate microwaves in the absence of an applied magnetic field. Also, if the applied magnetic field is too large, microwave generation ceases. Note that in the presence of a strong applied magnetic field, the vircator and reditron are essentially the same one-dimensional device. Both the oscillating virtual cathode and reflexing electron-bunching mechanisms are much less efficient when the incoming beam has a large energy divergence [13, 36]. It has been suggested that in the one-dimensional limit of a strong applied magnetic field, the initial and reflected electron streams undergo two-stream instability [32]. Because of this interaction, the initial beam distribution is significantly degraded and microwave generation ceases. The reditron avoids this problem by absorbing the reflected electron stream.

The operational domain of the reditron is in the moderate applied magnetic field regime. Physically, reflected electron discrimination depends on the incoming electron beam propagating through the narrow anode slit, while the reflected beam is simultaneously absorbed in the anode. In other words, the thick anode with a narrow slit acts as a thin foil to the incoming beam and as a collimator to the reflected beam. Using this approach, the incoming electron beam must gain sufficient transverse energy as it is reflected from the virtual cathode to ensure absorption by the anode. This should occur at a moderate applied magnetic field strength, and the exact magnitude of the field would depend on the details of the diode and applied voltage. Indeed, particle-in-cell plasma simulations of the reditron show the expected dependence on the applied magnetic field strength [26]. The experimental results for a 2.6 cm anode-cathode gap are shown in Figure 14.28. In this case, the relative power is defined as the power observed at a given magnetic field strength divided by the power observed at 5 kG. As expected, there is a strong dependence of the microwave power on the applied magnetic field strength.

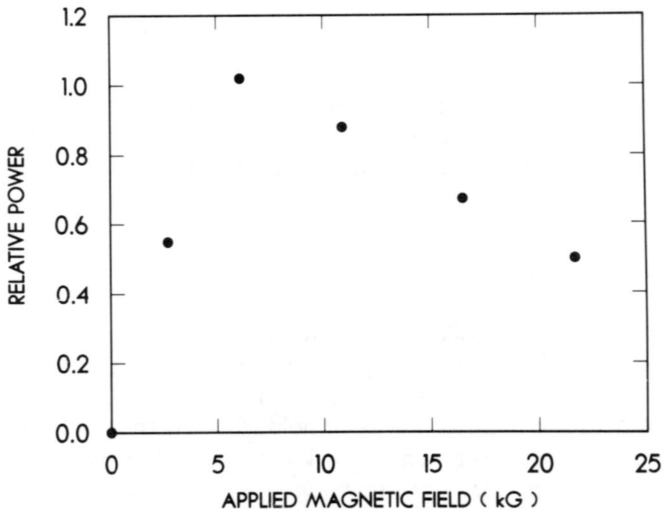

Fig. 14.28 Microwave Power as a Function of Applied Magnetic Field for the Reditron Experiment

14.7.3 Frequency Dependence

Because the oscillating virtual cathode is the microwave generation mechanism, the frequency dependence should be related to the beam plasma frequency. In particular, as the applied magnetic field is increased, or the anode-cathode gap spacing is decreased, the beam plasma frequency and related microwave frequency should increase. In Figure 14.29, the observed frequency dependence on applied magnetic field strength is shown for a 2.6 cm anode-cathode gap. In contrast to what might be

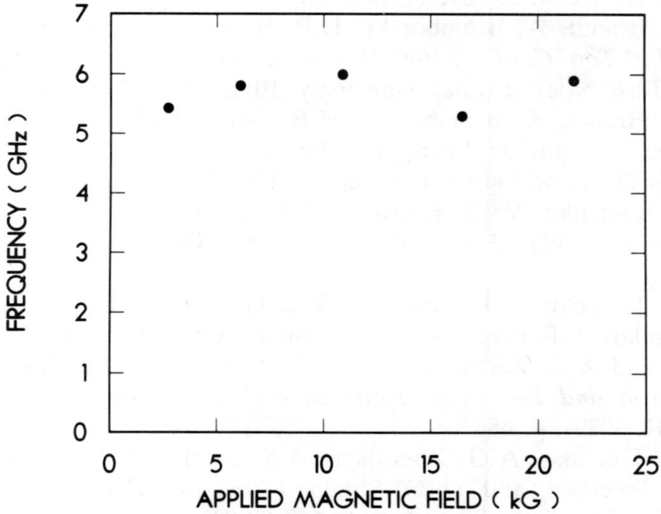

Fig. 14.29 Frequency as a Function of Applied Magnetic Field for the Reditron Experiment

expected, the frequency does not show any strong dependence on the magnetic field strength. However, the frequency was observed to decrease with increasing anode-cathode gap spacing.

ACKNOWLEDGEMENTS

This chapter would not have been possible without help from a large number of individuals. The author gratefully acknowledges discussions with B. Anderson, D. Bartsch, J. Benford, H. Brandt, A. Bromborsky, S. Burkhardt, C. Clark, H. Davis, B. DeVolder, B. Dugnan, S. Graybill, M. Hayworth, A. Kehs, T. Kwan, A. Peratt, D. Price, R. Scarpetti, C. Snell, R. Stringfield, D. Sullivan, H. Sze, D. Voss, and W. Woo. I also want to acknowledge A. Thode for numerous suggestions concerning the presentation of the material.

REFERENCES

1. R.A. Mahaffey, P. Sprangle, J. Golden, and C.A. Kapetanakos,

Phys. Rev. Lett. **39**, 843 (1977).
2. H.E. Brandt, A. Bromborsky, H.B. Burns, and R.A. Kehs, *Proc. 2nd Int. Top. Conf. on High Power Electron and Ion Beam Research and Technology* (Cornell University, Ithaca, NY, 1977), p. 649.
3. H.E. Brandt, A. Bromborsky, H.B. Burns, R.A. Kehs, and G.P. Lasche, "Gigawatt Microwave Emission from a Reflex Trode," Harry Diamond Laboratory Report HDL-TR-1917 (August 1980).
4. A.N. Didenko, Ya.E. Krasik, S.F. Perelygin, and G.P. Fomenko, *Pisma Zh. Tekh. Fiz.* **5**, 321 (1979); *Sov. Tech. Phys. Lett.* **5**, 128 (1979).
5. A.N. Didenko, G.P. Fomenko, I.Z. Gleizer, Ya.E. Krasik, G.V. Meknikov, S.F. Perelygin, Yu.G. Shtein, A.S. Sulakshin, V.I. Tsvetkov, and A.G. Zerlitsin, *Proc. 3rd Int. Top. Conf. on High Power Electron and Ion Beam Research and Technology* (Novosibirsk, USSR, 1979), p. 683.
6. A.N. Didenko, A.G. Zherlitsyn, A.S. Sulakshin, G.P. Fomenko, V.I. Tsvetkov, and Yu.G. Shtein, *Pisma Zh. Tekh. Fiz.* **9**, 1510 (1983); *Sov. Tech. Phys. Lett.* **9**, 647 (1983).
7. A.G. Zherlitsyn, S.I. Kuznetsov, G.V. Mel'nikov, G.P. Fomenko, and V.I. Tsvetkov, *Pisma Zh. Tekh. Fiz.* **11**, 1083 (1985); *Sov. Tech. Phys. Lett.* **11**, 450 (1985).
8. J.M. Buzzi, H.J. Doucet, H. Lamain, C. Rouille, J. Delvaux, and J.C. Jouys, *Le Journal de Physique Lettres* **39**, 1 (1978).
9. J.M. Buzzi, H.J. Doucet, B. Etlicher, P. Haldenwang, J. Cabe, J. Delvaux, J.C. Jouys, and C. Peugnet, *Proc. 2nd Int. Top. Conf. on High Power Electron and Ion Beam Research and Technology* (Cornell University, Ithaca, NY, 1977), p. 663.
10. M.C. Clark, Sandia National Laboratory, private communication.
11. H. Sze, J. Benford, T. Young, D. Bromley, and B. Harteneck, *IEEE Trans. Plasma Sc.* **PS-13**, 492 (1985).
12. R.D. Scarpetti and S.C. Burkhardt, *IEEE Trans. Plasma Sc.* **PS-13**, 506 (1985).
13. S.C. Burkhardt, R.D. Scarpetti, and R.L. Lundberg, *J. Appl. Phys.* **58**, 28 (1985).
14. H.A. Davis, R.R. Bartsch, L.E. Thode, E.G. Sherwood, and R.M. Stringfield, *Phys. Rev. Lett.* **55**, 2293 (1985).
15. D. Saunders, D. Voss, B. Miera, M.C. Clark, and B. Dungan, US Air Force Weapons Laboratory, private communication.
16. M.C. Clark and L.E. Thode, *Conference Record*, 1986 IEEE International Conference on Plasma Science (Saskatoon, Saskatchewan, 1986), p. 55.
17. S.C. Burkhardt, to be published in *J. Appl. Phys.* (1987).

18. M.C. Clark, Sandia National Laboratory, private communication.
19. H.A. Davis, R.R. Bartsch, L.E. Thode, E.G. Sherwood, and R.M. Stringfield, Los Alamos National Laboratory, private communication.
20. M.C. Clark, Sandia National Laboratory, private communication.
21. A. Bromborsky, Harry Diamond Laboratory, private communication.
22. M.C. Clark and L.M. Earley, Sandia National Laboratory report SAND86–0665 (July-September 1986), p. 33.
23. J. Benford, H. Sze, W. Woo, and B. Harteneck, *Phys. Rev. Lett.* **56**, 344 (1986).
24. J. Benford, D. Price, D. Bromley, and H. Sze, *Conference Record,* 1986 IEEE International Conference on Plasma Science (Saskatoon, Saskatchewan, 1986), p. 68.
25. D. Price, Physics International Company, private communication.
26. T.J.T. Kwan, *Phys. Rev. Lett.* **57**, 1895 (1986).
27. H.A. Davis, R.R. Bartsch, E.G. Sherwood, R.M. Stringfield, and T.J.T. Kwan, *Conference Record*, 1986 IEEE International Conference on Plasma Science (Saskatoon, Saskatchewan, 1986), p. 68.
28. D.J. Sullivan, *Proc. 3rd Int. Top. Conf. on High Power Electron and Ion Beam Research and Technology* (Novosibirsk, USSR, 1979), p. 769.
29. D.J. Sullivan and E.A. Coutsias, *Proc. 4th Int. Top. Conf. on High Power Electron and Ion Beam Research and Technology* (Palaiseau, France, 1981), p. 371.
30. D.J. Sullivan, *IEEE Trans. Nucl. Sc.* **30**, 3426 (1983).
31. E.A. Coutsias and D.J. Sullivan, *Phys. Rev. A* **27**, 1535 (1983).
32. T.J.T. Kwan and L.E. Thode, *Bull. APS* **27**, 937 (1981).
33. T.J.T. Kwan and L.E. Thode, *Conference Record*, 1982 IEEE International Conference on Plasma Science (Ottawa, Ontario, 1982), p. 79.
34. T.J.T. Kwan and L.E. Thode, *Bull. APS* **26**, 1017 (1982).
35. T.J.T. Kwan and L.E. Thode, *Conference Record*, 1983 IEEE International Conference on Plasma Science (San Diego, CA, 1983), p. 40.
36. T.J.T. Kwan, C.M. Snell, M.A. Mostrom, and L.E. Thode, *Proc. 10th Conf. Numerical Simulation of Plasma* (San Diego, CA, 1983), p. 2B1.
37. B.G. DeVolder, L.E. Thode, T.J.T. Kwan, and M.C. Clark, *Conference Record*, 1984 IEEE International Conference on Plasma Science (St. Louis, MO, 1984), p. 20.
38. T.J.T. Kwan, *Phys. Fluids* **27**, 228 (1984).

39. T.J.T. Kwan and L.E. Thode, *Phys. Fluids* **27**, 1570 (1984).
40. A.L. Peratt, C.M. Snell, and L.E. Thode, *IEEE Trans. Plasma Sc.* **PS-13**, 498 (1985).
41. W. Woo, *Phys. Fluids* **30**, 239 (1987).
42. H.E. Brandt, *IEEE Trans. Plasma Sc.* **PS-13**, 513 (1985).
43. H.E. Brandt, *Bull. APS* **31**, 1505 (1986).
44. A.N. Didenko and N.M. Philipenko, *Radiotekhnika i Elektronika* **30**, 820 (1985).
45. L. Thode, *Conference Record*, 1986 IEEE International Conference on Plasma Science (Saskatoon, Saskatchewan, 1986), p. 67.
46. H.G. Unger, *Bell Syst. Tech. J.* **37**, 899 (1958).
47. M.M. Cambell, D.J. Sullivan, and B.B. Godfrey, *CCUBE Users Manual*, Mission Research Corporation, Report No. AMRC-R-341 (1983).

Index